PROPERTY OF
UNIVERSITY OF COLORADO AT DENVER
MATHEMATICS DEPARTMENT
1250 14TH ST RM 600
DENVER, CO 80202
(303) 556-8442

ELEMENTARY MODERN ALGEBRA

Robert C. Thompson

University of California, Santa Barbara

SCOTT, FORESMAN AND COMPANY
Glenview, Illinois Brighton, England

Library of Congress Catalog Card Number 73-90592
ISBN: 0-673-07829-9
AMS 1970 Subject Classification 98A10

Printed in the United States of America.

Regional Offices of Scott, Foresman and Company
are located in Dallas, Texas;
Glenview, Illinois;
Oakland, New Jersey;
Palo Alto, California;
Tucker, Georgia; and
Brighton, England.

PREFACE ✻✻ CONTENTS

This book is intended for use in algebra courses ranging upward from introductory courses parallel to (or subsequent to) the first or second calculus course. The manuscript has been developed over a period of several years, and portions of it have been used with the following types of classes:

> Freshman and sophomore-level classes, of one-quarter and two-quarter lengths, composed principally of students majoring in mathematics or science, and sometimes including gifted students;
>
> A junior-senior level class of students majoring in mathematics and intending to become teachers of high school mathematics;
>
> A graduate-level class of mathematics teachers enrolled in a program leading to a Master of Education degree;
>
> A class of teachers of high school mathematics enrolled in an enrichment program.

There is sufficient material, and sufficient flexibility in its organization, to accommodate the needs of a variety of audiences, in courses ranging from one quarter to a full year.

In many ways this is a new type of algebra text. It is a book about integers, rational numbers, real numbers, complex numbers, and also about polynomials and rational functions in a single variable with coefficients from these number systems. (In Chapter 7 the coefficients may lie in a finite field.) These topics are studied in a setting of groups, rings, and fields. This certainly is standard material. Within this framework, however, there are a variety of new topics, quite suitable for inclusion in an elementary algebra text, but which have usually been omitted. Most of these are in optional sections or subsections. Equally important, a considerable effort is made to motivate, to show connections between apparently unconnected topics, and to show applications outside algebra, particularly in calculus, analysis, geometry, computer science, and engineering. A unifying theme throughout the book is the strong analogy between the theory of integers and rational numbers on the one hand, and the theory of polynomials and rational functions on the other. Many of the new topics are introduced in a natural way by appeal to this striking parallelism.

To avoid misunderstanding, let it be emphasized that this is an *elementary* textbook, with all the new topics fitted into a unified, elementary framework. Because most of the nontraditional material is presented in optional (sub)sections, instructors wishing to use this book for courses along strictly traditional lines will be able to do so without difficulty. However, if a substantial number of the optional (sub)sections are covered, the student's comprehension will be considerably enhanced.

Instructors often observe that students fail to transfer concepts and facts from one course to another. Frequently this failure is brought about by failure of the textbook to point to related material elsewhere in undergraduate mathematics. Seemingly different parts of mathematics are often related, even at the undergraduate level, and sometimes in surprising ways. I have attempted to create a text in which this is clearly evident. My experience has shown that superior students from the freshman level upward strongly appreciate this point. I have also attempted to create a text showing clearly how algebraic concepts can lead to deepened understanding of the most primitive mathematical objects (integers, rational numbers, polynomials, rational functions). In varying degree, this is done in other texts, but in none is it pursued as vigorously as in this book.

Although this is an algebra text, it is perhaps best described as a book about mathematics from an algebraic point of view. It is, in fact, a book about elementary mathematics from this algebraic viewpoint, even though it brings in some topics previously regarded as advanced.

Preliminary drafts of the manuscript were reviewed with great thoroughness by an anonymous reviewer. I wish to record here my appreciation for his critical but helpful comments.

Suggestions for use of the book and further discussion of its features will be found after the table of contents.

R. C. Thompson

1

BASIC CONCEPTS

2

THE COMPLEX NUMBER FIELD

3

INTEGERS AND POLYNOMIALS

4

INTEGRAL ARITHMETIC

5

POLYNOMIALS AND THEIR ZEROS

6

THE FIELD OF FRACTIONS. RATIONAL NUMBERS AND FUNCTIONS

7

MAPPINGS OF RINGS

8

SPECIAL TOPICS (optional)

SUGGESTIONS FOR USE

A feature of this text is that discussions are detailed, thorough, and clear. For this reason, depending on the audience, some instructors may wish to cover portions of the text as assigned reading, with minimal in-class discussion.

There are many exercises. They are classified into three types:

Type I are computational, usually requiring a numerical answer, but sometimes calling for the construction of a numerical example or counterexample. Most students should be able to work the majority of these exercises, and answers for nearly all are provided at the back of the book.

Type II are of a not-too-difficult theoretical nature, typically calling for proofs like those in the text, or for the construction of examples or counterexamples. Most students should be able to work the bulk of these exercises. Some answers are given at the back of the book.

Type III are theoretical, and of a broader scope and moderate-to-substantial difficulty. These exercises may require knowledge of optional sections of the text, and some of them provide additional theory supplementing the text. These exercises should appeal to gifted students. No answers are given.

Suggestions for various courses are in the table below. It was not practical to list the subsections (subtopics). Instructors will want to decide which subsections to include, depending on the pace and depth desired for a given set of students.

	Chapter	*Include sections (all or part)*	*Omit sections*
One-semester course			
	1	1, 2, 3, 4, 5, 6, 7	None
	2	1, 2, 3, 4, 6	5
	3	1, 2, 3, 4, 5, 6, 7	None
	4	1, 2, 5	3, 4
	5	1, 2, 3, 4, 8	5, 6, 7
	6	1, 2, 3, 4, 7	5, 6
	7	None	1–8
	8	None	1–6
Two-quarter course			
	1	1, 2, 3, 4, 5, 6, 7	None
	2	1, 2, 3, 4, 6	5
	3	1, 2, 3, 4, 5, 6, 7	None
	4	1, 2, 3, 5	4
	5	1, 2, 3, 4, 5, 8	6, 7
	6	1, 2, 3, 4, 5, 7	6
	7	1, 2, 3, 8	4, 5, 6, 7
	8	None	1–6
One-quarter course			
	1	1–7	Optional sections and subsections (chapters 1–4)
	2	1–6	
	3	1–7	
	4	1–5	
	5	Selected topics only	

TABLE OF NOTATION

$\mathbf{C}$	66	$\mathbf{Q}$	31	$\mathbf{R}$	56	$\mathbf{Z}$	30
$\mathfrak{R}[\lambda]$	106	$\mathfrak{R}[[\lambda]]$	111	$\mathfrak{F}(\lambda)$	265	$\mathbf{Z}_m$	306
$\mathfrak{R}_m$	306	$\mathfrak{F}[\lambda]_m$	306	$\mathfrak{R}^m$	337, 387	$\mid, \nmid$	121
$(\ )$	62, 98	$(\ )$	124	$\langle\ \rangle$	129, 429	$\geq$	7
$\equiv$	141	λ	103	$\lvert\ \rvert$	8, 72, 268	$\simeq$	343
$\lVert\ \rVert$	380	$\lvert\ \rvert_p$	381	$!$	34	$\binom{n}{m}$	158

1

BASIC CONCEPTS

1.1 INTRODUCTION

This chapter introduces the mathematical structures known as *set*, *group*, *ring*, and *field*. These structures occupy a central position in modern algebra, in both its beginning and advanced aspects, and appear as well in other branches of mathematics, physics, chemistry, computer science, engineering, communication theory, and elsewhere. They will occupy a central position in this book, and in this chapter we explain some of their simpler properties. The reader may find much of this explanation to be an orderly presentation of ideas he has already acquired, possibly in high school.

The chapter begins with the basic notions of set, order, and proof by induction, goes on to a discussion of groups, then to rings, and finally to fields. Each of these topics is used in the study of those that follow. In later chapters of this book, we shall often employ rings and fields; groups will also occur later, but slightly less often. Proof by induction is used at numerous places in the book.

1.2 SETS

A *set* is simply a collection of objects. When describing a set, a precise rule or statement must be given specifying when an object is in the set, and when two objects are the same, that is, are equal. (Illustrations of this remark will be seen in Example 1.4.8, in §2.5, and elsewhere.) The objects of which a set is composed are called the *elements* of the set. It is customary to use the following notation. If S is a set in which the elements are given by a certain rule, we write

$$S = \{x \mid x \text{ satisfies the rule}\}$$

and read "S is the set of all x such that x satisfies the rule." Thus, for example, if we wish to study the numbers 1, 2, 3, 4, 5, 6, 7, 8, we may introduce the set S defined by

$$S = \{x \mid x \text{ is a positive integer not exceeding } 8\}.$$

Sometimes it is more convenient just to list within braces the elements in the set. For the set just described we would then write

$$S = \{1, 2, 3, 4, 5, 6, 7, 8\}. \tag{1}$$

We write $x \in S$ if x is an element of S and $x \notin S$ if x is not an element of S. Read "$x \in S$" as "x belongs to S" or as "x is an element of S"; and "$x \notin S$" as "x does not belong to S" or as "x is not an element of S." For example, if S is the set (1) above, then $7 \in S$ but $9 \notin S$.

Associated with the notion of a set is that of a subset. If S and T are sets, and if each element in T is also in S, we say that T is a *subset* of S or that S *contains* T. Thus T is a subset of S if S contains all elements of T and perhaps other elements as well. If T is a subset of S, we write $T \subseteq S$ or $S \supseteq T$; read these formulas respectively as "T is contained in S" and "S contains T." As examples we have $\{1, 2\} \subseteq \{1, 2, 3, 4\}$ and $\{2, 4\} \supseteq \{2\}$. If $T \subseteq S$ and there is at least one element of S that is not in T, we say that T is a *proper subset* of S, and write this in symbols as $T \subset S$ or as $S \supset T$. Read these last two formulas respectively as "T is strictly contained in S" and "S strictly contains T." Thus in fact $\{1, 2\} \subset \{1, 2, 3, 4\}$ and $\{2, 4\} \supset \{2\}$.

Sets S and T are *equal* (written as $S = T$) if and only if $S \subseteq T$ and $T \subseteq S$. That is, $S = T$ if and only if S and T consist of precisely the same elements. If S and T do not consist of exactly the same elements, we write $S \neq T$; read this as "S is not equal to T."

A very useful set is the *empty set*. It is, by definition, the set consisting of no elements. We let the symbol $\varnothing$ denote it. Then, for any set S, we have $\varnothing \subseteq S$ since the condition "each element $x \in \varnothing$ is also in S" is satisfied (because there are no elements x in $\varnothing$ to fail to satisfy it.)

There are several other useful ideas associated with sets. First, let S and T be any sets. By $S \cap T$ we denote the set consisting of all elements belonging to *both* S and T. In symbols,

$$S \cap T = \{x \mid x \in S \text{ and } x \in T\}.$$

We call $S \cap T$ the *intersection* of S and T. By $S \cup T$ we denote the set consisting of all elements belonging to *at least one* of S and T. In symbols,

$$S \cup T = \{x \mid x \in S, \text{ or } x \in T, \text{ or both}\}.$$

We call $S \cup T$ the *union* of S and T. As examples we have

$$\{1, 2, 3, 5\} \cap \{2, 3, 6, 7\} = \{2, 3\}$$

and

$$\{1, 2, 3, 5\} \cup \{2, 3, 6, 7\} = \{1, 2, 3, 5, 6, 7\}.$$

Now let T be a subset of S. The *difference of S and T*, sometimes also called the *complement of T with respect to S*, is the subset of S consisting of all elements of S which are not in T. In symbols it is

$$\{x \mid x \in S \text{ and } x \notin T\},$$

and is denoted by $S - T$.

It is often advantageous when discussing sets to make use of *Venn diagrams*. These are illustrations of relationships between sets in which we depict a set S as the region contained within an oval (or other geometric figure), as in Figure 1.2.1.

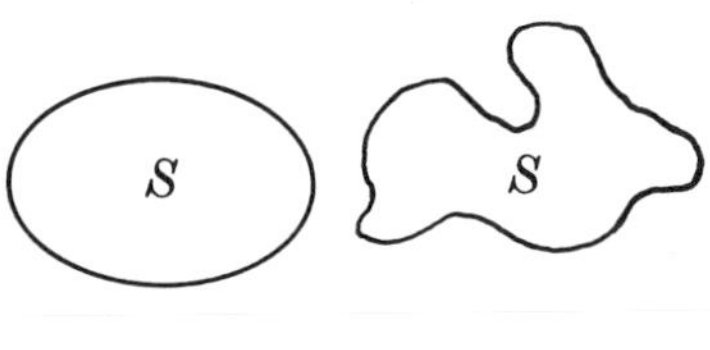

Figure 1.2.1

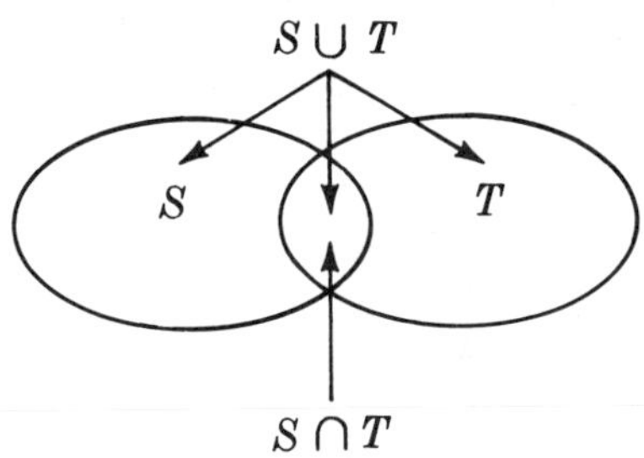

Figure 1.2.2

If sets S and T are illustrated (see Figure 1.2.2) as regions within a pair of ovals, then $S \cap T$ is the region belonging to both ovals, and $S \cup T$ is the region belonging to at least one of the ovals. If T happens to be a subset of S, the oval delineating T is drawn within the oval delineating S; see Figure 1.2.3. Then $S - T$ is the shaded region between the ovals, as shown in Figure 1.2.4.

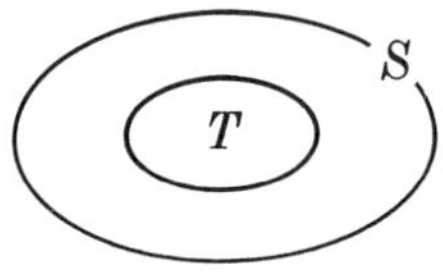

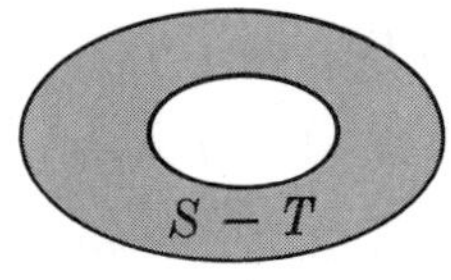

We say sets S, T are *disjoint* if $S \cap T = \varnothing$. Disjoint sets, therefore, are sets having no common element. For example, $\{1, 3, 5\}$ and $\{2, 4, 6\}$ are disjoint, but $\{1, 3, 5\}$ and $\{2, 3, 4\}$ are not. If T is a subset of S, obviously T and $S - T$ are disjoint.

Intersections and unions of more than two sets are often encountered. If $S, T, U, \ldots$ are sets, their intersection $S \cap T \cap U \cap \cdots$ and union $S \cup T \cup U \cup \cdots$ are defined by

$$S \cap T \cap U \cap \cdots = \{x \mid x \in \text{all of } S, T, U, \ldots\},$$

$$S \cup T \cup U \cup \cdots = \{x \mid x \in \text{at least one of } S, T, U, \ldots\}.$$

For example,

$$\{1, 2, 3, 5\} \cap \{2, 3, 4, 6\} \cap \{1, 3, 6, 7\} = \{3\},$$

$$\{1, 2, 3, 5\} \cup \{2, 3, 4, 6\} \cup \{1, 3, 6, 7\} = \{1, 2, 3, 4, 5, 6, 7\}.$$

Illustrations of $T \cap U \cap V$ and $T \cup U \cup V$ by Venn diagrams, when T, U, V are subsets of S, are shown in Figure 1.2.5.

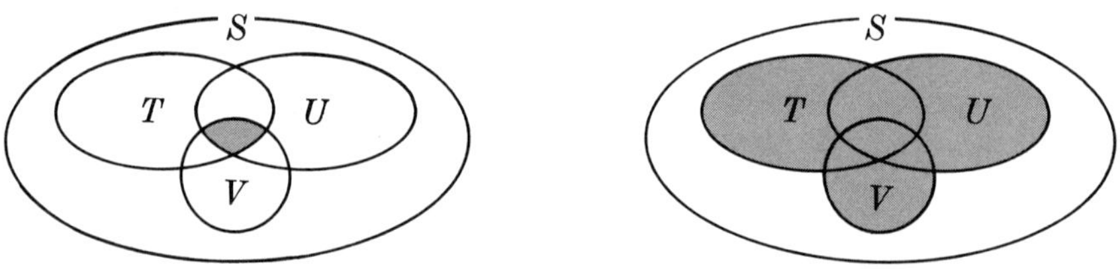

The shaded region is $T \cap U \cap V$ The shaded region is $T \cup U \cup V$

Figure 1.2.5

Partitions

If S is a set, and $T, U, V, \ldots$ are subsets of S, we say that $T, U, V, \ldots$ *partition* S if

(i) Any two of the subsets $T, U, V, \ldots$ are disjoint;

(ii) $T \cup U \cup V \cup \cdots = S$.

Thus $T, U, V, \ldots$ partition S if each element of S belongs to precisely one of these subsets. The subsets $T, U, V, \ldots$ are sometimes called the *components of the partition*.

Examples: (i) If T is a subset of S, then T and $S - T$ partition S.

(ii) If $S = \{1, 2, 3, \ldots, 100\}$ is the set of all natural numbers not exceeding 100, a partition of S with two components is obtained by taking T to be all even integers in S, and U to be all odd integers in S.

(iii) If S is the set of all (positive, negative, or zero) integers, a partition of S with three components is obtained by taking T to be all positive integers, $U = \{0\}$, V to be all negative integers.

An illustration using Venn diagrams of a set S partitioned into subsets T, U, V is shown in Figure 1.2.6.

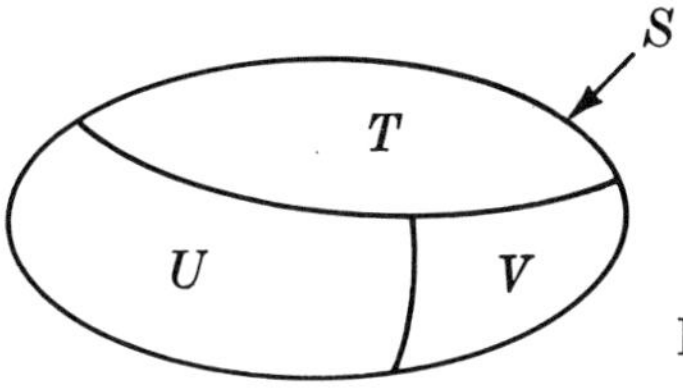

Figure 1.2.6

EXERCISES 1.2

I

1. List all subsets of $\{1, 2, 3\}$.

2. Is it true that $\{1, 2, 3, 4, 5\} = \{2, 3, 5, 1, 4\}$?

3. Find $\{1, 2, 4, 7, 9\} \cap \{2, 4, 10, 3\}$ and $\{1, 2, 4, 7, 9\} \cup \{2, 4, 10, 3\}$. Find $\{1, 5, 7\} \cap \{2, 4, 6\}$ and $\{1, 5, 7\} \cup \{2, 4, 6\}$.

4. Find all subsets of $S = \{1, 2, 3, 4\}$ containing exactly two elements, including 1. Find the complements with respect to S of each of these subsets.

5. List all subsets of $S = \{1, 2, 3, 4\}$ containing exactly two elements, not including 1. Show that the complement with respect to S of each of the subsets on your list contains the integer 1.

6. If $S = \{1, 2, 3, 4, 5, 6, 7, 8\}$, find the set V if T, U, V partition S, where $T = \{1, 3\}$, $U = \{2, 4, 7\}$.

7. With S as in Exercise 6, if $T = \{1, 7\}$, $U = \{3, 4\}$, find V if U, V partition $S - T$.

II

8. If $T, U, V, W, \ldots$ are subsets of S that partition S, show that $U, V, W, \ldots$ are subsets of $S - T$ that partition $S - T$.

A set is said to be *finite* if it consists of exactly m elements, where m is a nonnegative integer. In particular, the empty set $\varnothing$ is considered to be a finite set. A set which is not finite is said to be *infinite*.

9. Is it necessarily true that finite sets having the same number of elements are equal?

10. If S and T are finite sets, show that the number of elements in $S \cap T$ does not exceed that of S, and that the number of elements in $S \cup T$ is always at least that of S.

11. Is the set I of all positive integers a finite set? Let S be an arbitrary finite set. Show that $S \cup I$ is infinite but that $S \cap I$ is finite.

12. Let S be a set. If all proper nonempty subsets of S are equal, show that S consists of a single element.

13. By drawing Venn diagrams, show that $S \cap T = S$ if and only if $T \supseteq S$.

14. By drawing Venn diagrams, show that $S \cup T = S$ if and only if $T \subseteq S$.

15. Let S and T be sets. Show that $S \cap T = T \cap S$ and $S \cup T = T \cup S$.

III

16. For any finite set S, let $n(S)$ denote the number of elements of S. If S and T are finite sets, show that $n(S \cup T) + n(S \cap T) = n(S) + n(T)$. Drawing a Venn diagram may help you find the proof.

17. If T is a subset of S, where S is a finite set, show that $n(S - T) + n(T) = n(S)$.

18. Let S be a finite set, with n elements. How many subsets does S have? Prove your assertion.

19. Are there any finite sets S for which the number of subsets of S is not more than the number of elements of S?

20. Let A, B, C be sets. Prove that $A \cup (B \cap C) = (A \cup B) \cap (A \cup C)$ and $A \cap (B \cup C) = (A \cap B) \cup (A \cap C)$.

21. Let S be a finite set. If subsets T, U, V, ... partition S, show that $n(S) = n(T) + n(U) + n(V) + \cdots$, where $n(S)$ denotes the number of elements in the set S.

22. Let T, U, V be subsets of the finite set S. Suppose $S = T \cup U \cup V$. Find the number of elements in S, $n(S)$, if: $n(T \cap U \cap V) = 7$, $n(T \cap U) = 9$, $n(T \cap V) = 10$, $n(U \cap V) = 11$, $n(T) = 15$, $n(U) = 16$, $n(V) = 17$.

1.3 ORDER. PROOF BY INDUCTION

The reader is undoubtedly aware that the set of integers may be partitioned into three types of integers: the positive integers 1, 2, 3, ... (sometimes also called the *natural numbers*); zero; and the negative integers -1, -2, -3 .... Furthermore, sums or products of positive integers are again positive. For easy reference we list these properties as a pair of statements P1, P2, called the *order axioms.* Let a, b be integers.

P1 (The Law of Trichotomy) *One, and only one of the following statements concerning a is true:*
(1) *a is positive;* (2) *a is zero;* (3) *$-a$ is positive.*

P2 *If a and b are positive, so are $a + b$ and ab.*

Later [Chapter 2] we shall encounter P1 and P2 again, and then a and b may be other kinds of numbers.

If $-a$ is positive, we say a is *negative.* If a is positive or zero, we say a is *nonnegative.* The nonnegative integers are thus 0, 1, 2, 3, ... If a is negative or zero, we say a is *nonpositive.* The nonpositive integers are thus 0, -1, -2, -3, ...

Property P2 is sometimes called the *closure rule for the positive numbers.* This simply means that the sum or product of positive quantities is again positive.

In general, a number system for which there is a subset of positive elements with properties P1 and P2 is said to be *ordered.* The integers therefore are ordered. Although many number systems are ordered, not all are. This we shall see in Chapter 2. In an ordered number system one may compare elements by introducing symbols $>$, $\geq$, $<$, $\leq$ with meanings given by these rules:

$$\begin{aligned} a > b &\quad \text{means} \quad a - b \text{ is positive;} \\ a \geq b &\quad \text{means} \quad a - b \text{ is nonnegative;} \\ a < b &\quad \text{means} \quad b - a \text{ is positive;} \\ a \leq b &\quad \text{means} \quad b - a \text{ is nonnegative.} \end{aligned}$$

We read $a > b$ as "a is greater than b";
$a \geq b$ as "a is greater than or equal to b";
$a < b$ as "a is less than b";
$a \leq b$ as "a is less than or equal to b".

Formulas involving $>$, $\geq$, $<$, $\leq$ are called *inequalities.* For additional clarity we may read the inequality $a > b$ as "a is strictly greater than b," and the inequality $a < b$ as "a is strictly less than b." A nonstrict inequality is one in which equality may occur.

The following are some basic rules for manipulating inequalities:

(i) (*Law of trichotomy*) One, and only one, of the following statements concerning a, b is true:

(1) $a > b$; (2) $a = b$; (3) $a < b$.

(ii) $a > b$ and $c > d$ imply $a + c > b + d$.

(iii) $a < b$ and $c < d$ imply $a + c < b + d$.

(iv) $a > b$ implies $ac > bc$ when c is positive.

(v) $a > b$ implies $ac < bc$ when c is negative.

(vi) (The property of *transitivity*) If $a > b$ and $b > c$, then $a > c$.

If everywhere in (ii), (iii), (iv), (v), and (vi) we replace $>$ and $<$ with $\geq$ and $\leq$ respectively, we obtain another valid set of implications. The proofs of all of these facts are easy. We give the proof of (ii), and ask the reader to supply the other proofs.

Proof of (ii) The inequalities $a > b$ and $c > d$ mean $a - b$ and $c - d$ are positive. Since the sum of positive quantities is positive (property P2), $(a - b) + (c - d)$ is also positive. That is, $(a + c) - (b + d)$ is positive, and this implies $a + c > b + d$.

Using P1, we introduce the *absolute value of a*, sometimes also called the *norm* of a, and denoted by $|a|$. It is defined by

$$|a| = \begin{cases} a & \text{if } \quad a \text{ is positive,} \\ 0 & \text{if } \quad a \text{ is zero,} \\ -a & \text{if } -a \text{ is positive.} \end{cases}$$

Thus always $|a|$ is nonnegative. The most important properties of the absolute value (or norm) are:

$$-|a| \leq a \leq |a|, \tag{1}$$

$$|a + b| \leq |a| + |b|, \tag{2}$$

$$|ab| = |a|\,|b|. \tag{3}$$

Proof of (1): We consider three cases: a positive, a zero, and a negative. When a is positive, $|a| = a$ and (1) becomes $-a \leq a$ and $a \leq a$. These are evidently correct, the first because $a - (-a) = a + a$ is positive, being the sum of positive quantities. The other two cases are handled similarly, using $|a| = a$ if a is zero and $|a| = -a$ if a is negative. □*

* The symbol □ signifies that the end of a proof has been reached.

Proof of (2): Applying (1) to both a and b, we have

$$-|a| \leq a \leq |a|,$$
$$-|b| \leq b \leq |b|.$$

Adding these inequalities, we deduce

$$-(|a| + |b|) \leq a + b \leq |a| + |b|. \tag{4}$$

From (4) we obtain $a + b \leq |a| + |b|$ and $-(a + b) \leq |a| + |b|$. Since $|a + b|$ equals either $a + b$ or $-(a + b)$, it is evident that (2) holds. □

The proof of (3) is straightforward and so is omitted.

Often statements involving absolute values may be reexpressed as inequalities. For example, for fixed a, fixed positive b, and variable x, the inequality

$$|x - a| \leq b \tag{5}$$

is equivalent to $-b \leq x - a \leq b$ and hence to

$$a - b \leq x \leq a + b. \tag{6}$$

The proof of the equivalence between (5) and (6) is left as an exercise. The reader should be aware that two steps are required: (i) assuming x satisfies (5), prove that it satisfies (6); (ii) assuming x satisfies (6), prove that it satisfies (5). See Exercise 16 below for some of the details.

The following is a list of several important properties of the absolute value, numbered for easy reference as V1, V2, V3, V4.

V1 $|a|$ *is nonnegative for all* a;

V2 $|a| = 0$ *if and only if* $a = 0$;

V3 $|ab| = |a|\,|b|$ *for all* a, b;

V4 $|a + b| \leq |a| + |b|$ *for all* a, b.

We comment that V4 is usually called the *triangle inequality;* the reason for this name will appear in Chapter 2.

Proof by induction

It is an intuitively clear fact that a *nonempty set* S *of positive integers contains a smallest integer*. That is, there is an integer m in S such that $m \leq s$ for every integer s in S. This fact is called the *well-ordering principle,* and the reader should accept it as an axiom. We employ it to explain a widely used method of proof called *proof by induction.*

Let $P(n)$ be a statement depending on a positive integer n, which we hope to prove is true for every $n \geq 1$. However, we are given only the following two facts:

(a) The statement $P(1)$ is true;

(b) Whenever $k - 1$ is a positive integer such that $P(k - 1)$ is true, then $P(k)$ is also true.

The first method of proof by induction is a general principle (called the *First Principle of Induction*) asserting that *if both* (a) *and* (b) *are satisfied, then* $P(n)$ *is true for every positive integer* n. To see why this sweeping conclusion should follow from (a) and (b), let S be the set consisting of all integers n for which $P(n)$ is false. If $P(n)$ is false for at least one n, then S will not be empty and so, by the well-ordering principle, will contain a smallest integer. Let k denote the smallest integer in S. Then $P(k)$ is false since k is in S. Because P(1) is true, we must have $k > 1$. Hence $k - 1$ is a positive integer and cannot be in S because k is the smallest integer in S. Therefore $P(k - 1)$ is true. But now (b) implies that $P(k)$ is true. We have produced the contradiction that $P(k)$ is both true and false. The only escape from this contradiction is that S must be the empty set, meaning that $P(n)$ is true for every positive integer n.

The truth of the statement $P(k - 1)$, which in part (b) is to imply the truth of $P(k)$, is called the *induction hypothesis*.

Example 1.3.1

Let $P(n)$ be the statement

$$1 + 2 + 3 + \cdots + n = \frac{n(n+1)}{2}. \tag{7}$$

Thus P(1) is the statement

$$1 = \frac{1(1+1)}{2}$$

and this is certainly true. Hence (a) is verified. We next verify (b). Suppose $k - 1$ is a positive integer such that $P(k - 1)$ is true, so that we have

$$1 + 2 + \cdots + (k-1) = \frac{(k-1)k}{2}.$$

(This is the induction hypothesis.) Add k to each side. After simplification, the right-hand side $(k - 1)k/2 + k$ becomes $k(k + 1)/2$, and therefore we get

$$1 + 2 + \cdots + (k-1) + k = \frac{k(k+1)}{2}.$$

This equation informs us that $P(k)$ is true. Hence property (b) is verified. The first principle of induction now shows that $P(n)$ is true for every positive integer n. That is, for any positive integer n, equation (7) is correct. □

Example 1.3.2
Let x be a nonnegative number and take $P(n)$ to be the statement

$$(1 + x)^n \geq 1 + nx. \tag{8}$$

Clearly $P(1)$ is true since each side of (8) equals $1 + x$ when $n = 1$. Thus (a) is verified. We next verify (b). Suppose $P(k - 1)$ is true, where $k - 1$ is a positive integer, so that

$$(1 + x)^{k-1} \geq 1 + (k - 1)x. \tag{9}$$

(This is the induction hypothesis.) The rules for manipulating inequalities (discussed above) permit us to multiply each side of (9) by $1 + x$, yielding

$$(1 + x)^k \geq (1 + (k - 1)x)(1 + x) = 1 + kx + (k - 1)x^2 \geq 1 + kx.$$

Therefore $(1 + x)^k \geq 1 + kx$, and so $P(k)$ is true. Thus (b) has been verified. The first principle of induction now implies that $P(n)$ is true for every positive integer n; in other words, (8) is a correct statement for any positive integer n and any nonnegative number x. □

NOTE Sometimes we replace (a) and (b) with

(a′) There is an integer t such that $P(t)$ is true;

(b)′ whenever $k - 1 \geq t$ and $P(k - 1)$ is true, then $P(k)$ is true. It then follows, by an argument similar to the one presented above, that $P(n)$ is true for all integers $n \geq t$. (See Exercise 15 below for the modified version of the well ordering principle needed for this argument.)

Example 1.3.3
Let x be a positive number. Prove that $(1 + x)^n > 1 + nx$ for all integers $n \geq 2$.

SOLUTION Let $P(n)$ be the statement $(1 + x)^n > 1 + nx$. Then $P(2)$ is true, as a simple calculation shows. This verifies part (a′). Let $k - 1 \geq 2$. The proof presented in Example 1.3.2 shows [after noticing that $(k - 1)x^2 > 0$] that if $P(k - 1)$ is true, then so is $P(k)$. Hence (b') is verified. It now follows that $P(n)$ is true for every integer $n \geq 2$ and every positive number x. □

Another version of the principle of induction (called the *Second Principle of Induction*) goes as follows. Suppose that

(a$''$) the statement $P(1)$ is true;

(b$''$) whenever $k - 1$ is a positive integer such that $P(1)$, $P(2)$, . . ., $P(k - 1)$ are *all* true, then $P(k)$ is also true.

The *second induction principle* asserts that if both (a$''$) and (b$''$) are satisfied, *then* $P(n)$ *is true for every positive integer* n. To see this, one again resorts to the well-ordering principle. If there are some values of n for which $P(n)$ is false, let k be the smallest of these values. Because $P(1)$ is true, we must have $k > 1$. Thus $P(1)$, $P(2)$, . . ., $P(k - 1)$ are all true (because k is the *smallest* integer for which $P(k)$ is false) and hence, by (b''), $P(k)$ is also true. We now have the contradiction that $P(k)$ is both true and false. This contradiction shows that no value of n can exist for which $P(n)$ is false, and hence $P(n)$ is true for every $n \geq 1$.

In this form of the principle of induction, the truth of all the statements $P(1)$, . . ., $P(k - 1)$ [from which the truth of the statement $P(k)$ is to follow] is called the *induction hypothesis*.

Of course we may modify the second principle of induction as follows. Suppose that

(a$'''$) the statement $P(t)$ is true;

(b$'''$) whenever k is an integer such that $k > t$ and $P(t)$, $P(t + 1)$, $P(t + 2)$, . . ., $P(k - 1)$ are all true, then $P(k)$ is also true.

It then follows that $P(n)$ is true for all integers $n \geq t$.

Remarks

1. Without supplying proofs, we state that the first principle of induction, the second principle of induction, and the well-ordering principle are logically equivalent in this sense: assuming one of these principles as an axiom, the other two may be deduced from it. In our discussion above we took the well-ordering principle as an axiom (because of its intuitive geometric appeal) and deduced from it the two induction principles, but it would have been possible to adopt instead either principle of induction as an axiom and from it deduce the other induction principle and the well-ordering principle. (See Exercises 27 and 28 below.) Thus these three principles are simply different versions of a single concept. In more advanced textbooks (for example, the Thurston book listed on page 455) the reader will find a full development of the integers and their properties, based on a concise set of axioms, one of which is the first principle of induction.

2. When we say we are using mathematical induction or that we are giving a proof by induction, we mean that we are using one of our principles of induction. Of course, any such proof must contain *two* distinct steps: verification of part (a) [or (a′), (a″), (a‴)], *and* verification of part (b) [or (b′), (b″), (b‴) respectively].

3. We employ two induction principles because sometimes one is easier to apply than the other. An example in which the second principle is the easier to apply will occur in §1.4 when we establish the generalized associative law.

EXERCISES 1.3

Whenever possible, use proof by induction in these exercises.

I

1. Find all integers x satisfying $x^2 - 3x + 2 \leq 0$.
2. Find all integers x satisfying $|x - 1| + |2x + 1| \leq 5$.
3. Find all integers x satisfying $|x^2 - x| \leq 3$.
4. Find all integers x satisfying $|x|^5 + |x^5 - x| \leq 4$.
5. Suppose numbers a, b, c, d satisfy $a > b$ and $c > d$. Does it follow that $a - c > b - d$ or that $ac > bd$ or that $a/c > b/d$? Give either proofs or examples in which these conclusions are false.
6. Let a and b be integers, with a positive, and take $P(n)$ to be the statement $an + b > 0$. Show that $P(k)$ is true whenever $P(k - 1)$ is true. Does it follow that $P(n)$ is true for every positive integer n? Give an example (by selecting special values for a and b) in which $P(10{,}000)$ is false but $P(n)$ is true for all $n > 10{,}000$. (The lesson to be learned from this exercise is that without part (a) [or (a′), (a″), (a‴)], the principle of induction is invalid.)

II

7. Prove that $2^n > n$ for each nonnegative integer n.
8. Prove that $n^3 + 5n$ is an even integer, for each positive integer n. (An integer is even if it is of the form $2m$ for some integer m.)
9. Prove that $(1 + x)^n > 1 + n(x + x^2)$ for all integers $n \geq 3$, if $x > 0$.
10. Prove that $2^n < 1 \cdot 2 \cdot 3 \cdots n$ for all integers $n \geq 4$.
11. Write out proofs of the rules (i)–(vi) for manipulating inequalities. Also write out proofs for the corresponding rules in which $>$ and $<$ are replaced by $\geq$ and $\leq$ respectively.

12. By applying the triangle inequality to $a = \alpha - \beta$ and $b = \beta$, prove that $|\alpha - \beta| \geq | \, |\alpha| - |\beta| \, |$.

13. Show that $|a + b| = |a| + |b|$ if and only if a, b are both nonnegative or both nonpositive.

14. Prove that

$$\frac{1}{1 \cdot 2} + \frac{1}{2 \cdot 3} + \frac{1}{3 \cdot 4} + \cdots + \frac{1}{n(n+1)} = \frac{n}{n+1}.$$

15. (a) Use the well-ordering principle, as stated in the text, to establish the following generalization: *If S is a nonempty set of integers for which there is a fixed integer M with $M \leq s$ for every s in S, then S contains a smallest integer*, i.e., *there is an integer m in S such that $m \leq s$ for all s in S.* (Note: We do not require M to be in S, but do require m to be in S.) Hint: Let set S_1 be given by

$$S_1 = \{s - M + 1 \mid s \in S\}.$$

Show that S_1 contains only positive integers. Apply the well-ordering principle to S_1; if m_1 is the smallest integer in S_1, take $m = m_1 + M - 1$.

(b) Use this generalized well-ordering principle to justify the conclusions in the text just below (b′) and (b‴) "that $P(n)$ is true for all integers $n \geq t$."

16. (a) Deduce this: If $|a| \leq b$, then $-b \leq a \leq b$. Hint: Consider the cases $|a| = a$ or $|a| = -a$.

(b) [The converse of part (a)]: If $-b \leq a \leq b$, deduce that $|a| \leq b$.

17. Is the law of trichotomy valid for $\geq$? That is, if a, b are integers, must exactly one of these statements be true: $a \geq b$; $a = b$; $a \leq b$? Must at least one be true?

III

18. An integer $p > 1$ is a (positive) *prime* if it cannot be factored as $p = ab$, where a and b are integers with $a > 1$ and $b > 1$.

(a) Prove, by the second induction principle, that an integer $n > 1$ may always be written as a product of primes. (This is part of *the Fundamental Theorem of Arithmetic.* See §3.5.) Hint: A positive integer n (with $n > 1$) is either prime or else $n = ab$ where $1 < a < n$ and $1 < b < n$. Apply the induction hypothesis to a and b.

(b) Rework this proof using the well-ordering principle in place of the second induction principle.

19. (i) Let T be the set of positive integers of the form $5m + 1$ where m is a positive integer. Prove that if $a \in T$ and $b \in T$, then $ab \in T$. Hint: If $a = 5m + 1$ and $b = 5m_1 + 1$, then $ab = 5m_2 + 1$ for $m_2 = 5m_1m_2 + m_1 + m_2$.
(ii) An integer n in T is said to be *prime in T* if there do not exist integers a, b in T with $n = ab$. Use the second induction principle or the well-ordering principle to prove that an integer in T may be written as a product of primes in T. (See the hint in Exercise 18.)
(iii) Show that the factorization into primes in T can sometimes be carried out in more than one way. Consider $n = 1296$.)

20. By considering the graph of $y = ax^2 + bx + c$ (where x, y are variable numbers and a, b, c are constants with $a > 0$), show that if y is nonnegative for every x, then $b^2 - 4ac \leq 0$.

21. Let a_1, a_2, b_1, b_2 be real numbers. Set $y = (xa_1 + b_1)^2 + (xa_2 + b_2)^2$. Use Exercise 20 to deduce that $(a_1b_1 + a_2b_2)^2 \leq (a_1^2 + a_2^2)(b_1^2 + b_2^2)$. Hence show that

$$|a_1b_1 + a_2b_2| \leq (a_1^2 + a_2^2)^{1/2}(b_1^2 + b_2^2)^{1/2}.$$

(This inequality is called the *Cauchy-Schwarz inequality.*)

22. Let a and b be positive numbers. The quantity $\sqrt{ab}$ is called the *geometric mean* of a and b, whereas the quantity $\frac{1}{2}(a + b)$ is called the *arithmetic mean.* Prove: $\sqrt{ab} \leq \frac{1}{2}(a + b)$. Hint: Start from $(\sqrt{a} - \sqrt{b})^2 \geq 0$.

23. If $x^2 + bx + c < 0$, show that $b^2 - 4c > 0$ and $-b - \sqrt{b^2 - 4c} < 2x < -b + \sqrt{b^2 - 4c}$.

24. Find all integers x for which both $x^2 + x \leq 12$ and $x^2 + 4 \geq 5x$.

25. If an integer x satisfies $2x^2 - 4x \geq 5$, show that in fact $x^2 \geq 2x + 3$. Does this conclusion follow if x is not an integer?

26. Describe all numbers x (integers or not) for which $|x| + |x - 1| = 1$.

27. Take the first induction principle as an axiom, and from it deduce the second induction principle.

28. Take the second induction principle as an axiom, and from it deduce the well-ordering principle and thence the first induction principle.

29. Use the well-ordering principle to prove that there is no integer x between 0 and 1, i.e., such that $0 < x < 1$. Hint: If there is such an integer, by the well-ordering principle there is a smallest one. Consider x^2.

1.4 GROUPS

Let S be a nonempty set. We lay the foundation for the introduction of groups by first explaining the concept of a *binary operation on* S. We shall use $\circ$ to denote it, and it is postulated to have this property: If a, b are elements of S, then $a \circ b$ is also an element of S. We read "$a \circ b$" as "a circle b" and call $\circ$ "operation circle." The idea inherent in a binary operation is that any elements a and b in S may be combined to produce an element which is *again in* S. We express the fact that operation $\circ$ combines elements of S to produce an element of S by saying that *S is closed under operation* $\circ$.

Example 1.4.1
Take S to be the set of nonnegative integers, and let operation circle be addition. Then $a \circ b$ is simply the sum of a and b. It is customary to use $+$ instead of $\circ$ for this operation, and write $a + b$ instead of $a \circ b$.

Example 1.4.2
Take S to be the set of all integers (that is, both positive and negative integers, and zero), and again let $\circ$ be addition, so that $a \circ b = a + b$. This example differs from Example 1.4.1 in that a different set S is involved.

Example 1.4.3
Take S to be the set of positive integers, and let operation circle be multiplication. Then $a \circ b$ is simply the product of a and b. Often the symbols $\times$ or $\cdot$ are used in place of $\circ$, when $\circ$ is multiplication, but most often the operation symbol is simply omitted and $a \circ b$ is written as ab.

Example 1.4.4
Let S be the set of all integers, as in Example 1.4.2, but now let operation circle be subtraction. The minus symbol $-$ is usually used for this operation, and then $a \circ b$ is written as $a - b$. Note that this operation has the property that $a \circ b$ is usually *not* equal to $b \circ a$, and that $(a \circ b) \circ c$ and $a \circ (b \circ c)$ are usually *not* equal, where c is also an element of S. Furthermore, there is no element e in S such that $e \circ a = a \circ e = a$ for every a in S. This is because $a = e \circ a$ implies $a = e - a$, and there is no e such that this holds for every a.

Example 1.4.5
Let S be the set of all nonzero fractions, that is, numbers of the form m/n where m, n are nonzero integers. (The denominator may be 1 so that the fractions include the integers.) Take $\circ$ to be the division operation; that is, for a and b in S we take $a \circ b = a/b$. The reader should convince

himself that $a \circ b$ is again an element of S. (For this he must show that a/b can be written in the form m/n, where m and n are integers.) Notice that, for this operation, $a \circ b$ is usually not equal to $b \circ a$ and that $(a \circ b) \circ c$ is usually not $a \circ (b \circ c)$, where c is also an element of S. Furthermore, there is no element e in S with the property that $e \circ a = a$ for every a in S. This is because $e \circ a = e/a$ and $e/a = a$ is not true for every a, no matter what e is.

Example 1.4.6

Let S be the set in Example 1.4.5, but now let $\circ$ be multiplication, so that $a \circ b$ is the product ab.

With these examples as illustrations, we now introduce groups.

DEFINITION 1.4.7

A group $\mathcal{G}$ *is a nonempty set endowed with a binary operation* $\circ$ *satisfying the following four postulates:*

G1 *If* g_1 *and* g_2 *are elements of* $\mathcal{G}$, $g_1 \circ g_2$ *is an element of* $\mathcal{G}$.

G2 *For all* g_1, g_2, g_3 *in* $\mathcal{G}$,

$$(g_1 \circ g_2) \circ g_3 = g_1 \circ (g_2 \circ g_3). \tag{1}$$

G3 *There is an element in* $\mathcal{G}$, *denoted by* e, *such that*

$$e \circ g = g \circ e = g \tag{2}$$

for every element g *in* $\mathcal{G}$.

G4 *If* g *is an element of* $\mathcal{G}$, *there is an element* h *in* $\mathcal{G}$ *(*h *depends on* g*) such that*

$$h \circ g = g \circ h = e, \tag{3}$$

where e *is the element described in Axiom* G3.

Some comments are in order concerning these axioms. Axiom G1 merely states that $\circ$ is a binary operation on the set $\mathcal{G}$. We call G1 the *closure* axiom, signifying that $\circ$ applied to elements of $\mathcal{G}$ always produces a result in $\mathcal{G}$. Axiom G2 is called the *associative law*. The element e described in Axiom G3 is called the *identity* or *identity element* of $\mathcal{G}$. It is not immediately clear that $\mathcal{G}$ has only one identity element. This, however, is true and will be proved below. If g is an element of $\mathcal{G}$, the element h in Axiom G4 is called the *inverse* of g. It is not immediately obvious that g has only one inverse. However, this is also true and will be proved below.

We now look at the examples above. Let $\mathcal{G}$ be, in turn, each of the sets S in these six examples.

In Example 1.4.1, the closure and associative Axioms G1 and G2 are fulfilled, and so is Axiom G3 if we take e to be the integer zero. In fact this is the only possible e that will satisfy Axiom G3 (Exercise 17 below). But then there is no element h that will satisfy $g \circ h = e$ since negative integers are not in S. We therefore do not have a group.

In Example 1.4.2 all axioms G1–G4 are satisfied; the integer zero plays the role of the identity e, and the inverse h of the element g is $-g$. These elements are in S since S now contains all integers. Thus we obtain a group.

Axioms G1 and G2 are satisfied in Example 1.4.3, and so is G3 if we take e to be the integer 1. In fact this is the only possible e that will satisfy Axiom G3 (Exercise 17 below). However, G4 is not satisfied since, for example, there is no integer h such that $2 \circ h = 2h = 1$. Thus Example 1.4.3 is not an example of a group.

We saw above that in Examples 1.4.4 and 1.4.5 the associative law, Axiom G2, is not satisfied. Moreover, Axiom G3 is not satisfied either in these examples. These examples are therefore not examples of groups.

Axioms G1–G4 are all satisfied in Example 1.4.6 if we take the identity e to be the integer 1, and the inverse h of the element g to be the number $1/g$. (Note that if g is a fraction, say m/n where m and n are integers, then $1/g = n/m$ is also a fraction and hence is in S.) Thus in Example 1.4.6 we do have an example of a group.

The groups in Examples 1.4.2 and 1.4.6 satisfy the following rule:

G5 $g_1 \circ g_2 = g_2 \circ g_1$ *for all* g_1, g_2.

If G5 holds for elements g_1, g_2 in $\mathcal{G}$, we say that $\circ$ is a commutative operation and that the group $\mathcal{G}$ is *commutative*. (More generally, an operation $\circ$ on a set S is said to be *commutative* if G5 holds for elements g_1, g_2 in S. In Examples 1.4.4 and 1.4.5 we encountered noncommutative operations.) We call G5 the *commutative law*. Not all groups are commutative, as the next example shows.

Example 1.4.8 (Optional)

If a, b are integers, with $a = 1$ or -1, consider the *function* f given by

$$f(x) = ax + b \qquad \text{for all integers } x.$$

If c, d are also integers, with $c = 1$ or -1, let g be the function given by

$$g(x) = cx + d \qquad \text{for all integers } x.$$

We *define* equality of these functions in this way: $f = g$ *if and only if* $f(x) = g(x)$ *for every integer* x. Notice that $f = g$ if and only if $a = c$ and $b = d$ [Proof: If $f = g$, then $f(x) = g(x)$ for every x, so that $ax + b = cx + d$ for every x, and therefore $(a - c)x = d - b$ for every x. If $a - c \neq 0$ this equation is valid for at most one integer x. Therefore $a - c = 0$, and hence $d - b = 0$. Conversely, if $a = c$, $b = d$, obviously $f(x) = g(x)$ for every integer x, so that $f = g$.] We let $\mathcal{G}$ be the set of all functions of the form of f above, in which a is 1 or -1 and b is any integer. Notice how carefully we defined equality of elements of $\mathcal{G}$—see the comment in

§1.2 to the effect that the description of a set must state in precise terms when two of its elements are equal. On $\mathcal{G}$ we define an operation $\circ$ by specifying that $f \circ g$ is the function whose value at x is obtained by evaluating g at x, then evaluating f on the result. That is,

$$(f \circ g)(x) = f(g(x)) \qquad \text{for all integers } x.$$

By direct calculation we find that

$$(f \circ g)(x) = acx + ad + b \qquad \text{for all integers } x.$$

The verification that $\mathcal{G}$ is a group under this operation is left to the reader. We merely remark that the identity e is the function defined by

$$e(x) = x \qquad \text{for all integers } x,$$

and that the inverse of the function f above is the function h given by

$$h(x) = (1/a)x - b/a \qquad \text{for all integers } x.$$

[Note that $1/a$ and b/a are integers since a is 1 or -1.] This group is not commutative because, if we take $a = -1$, $b = c = d = 1$, we get

$$(f \circ g)(x) = -x, \qquad (g \circ f)(x) = -x + 2,$$

and clearly $(f \circ g)(x) \neq (g \circ f)(x)$ for any x, let alone for every x.

The operation $\circ$ introduced in Example 1.4.8 is called *composition* and will be studied in some detail in Chapter 7. Other examples of groups, including noncommutative groups, will be seen in §§2.5 and 7.7. In passing we remark that commutative groups are often called *abelian* groups.

Although the operation $\circ$ may sometimes be addition, when we are speaking of an abstract group $\mathcal{G}$ with operation $\circ$, we shall call $g_1 \circ g_2$ the "product" of g_1 and g_2.

We now state and prove several basic facts about groups which follow from the axioms and which will be needed in §§1.5 and 1.6. For easy reference we number them as G6, G7, etc. Because we do not assume the commutative law G5, the order of factors in any product must receive careful attention, and may be changed only when proper justification is provided.

The closure Axiom G1 permits us to use operation $\circ$ to combine only pairs of elements of $\mathcal{G}$. Thus, if g_1, g_2, g_3 are elements of $\mathcal{G}$ and we wish to calculate $g_1 \circ g_2 \circ g_3$, we have to insert parentheses so that the calculation involves only combinations of pairs of group elements. This we may do in two ways:

$$(g_1 \circ g_2) \circ g_3 \qquad \text{or} \qquad g_1 \circ (g_2 \circ g_3). \tag{4}$$

By the associative law (Axiom G2), these two ways yield the same result. Now suppose g_1, g_2, g_3, g_4 are elements of $\mathcal{G}$. To calculate $g_1 \circ g_2 \circ g_3 \circ g_4$ we again have to insert parentheses such that operation $\circ$ is applied only to pairs of group elements. This we may do in a variety of ways, for example, $((g_1 \circ g_2) \circ g_3) \circ g_4$, or $(g_1 \circ g_2) \circ (g_3 \circ g_4)$, or $g_1 \circ (g_2 \circ (g_3 \circ g_4))$. It requires a theorem to establish that these different ways of calculating $g_1 \circ g_2 \circ g_3 \circ g_4$ yield the same result. This theorem is called the *generalized associative law*. We state it as fact G6.

G6 *Let $g_1, g_2, \ldots, g_n$ be elements of $\mathcal{G}$. Then all meaningful ways of calculating $g_1 \circ g_2 \circ \cdots \circ g_n$ yield the same result.*

REMARK By "meaningful way of calculating $g_1 \circ g_2 \circ \cdots \circ g_n$" we mean that parentheses are inserted so that the calculation involves operation $\circ$ applied only to pairs of group elements.

PROOF We shall prove that no matter how we insert parentheses to compute $g_1 \circ \cdots \circ g_n$, the value we obtain equals the value obtained by inserting the parentheses in the following specific way:

$$(\cdots(((g_1 \circ g_2) \circ g_3) \circ g_4) \cdots \circ g_{n-1}) \circ g_n. \tag{5}$$

For example, when n is 1 or 2, the product $g_1 \circ \cdots \circ g_n$ is g_1 or $g_1 \circ g_2$ and hence is that given by (5). The smallest value of n for which we can compute $g_1 \circ \cdots \circ g_n$ in more than one way is $n = 3$, and then the only ways of inserting parentheses are those displayed in (4) [of which the first is (5)]; by the associative law G2 these yield equal results. Thus if n is 1, 2, or 3, the product $g_1 \circ \cdots \circ g_n$ equals that displayed in (5).

We shall use the second principle of induction. The statement $P(n)$ is to be this: for any elements $g_1, \ldots, g_n$, the calculation of $g_1 \circ \cdots \circ g_n$ by insertion of parentheses in any meaningful way yields a result equal to that displayed in (5). By the paragraph above, $P(1)$ is true. The first step in the proof by induction is therefore completed. We now turn to the second step. Let $k > 1$. We phrase the induction hypothesis [that $P(1), \ldots, P(k-1)$ are all true] in this way:

$$b_1 \circ b_2 \circ \cdots \circ b_l = (\cdots((b_1 \circ b_2) \circ b_3) \cdots) \circ b_l \tag{6}$$

for any elements $b_1, b_2, \ldots, b_l$, whenever $1 \leq \ell < k$. Consider a product $g_1 \circ g_2 \circ \cdots \circ g_k$. We wish to prove that $P(k)$ is true, i.e., that

$$g_1 \circ g_2 \circ \cdots \circ g_k = (\cdots((g_1 \circ g_2) \circ g_3) \cdots) \circ g_k \tag{7}$$

whenever parentheses are meaningfully inserted into the left-hand side. After parentheses are inserted, $g_1 \circ \cdots \circ g_k$ must become $c \circ d$ where $c = g_1 \circ g_2 \circ \cdots \circ g_r$ and $d = g_{r+1} \circ g_{r+2} \circ \cdots \circ g_k$. Here r is some positive integer smaller than k. Using (6) with $\ell = r$ and $b_1 = g_1$, $b_2 = g_2, \ldots, b_r = g_r$, we see that $c = (\cdots((g_1 \circ g_2) \circ g_3)\cdots) \circ g_r$. Using (6) with $\ell = k - r$ and $b_1 = g_{r+1}$, $b_2 = g_{r+2}$, . . ., $b_{k-r} = g_k$, we get

$$d = (\cdots((g_{r+1} \circ g_{r+2}) \circ g_{r+3})\cdots) \circ g_k.$$

The argument now splits into two cases, depending on the value of r. If $r = k - 1$, then $d = g_k$ and $c \circ d = (c) \circ g_k$, so that

$$c \circ d = (\cdots((g_1 \circ g_2) \circ g_3)\cdots) \circ g_k.$$

Thus in this case, $c \circ d$ has the form on the right-hand side of (7). If $r < k - 1$, then $d = \delta \circ g_k$ where $\delta = g_{r+1} \circ g_{r+2} \circ \cdots \circ g_{k-1}$. Thus $c \circ d = c \circ (\delta \circ g_k) = (c \circ \delta) \circ g_k$, using the associative law, Axiom G2. Using (6) with $\ell = k - 1$, we have $c \circ \delta = (\cdots((g_1 \circ g_2) \circ g_3)\cdots) \circ g_{k-1}$. We again see that $c \circ d = (c \circ \delta) \circ g_k$ is given by (7).

Therefore, in all cases, no matter how we compute $g_1 \circ \cdots \circ g_k$, we have proved that the result obtained equals that given by (7). Therefore $P(k)$ is true. This completes the second step in the proof by induction, so that we now know $P(n)$ to be true for every natural number n. That is: any meaningful calculation of $g_1 \circ \cdots \circ g_n$ gives a result equal to the product displayed in (5). The generalized associative law is proved. □

REMARK The above proof made use of Axioms G1 and G2 only, and did not use Axioms G3 and G4. We therefore have: *Whenever the closure and associative laws* (Axioms G1 *and* G2) *are valid, then the generalized associative law is also valid.* We shall use this remark in §1.5.

We next prove the uniqueness, mentioned above, of the elements e and h.

G7 *The identity element e of $\mathcal{G}$ is unique.*

PROOF Suppose that an element e' in $\mathcal{G}$ also satisfies G3, so that

$$e' \circ g = g \circ e' = g \tag{8}$$

for all g in $\mathcal{G}$. We wish to prove that $e = e'$. Taking g to be e' in (2), and taking g to be e in (8), we get $e' = e \circ e' = e$. This shows that $e = e'$. □

G8 *Let g be an element of $\mathcal{G}$. Then the inverse h of g is unique.*

PROOF Suppose that h' satisfies

$$h' \circ g = g \circ h' = e. \tag{9}$$

We wish to show that $h' = h$. To do this we calculate as follows:

$$\begin{aligned} h' &= h' \circ e && \text{by G3} \\ &= h' \circ (g \circ h) && \text{by G4} \\ &= (h' \circ g) \circ h && \text{by G2} \\ &= e \circ h && \text{by (9)} \\ &= h. && \text{by G3} \end{aligned}$$

Thus $h' = h$, as desired. □

NOTATION We let g^{-1} denote the inverse of g. Thus g^{-1} is the unique element in $\mathcal{G}$ satisfying $g^{-1} \circ g = g \circ g^{-1} = e$. The use of the superscript $^{-1}$ to denote the inverse of g is a notational convenience that avoids introducing a new letter each time we talk about inverses. Moreover, we get the following inversion rules.

G9 *Let f and g be elements of $\mathcal{G}$. Then:*

(a) $(f \circ g)^{-1} = g^{-1} \circ f^{-1}$;

(b) $(g^{-1})^{-1} = g$.

PROOFS (a) Let $h = g^{-1} \circ f^{-1}$. Then

$$\begin{aligned} (f \circ g) \circ h &= (f \circ g) \circ (g^{-1} \circ f^{-1}) = f \circ (g \circ g^{-1}) \circ f^{-1} \\ &= f \circ e \circ f^{-1} = f \circ f^{-1} = e \end{aligned}$$

and

$$\begin{aligned} h \circ (f \circ g) &= (g^{-1} \circ f^{-1}) \circ (f \circ g) = g^{-1} \circ (f^{-1} \circ f) \circ g \\ &= g^{-1} \circ e \circ g = g^{-1} \circ g = e. \end{aligned}$$

In each of these calculations we used, in order, the definition of h, G6, G4, G3, G4. Thus h satisfies $(f \circ g) \circ h = e$ and $h \circ (f \circ g) = e$, so that it meets the requirements of $(f \circ g)^{-1}$. Hence $h = (f \circ g)^{-1}$. This proves (a).

(b) We have $g^{-1} \circ g = g \circ g^{-1} = e$. The requirements of the inverse x of g^{-1} are $g^{-1} \circ x = x \circ g^{-1} = e$, and these equations say that g fulfills these requirements. Therefore $g = (g^{-1})^{-1}$. □

In many groups the operation $\circ$ is addition, denoted by $+$. When this is the case we denote the identity element of $\mathcal{G}$ by 0 (zero) and the inverse of g by $-g$ (minus g). Thus, when the group operation is $+$, the symbol 0 indicates the unique element of $\mathcal{G}$ for which $0 + g = g + 0 = g$ for all g in $\mathcal{G}$, and the symbol $-g$ indicates the unique element of $\mathcal{G}$ for which $(-g) + g = g + (-g) = 0$. The notation $+$ for the operation $\circ$ is commonly used only for commutative groups, and we say that a commutative group in which $\circ$ is $+$ is an *additive* group. In such a group $g_1 + g_2 = g_2 + g_1$ holds for all group elements g_1 and g_2, and it is customary to write $g_1 + (-g_2)$ more simply as $g_1 - g_2$. Notice that in an additive group the inversion rules G9 yield the following *rules of the minus sign:*

$$-(f + g) = (-f) + (-g), \tag{10}$$

$$-(-g) = g. \tag{11}$$

Let p be a positive integer, and let g be an element of $\mathcal{G}$. We define positive, zero, and negative powers of g by

$$g^p = \underbrace{g \circ g \circ g \circ \cdots \circ g}_{p \text{ factors}}, \tag{12}$$

$$g^0 = e, \tag{13}$$

$$\begin{aligned} g^{-p} &= \underbrace{g^{-1} \circ g^{-1} \circ g^{-1} \circ \cdots \circ g^{-1}}_{p \text{ factors}} \\ &= (g^{-1})^p \qquad \text{by (12)}. \end{aligned} \tag{14}$$

We now give the familiar *laws of exponents.*

G10 *Let m and n be integers and g an element of $\mathcal{G}$. Then*

$$g^m \circ g^n = g^{m+n}, \tag{15}$$

$$(g^m)^n = g^{mn}. \tag{16}$$

Proof of (15). If $m = 0$ or $n = 0$ it is easy [using (13)] to verify (15). So assume m and n are both nonzero. The proof splits into six cases as follows: m, n both positive; m, n both negative; m positive, n negative, and $m + n$ nonnegative; m positive, n negative, and $m + n$ negative; m negative, n positive, and $m + n$ nonnegative; m negative, n positive, and $m + n$ negative. We give the proof in two of the cases; the reader may supply for himself the proof in the other cases. If m, n are both positive, then

$$\begin{aligned} g^m \circ g^n &= \underbrace{(g \circ \cdots \circ g)}_{m \text{ factors}} \circ \underbrace{(g \circ g \circ \cdots \circ g)}_{n \text{ factors}} \\ &= \underbrace{g \circ g \circ \cdots \circ g}_{m + n \text{ factors}}, \qquad \text{by generalized associativity,} \\ &= g^{m+n}. \end{aligned}$$

If m is positive and n is negative with $m + n$ negative, then [using (12), (14), and the generalized associative law] we have

$$\begin{aligned} g^m \circ g^n &= \underbrace{(g \circ g \circ \cdots \circ g)}_{m \text{ factors}} \circ \underbrace{(g^{-1} \circ g^{-1} \circ \cdots \circ g^{-1})}_{-n \text{ factors}} \\ &= \underbrace{(g \circ \cdots (g \circ (g \circ (g}_{m \text{ factors}} \circ \underbrace{g^{-1}) \circ g^{-1}) \circ g^{-1}) \cdots \circ g^{-1})}_{m \text{ factors}} \circ \underbrace{g^{-1} \circ g^{-1} \circ \cdots \circ g^{-1}}_{-n - m \text{ factors}}. \end{aligned}$$

Since $g \circ g^{-1} = e$ and $g \circ e \circ g^{-1} = g \circ g^{-1}$, we may start at the innermost parentheses and delete $g \circ g^{-1}$, then delete a second $g \circ g^{-1}$, a third $g \circ g^{-1}$, etc. Continuing, each g cancels against one g^{-1}, and we eventually get

$$g^m \circ g^n = \underbrace{g^{-1} \circ g^{-1} \circ \cdots \circ g^{-1}}_{-n-m \text{ factors}} = g^{m+n},$$

using (14). □

Proof of (16). By taking $m + n = 0$ in (15), we get $g^m \circ g^{-m} = e = g^{-m} \circ g^m$. Hence

$$(g^m)^{-1} = g^{-m} \tag{17}$$

for any integer exponent m. We shall use this below.

If $m = 0$ or $n = 0$, it is evident from (13) that (16) is valid. We may therefore assume in the remainder of the proof that m and n are both nonzero. The proof now divides into several cases. If m and n are both positive, we have

$$(g^m)^n = \underbrace{\underbrace{(g \circ \cdots \circ g)}_{m \text{ factors}} \circ \underbrace{(g \circ \cdots \circ g)}_{m \text{ factors}} \circ \cdots \circ \underbrace{(g \circ \cdots \circ g)}_{m \text{ factors}}}_{n \text{ blocks of factors}}.$$

There are altogether mn occurrences of g on the right-hand side here, so that removing parentheses produces

$$\underbrace{g \circ \cdots \circ g}_{mn \text{ factors}} = g^{mn}.$$

This proves (16) for any g and any positive exponents m and n. Now let $m = -p$ be negative and n be positive. Set $h = g^{-1}$. Then we have

$$(g^m)^n = (g^{-p})^n = ((g^{-1})^p)^n = (h^p)^n = h^{pn} = (g^{-1})^{pn} = g^{-pn} = g^{mn}.$$

In order here we used: $m = -p$, (14), the definition of h, the already proved case of (16) applied to h, the definition of h, (14) with pn in place of p, and $m = -p$. Therefore (16) is proved for any positive or negative m and any positive n. To complete the proof we let m be any integer and $n = -p$ be negative. Set $h = g^m$. Then:

$$(g^m)^n = h^n = h^{-p} = (h^{-1})^p = ((g^m)^{-1})^p = (g^{-m})^p = g^{-mp} = g^{mn}.$$

In order here we used: the definition of h, $n = -p$, (14) applied to h, the definition of h, (17), the already proved cases of (16), and $n = -p$. This ends the proof. □

There is one more property, valid in any group, that we shall require.

G11 *Let f, g be elements of the group $\mathcal{G}$. Then:*

(a) *the equation $f \circ x = g$ in unknown x has exactly one solution in $\mathcal{G}$;*
(b) *the equation $x \circ f = g$ in unknown x has exactly one solution in $\mathcal{G}$.*

PROOF (a) It is easy to verify that $x = f^{-1} \circ g$ satisfies the equation $f \circ x = g$. Conversely, if x satisfies $f \circ x = g$, then $f^{-1} \circ (f \circ x) = f^{-1} \circ g$, hence $(f^{-1} \circ f) \circ x = f^{-1} \circ g$, yielding $x = f^{-1} \circ g$ since $f^{-1} \circ f = e$. Thus $f^{-1} \circ g$ is the only solution.

(b) By a similar argument it is easy to show that $x = g \circ f^{-1}$ is the only solution of $x \circ f = g$. □

In an additive group, G11 says: *there is exactly one element* x, *namely* $x = g - f$, *satisfying the equation* $f + x = g$.

We may summarize our discussion of groups this way: A group is a set endowed with a binary operation for which certain simple rules (closure, associativity, existence of identity and inverses) are valid. An important type of group is the commutative group (also called abelian group) in which the commutative law G5 is also valid.

In §1.6 we shall encounter groups in which the group operation is multiplication. Groups of this type are called *multiplicative groups.*

Because we often encounter sets satisfying some, but not all, of the group axioms G1, G2, G3, G4, we introduce two additional concepts, that of *semigroup* and *monoid.*

DEFINITION 1.4.9

Let S *be a nonempty set, endowed with a binary operation* $\circ$.

(i) *If the closure and associative laws*

$$\text{G1:}\quad g_1 \circ g_2 \in S; \qquad \text{G2:}\quad g_1 \circ (g_2 \circ g_3) = (g_1 \circ g_2) \circ g_3,$$

are valid for all elements g_1, g_2, g_3 *of* S, *we call* S *a* semigroup.

(ii) *If* S *is a semigroup and contains also an element* e *satisfying*

$$\text{G3:}\quad e \circ g = g \circ e = g \qquad \text{for all } g \in S,$$

we call S *a* monoid.

A semigroup or monoid S satisfying

$$\text{G5:}\quad g_1 \circ g_2 = g_2 \circ g_1 \qquad \text{for all } g_1, g_2 \in S,$$

is said to be commutative. Examples of commutative semigroups which are not monoids are:

(i) the positive integers, $\circ$ being addition;
(ii) the positive integers > 1, $\circ$ being multiplication.

Examples of commutative monoids which are not groups are:

(iii) the nonnegative integers, $\circ$ being addition;
(iv) the positive integers, $\circ$ being multiplication.

(*Optional*) If we modify Example 1.4.8 and consider all functions $f(x) = ax + b$ in which a is allowed to be any even integer and b any integer, we obtain a noncommutative semigroup which is not a monoid. If we allow a, b to be any integers, we obtain a noncommutative monoid which is not a group.

EXERCISES 1.4

I

1. Solve the following equations for g:
   (a) $3 \circ g = 2$ in the group of Example 1.4.2;
   (b) $3 \circ g = 2$ in the group of Example 1.4.6;
   (c) [*optional*] If $f(x) = -x + 2$ and $h(x) = x - 3$, for all integers x, solve $f \circ g = h$ and $g \circ f = h$ for g in the group of Example 1.4.8.

2. In the groups of Examples 1.4.2 and 1.4.6, compute $2 \circ 4 \circ 2 \circ 3 \circ 6$.

3. Find 2^{-1}, 3^{-1}, 4^{-1}, 5^{-1} in the groups of Examples 1.4.2 and 1.4.6.

4. Find all elements g in the groups of Examples 1.4.2 and 1.4.6 satisfying $g \circ g \circ g \circ g = e$. [*Optional*] Do the same in the group of Example 1.4.8.

5. Let n be a positive integer. Solve the equation $g^n = e$ for g in the groups of Examples 1.4.2 and 1.4.6.

6. Let $\mathcal{G}$ be the set of all integers. Under operation $\circ$, as below, is $\mathcal{G}$ a semigroup? a monoid? a group? Is it commutative?
   (a) $a \circ b = a + b - 1$;
   (b) $a \circ b = |a| + |b|$;
   (c) $a \circ b = a$;
   (d) $a \circ b = -ab$.

7. If $\mathcal{G}$ is the set of all fractions (including zero and negative fractions), is $\mathcal{G}$ a semigroup, monoid, or a group under the operations $\circ$ in Exercise 6? Repeat, on the set of all nonzero fractions.

8. If $\circ$ is the operation of Exercise 6(a), and if $a \circ b = 0$, does it follow that a or b is 0? Here 0 denotes the zero integer.

9. Do the integers 1, -1 under operation multiplication form a group? What if the operation is addition?

10. Do the (positive, negative, or zero) even integers under operation addition form semigroup, monoid, or a group? What about the odd integers? (An even integer has the form $2s$, an odd integer has the form $2s + 1$, for s an integer.)

II

11. Let $g_1, \ldots, g_n$ be elements of a group $\mathcal{G}$. Generalize the rule G9(a) and show that $(g_1 \circ \cdots \circ g_n)^{-1} = g_n^{-1} \circ \cdots \circ g_1^{-1}$. Imitate the proof of G9(a).

12. Let $\mathcal{G}$ be a commutative group, the group operation $\circ$ being $+$. In this type of group it is customary to denote the mth power of g by mg (instead of g^m). Thus, if p is a positive integer, we have

$$pg = g + g + \cdots + g, \qquad (p \text{ terms}),$$

$$0g = 0, \qquad (\text{the first } 0 \text{ is the zero integer, the second zero is the } e \text{ element of } \mathcal{G}),$$

$$\begin{aligned}(-p)g &= (-g) + (-g) + \cdots + (-g) \qquad (p \text{ terms})\\ &= p(-g).\end{aligned}$$

As an application of the laws of exponents (G10), show that if m and n are integers (positive, negative, or zero) then

$$mg + ng = (m + n)g,$$

$$m(ng) = (mn)g.$$

13. Show that an element g in a group $\mathcal{G}$ satisfies $g^2 = e$ if and only if $g = g^{-1}$.

14. Show that if every element g in a group $\mathcal{G}$ satisfies $g^2 = e$, then $\mathcal{G}$ is a commutative group. Hint: Use $(g_1 \circ g_2)^2 = e$.

15. Let $g_1, g_2, \ldots, g_n$ be elements of an additive commutative group $\mathcal{G}$. Prove that

$$-(g_1 + g_2 + \cdots + g_n) = (-g_1) + (-g_2) + \cdots + (-g_n).$$

(Use induction on n or Exercise 11.)

16. Let $\mathcal{G}$ be a commutative group with operation $\circ$. Thus $g_1 \circ g_2 = g_2 \circ g_1$ for all elements g_1, g_2 of $\mathcal{G}$. Convince yourself of the validity of the following generalization of this formula: let $g_1, g_2, \ldots, g_n$ be elements of $\mathcal{G}$, and let $i_1, i_2, \ldots, i_n$ denote the integers $1, 2, \ldots, n$, but not necessarily in this order. Then

$$g_{i_1} \circ g_{i_2} \circ \cdots \circ g_{i_n} = g_1 \circ g_2 \circ \cdots \circ g_n.$$

This is known as the *generalized commutative law.*

17. In Example 1.4.1, show that $e = 0$ is the only element of S satisfying Axiom G3. Do this also for Example 1.4.2. In Example 1.4.3, show that $e = 1$ is the only element of S satisfying Axiom G3. Repeat for Example 1.4.6.

III

18. Find an example of a group $\mathcal{G}$ in which $g^2 = e$ for every g in $\mathcal{G}$.

19. Find an example of a group $\mathcal{G}$ in which e is the only element satisfying an equation $g^n = e$ with the exponent n nonzero.

20. Let $\mathcal{G}$ be a set consisting of the integers 0, 1, 2. Define operation $\circ$ on $\mathcal{G}$ by these rules: $0 \circ 0 = 0$, $0 \circ 1 = 1 \circ 0 = 1$, $0 \circ 2 = 2 \circ 0 = 2$, $1 \circ 1 = 2$, $1 \circ 2 = 2 \circ 1 = 0$, $2 \circ 2 = 1$. Verify that $\mathcal{G}$ with this operation $\circ$ is a commutative group. Find 1^{-1} and 2^{-1} in this group.

21. Let n be a positive integer. Solve the equation $g^n = e$ for g in the group of Example 1.4.8.

22. Suppose, for each pair of elements f, g in a group $\mathcal{G}$, the solution of $f \circ x = g$ is also a solution of $x \circ f = g$. Prove or disprove: $\mathcal{G}$ is a commutative group.

23. Define functions $f_1, f_2, f_3, f_4, f_5, f_6$ by

$$f_1(x) = x, \qquad f_2(x) = \frac{1}{x}, \qquad f_3(x) = \frac{-1}{x+1},$$

$$f_4(x) = \frac{-x-1}{x}, \qquad f_5(x) = -x-1, \qquad f_6(x) = \frac{-x}{x+1}.$$

Here x is any integer or fraction other than 0 or -1. Define equality of functions and operation $\circ$ on functions in the manner of Example 1.4.8; e.g., functions f, g are equal if and only if $f(x) = g(x)$ for all x, and $(f \circ g)(x) = f(g(x))$. Show that $f_1, f_2, f_3, f_4, f_5, f_6$ constitute a group having exactly six elements. Is this group commutative?

24. Define functions g_1, g_2, g_3, g_4 for $x \neq 0$ by

$$g_1(x) = x, \qquad g_2(x) = \frac{1}{x}, \qquad g_3(x) = -x, \qquad g_4(x) = -\frac{1}{x}.$$

Show that under the operation $\circ$ of the exercise above these four functions form a group. How many elements g in this group satisfy $g^2 = e$? Is this group commutative?

1.5 RINGS

Most number systems involve two binary operations: one called addition and denoted by $+$, the other called multiplication and denoted by $\times$, or $\cdot$, or simply by juxtaposition (that is, by writing ab instead of $a \times b$ or $a \cdot b$). We now introduce an algebraic structure, called a *ring*, based on this simple observation.

DEFINITION 1.5.1
A ring $\mathcal{R}$ *is a nonempty set endowed with two binary operations, called* addition *(denoted by* $+$*) and* multiplication *(denoted by juxtaposition), such that the following axioms are satisfied:*

R1 $\mathcal{R}$, *with operation* $+$, *is a commutative group.*

R2 ab *is in* $\mathcal{R}$ *whenever* a *and* b *are in* $\mathcal{R}$.

R3 *For all* a, b, c *in* $\mathcal{R}$, $(ab)c = a(bc)$.

R4 *For all* a, b, c, *in* $\mathcal{R}$,

$$a(b + c) = ab + ac,$$
$$(b + c)a = ba + ca.$$

Some words of explanation are in order concerning these axioms. Written out in full, Axiom R1 states that:

R1(i) $a + b$ *is in* $\mathcal{R}$ *whenever* a *and* b *are elements of* $\mathcal{R}$.
R1(ii) $(a + b) + c = a + (b + c)$ *for all* a, b, c *in* $\mathcal{R}$.
R1(iii) $\mathcal{R}$ *contains an element* 0 *such that* $0 + a = a + 0 = a$ *for all* a *in* $\mathcal{R}$.
R1(iv) *If* a *is an element of* $\mathcal{R}$, *there is an element* $-a$ *in* $\mathcal{R}$ *such that* $a + (-a) = (-a) + a = 0$.

Axioms R1(i)-(iv) are the group axioms G1–G4 rewritten in additive form. We also have

R1(v) $a + b = b + a$ *for all* a, b *in* $\mathcal{R}$.

This is just G5. We call R1(i) the *closure law* for addition and R1(ii) the *associative law* for addition. The element 0 is called the *zero element* of $\mathcal{R}$ and $-a$ is called the *additive inverse* (or, sometimes, the *negative*) of a. We call R1(v) the *commutative law* for addition.

Axiom R2 is simply a restatement of the fact that multiplication is a binary operation, just as R1(i) is simply a restatement of the fact that addition is a binary operation. We call R2 the *closure law for multiplication* and R3 the *associative law for multiplication*. Together R2 and R3 imply that $\mathcal{R}$ is a semigroup under multiplication. The only axiom in which both addition and multiplication appear is Axiom R4; the formulas in R4 are called the *distributive laws*.

Notice that the following rule R5 is *not* part of the axioms:

R5 $ab = ba$ *for all* a, b *in* $\mathcal{R}$.

Rings in which R5 is valid are called *commutative rings*. Thus a ring is commutative *if and only if the multiplication operation is commutative.* Furthermore, we do *not* require that there be an element e such that

R6 $ea = ae = a$ *for all* a *in* $\mathcal{R}$.

If $\mathcal{R}$ contains an element e such that R6 holds, we say $\mathcal{R}$ is a *ring with identity* and that e is the *identity element of* $\mathcal{R}$. Nearly all rings encountered in this book will be commutative rings and many, though not all, will be rings with identity.*

We now give a number of examples.

Example 1.5.2

Let $\mathcal{R}$ be the set of integers, including zero and both positive and negative integers, and take the addition and multiplication operations in $\mathcal{R}$ be the usual operations of addition and multiplication of integers. Then $\mathcal{R}$ is a commutative ring with identity. We call this ring the *ring of integers* and we shall always use the symbol $\mathbf{Z}$ to denote it. We use $\mathbf{Z}$ because it is the first letter of the German word *Zahl*, meaning "number." In Chapters 3 and 4 we shall study this ring in great detail.

Example 1.5.3

A *rational number* is, by definition, a number of the form m/n, where $m, n \in \mathbf{Z}$ and $n \neq 0$. Thus the rational numbers are the numbers commonly called fractions, including positive fractions, negative fractions, and zero. An integer is automatically a rational number since the denominator n may be 1. A rational number may have more than one expression as a fraction m/n with integer numerator m and integer denominator n. For example, m and n may have a common factor which, if cancelled, yields another fraction equal to m/n. If m, m', n, n' are integers, with n, n' nonzero, we therefore *define* equality of fractions by specifying that

$$\frac{m}{n} = \frac{m'}{n'} \quad \text{if and only if} \quad mn' = nm'. \tag{1}$$

Letting m'', n'' also be integers, $n'' \neq 0$, from this definition we get the usual properties of equality:

(i) $\dfrac{m}{n} = \dfrac{m}{n}$;

(ii) If $\dfrac{m}{n} = \dfrac{m'}{n'}$, then $\dfrac{m'}{n'} = \dfrac{m}{n}$;

(iii) If $\dfrac{m}{n} = \dfrac{m'}{n'}$ and $\dfrac{m'}{n'} = \dfrac{m''}{n''}$, then $\dfrac{m}{n} = \dfrac{m''}{n''}$.

*** In many texts the identity e is required to be different from the element 0. We shall not impose this condition on e. It will turn out, however, that in all examples of interest, e is different from 0. (See Exercise 10 at the end of this section.) We also note that some texts require that a ring have an identity element.**

The proofs are easy:

Proof of (i). From the definition of fraction equality, (i) holds if and only if $mn = nm$, and this is clearly true. □

Proof of (ii). From $m/n = m'/n'$ we get $mn' = nm'$; hence $m'n = n'm$, and therefore $m'/n' = m/n$. □

Proof of (iii). From $m/n = m'/n'$ and $m'/n' = m''/n''$, we get $mn' = nm'$ and $m'n'' = n'm''$. Multiplying the first equation of this last pair by n'' and the second equation by n gives $mn'n'' = nm'n'' = nn'm''$. Hence $n'(mn'' - nm'') = 0$. Because $n' \neq 0$, we get $mn'' - nm'' = 0$, yielding $mn'' = nm''$, and hence $m/n = m''/n''$. □

We denote the set of rational numbers by **Q**, using **Q** because it is the first letter of "quotient," and so indicates that an element of **Q** is a quotient of integers.

If

$$a = \frac{m}{n}, \qquad b = \frac{m'}{n'} \tag{2}$$

are rational numbers, we define addition and multiplication in the usual way:

$$a + b = \frac{mn' + nm'}{nn'}, \qquad ab = \frac{mm'}{nn'}. \tag{3}$$

Since the right-hand sides of equations (3) are quotients of integers, both $a + b$ and ab are rational numbers. The closure axioms R1(i) and R2 are therefore satisfied in **Q**. However, because fractions a and b may be represented as quotients of integers in more than one way, the definitions (3) involve a logical difficulty which must be cleared up before proceeding. If we write a and b as fractions in another way, as

$$a = \frac{m_1}{n_1}, \qquad b = \frac{m_1'}{n_1'}, \tag{4}$$

where m_1, n_1, m_1', n_1' are integers, the definitions (3) of $a + b$ and ab, when applied to the representations (4), yield

$$a + b = \frac{m_1n_1' + n_1m_1'}{n_1n_1'}, \qquad ab = \frac{m_1m_1'}{n_1n_1'}. \tag{5}$$

It must be shown that the calculations in (3) yield the same results as the calculations in (5). According to our definition of fraction equality, this will be the case if and only if

$$(mn' + nm')n_1n_1' = nn'(m_1n_1' + n_1m_1'), \tag{6}$$

$$(mm')(n_1n_1') = (nn')(m_1m_1'). \tag{7}$$

From (2) and (4) we get

$$mn_1 = nm_1 \qquad \text{and} \quad m'n_1' = n'm_1'. \tag{8}$$

If we can show that (6) and (7) follow from (8), we will know that $a + b$ and ab depend only on a and b themselves and not on the way they are written as fractions.

It is easy to show that (6) and (7) follow from (8). To get (6), multiply the first and second equations in (8) by $n'n_1'$ and nn_1 respectively, then add. To get (7), multiply together equations (8).

It is now straightforward to verify axioms R1–R4 for fractions. The verifications are left as exercises. We merely note that 0 is in **Q** since $0 = 0/1$ is a quotient of integers, and if $a = m/n$, then $-a = (-m)/n$ is also a quotient of integers and therefore in **Q**. As an example of the verifications of the other axioms, we verify the first formula in the distributive law R4. If $c = m''/n''$, using the definitions of sum and product given in (3), we compute that

$$ab + ac = \frac{(mm')(nn'') + (nn')(mm'')}{(nn')(nn'')},$$

$$a(b + c) = \frac{m(m'n'' + n'm'')}{n(n'n'')}.$$

The fractions appearing on the right-hand sides of this pair of equations are equal, since

$$\{(mm')(nn'') + (nn')(mm'')\}n(n'n'') = (nn')(nn'')\{m(m'n'' + n'm'')\},$$

as one sees by multiplying out both sides.

Thus **Q** is a ring. It is commutative because $m/n \cdot m'/n' = (mm')/(nn') = m'/n' \cdot m/n$, and a ring with identity because it contains the integer 1.

Notice that the addition and multiplication operations in **Q**, when applied to integers (i.e., to fractions of the form $m/1$) reproduce the addition and multiplication operations for integers. To see this, take $n = n' = 1$ in (2) and (3), thereby obtaining

$$\frac{m}{1} + \frac{m'}{1} = \frac{m + m'}{1}, \qquad \frac{m}{1} \cdot \frac{m'}{1} = \frac{mm'}{1}.$$

In §6.3 we shall repeat this discussion in greater detail and generality.

For use later, observe that the *nonzero* rational numbers form a multiplicative group. We leave the verifications of group axioms G1, G2, G3, to the reader, giving here only the verification of the axiom about inverses, G4: If m/n is a nonzero fraction, as above, the nonzero fraction n/m is its multiplicative inverse. This verifies G4.

Example 1.5.4

Let $\mathcal{R}$ consist of all the even integers. (An integer z is *even* if and only if $z = 2t$ for some integer t. It is *odd* if and only if $z = 2t + 1$ for some integer t.) The even integers include, of course, the negative even integers and zero. Endow $\mathcal{R}$ with the usual operations of addition and multiplication of integers. Notice that if $z_1 = 2t_1$ and $z_2 = 2t_2$ are even integers $(t_1, t_2 \in \mathbf{Z})$, then $z_1 + z_2 = 2(t_1 + t_2)$, $z_1z_2 = 2(2t_1t_2)$, $-z_1 = 2(-t_1)$, and $0 = 2 \cdot 0$ are all even integers. It is now easy to check that $\mathcal{R}$ is a commutative ring. It does not have an identity element, however, since the integer 1 is odd and hence is not in $\mathcal{R}$.

Example 1.5.5

Let $\mathcal{G}$ be any commutative group with $+$ as its operation. We may make $\mathcal{G}$ into a ring by defining multiplication in the following way: for any elements g_1 and g_2 of $\mathcal{G}$, define $g_1 \cdot g_2 = 0$. Here 0 is, of course, the (additive) identity of $\mathcal{G}$. It is a straightforward exercise (that we shall leave to the reader) to verify that $\mathcal{G}$, with these $+$ and $\cdot$ operations, is a commutative ring.

An example of a noncommutative ring may be found in Exercise 14 below. The reader familiar with matrices will observe that the set of 2×2 matrices with integers as entries furnishes another example of a noncommutative ring. Throughout this book we shall chiefly be interested in commutative rings.

We now discuss some basic properties common to all rings. We give proofs here so that we will know, without further discussion, that these properties hold for the rings to be encountered in Chapters 2, 3, and in later chapters. We list these properties as R7, R8, etc. Although our principal interest is in commutative rings, we do not assume commutativity in R7, R8, etc. Thus the order of the factors in any product should receive careful attention and should not be changed without adequate justification.

We first inquire about sums or products of more than two elements of a ring $\mathcal{R}$. The hypotheses permit addition or multiplication of pairs of elements of $\mathcal{R}$ only. Whenever more than two elements are to be added or multiplied, parentheses must be inserted to reduce the calculation to repeated addition or multiplication of pairs of elements. As in group theory, we need a generalized associative law to be certain that the result of the calculation will not depend on how the parentheses are inserted. We already know this to be the case for addition since $\mathcal{R}$, with operation $+$, is a group. By taking the operation $\circ$ to be multiplication in the remark following the proof of G6, we see that the generalized associative law for multiplication follows from the closure and associativity axioms R2 and R3. We state these conclusions as R7.

R7 (The generalized associative laws) *Let $r_1, r_2, \ldots, r_n$ be elements of $\mathcal{R}$. Then all calculations of $r_1 + r_2 + \cdots + r_n$ by meaningful insertion of parentheses yield equal results. This is also true of $r_1 r_2 \cdots r_n$.*

R8 (The generalized commutative laws) *Let $r_1, r_2, \ldots, r_n$ be elements of a ring $\mathcal{R}$, and let $i_1, i_2, \ldots, i_n$ denote the integers* 1, 2, . . ., *n in any other order. Then*

$$r_1 + r_2 + \cdots + r_n = r_{i_1} + r_{i_2} + \cdots + r_{i_n}. \tag{9}$$

If $\mathcal{R}$ is a commutative ring, then also

$$r_1 r_2 \cdots r_n = r_{i_1} r_{i_2} \cdots r_{i_n}. \tag{10}$$

PROOF A moment's thought will convince the reader that (9) follows from the commutative rule for addition and (10) follows from the commutative rule for multiplication. □

It is worthwhile to have a shorthand notation for sums such as (9). We use the Greek letter Σ (capital sigma) for this purpose, by defining

$$\sum_{i=1}^{n} r_i = r_1 + r_2 + \cdots + r_n.$$

Notice that $\sum_{i=1}^{n} r_i = \sum_{j=1}^{n} r_j$ since both mean $r_1 + r_2 + \cdots + r_n$. For the same reason,

$$\sum_{i=1}^{n} r_i = \sum_{i=1}^{n-1} r_i + r_n$$

if n is a positive integer. In a commutative ring we use the Greek letter Π (capital pi) similarly as a shorthand notation for products by defining

$$\prod_{i=1}^{n} r_i = r_1 r_2 \cdots r_n.$$

A frequently occurring product involving integers is the quantity $n!$ (read this as "n factorial") defined by

$$n! = \prod_{i=1}^{n} i = 1 \cdot 2 \cdot 3 \cdots n.$$

Often the additional symbol $0!$ is needed. It is defined by $0! = 1$. It is then an easy exercise to show that

$$n! = n \cdot (n - 1)!$$

for any positive integer n.

The two formulas in the distributive axiom R4 show how parentheses are removed when a ring element is multiplied onto a sum of two terms.

It is natural to ask how parentheses are removed when sums of several terms are multiplied. The next result, called the *generalized distributive law,* provides the method.

R9 (The generalized distributive law) *Let* $r_1, r_2, \ldots, r_m, s_1, s_2, \ldots, s_n$ *be elements of a ring* $\mathcal{R}$. *Then*

$$\left(\sum_{i=1}^{m} r_i\right)\left(\sum_{j=1}^{n} s_j\right) = \sum_{i=1}^{m}\sum_{j=1}^{n} r_i s_j. \tag{11}$$

The right-hand side of (11) consists of all terms $r_i s_j$ added together as i assumes the values $1, 2, \ldots, m$ and j assumes the values $1, 2, \ldots, n$. We may express (11) this way: parentheses are removed by multiplying each term of the first factor onto each term of the second factor.

PROOF Let $N = m + n$, and let $P(N)$ be the statement that the two sides of (11) are equal for any N ring elements $r_1, \ldots, r_m, s_1, \ldots, s_n$. The smallest value possible for N is 2 (when $m = 1$, $n = 1$), and in this case it is easy to see that each side of (11) is $r_1 s_1$. Thus $P(2)$ is true. We employ the second principle of induction. Because $P(2)$ is true, the first step in the induction proof is completed. We now give the verification of the second step. Let $k > 2$ and assume (as induction hypothesis) the truth of all of $P(2), \ldots, P(k - 1)$. We wish to show that $P(k)$ is true. Since $k > 2$ and $m + n = k$, at least one of m, n is larger than 1, say $n > 1$. Let

$$a = \sum_{i=1}^{m} r_i, \qquad b = \sum_{j=1}^{n-1} s_j, \qquad c = s_n.$$

Then

$$\begin{aligned}
\left(\sum_{i=1}^{m} r_i\right)\left(\sum_{j=1}^{n} s_j\right) &= \left(\sum_{i=1}^{m} r_i\right)\left(\sum_{j=1}^{n-1} s_j + s_n\right) \\
&= a(b + c) \\
&= ab + ac \\
&= \left(\sum_{i=1}^{m} r_i\right)\left(\sum_{j=1}^{n-1} s_j\right) + \left(\sum_{i=1}^{m} r_i\right) s_n \\
&= \sum_{i=1}^{m}\sum_{j=1}^{n-1} r_i s_j + \sum_{i=1}^{m} r_i s_n \\
&= \sum_{i=1}^{m}\sum_{j=1}^{n} r_i s_j.
\end{aligned}$$

We used the distributive law R4 here to pass from the second line to the third, and used $P(m + n - 1)$ and $P(m + 1)$ to go from the fourth to the fifth line. (Note: $m + n - 1 < k$ and $m + 1 < k$.) The passage to the last line amounts to observing that if we add all terms r_is_j with $i = 1, \ldots, m$, $j = 1, \ldots, n - 1$, to all terms r_is_n with $i = 1, \ldots, m$, we get all terms r_is_j with $i = 1, \ldots, m$ and $j = 1, \ldots, n$. Thus $P(k)$ is true, completing the proof of the generalized distributive law. □

As applications of G7 and G8 we have the following ring-theoretic facts R10 and R11.

R10 *The element* 0 *of* $\mathcal{R}$ *is unique.*

PROOF The element 0 is the identity in the group consisting of $\mathcal{R}$ and operation $+$. □

R11 *Let* $r \in \mathcal{R}$. *Then* $\mathcal{R}$ *contains exactly one element, denoted by* $-r$, *satisfying* $(-r) + r = r + (-r) = 0$.

PROOF The element $(-r)$ is the inverse of r in the group consisting of $\mathcal{R}$ with operation $+$. □

When we apply G11 to the group consisting of $\mathcal{R}$ with operation $+$, we get: *If* r_1, r_2 *are elements of* $\mathcal{R}$, *the unique element* x *in* $\mathcal{R}$ *satisfying* $r_1 + x = r_2$ is $x = r_2 - r_1$.

We next develop a familiar property of 0.

R12 *For any element* r *in the ring* $\mathcal{R}$,

$$r0 = 0r = 0. \tag{12}$$

PROOF Since $0 + a = a$ for any element a of $\mathcal{R}$, setting $a = 0$ yields $0 + 0 = 0$. Let $b = r\cdot 0$. Using distributivity (Axiom R4), we see that

$$b = r\cdot 0 = r\cdot(0 + 0) = r\cdot 0 + r\cdot 0 = b + b.$$

Thus $b = b + b$. Adding $-b$ to each side then yields $0 = b$. Hence $r\cdot 0 = 0$ and similarly one shows that $0\cdot r = 0$. □

We now develop the rules of the minus sign in a ring $\mathcal{R}$.

R13 *Let* a *and* b *be elements of a ring* $\mathcal{R}$. *Then:*

$$-(a + b) = (-a) + (-b), \tag{13}$$

$$-(-a) = a, \tag{14}$$

$$(-a)b = a(-b) = -(ab), \tag{15}$$

$$(-a)(-b) = ab. \tag{16}$$

PROOF Since $\mathcal{R}$ with operation $+$ is a group, formulas (13) and (14) follow immediately from formulas (10) and (11) of §1.4. Let us now prove (15). Using R12 and the distributive law, we obtain: $0 = 0 \cdot b = (a + (-a))b = ab + (-a)b$. Thus $0 = ab + (-a)b$. Adding $-(ab)$ to each side gives $-(ab) = (-a)b$. This proves part of (15), and the rest of (15) is proved similarly. We now prove (16) by using (15) twice, and then (14) in this way: $(-a)(-b) = -(a(-b)) = -(-(ab)) = ab$. □

As the last of our basic properties of rings we have:

R14 *Let $\mathcal{R}$ be a ring with identity e. Then e is unique and moreover $(-e)r = r(-e) = -r$ for all r in $\mathcal{R}$.*

PROOF The identity e satisfies $er = re = r$ for all r. Suppose that the element e' of $\mathcal{R}$ satisfies $e's = se' = s$ for all s in $\mathcal{R}$. We wish to prove that $e = e'$. Setting $r = e'$ and $s = e$ in these equations produces $ee' = e = e'$. As for the second part of R14, we use R13 to get $(-e)r = -(er) = -r$. □

NOTATION Often the symbol 1 is used for the identity element. This is consistent with the fact that the identity element in the rings **Z** and **Q** is the number one.

We may summarize our discussion of rings in this way: A ring is a set endowed with two operations, addition and multiplication, in which several simple axioms (R1–R4) are valid. A number of natural properties (R5–R14) are consequences of these axioms. In particular, a ring is endowed with subtraction, i.e., inverses under addition.

EXERCISES 1.5

I

1. Which of the following subsets of **Z** are rings? rings with identity? The operations are to be the usual addition and multiplication of integers.
   (a) The nonnegative integers.
   (b) The subset consisting of the single integer 0.
   (c) All the integers, positive, negative, or zero, which are multiples of 3.
   (d) $\{0, 1, -1\}$.

2. Consider the group **Z** with operation $+$. On **Z** define a new multiplication law $\times'$ by specifying $z_1 \times' z_2 = 0$ for all integers z_1 and z_2. When endowed with addition $+$ and multiplication $\times'$, is **Z** a ring? Is it a ring with identity? Is it commutative?

3. Let $S = \{0, 1\}$ be a set consisting of the integers zero and 1. On S define the operations of addition and multiplication by these rules:

$0 + 0 = 0$, $0 + 1 = 1 + 0 = 1$, $1 + 1 = 0$, $0 \cdot 0 = 0 \cdot 1 = 1 \cdot 0 = 0$, $1 \cdot 1 = 1$. Prove or disprove: S, with these operations, is a ring. Is it commutative? Is it a ring with identity? Find -1. How are your answers changed if we replace $1 \cdot 1 = 1$ with $1 \cdot 1 = 0$?

4. Which of the following subsets of **Q** are rings? rings with identity? commutative rings?
   (a) All fractions representable in the form m/n where $m, n \in \mathbf{Z}$ and n is an odd integer.
   (b) All fractions representable in the form m/n where $m, n \in \mathbf{Z}$ and n is a nonzero even integer.
   (c) All fractions representable in the form m/n where $m, n \in \mathbf{Z}$ and n is a power of 2, that is, $n = 2^t$ for some nonnegative integer t (which may be different for different fractions).
   (d) The same as (c) except that $n = 3^t$ for some nonnegative integer t.
   (e) All fractions representable in the form m/n where $m, n \in \mathbf{Z}$, $n \neq 0$, and m is even.

5. An operation $\circ$ is *left distributive* with respect to addition if $a \circ (b + c) = a \circ b + a \circ c$ for all a, b, c. It is *right distributive* with respect to addition if $(b + c) \circ a = b \circ a + c \circ a$ for all a, b, c. On **Z**, is subtraction left or right distributive with respect to addition? On the positive rational numbers, is division (i.e., $a \circ b = a/b$) left or right distributive with respect to addition?

6. (a) Give an example of a ring $\mathcal{R}$ in which the nonzero elements of $\mathcal{R}$, under multiplication, form a group.
   (b) Give also an example of a ring $\mathcal{R}$ in which the nonzero elements of $\mathcal{R}$, under multiplication, do not form a group.

7. Find an example of a ring $\mathcal{R}$ in which the following property does not hold: If $rs = rt$ and $r \neq 0$, then $s = t$. Find an example of another ring $\mathcal{R}$ in which this property does hold.

II

8. Let a/b be a rational number with $0 < a < b$. Let x be positive. Prove or disprove:

$$\frac{a}{b} < \frac{a + x}{b + x}. \qquad \text{Prove also that } \frac{1}{b} < \frac{1}{a}.$$

9. Let r be an element of a ring $\mathcal{R}$, and let m be a positive integer. Define r^m by $r^m = r \cdot r \cdots r$ (m factors). If also n is a positive integer, prove the *laws of exponents*:

$$r^m r^n = r^{m+n}, \qquad (r^m)^n = r^{mn}.$$

If $\mathcal{R}$ is a ring with identity e, show that these exponent rules hold for nonnegative integers m, n if we define $r^0 = e$. Follow the proof in G10.

10. Let $\mathcal{R}$ be a ring with identity e. Suppose $\mathcal{R}$ contains at least two elements. Prove that $e \neq 0$.

11. Let $\mathcal{G}$ consist of the rational numbers representable in the form m/n, where m and n are odd integers. Endow $\mathcal{G}$ with the operation of multiplication. Is $\mathcal{G}$ a group?

12. Let n be a positive integer. Prove that $n! = n\cdot(n-1)!$

13. Assuming the validity of R11 (but not necessarily R10), deduce R10.

III

14. Let S be a set consisting of four letters a, b, c, d. Define operations of addition and multiplication on the elements of S as shown in these tables:

$+$	a	b	c	d
a	a	b	c	d
b	b	a	d	c
c	c	d	a	b
d	d	c	b	a

$\cdot$	a	b	c	d
a	a	a	a	a
b	a	b	b	a
c	a	c	c	a
d	a	d	d	a

Here the entry at the intersection of the row labelled x and the column labelled y denotes the sum $x + y$ in the first table and the product xy in the second table, e.g. $c + d = b$ and $cd = a$. Verify that these operations make S into a noncommutative ring. (This is the noncommutative ring having the fewest possible number of elements. See Exercise 21 of §7.5 for a simpler representation of it.)

15. Show that any ring $\mathcal{R}$ having exactly one element, or exactly two elements, must be commutative. Do this by finding all rings which have exactly one, or exactly two, elements.

16. Repeat Exercise 15 for any ring having exactly three elements.

17. Let S be a set, and let $\mathcal{R}$ be the collection of subsets of S, so that an element of $\mathcal{R}$ is a subset of S. Endow $\mathcal{R}$ with operations of addition and multiplication this way: If $r, s \in \mathcal{R}$ then

$$r + s = r \cup s - r \cap s,$$

$$r\cdot s = r \cap s.$$

(Note: The minus symbol $-$ here denotes the difference of sets, as in §1.2.) Prove that $\mathcal{R}$, under these operations, is a commutative ring with identity. Show that every element r of $\mathcal{R}$ satisfies $r^2 = r$. If S is a finite set with n elements, show that $\mathcal{R}$ is a finite ring with 2^n elements. Prove that $r + r = 0$ for any element r of $\mathcal{R}$. Hence show that each element of $\mathcal{R}$ is its own additive inverse. Do the nonzero elements of $\mathcal{R}$ form a group? Prove: If r is neither 0 nor the identity element of $\mathcal{R}$, then a nonzero element s of $\mathcal{R}$ exists such that $rs = 0$.

18. Let n be a positive integer and not a perfect square (i.e., there is no integer t such that $t^2 = n$.) Let $\mathcal{R}$ consist of all numbers of the form $x + y\sqrt{n}$ where x, y are in $\mathbf{Z}$. Endow $\mathcal{R}$ with the usual operations of addition and multiplication of numbers. Is $\mathcal{R}$ a ring? a ring with identity?

1.6 INTEGRAL DOMAINS AND FIELDS

Let $\mathcal{R}$ be a commutative ring, and by $\mathcal{R}^*$ denote the set of nonzero elements of $\mathcal{R}$. Under operation multiplication, how near is $\mathcal{R}^*$ to being a group? In general, not very near. It may happen that $\mathcal{R}^*$ is empty. (This is the case for the ring whose only element is zero.) If $\mathcal{R}$ has nonzero elements, group axiom G1 (closure) may not hold for $\mathcal{R}^*$ since the product of nonzero elements may be zero. (This happens in the ring of Example 1.5.5.) The associative law G2 must hold for the elements of $\mathcal{R}^*$ since it is the special case of R3, the associative law for multiplication of ring elements, in which a, b, c are nonzero. But Axiom G3 (about the identity) need not hold (an example is the ring of *even* integers, which has no multiplicative identity element) and neither need Axiom G4 (about inverses), as the example $\mathbf{Z}$ shows. However, for $\mathbf{Z}$, the set $\mathbf{Z}^*$ is a monoid, i.e., satisfies axioms G1, G2, and G3 of a multiplicative group, and for $\mathbf{Q}$ the set $\mathbf{Q}^*$ is a group. Motivated by the examples of $\mathbf{Z}$ and $\mathbf{Q}$, we introduce two special types of rings, *integral domains* and *fields*, which will occupy a prominent position in this book.

DEFINITION 1.6.1

A commutative ring $\mathcal{I}$ is an integral domain *if*

(i) *the set $\mathcal{I}^*$ of nonzero elements is nonempty;*
(ii) *$\mathcal{I}^*$, under operation multiplication, is a monoid.*

DEFINITION 1.6.2
A commutative ring $\mathfrak{F}$ is a field *if the set $\mathfrak{F}^*$ of nonzero elements is a group under multiplication.*

Thus a field is *always* an integral domain.

Example. The ring **Z** of integers is an integral domain but not a field, and the ring **Q** of rational numbers is a field.

In Chapters 2 and 3 we shall encounter other important examples of integral domains and fields. Polynomials, to be studied in Chapter 3, provide another example of an integral domain, and the number systems known as the real numbers and the complex numbers, both of which will be described in Chapter 2, are additional examples of fields. In the next few pages, we give an introductory discussion of integral domains and fields. This discussion is given now so that it can be applied to any integral domain or field that may be encountered later in this book (or, for that matter, in any book).

To facilitate the discussion we introduce a technical term, that of *zero divisor*.

DEFINITION 1.6.3
Let $\mathfrak{R}$ be a commutative ring. A nonzero element r in $\mathfrak{R}$ is a zero divisor *if $rr_1 = 0$ for some nonzero element r_1 of $\mathfrak{R}$.*

In some rings zero divisors are extremely plentiful. For example, each nonzero element of the ring of Example 1.5.5 is a zero divisor. In the ring of Exercise 17 of §1.5, each element other than zero or the multiplicative identity is a zero divisor. In Example 1.6.6 (below) we shall see still another ring with numerous zero divisors.

THEOREM 1.6.4
A commutative ring is an integral domain if and only if:
I1 *It has an identity e with $e \neq 0$.*
I2 *It has no zero divisors.*

PROOF Since this is an "if and only if" theorem, it requires proof in two directions: we must show that an integral domain satisfies I1 and I2, and we must also show that a commutative ring in which I1 and I2 hold is an integral domain.

Let $\mathfrak{I}$ be an integral domain. Then $\mathfrak{I}^*$ is a monoid and therefore is nonempty and satisfies group axioms G1, G2, G3. Thus $\mathfrak{I}^*$ is closed under multiplication, meaning that nonzero elements of $\mathfrak{I}$ have a nonzero product. Therefore $\mathfrak{I}$ has no zero divisors, verifying I2. By G3, $\mathfrak{I}^*$ contains an element e with $es = se = s$ for all $s \in \mathfrak{I}^*$. Since $e0 = 0e = 0$ (by R12, applied to e), we see that $es = se = s$ for all s in $\mathfrak{I}$. Thus e is an identity for $\mathfrak{I}$ and is nonzero because it is in $\mathfrak{I}^*$. Therefore I1 is verified.

Conversely, let $\mathfrak{R}$ be a commutative ring in which I1, I2 hold. We wish to prove that $\mathfrak{R}$ is an integral domain, and for this, we must show that $\mathfrak{R}^*$ is a monoid under multiplication. By I1, $\mathfrak{R}$ contains a nonzero element, namely e. Thus $\mathfrak{R}^*$ is nonempty. By I2, $\mathfrak{R}$ has no zero divisors, so that nonzero elements of $\mathfrak{R}$ have a nonzero product, meaning that $\mathfrak{R}^*$ is closed under multiplication. This verifies G1 for $\mathfrak{R}^*$. No verification of associativity is needed for $\mathfrak{R}^*$ since associativity holds for all elements of $\mathfrak{R}$, by R3. Thus G2 holds for $\mathfrak{R}^*$. No verification of G3 is needed for $\mathfrak{R}^*$ either, since, by I1, the element e is an identity for $\mathfrak{R}^*$. Thus $\mathfrak{R}^*$ is a monoid. □

Integral domains have a very important property which the reader undoubtedly has encountered previously, namely: a nonzero factor may be cancelled from an equation. We state this as Theorem 1.6.5.

THEOREM 1.6.5

If a, b, c are elements of a commutative ring without zero divisors, this cancellation law is valid:

Cancellation law: $ab = ac$ *implies* $b = c$, *provided* $a \neq 0$.

In particular, this cancellation law is valid when a, b, c are elements of an integral domain.

PROOF If $ab = ac$, then $ab - ac = 0$, hence $ab + a(-c) = 0$ (by R13) and so, by the distributive law, $a(b - c) = 0$. Here we have a product equal to zero. Since a, $b - c$ are elements of a ring without zero divisors, either a or $b - c$ must be zero. By hypothesis, a is nonzero. Therefore $b - c = 0$, so that $b = c$. □

Notice that this cancellation law is *not* valid in every ring. For example, in the ring of Example 1.5.5, we know $g_1g_2 = g_1g_3$ for any elements g_1, g_2, g_3 (since $g_1g_2 = 0$ and $g_1g_3 = 0$). If we cancelled g_1, the resulting equation is $g_2 = g_3$, which is not true if g_2 and g_3 are different. *Thus, we are permitted to cancel a nonzero common factor from an equation involving ring elements without giving justification only when the ring is known to have no zero divisors.*

Example 1.6.6

We pause to give another example of a ring containing numerous zero divisors and for which the cancellation law mentioned above does not hold. Take an XY axis system (Figure 1.6.1) and take $\mathfrak{R}$ to be all points (x, y) in the plane of this axis system. This is just the Cartesian coordinate plane encountered in analytic geometry and calculus; thus points (x, y) and (x', y') are equal if and only if $x = x'$ and $y = y'$. We make this coordinate plane into a ring by defining point addition and multiplication in this way:

$$(x, y) + (x', y') = (x + x', y + y'),$$

$$(x, y) \cdot (x', y') = (xx', yy').$$

Thus we add or multiply points by adding or multiplying their coordinates. We leave as an exercise for the reader the straightforward verification that $\mathcal{R}$ is a commutative ring with identity, merely observing that the origin (0, 0) is the zero element and (1, 1) is the multiplicative identity. Now notice that

$$(x, 0) \cdot (0, y) = (0, 0),$$

for any x and y. That is, *any* point on the X or Y axes (except the origin) is a zero divisor.

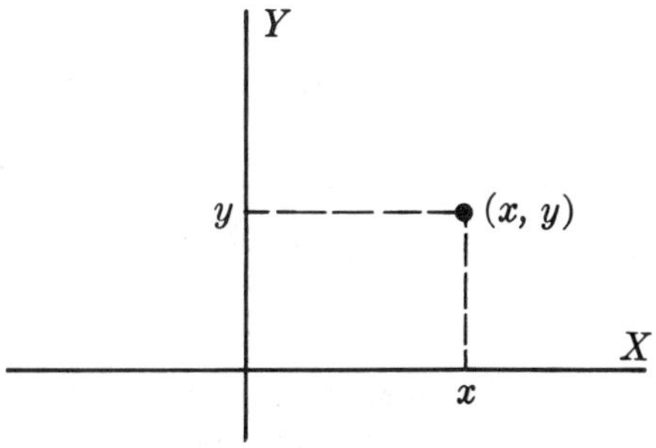

Figure 1.6.1

In this ring, $ab = ac$ does not imply $b = c$ even if a is nonzero. As an example, take $a = (1, 0)$, $b = (0, 1)$, $c = (0, 2)$.

In Chapter 2 we shall see that if we take a different definition of multiplication, the points of this coordinate plane will then form an integral domain, and in fact, even a field.

We turn now to fields. Because a field is an integral domain, the cancellation law of Theorem 1.6.5 is valid in any field. Parallel to Theorem 1.6.4, we have the following characterization of a field.

THEOREM 1.6.7

A commutative ring $\mathfrak{F}$ is a field if and only if:

F1 *It has an identity e with $e \neq 0$;*

F2 *Each nonzero element of $\mathfrak{F}$ has a multiplicative inverse in $\mathfrak{F}$.*

PROOF Suppose $\mathfrak{F}$ is a field. Then $\mathfrak{F}$ is an integral domain. By I1 there must be an identity e in $\mathfrak{F}$ and e is nonzero. Therefore F1 holds. According to Definition 1.6.2, $\mathfrak{F}^*$ is a group, so that each nonzero element of $\mathfrak{F}$ has a multiplicative inverse in $\mathfrak{F}$. (This is group axiom G4 applied to the group $\mathfrak{F}^*$.) Hence F2 holds.

Now suppose that both F1 and F2 hold. We wish to prove that $\mathfrak{F}$ is a field. According to Definition 1.6.2 we must show that the set $\mathfrak{F}^*$ of nonzero elements forms a group under multiplication. Certainly $\mathfrak{F}^*$ is nonempty since, by F1, the element e is nonzero. Three of the group axioms are obviously satisfied by $\mathfrak{F}^*$: associativity (G2) holds because it holds for all elements of $\mathfrak{F}$ (since $\mathfrak{F}$ is a ring) and hence holds when these elements happen to be nonzero; G3 holds because the existence of the identity e is postulated in F1; and G4 holds because the existence of inverses is postulated in F2. Therefore, only the closure law G1 requires verification. For this we must show that if a and b are elements of $\mathfrak{F}^*$ then so is ab. If ab were not an element of $\mathfrak{F}^*$ it would have to be zero, since zero is the only element of $\mathfrak{F}$ not in $\mathfrak{F}^*$. Since a is nonzero, there is (by F2) an element α in $\mathfrak{F}$ such that $\alpha a = e$. If $ab = 0$ we would have

$$0 = \alpha 0 = \alpha(ab) = (\alpha a)b = eb = b.$$

This contradicts the fact that b belongs to $\mathfrak{F}^*$ and hence is nonzero. Thus $\mathfrak{F}^*$ is closed under multiplication and therefore is a group. Consequently $\mathfrak{F}$ is a field. □

We use the same notation for inverses in the group $\mathfrak{F}^*$ as in an abstract group. Thus the inverse of a nonzero element a of $\mathfrak{F}$ is denoted by a^{-1}, and so $x = a^{-1}$ is the unique solution of the equation $ax = e$. The laws of exponents, formulas (15) and (16) of G10 in §1.4, then hold for every nonzero g in $\mathfrak{F}$ and for all integer exponents m and n. They also hold if $g = 0$, provided m and n are positive integers. It is sometimes convenient to have a meaning for the symbol 0^0. (The first 0 is the zero element of $\mathfrak{F}$, the second 0 is the integer zero.) We *define* 0^0 by $0^0 = e$. It is then easy to check that the exponent laws G10 hold when $g = 0$ and the exponents m, n are any nonnegative integers.

If a, b are elements of a field $\mathfrak{F}$, with a nonzero, the solution of $ax = b$ is $x = a^{-1}b$. If $\mathfrak{F}$ is $\mathbf{Q}$, the solution is $x = b/a$. Thus $a^{-1}b$ is the abstract generalization of the notion of a fraction with numerator b and denominator a.

The major distinctions between rings, integral domains, and fields may be summarized this way: In a ring, one may add, subtract, and multiply, but usually not divide. In an integral domain, one may add, subtract, multiply, and cancel a common nonzero factor from an equation, but apart from this, usually not divide. In a field, one may add, subtract, multiply, and divide (except by zero.)

Example 1.6.8
We now give an example of an integral domain $\mathfrak{I}$ and a field $\mathfrak{F}$. Let $\mathfrak{I}$ consist of all

$$x + y\sqrt{2}, \qquad \text{where } x, y \in \mathbf{Z},$$

and let $\mathfrak{F}$ consist of all

$$x + y\sqrt{2}, \qquad \text{where } x, y \in \mathbf{Q}.$$

Thus $\mathfrak{I} \subset \mathfrak{F}$. Let $x, y, x', y' \in \mathbf{Q}$. It will be proved in §2.2 that $\sqrt{2}$ is not a rational number. Because of this, the condition for equality of elements $x + y\sqrt{2}$, $x' + y'\sqrt{2}$ of $\mathfrak{F}$ is

$$x + y\sqrt{2} = x' + y'\sqrt{2} \quad \text{if and only if } x = x',\ y = y'.$$

The sum and product of elements of $\mathfrak{F}$ are given in this way:

$$(x + y\sqrt{2}) + (x' + y'\sqrt{2}) = (x + x') + (y + y')\sqrt{2},$$
$$(x + y\sqrt{2}) \cdot (x' + y'\sqrt{2}) = (xx' + 2yy') + (xy' + yx')\sqrt{2},$$

(i.e., to multiply, multiply out parentheses in the normal way, use $(\sqrt{2})^2 = 2$, then collect terms.) We leave as exercises the verifications that $\mathfrak{I}$ and $\mathfrak{F}$ are commutative rings with identity. In fact, $\mathfrak{I}$ is an integral domain and $\mathfrak{F}$ a field. To verify F2, observe that if $x + y\sqrt{2} \neq 0$, its multiplicative inverse is

$$x/(x^2 - 2y^2) - y/(x^2 - 2y^2) \cdot \sqrt{2}. \tag{1}$$

(Note that $x^2 - 2y^2 \neq 0$, since otherwise 2 is the square of the rational number x/y, which is impossible.) The number (1) is therefore in $\mathfrak{F}$ since $x/(x^2 - 2y^2)$, $-y/(x^2 - 2y^2)$ are rational numbers. Consequently, $\mathfrak{F}$ is a field. Since a field $\mathfrak{F}$ cannot contain zero divisors, its subset $\mathfrak{I}$ cannot either, so that $\mathfrak{I}$ is an integral domain. Of course, $\mathfrak{F}$ is also an integral domain; however, $\mathfrak{I}$ is not a field.

Other examples of integral domains and fields will be seen in Chapters 2 and 3.

Subrings, subfields, subgroups

A subset of a ring which itself is a ring using the same addition and multiplication operations is called a *subring*. An example of a subring of $\mathbf{Z}$ is the ring of even integers, and $\mathbf{Z}$ is a subring of the ring $\mathfrak{I}$ in the above example, and of $\mathbf{Q}$. A subring of a field which is itself a field is called a *subfield*. It can be proved (Exercise 5) that no subring of $\mathbf{Z}$ is a field and that the only subfield of $\mathbf{Q}$ is $\mathbf{Q}$ itself. However, in Chapter 2, we shall encounter several fields containing $\mathbf{Q}$ as a subfield. (The field $\mathfrak{F}$ in the example above contains $\mathbf{Q}$ as a subfield.) The corresponding concept for groups is that of *subgroup:* this is a subset of a group which itself is a group using the same operation. The *set* $\mathbf{Z}$, under operation $+$, is a group containing many subgroups (Exercise 14), and the *set* $\mathbf{Q}^*$, under operation multiplication, is also a group containing many subgroups.

In later chapters, we shall need the following two results giving criteria for determining when a subset of a ring is a subring and when a subset of a group is a subgroup.

THEOREM 1.6.9

Let $\mathcal{R}$ be a ring, and suppose $\mathcal{S}$ is a nonempty subset of $\mathcal{R}$. Then $\mathcal{S}$ is a subring of $\mathcal{R}$ if $s_1 + s_2$, s_1s_2, and $-s_1$ are in $\mathcal{S}$ whenever s_1 and s_2 are elements of $\mathcal{S}$.

Thus a nonempty subset $\mathcal{S}$ of $\mathcal{R}$ is a subring if it is closed under addition and multiplication and contains also the additive inverse of each of its elements (i.e., closure under subtraction). The proof is easy. We must check that $\mathcal{S}$ satisfies all axioms R1–R4. The closure axioms R1(i) and R2 are satisfied by hypothesis. The associativity, commutativity, distributivity axioms R1(ii), R1(v), R3, R4 need no verification since these axioms hold for all elements of $\mathcal{R}$, hence hold when the elements happen to be in $\mathcal{S}$. Let s be an element of $\mathcal{S}$. Then, by hypothesis, $-s \in \mathcal{S}$ and hence $0 = s + (-s)$ is also in $\mathcal{S}$. Axiom R1(iii) is therefore verified. By hypothesis, R1(iv) is satisfied. □

THEOREM 1.6.10

Let $\mathcal{G}$ be a group with operation $\circ$, and suppose $\mathcal{H}$ is a nonempty subset of $\mathcal{G}$. Then $\mathcal{H}$ is a subgroup if $h_1 \circ h_2$ and h_1^{-1} are in $\mathcal{H}$ whenever h_1 and h_2 are elements of $\mathcal{H}$.

PROOF The associative law is valid for elements of $\mathcal{H}$ since it holds for all elements of $\mathcal{G}$, and if h is an element of $\mathcal{H}$, $e = h \circ h^{-1} \in \mathcal{H}$. The remaining details of the proof are left as an exercise. □

The reader will find below, in Exercises 11 and 13, slightly different versions of the assertions in Theorems 1.6.9 and 1.6.10.

EXERCISES 1.6

I

1. Which of the rings in Exercise 1 of §1.5 are integral domains? fields? Answer the same questions for the ring in Exercise 3 of §1.5.
2. Let a be a nonzero element of a field. Can it happen that $-a = a^{-1}$, i.e., that the additive and multiplicative inverses of a are the same? (See Exercise 3 of §1.5. Another example will be seen when we study the *complex number field* in Chapter 2.)

3. Let $\mathcal{R}$ consist of all numbers of the form $x + y\sqrt{3}$ where $x, y \in \mathbf{Q}$. Endow $\mathcal{R}$ with the usual operations of addition and multiplication of numbers. Is $\mathcal{R}$ an integral domain? a field? (You may assume that no rational number t exists such that $t^2 = 3$.)

4. Find all subrings of $\mathbf{Z}$ which are integral domains.

5. Find all subfields of $\mathbf{Q}$.

6. Find all zero divisors in the ring of Example 1.6.6.

II

7. Let a and b be elements of an integral domain, and suppose $a^2 = b^2$. Prove: Either $a = b$ or $a = -b$. Give full details in your proof.

8. Let a be a nonzero integer, b any integer. The equation $ax = b$ always has a solution, namely $x = b/a$. Why does this not prove that $\mathbf{Z}$ is a field?

9. Let S be a nonempty set endowed with operations of addition and multiplication. Prove that S is a field if and only if the following three properties are fulfilled:
   (i) S, under operation addition, is a commutative group.
   (ii) S^*, under operation multiplication, is a commutative group.
   (iii) $a(b + c) = ab + ac$ for all a, b, c in S.
   Here S^* is the set of nonzero elements of S.

10. Suppose $\mathfrak{F}$ is a commutative ring with identity e, with $e \neq 0$, and suppose, whenever a is a nonzero element of $\mathfrak{F}$, there is an x in $\mathfrak{F}$ such that $ax = e$. Prove that $\mathfrak{F}$ is a field.

11. Let $\mathcal{R}$ be a ring and $\mathcal{S}$ a nonempty subset of $\mathcal{R}$. Suppose $\mathcal{S}$ is closed under subtraction and multiplication, i.e., $s_1 - s_2$ and $s_1 s_2$ are in $\mathcal{S}$ whenever s_1 and s_2 are elements of $\mathcal{S}$. Deduce that $\mathcal{S}$ is a subring of $\mathcal{R}$. In fact show this: closure of $\mathcal{S}$ under addition follows from closure of $\mathcal{S}$ under subtraction.

12. [Continuation of 11] Show by an example that closure of $\mathcal{S}$ under addition and multiplication does not imply the closure of $\mathcal{S}$ under subtraction.

13. Let $\mathcal{G}$ be a group, with operation $\circ$, and $\mathcal{H}$ a nonempty subset of $\mathcal{G}$. If $h_1 \circ h_2^{-1}$ is in $\mathcal{H}$ whenever h_1 and h_2 are elements of $\mathcal{H}$, show that $\mathcal{H}$ is a subgroup of $\mathcal{G}$.

14. Find several examples of subgroups of the additive group of integers. Find several examples of subgroups of the multiplicative group of nonzero rational numbers.

III

15. Find, to within change of notation, all fields having exactly two elements.

16. Repeat 15 for all fields having exactly three elements.

17. Describe all sets S for which the ring of Exercise 17, §1.5, is an integral domain or a field.

18. Show that a field $\mathfrak{F}$ can contain at most four elements x for which $x^4 = e$. What about $x^3 = e$ and $x^6 = e$?

19. The best justification for the definition $0^0 = e$ is that it simplifies statements and formulas to appear in later chapters. Give a further justification by using methods from calculus to evaluate the limit approached by x^x as the positive number x decreases toward zero. (The value of the limit is 1.)

20. Modify Example 1.6.6 by taking this to be the multiplication of points:
$$(x, y) \cdot (x', y') = (xx' + 2yy', xy' + yx').$$
With this multiplication and the addition in Example 1.6.6, show that the points of the plane form a field. Is there any connection between this field and the field in Example 1.6.8?

1.7 REVIEW

This chapter introduced the basic algebraic structures of groups, rings, fields. A group is endowed with only one operation, which may be addition, or multiplication, or an abstract operation $\circ$. The group axioms require that a group have an identity element and an inverse for each of its elements. A ring is endowed with two operations, addition and multiplication. Under addition a ring is a commutative group. This implies that it has an additive identity (denoted by 0) and that each of its elements has an additive inverse—i.e., a ring is also endowed with subtraction. The multiplication operation need not be commutative. If it is, the ring is said to be commutative. Most of the rings encountered in this book will be commutative rings. A ring need not contain a multiplicative identity e and, even if it does contain e, it need not contain a multiplicative inverse a^{-1} for a nonzero element a. A special type of ring is the integral domain. An integral domain always contains a multiplicative identity element, the product of two nonzero elements of an integral domain is nonzero, and a common nonzero factor may be cancelled from an equation involving elements in an integral domain. A field is an integral domain which contains a multiplicative inverse a^{-1} for each of its nonzero elements a, i.e., one may divide, except by zero. The ring **Z** is an integral domain, but not a field, and **Q** is a field. A subgroup is a nonempty subset of a group which is itself a group (using the same operation). A subring is a nonempty subset of a ring which is itself a ring (using the same operations), and a subfield is a nonempty subset of a field which is itself a field (using the same operations). A nonempty subset of a ring is a subring if it satisfies closure under both addition and multiplication and contains the additive inverse of each of its elements. A nonempty subset of a group is a subgroup if it satisfies closure and contains the inverse of each of its elements.

REVIEW EXERCISES 1.7

1. If the nonzero elements of a ring form a multiplicative commutative group, what conclusion can you draw about the ring?

2. Let a, b, c be elements of a ring. If $ab = ac$, does it follow that $b = c$? What if a is nonzero?

3. Let $\mathcal{R}$ be a ring. Suppose, for each nonzero element a in $\mathcal{R}$ and each b in $\mathcal{R}$, one may find an element x in $\mathcal{R}$ for which $ax = xa = b$. What conclusion can you draw about $\mathcal{R}$?

4. Give an example of a field with the following property: it has a subring which is not a subfield.

5. Give two examples of a field $\mathfrak{F}$ whose only subfield is $\mathfrak{F}$ itself.

6. Give several examples of groups with operation addition, which are contained in the field **Q** and give several examples of groups, with operation multiplication, which are also contained in **Q**.

7. Is it true that every commutative ring with identity is an integral domain?

8. Give an example of an integral domain which is
   (a) a field; (b) not a field.

9. Let $\mathfrak{F}$ be a field. Every equation $ax + b = 0$ (where $a, b \in \mathfrak{F}$, $a \neq 0$) has a solution x in $\mathfrak{F}$. Prove this. Is it true that every equation $ax^2 + bx + c = 0$ (where $a, b, c \in \mathfrak{F}$, $a \neq 0$) has a solution x in $\mathfrak{F}$?

2

THE COMPLEX NUMBER FIELD

2.1 INTRODUCTION

Most readers will have encountered at least once the interesting number i satisfying $i^2 = -1$. Very likely, however, many readers will not have seen a systematic development of the number system (called the *complex number system*) of which i is a member. This development will now be carried out, beginning with rational numbers, then proceeding to a short discussion of real numbers, and finally to a thorough discussion of complex numbers. These number systems, as we shall see, are all fields.

2.2 RATIONAL AND REAL NUMBERS

As we saw in §1.5, the rational number field **Q** consists of fractions m/n in which m and n are integers with $n \neq 0$. The ring **Z** is contained in **Q** since the denominator n may be 1. If m_1, n_1 are also integers, $n_1 \neq 0$, fractions m/n, m_1/n_1 may be equal without having equal numerators m, m_1 or equal denominators n, n_1. The definition of equality of fractions is that $m/n = m_1/n_1$ when, and only when, $mn_1 = nm_1$.

A nonzero rational number $a = m/n$ is said to be *positive* if the integers m and n have the same sign (i.e., both positive or both negative) and to be *negative* if m and n have opposite signs (i.e., one positive, one negative). If $m/n = m_1/n_1$, then $mn_1 = nm_1$ so that, if m, n have the same (opposite) signs, then m_1, n_1 also have the same (opposite, respectively) signs. Thus the positivity of a rational number depends only on the rational number itself and does not depend on which (of the many possible) representations of it as a quotient of integers is used to determine the positivity. Since $m/n = -m/-n$, a positive fraction may always be written as a fraction with positive numerator and denominator. If $a = m/n$ and $b = m_1/n_1$ are positive fractions, it follows from $a + b = (mn_1 + nm_1)/nn_1$ and $ab = mm_1/nn_1$ that $a + b$ and ab are also positive. This means that the order axioms P1 and P2 of §1.3 are valid for rational numbers. Consequently, the discussion of order, inequalities, and absolute value given in §1.3 for integers applies equally well to rational numbers. This should be checked by rereading the first part of §1.3, taking a, b, c, d in that discussion to be rational numbers. In particular, the triangle inequality

$$|a + b| \leq |a| + |b| \tag{1}$$

is valid when a and b are rational numbers. The reason for calling (1) the "triangle inequality" will be seen later in this chapter. Some important properties of absolute values are:

V1 $|a| \geq 0$ *for all* a;

V2 $|a| = 0$ *if and only if* $a = 0$;

V3 $|ab| = |a|\,|b|$ *for all* a, b;

V4 $|a + b| \leq |a| + |b|$ *for all* a, b.

One might hope that rational numbers would be sufficient for most purposes and that no additional types of numbers would be needed. That this hope is not realized is shown by the following lemma.

LEMMA 2.2.1
There is no rational number whose square is 2. In other words: $\sqrt{2}$ *is not a rational number.*

PROOF We give a proof by contradiction, supposing there is a rational number a satisfying $a^2 = 2$, and from this deducing a contradiction.

Let $a = m/n$ where m and n are integers. If m and n are both even, we cancel a factor 2 from the numerator and denominator of a. (See Exercise 17.) If the new numerator and denominator are again both even, we cancel another factor 2. After perhaps several cancellations of 2, we arrive at a fraction equal to a in which the numerator and denominator are not both even. This must eventually happen since each cancellation of 2 diminishes the size of the numerator and denominator; see Exercise 18. We therefore assume $a = m/n$ where m, n are integers, at least one of which is odd. If $a^2 = 2$ we obtain $m^2 = 2n^2$. The right-hand side of this equation is even, hence m^2 is even, and hence m is even (because the square of an odd integer is odd; see Exercise 13). Thus $m = 2t$ for some integer t. From $m^2 = 2n^2$ and $m = 2t$ we get $n^2 = 2t^2$. Therefore n^2 is even and hence n is even. Thus both m and n are even, contradicting the fact that at least one is odd. The assumption that $(m/n)^2 = 2$ must therefore be incorrect, implying that no rational number a exists satisfying $a^2 = 2$. □

Many other frequently encountered numbers can be proved to be not rational. Examples are $\sqrt[3]{2}$, $\sqrt[4]{2}$, $\sqrt{3}$, $\sqrt{5}$, $\sqrt{3} + \sqrt{5}$, e, π. Here $e = 2.71828\ldots$ is the natural logarithm base encountered in calculus, and π is the number giving the circumference of the circle of diameter 1. For some of these numbers the proof that the number is not rational is as easy as that for $\sqrt{2}$. For the others, slightly more sophisticated ideas are needed. Below, we shall prove that e is not rational. In Chapters 3 and 5 (Examples 3.5.6 and 5.4.11) we shall give general methods for proving that $\sqrt{3}$, $\sqrt{5}$, $\sqrt{3} + \sqrt{5}$, and many similar types of numbers are not rational. Often the rational number $22/7$ is used as a substitute for π. This is justified on the basis that $22/7$ is a rational number reasonably close to π. In fact, correct to four decimal places, $22/7 = 3.1429$, whereas $\pi = 3.1416$. It can be proved that π is not rational; see the Niven book cited on page 455. In §8.5 we shall give methods for calculating rational numbers closely approximating π. A number which is not rational, that is, not in **Q**, is called an *irrational number*.

Since $\sqrt{2}$, π, $\sqrt{3}$, etc., are irrational, **Q** does not contain many numbers needed in mathematics. A larger number system must therefore be invented. This larger number system is called the *real number system*, and we now give a short description of it. Afterwards we discuss the possibility of giving a more complete description.

Let a straight line (an X axis) be given, and on it mark two points, which we call 0 and 1. The distance between these two points becomes the unit of distance, and using it we mark the integer points $0, 1, 2, 3, \ldots, -1, -2, -3, \ldots$.

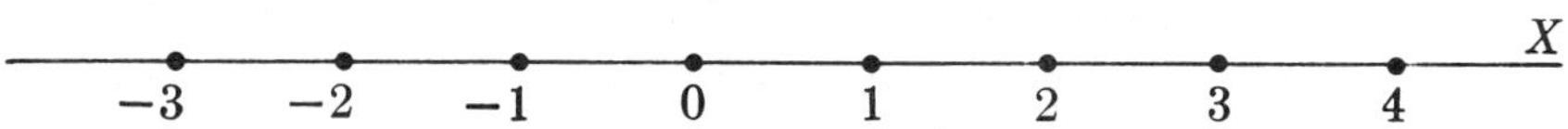

Figure 2.2.1

Having done this, we plot other rational numbers, e.g. 4/3 is one-third of the way from 1 to 2. Thus each rational number may be regarded as a point on this X axis. However, an irrational number such as $\sqrt{2}$ is not yet plotted.

If we are asked to compute a numerical value for $\sqrt{2}$ (or $\sqrt{3}$, π, e, or other irrational number), how do we respond? We work out its value *accurate to a specified number of decimals.* In other words, we compute an approximate value, and with sufficient labor we make the degree of precision of this approximation as good as is desired by computing enough digits to the right of the decimal point. Let us discuss this in detail for $\sqrt{2}$.

Let α_k be the largest k-decimal place number* satisfying $\alpha_k^2 \leq 2$, and let

$$\beta_k = \alpha_k + \underbrace{.000\ldots001}_{k \text{ decimal places}} = \alpha_k + 10^{-k}.$$

Thus β_k, being the sum of two k-decimal place numbers, is a k-decimal place number, and since it is larger than α_k, it satisfies $\beta_k^2 > 2$. We can compute α_k, β_k, for $k = 1, 2, 3, \ldots$ recursively, as follows. Since $1.4^2 < 2$, $1.5^2 > 2$, we have

$$\alpha_1 = 1.4, \quad \beta_1 = 1.5, \quad \alpha_1^2 < 2, \quad \beta_1^2 > 2, \quad \beta_1 = \alpha_1 + 10^{-1}.$$

We then calculate α_2, β_2 by squaring each of the two-decimal place numbers

$$\alpha_1 = 1.40, 1.41, 1.42, 1.43, \ldots, 1.48, 1.49, 1.50 = \beta_1 \tag{2}$$

to find the largest whose square is ≤ 2. This will be α_2, and then β_2 is the next number on the list (2). Thus

$$\alpha_1 \leq \alpha_2, \quad \alpha_2^2 \leq 2, \quad \beta_2 = \alpha_2 + 10^{-2}, \quad \beta_2 \leq \beta_1, \quad \beta_2^2 > 2.$$

We find that $\alpha_2 = 1.41$, $\beta_2 = 1.42$. Repeating this procedure, we calculate α_3, β_3 by squaring each of the three decimal place numbers

$$\alpha_2 = 1.410, 1.411, 1.412, 1.413, \ldots, 1.418, 1.419, 1.420 = \beta_2, \tag{3}$$

to find the largest whose square is ≤ 2. This will be α_3, and then β_3 is the next number on the list (3). Thus

$$\alpha_2 \leq \alpha_3, \quad \alpha_3^2 \leq 2, \quad \beta_3 = \alpha_3 + 10^{-3}, \quad \beta_3 \leq \beta_2, \quad \beta_3^2 > 2.$$

In general, if we have found the largest k-decimal place number α_k with $\alpha_k^2 \leq 2$, so that $\beta_k^2 > 2$ with $\beta_k = \alpha_k + 10^{-k}$, we calculate α_{k+1}, β_{k+1} by squaring each of the $k + 1$ decimal place numbers

$$\alpha_k = \alpha_k + 0 \cdot 10^{-(k+1)}, \alpha_k + 1 \cdot 10^{-(k+1)}, \alpha_k + 2 \cdot 10^{-(k+1)}, \\ \ldots, \alpha_k + 9 \cdot 10^{-(k+1)}, \alpha_k + 10 \cdot 10^{-(k+1)} = \beta_k, \tag{4}$$

* We say α_k has k decimal places if it has the form $1.d_1d_2 \ldots d_k$ where $d_1, d_2, \ldots, d_k$ are digits, each being 0, 1, 2, ..., or 9. It may happen that $d_k = 0$. (Other digits may also be 0.) Here, and in many other places in the book, the bold point **.** denotes the decimal point.

to find the largest whose square is ≤ 2. This will be α_{k+1}, and then the next number on the list (4) is β_{k+1}. Thus

$$\alpha_k \leq \alpha_{k+1}, \quad \alpha_{k+1}{}^2 \leq 2, \quad \beta_{k+1} = \alpha_{k+1} + 10^{-(k+1)}, \quad \beta_{k+1} \leq \beta_k, \quad \beta_{k+1}{}^2 > 2.$$

The first few values obtained by this procedure are

$$\alpha_1 = 1.4, \quad \alpha_2 = 1.41, \quad \alpha_3 = 1.414, \quad \alpha_4 = 1.4142, \quad \alpha_5 = 1.41421,$$
$$\beta_1 = 1.5, \quad \beta_2 = 1.42, \quad \beta_3 = 1.415, \quad \beta_4 = 1.4143, \quad \beta_5 = 1.41422.$$

Each of these numbers (and in fact, any k-decimal place number for any k) is rational since, for example,

$$\alpha_1 = \frac{14}{10}, \quad \alpha_2 = \frac{141}{100}, \quad \alpha_3 = \frac{1414}{1000}, \quad \alpha_4 = \frac{14142}{10000}, \quad \alpha_5 = \frac{141421}{100000}, \text{ etc.}$$

In this way, therefore, we calculate *rational numbers* for which*

$$\alpha_1 \leq \alpha_2 \leq \alpha_3 \leq \alpha_4 \leq \cdots \quad \cdots \leq \beta_4 \leq \beta_3 \leq \beta_2 \leq \beta_1,$$
$$\alpha_k{}^2 < 2, \quad \beta_k{}^2 > 2, \quad \beta_k = \alpha_k + 10^{-k}, \quad k = 1, 2, 3, \ldots.$$

(We do not have $\alpha_k{}^2 = 2$ because α_k is a rational number and there is no rational number whose square is 2.)

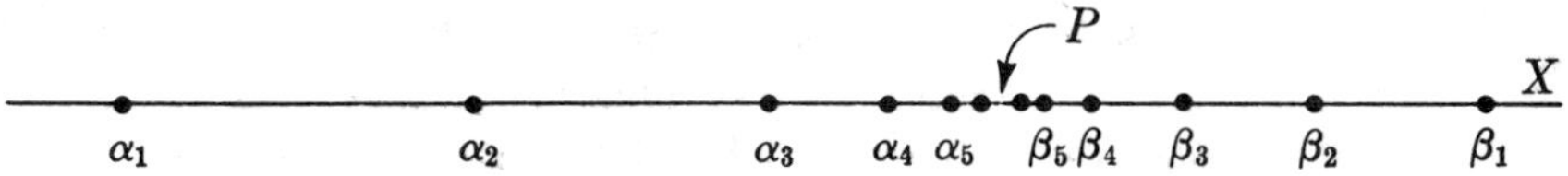

For clarity, this figure is not drawn to scale.

Figure 2.2.2

If we plot the rational numbers $\alpha_1, \alpha_2, \alpha_3, \ldots, \beta_1, \beta_2, \beta_3, \ldots$ on the X axis, the α's move progressively to the right, the β's progressively to the left, every α is to the left of every β, and (since $\beta_k = \alpha_k + 10^{-k}$) α_k, β_k become progressively nearer each other as k increases. Consequently—see Figure 2.2.2—a unique point P exists to the right of every α and to the left of every β. This point P is to be regarded as the location of $\sqrt{2}$ on the X axis. Point P can be closely approximated by rational numbers, the procedure above giving a method for calculating rational numbers α_k, β_k arbitrarily near P. Notice that since the rational numbers α_k cluster near P for large k, the distance $|\alpha_i - \alpha_j|$ between two of these numbers will be small when the subscripts i, j are large. A similar statement holds for the β's.

* More efficient methods for calculating rational numbers closely approximating $\sqrt{2}$ will be seen in Chapter 5. See also Exercise 23.

In general we may regard real numbers as numbers which can be calculated to any desired degree of accuracy using sufficiently many decimal places, i.e., can be approximated arbitrarily closely by rational numbers. We therefore *define a real number to be a number approximable to any specified degree of accuracy by rational numbers.* Any rational number, being approximable with zero error by a rational number (namely, itself) meets this definition, so that the real numbers include the rational numbers. As with $\sqrt{2}$, the rational approximations of a real number may be used to locate its position on the X axis. In geometric language, the real numbers therefore consist of points on the X axis; these points include all points at which rational numbers are located, together with other points at which irrational numbers are located, and arbitrarily close to a point giving an irrational number is a rational point.

This description of the real numbers as points on the X axis is adequate to handle the topics to be studied in this book, and the reader who accepts it will not experience conceptual difficulty. In advanced work, a mathematically more precise description of the real numbers is needed, and we now discuss it briefly. To construct the real numbers in a rigorous way, one begins with sequences

$$\alpha_1, \alpha_2, \alpha_3, \ldots$$

of rational numbers which have the clustering property mentioned above: for large subscripts i, j, the distance $|\alpha_i - \alpha_j|$ between terms α_i, α_j in the sequence is small. Sequences of rational numbers with this property are called *Cauchy sequences,* and by studying them, real numbers can be described with acceptable standards of mathematical precision. The reader will find this description in Chapter 3 of the book *Elements of Real Analysis* by D. Sprecher (Academic, 1970). He may consult this book now, or later, or accept the description of real numbers as points on the X axis.

Throughout this book we shall use the symbol $\mathbf{R}$ to denote the real numbers.

We now state without proof the following fundamental theorem. (The proof is in the Sprecher book.)

THEOREM 2.2.2

The set $\mathbf{R}$ *of real numbers may be endowed with operations of addition and multiplication such that it becomes a field containing* $\mathbf{Q}$ *as a subfield. Furthermore, the real numbers partition into three subsets (the positive real numbers,* 0, *the negative real numbers) such that the order axioms* P1 *and* P2 *of* §1.3 *hold.*

Because P1 and P2 hold, we may study absolute values and inequalities involving real numbers. The discussion of these ideas in the first part of §1.3 applies without change to real numbers. In particular, the triangle inequality

$$|a + b| \leq |a| + |b| \qquad (5)$$

is valid when a and b are real numbers.

In fact, the absolute value of real numbers satisfies the four properties V1, V2, V3, V4 listed earlier in this section. (The same properties V1–V4 are also listed in §1.3.)

The geometric series

We now derive some results needed later in the book to study rational and real numbers, and in particular, needed below to prove that e is irrational.

The expression $1 + x + x^2 + \cdots + x^{k-1}$ (where k is a natural number) is called a *finite geometric series.* When $x = 1$, this series obviously equals k. We now give a formula for it when $x \neq 1$.

***LEMMA* 2.2.3**
Let x be an element of a field $\mathfrak{F}$, with $x \neq 1$. Set

$$S_k = 1 + x + x^2 + \cdots + x^{k-1}.$$

Then

$$S_k = \frac{1 - x^k}{1 - x}. \qquad (6)$$

PROOF Multiply S_k by x. Writing out both S_k and xS_k, we have

$$\begin{aligned} S_k &= 1 + x + x^2 + \cdots + x^{k-2} + x^{k-1}, \\ xS_k &= \phantom{1 + {}} x + x^2 + \cdots + x^{k-2} + x^{k-1} + x^k. \end{aligned}$$

Subtracting the second line from the first clearly causes most powers of x to cancel, producing the equation

$$(1 - x)S_k = 1 - x^k.$$

This gives the result. □

***COROLLARY* 2.2.4**
If the number x satisfies $|x| < 1$, then

$$1 + x + x^2 + x^3 + \cdots = \frac{1}{1 - x}. \qquad (7)$$

[*The left-hand side here is called an* infinite series *and is understood to mean*

$$\lim_{k\to\infty} (1 + x + x^2 + \cdots + x^{k-1}) = \lim_{k\to\infty} S_k.] \qquad (8)$$

PROOF From (6) we have

$$S_k = \frac{1}{1-x} - \frac{x^k}{1-x}.$$

For $|x| < 1$ it is clear that x^k approaches zero as k grows large. Thus we get

$$\lim_{k\to\infty} S_k = \frac{1}{1-x}. \qquad \square$$

REMARK (OPTIONAL) We interrupt the discussion to give a more detailed explanation of this limit. The meaning to be conveyed by (7) and (8) is that the sum of a finite number of terms on the left-hand side of (7), say

$$S_k = 1 + x + x^2 + \cdots + x^{k-1},$$

approximates $1/(1-x)$ with a specified precision, provided only that k is sufficiently large. That is, if a tolerance ϵ (a positive number) is given, we can guarantee that S_k approximates $1/(1-x)$ within this tolerance, provided only that k is large enough. (See Figure 2.2.3.) That this is the case is proved very easily. Because

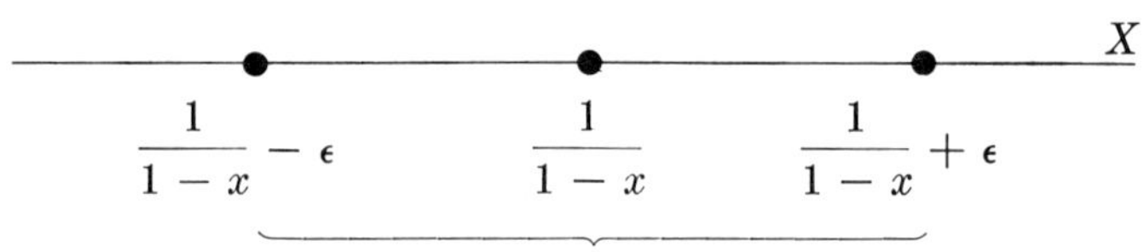

S_k lies in this interval, for all $k \geq K$.

Figure 2.2.3

$|x| < 1$, a high power of $|x|$ will be a number close to zero and hence there will be a natural number K such that $|x|^K < \epsilon\,|1-x|$, i.e.,

$$\left|\frac{x^K}{1-x}\right| < \epsilon.$$

For any $k \geq K$, we then have

$$\left|\frac{x^k}{1-x}\right| = |x^{k-K}| \left|\frac{x^K}{1-x}\right| \leq \left|\frac{x^K}{1-x}\right| < \epsilon,$$

(because $|x^{k-K}| = |x|^{k-K} \leq 1$ since $|x| < 1$). Using (6), for any $k \geq K$, we therefore have

$$\left|S_k - \frac{1}{1-x}\right| = \left|\frac{1-x^k}{1-x} - \frac{1}{1-x}\right| = \left|\frac{x^k}{1-x}\right| < \epsilon. \qquad \square$$

Other infinite series, besides the one on the left-hand side of (7), will be encountered later in the book, in §6.5 and Chapter 8. The series

$1 + x + x^2 + \cdots$ on the left-hand side of (7) is called the *infinite geometric series* or (usually) simply the *geometric series.*

The irrationality of *e* (Optional)

We now prove the irrationality of e that was stated above. Although this result will not be used elsewhere in the book, we include the proof because it is not difficult and the result is interesting, and because it provides an application of Corollary 2.2.4. The reader who wishes to omit this proof may do so. The proof is based on the following formula from calculus:

$$e = 1 + \frac{1}{2!} + \frac{1}{3!} + \frac{1}{4!} + \cdots = \sum_{k=0}^{\infty} \frac{1}{k!}.$$

We now make the hypothesis (contrary to what we wish to be true) that e is a rational number, say m/n, where m, n are positive integers. Then

$$\frac{m}{n} = \sum_{k=0}^{\infty} \frac{1}{k!} = \sum_{k=0}^{n} \frac{1}{k!} + \sum_{k=n+1}^{\infty} \frac{1}{k!}.$$

Multiplying each side by $n!$, we get

$$m(n-1)! = \left(\sum_{k=0}^{n} \frac{n!}{k!}\right) + \sum_{k=n+1}^{\infty} \frac{n!}{k!}.$$

In this expression the left-hand side is an integer, and each term in the sum in parentheses on the right-hand side is also an integer, since $n!/k! = (k+1)(k+2)\cdots n$ if $k \leq n$. Therefore the number E defined by

$$E = \sum_{k=n+1}^{\infty} \frac{n!}{k!}$$

is the difference of two integers and so is an integer. We shall produce a contradiction by proving that E is a number lying strictly between 0 and 1. Clearly $E > 0$ since it is a sum of positive terms. For $k > n$, we have

$$\frac{n!}{k!} = \frac{1}{(n+1)(n+2)\cdots(k)} < \frac{1}{(n+1)(n+1)\cdots(n+1)}$$

$$= \frac{1}{(n+1)^{k-n}}. \tag{9}$$

Here we used the fact that a fraction is increased if its denominator is decreased. Thus

$$E = \sum_{k=n+1}^{\infty} \frac{n!}{k!} < \sum_{k=n+1}^{\infty} \frac{1}{(n+1)^{k-n}} = \frac{1}{n+1} \sum_{k=n+1}^{\infty} \frac{1}{(n+1)^{k-1-n}}$$

$$= \frac{1}{n+1} \sum_{j=0}^{\infty} \frac{1}{(n+1)^j} = \frac{1}{n+1} \frac{1}{1 - \dfrac{1}{n+1}} = \frac{1}{n} \leq 1.$$

In this order: we used (9) for $k = n + 1, n + 2, \ldots$, then changed the summation index from k to j by the substitution $k - 1 - n = j$, and finally used Corollary 2.2.4 with $x = 1/(n + 1)$.

Thus $0 < E < 1$. This cannot be since E is an integer. Therefore e is irrational. □

EXERCISES 2.2

I

1. Give a shorter formula for $3^{-1} + 3^{-2} + 3^{-3} + \cdots + 3^{-11}$.

2. Give a shorter formula for $4^2 + 4^3 + \cdots + 4^8$.

3. Find sequences $\alpha_1, \alpha_2, \alpha_3, \ldots, \beta_1, \beta_2, \beta_3, \ldots$ of rational numbers, other than the sequences listed in the text, for which $\alpha_1 \le \alpha_2 \le \alpha_3 \le \cdots$, $\beta_1 \ge \beta_2 \ge \beta_3 \ge \cdots$, $\alpha_k^2 < 2 < \beta_k^2$ for all $k = 1, 2, 3, \ldots$, and $\beta_k - \alpha_k$ approaches zero as k grows large.

4. Describe all real numbers x for which $|x| + |x-1| \le 1$. Does every real number satisfying this inequality satisfies the equation $|x| + |x - 1| = 1$?

5. Describe all real numbers x for which $x^2 + 3x \le 4$. Do the same for $x^2 - 2x \ge -2$.

6. Let α and β be rational numbers, each having k decimal places, i.e., each has the form of an integer plus $.d_1d_2 \ldots d_k$ where the digits $d_1, d_2, \ldots, d_k$ are $0, 1, \ldots$, or 9. If $\alpha \ne \beta$, is it true that $|\alpha - \beta| \ge 10^{-k}$?

7. For which nonnegative real numbers x do we have $x^{1/2} \ge \frac{1}{2}x$? What about $x^{1/2} \le \frac{1}{2}x$?

8. Describe all real numbers x satisfying both $x^2 - 5x + 6 \ge 0$ and $x^2 - 3x + 2 \le 0$.

II

9. By imitating the proof that $\sqrt{2}$ is irrational, prove that $\sqrt{6}$, $\sqrt{10}$, $\sqrt{14}$ are all irrational. What about $\sqrt{4}$? $\sqrt{8}$? $\sqrt{12}$?

10. Let ABC be a right-angled triangle with the right angle at B. Suppose that AB and BC each have length 1. Show that the length AC is not a rational number.

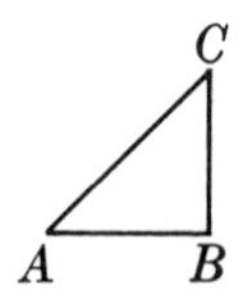

11. (a) Let t be an integer. Using the fact that t must have one of the three forms $3s$, $3s + 1$, $3s + 2$ (where s is an integer), show that t^2 is a multiple of 3 only if t is a multiple of 3.
(b) Using (a), prove that $\sqrt{3}$ is irrational. Hint: Imitate the proof of Lemma 2.2.1, substituting 3 into the role played by 2 in the proof.

12. Use the triangle inequality to prove that $|a - b| \geq \big| |a| - |b| \big|$ if a, b are real numbers.

13. The square of an even integer is even, and the square of an odd integer is odd. Prove this. (Hint: Square $2s$, $2s + 1$.)

14. Using the fact that $\sqrt{2}$ is irrational, prove that $1 + \sqrt{2}$, $\sqrt{2} + 3$, $3\sqrt{2}$, $1/\sqrt{2}$, $\sqrt{3} + \sqrt{6}$ are also irrational numbers.

15. If $x = -1$, show that $\lim_{k\to\infty} (1 + x + x^2 + \cdots + x^{k-1})$ does not exist. What if $x = 1$?

16. If a and b are irrational numbers, does it follow that $a + b$ and ab are also irrational? Do the irrational numbers form a ring? Do they form a group under addition or under multiplication?

17. If m, n, k are nonzero integers, show that $km/kn = m/n$.

18. Using the well-ordering principle, show that in the proof of Lemma 2.2.1 only finitely many cancellations of 2 from the numerator and denominator of a are possible.

III

19. Let $\alpha_1, \alpha_2, \alpha_3, \ldots$ be the numbers constructed in the text approximating $\sqrt{2}$. Prove that this sequence of numbers has the following property: If ϵ is a positive rational number, there is an integer K (dependent on ϵ) such that $|\alpha_i - \alpha_j| < \epsilon$ for all $i, j \geq K$.

20. Are $\sqrt{2} + \sqrt{3}$, $\sqrt{2} - \sqrt{3}$, $\sqrt{2 + \sqrt{2}}$ rational or irrational numbers?

21. Let $n = 2k$ where k is an odd integer. If t is an integer, with $t > 1$, show that $n^{1/t}$ is an irrational number.

22. The number e is irrational. What about $e^{1/2}$?

23. Let x be a real number, with $x > \sqrt{2}$. Show that $\frac{1}{2}(x + 2/x)$ lies between $\sqrt{2}$ and x. Recursively define numbers $x_0, x_1, x_2, \ldots$ by $x_0 = x$; then $x_{n+1} = \frac{1}{2}(x_n + 2/x_n)$ for $n = 0, 1, 2, \ldots$. Show that x_n steadily decreases and approaches $\sqrt{2}$ as n increases. If x_0 is a rational number, conclude that $x_0, x_1, x_2, x_3, \ldots$ are rational numbers approximating $\sqrt{2}$ with steadily increasing precision.

24. Let x be a real number. Define numbers $x_0, x_1, x_2, \ldots,$ recursively by $x_0 = x$; then $x_{n+1} = 1 + \frac{1}{4}x_n$. Show that x_n approaches a certain number a as n grows large. What is a? If x_0 is rational, prove that $x_0, x_1, x_2, \ldots$ are all rational. In this way obtain rational numbers which ever more closely approximate a. Is a itself rational?

25. What fraction is x if, as a decimal, $x = .12121212\ldots$? (12 repeats indefinitely.)

26. Prove that any decimal in which the digits have a repeating pattern (for example, x in the exercise above) is a rational number.

2.3 CONSTRUCTION OF THE COMPLEX NUMBERS

We begin with a Cartesian coordinate plane. This consists of two straight lines perpendicular to each other, one called the X axis, the other the Y axis. Mark off coordinates on the X axis, as in §2.2, and similarly put coordinates on the Y axis; see Figure 2.3.1. The number zero is placed on each axis at the intersection of the axes. (The intersection of the axes is called the *origin* and so is denoted by O.) Let P be any point in the plane of this XY axis system. Dropping straight lines L_1 and L_2 from P parallel to the Y and X axes, respectively, as shown in Figure 2.3.1, we obtain coordinates x_0 and y_0 for P by determining where L_1 intersects the X axis and where L_2 intersects the Y axis.

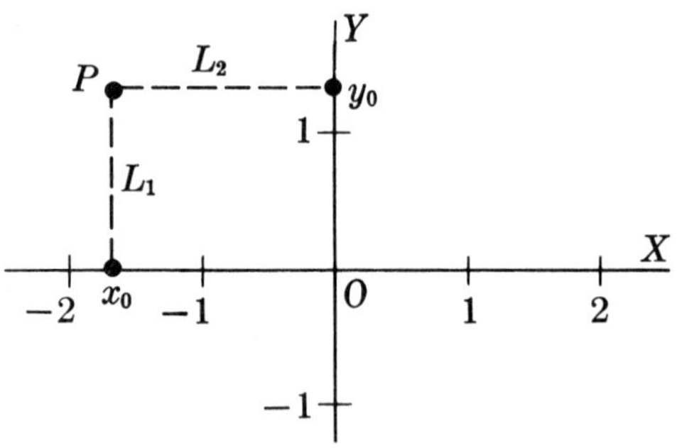

Figure 2.3.1

The coordinates are written (x_0, y_0), with the X coordinate x_0 always placed first. We call (x_0, y_0) the *coordinate pair*, or *ordered pair*, for P. The pair (x_0, y_0) uniquely determines P in the sense that no other point P' can have the same coordinate pair. That is, if (x_0', y_0') is the coordinate pair for point P', then $P = P'$ if and only if $x_0 = x_0'$ and $y_0 = y_0'$. Often we simply say that P is the point (x_0, y_0), and write $P = (x_0, y_0)$ or $P(x_0, y_0)$ when we wish the coordinates of P to appear explicitly. The coordinate system constructed this way is called the *Cartesian coordinate plane* or, sometimes, just the XY plane.

The points on the X axis (the real numbers **R**) are endowed with operations of addition and multiplication such that they form a field. We shall now endow the points in the Cartesian coordinate plane with addition and multiplication such that a field is obtained. We begin with addition, first introducing it algebraically, then explaining it geometrically. Let $P_1 = (x_1, y_1)$ and $P_2 = (x_2, y_2)$ be points in the Cartesian plane. We *define* addition of points by this rule:

$$(x_1, y_1) + (x_2, y_2) = (x_1 + x_2, y_1 + y_2). \tag{1}$$

Thus points add by addition of their coordinates.

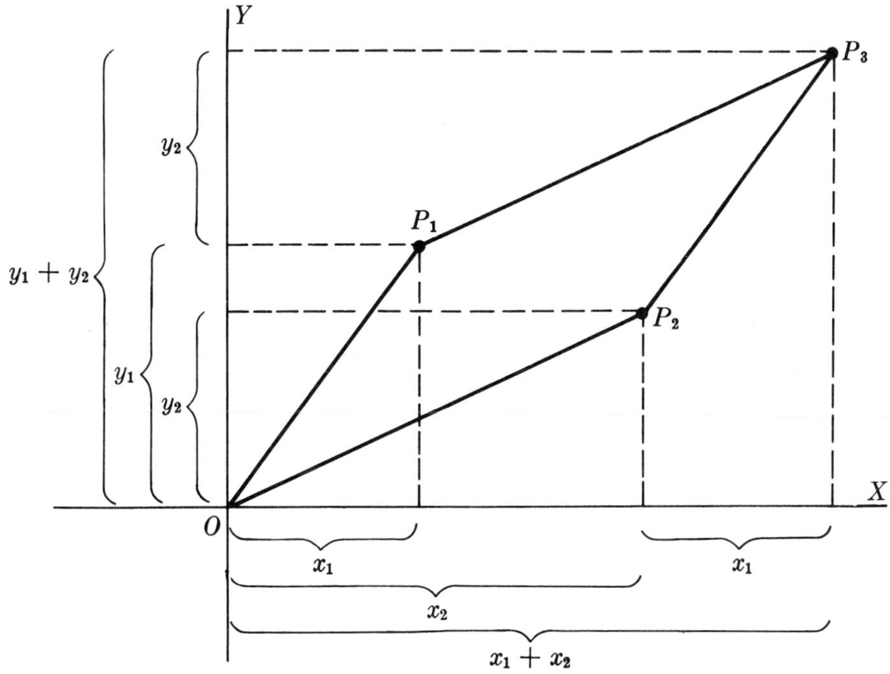

Figure 2.3.2

The geometric meaning of this rule may be obtained from Figure 2.3.2. Let P_3 be the fourth vertex of the parallelogram whose other three vertices are O, P_1, P_2. Because P_2P_3 is parallel to and has the same length as OP_1, the projections of OP_1 and P_2P_3 downwards onto the X axis have the same length. From this it follows that the X coordinate of P_3 is $x_1 + x_2$.

Similarly the horizontal projections of OP_2 and P_1P_3 onto the Y axis have the same lengths. From this we see that the Y coordinate of P_3 is $y_1 + y_2$. Hence $P_3 = (x_1 + x_2, y_1 + y_2)$. Since the definition of point addition gives $(x_1 + x_2, y_1 + y_2) = (x_1, y_1) + (x_2, y_2)$, we see that $P_3 = P_1 + P_2$. Thus: *the sum of points P_1 and P_2 is the fourth vertex of the parallelogram whose other three vertices are O, P_1, P_2.* This fact is called the *parallelogram law.* (Note: When O, P_1, P_2 are collinear, that is, on the same straight line, this law requires amplification since it is not clear how to construct a parallelogram using collinear points. See Exercise 11 below.)

We claim that the points of the Cartesian plane form a commutative group under this method of addition. Letting $P_1 = (x_1, y_1)$, $P_2 = (x_2, y_2)$, $P_3 = (x_3, y_3)$, we verify the axioms of an additive commutative group.

Closure (G1): $(x_1, y_1) + (x_2, y_2) = (x_1 + x_2, y_1 + y_2)$ is a point in the plane.

Associativity (G2):

$$\begin{aligned}(P_1 + P_2) + P_3 &= \big((x_1, y_1) + (x_2, y_2)\big) + (x_3, y_3)\\ &= (x_1 + x_2, y_1 + y_2) + (x_3, y_3)\\ &= (x_1 + x_2 + x_3, y_1 + y_2 + y_3)\\ &= (x_1, y_1) + (x_2 + x_3, y_2 + y_3)\\ &= (x_1, y_1) + \big((x_2, y_2) + (x_3, y_3)\big)\\ &= P_1 + (P_2 + P_3).\end{aligned}$$

Identity element (G3):
The additive identity is the origin $O = (0, 0)$.

Inverse element (G4):
The additive inverse of (x_1, y_1) is $(-x_1, -y_1)$.

Commutativity (G5):

$$\begin{aligned}P_1 + P_2 &= (x_1, y_1) + (x_2, y_2) = (x_1 + x_2, y_1 + y_2)\\ &= (x_2 + x_1, y_2 + y_1) = (x_2, y_2) + (x_1, y_1) = P_2 + P_1.\end{aligned}$$

The first ring axiom, R1, namely the axioms of an additive commutative group, is now verified.

Next we define multiplication of points, doing it algebraically. At first sight this multiplication may seem strange. The motivation behind it will appear later in this section, and the geometric interpretation will be seen in §2.4. We define the product of points $P_1 = (x_1, y_1)$ and $P_2 = (x_2, y_2)$ by this rule:

$$P_1P_2 = (x_1x_2 - y_1y_2, x_1y_2 + y_1x_2).$$

Clearly, the product of points is a point. Thus the closure Axiom R2 is satisfied. By direct calculation we shall verify Axioms R3 and R4, the associative and distributive laws. For this let $P_3 = (x_3, y_3)$. Then

$$\begin{aligned}
P_1(P_2P_3) &= (x_1, y_1)(x_2x_3 - y_2y_3, x_2y_3 + y_2y_3) \\
&= \big(x_1(x_2x_3 - y_2y_3) - y_1(x_2y_3 + y_2x_3), \\
&\qquad\qquad x_1(x_2y_3 + y_2x_3) + y_1(x_2x_3 - y_2y_3)\big) \\
&= (x_1x_2x_3 - x_1y_2y_3 - y_1x_2y_3 - y_1y_2x_3, \\
&\qquad\qquad x_1x_2y_3 + x_1y_2x_3 + y_1x_2x_3 - y_1y_2y_3) \\
&= \big((x_1x_2 - y_1y_2)x_3 - (x_1y_2 + y_1x_2)y_3, \\
&\qquad\qquad (x_1x_2 - y_1y_2)y_3 + (x_1y_2 + y_1x_2)x_3\big) \\
&= (x_1x_2 - y_1y_2, x_1y_2 + y_1x_2)(x_3, y_3) \\
&= (P_1P_2)P_3.
\end{aligned}$$

This verifies the associative law for multiplication. The verification of the distributive law, i.e., of $P_1(P_2 + P_3) = P_1P_2 + P_1P_3$, is done similarly by direct calculation using the definitions of addition and multiplication of points, and is left to the reader. Check that each side of the equation $P_1(P_2 + P_3) = P_1P_2 + P_1P_3$ equals the point

$$(x_1x_2 + x_1x_3 - y_1y_2 - y_1y_3, x_1y_2 + x_1y_3 + y_1x_2 + y_1x_3).$$

The verification of $(P_2 + P_3)P_1 = P_2P_1 + P_3P_1$ is similar.

Thus points of the Cartesian plane form a ring under their addition and multiplication operations. We next show that this ring is a field. First we check that $P_1P_2 = P_2P_1$ by calculating that each side of this equation is the point $(x_1x_2 - y_1y_2, x_1y_2 + x_2y_1)$. Thus the ring is commutative. The remainder of the proof consists of a verification of the conditions F1, F2 of Theorem 1.6.7, i.e., of the existence of a multiplicative identity and of multiplicative inverses of nonzero elements.

To verify F1, observe (by direct multiplication of points) that

$$(1, 0)(x, y) = (x, y)$$

when (x, y) is any point. Thus the point $(1, 0)$ is a multiplicative identity. Since it plainly is different from the zero point $(0, 0)$, the verification of F1 is complete.

If (x, y) is a nonzero point, then at least one of the real numbers x or y is nonzero, and hence $x^2 + y^2$ is a nonzero real number. By multiplying out the left-hand side, it is easily checked that

$$(x, y)\left(\frac{x}{x^2 + y^2}, -\frac{y}{x^2 + y^2}\right) = (1, 0).$$

Therefore the point

$$\left(\frac{x}{x^2 + y^2}, -\frac{y}{x^2 + y^2}\right)$$

is the multiplicative inverse of (x, y). This verifies F2.

We have completed the proof that under the addition and multiplication operations introduced above, the points of the Cartesian plane form a field. This field is called the *field of complex numbers*. We use **C** to denote it. An individual point often will be called a *complex number*, and the Cartesian plane itself will be called the *complex plane*. A notational device often used in this and other books is to let z denote the complex number (x, y).

THEOREM 2.3.1
Under the definitions of addition and multiplication given above, the points of the complex plane **C** *form a field.*

We next proceed to simplify our notation for points. Any point of the form $(x, 0)$ lies on the X axis. We agree to simplify the notation for it by writing x instead of $(x, 0)$. We call the X axis the *real axis*, and this simplification of notation means that any point on the real axis may be regarded either as a real number x or as a point $(x, 0)$ in the complex plane. Under this natural simplification the additive and multiplicative identities $(0, 0)$ and $(1, 0)$ become 0 and 1.

Notice that equality, addition, and multiplication of real numbers x_1 and x_2 is the same whether we regard them as members of the real field **R** or as members of the complex field **C**. To prove this merely observe that points $(x_1, 0)$ and $(x_2, 0)$ satisfy

$$\begin{aligned} (x_1, 0) &= (x_2, 0) \quad \text{if and only if } x_1 = x_2. \\ (x_1, 0) + (x_2, 0) &= (x_1 + x_2, 0), \\ (x_1, 0)(x_2, 0) &= (x_1 x_2, 0). \end{aligned}$$

Thus, by virtue of the fact that a real number x is also a point $(x, 0)$, we see that the real numbers **R** are contained in the complex numbers **C**, and furthermore, that the addition and multiplication operations in **C** when applied to elements of **R** coincide with the addition and multiplication operations in **R**. This means:

R *is a subfield of* **C**.

Observe that

$$x(x_1, y_1) = (xx_1, xy_1) \tag{2}$$

if x is a real number and (x_1, y_1) is a complex number. This is because x denotes the point $(x, 0)$, and the point product $(x, 0)(x_1, y_1)$ equals

(xx_1, xy_1). When x is a nonnegative real number, the geometric meaning of (2) is that $x(x_1, y_1)$ lies on the straight line passing through the origin and (x_1, y_1), with the distance from the origin increased by the factor x. (See Figure 2.3.3.) When x is negative, $x(x_1, y_1)$ is still on the straight line passing through the origin and (x_1, y_1), but is in the quadrant opposite to (x_1, y_1), with the distance from the origin increased by the factor $|x|$. The reader may draw a figure illustrating this for himself.

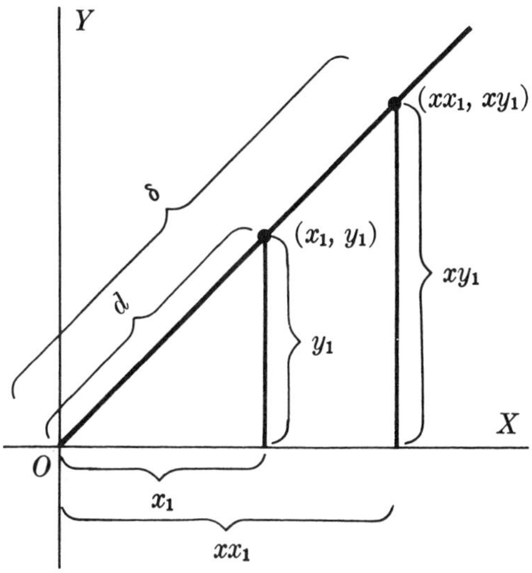

$$d = \sqrt{x_1^2 + y_1^2}$$

$$\delta = \sqrt{(xx_1)^2 + (yy_1)^2}$$
$$= xd$$

Figure 2.3.3

We will see in §2.4 that this geometric interpretation of a product of a real number and a complex number is a special case of the geometric interpretation of a product of two complex numbers.

Continuing our simplification of notation, we give the name i to the point $(0, 1)$ lying on the Y axis. Notice that

$$i^2 = (0, 1)(0, 1) = (-1, 0) = -1,$$

using first the multiplication rule for points, then our simplification of notation. Thus i has the property (not possessed by any real number) that its square is -1. For this reason we call i an *imaginary number* and the Y axis the *imaginary axis*. Of course, there is nothing imaginary about i; it is simply a point on the Y axis.

Now let (x, y) be any point. It is straightforward to check that

$$(x, y) = (x, 0) + (y, 0)(0, 1)$$

by direct application to the right-hand side of the rules for adding and multiplying points. Using the notation described above, we replace $(x, 0)$ with x, $(y, 0)$ with y, and $(0, 1)$ with i, obtaining

$$(x, y) = x + yi.$$

In other words: *a complex number* (x, y) *may, at our pleasure, be written* either as the coordinate pair (x, y) *or as* $x + yi$.

Observe the following elementary but basic fact: If x, y, x_1, y_1 are real numbers, *the complex numbers*

$$x + yi \qquad \text{and} \qquad x_1 + y_1 i$$

are equal if and only if

$$x = x_1 \qquad \text{and} \qquad y = y_1.$$

This is because

$$x + yi = (x, y), \qquad x_1 + y_1 i = (x_1, y_1),$$

and

$$(x, y) = (x_1, y_1) \quad \text{if and only if} \quad x = x_1 \quad \text{and} \quad y = y_1.$$

TERMINOLOGY When $z = x + yi$, with real x and y, we call x the *real part* of z, and y the *imaginary part* of z.

Thus: *complex numbers are equal*
when, and only when,
their real parts are equal
and their imaginary parts are equal.

Let $z_1 = x_1 + y_1 i$ and $z_2 = x_2 + y_2 i$ be any complex numbers, with real x_1, y_1, x_2, y_2. We can now compute the product $z_1 z_2$ in this easy way: Multiply out brackets using the distributive laws in the normal way, then use the fact that $i^2 = -1$, as follows:

$$\begin{aligned} z_1 z_2 &= (x_1 + y_1 i)(x_2 + y_2 i) = x_1 x_2 + x_1 y_2 i + y_1 x_2 i + y_1 y_2 i^2 \\ &= (x_1 x_2 + i^2 y_1 y_2) + (x_1 y_2 + x_2 y_1) i \\ &= (x_1 x_2 - y_1 y_2) + (x_1 y_2 + x_2 y_1) i. \end{aligned}$$

If we rewrite this last equation in terms of pairs, it says:

$$(x_1, y_1)(x_2, y_2) = (x_1 x_2 - y_1 y_2, x_1 y_2 + y_1 x_2). \tag{3}$$

This equation is just our definition of point multiplication. The above derivation of (3) reveals the algebraic foundation for this definition: In order to obtain a field containing a quantity i satisfying $i^2 = -1$, the definition of point multiplication should be that given by (3).

In §2.2 we saw that the real number field is *ordered*. The complex number field, however, is *not ordered*. See Exercise 13. The quite easy proof may be supplied by the reader.

REMARK This remark is addressed to readers familiar with *vectors* and may be omitted by other readers. Let us regard the complex number $z = (x, y)$ as the terminal point of a vector (arrow) $\vec{v}$

emanating from the origin. (See Figure 2.3.4.) If $z_1 = (x_1, y_1)$, $z_2 = (x_2, y_2)$ are complex numbers, let $\vec{v_1} = \overrightarrow{Oz_1}$, $\vec{v_2} = \overrightarrow{Oz_2}$ be vectors from the origin to z_1, z_2, respectively. The rules (1), for addition of complex numbers z_1, z_2, and (2), for multiplication of a complex number z_1 by a real number x, translate directly into the usual rules for computing the sum $\vec{v_1} + \vec{v_2}$ of vectors $\vec{v_1}$, $\vec{v_2}$ and product $x\vec{v_1}$ of vector $\vec{v_1}$ by real number x. This is evident from the parallelogram law and the geometric description above of the product $x(x_1, y_1)$. Complex number theory therefore contains some elements of the theory of vectors. However, the reader should be cautioned that the general multiplication rule for complex numbers is neither of the usual products for vectors, i.e., not the dot product (also called "inner product"), and not the cross product. See Exercise 17 for the connection between these multiplications.

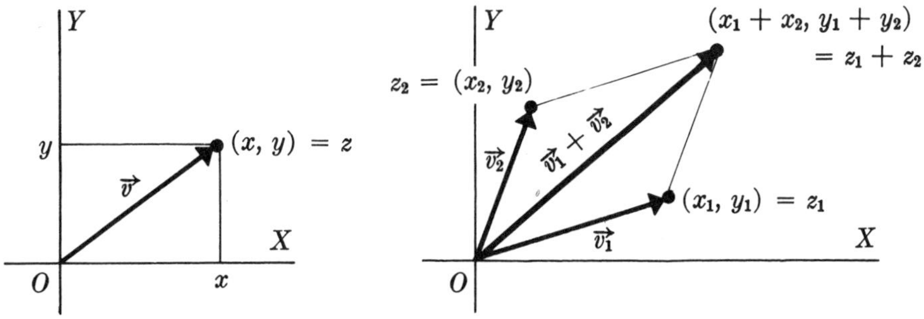

Figure 2.3.4

Complex number addition may be used to carry out translations of the complex plane. Let u be a fixed complex number and z be any complex number. By the parallelogram law, $z + u$ is obtained by translating z in the amount and direction indicated by the arrow from O to u. (See Figure 2.3.5.) Thus, if z_1, z_2, z_3 are vertices of a triangle T, $z_1 + 1$, $z_2 + 1$, $z_3 + 1$ will be vertices of a triangle congruent to T. (See Figure 2.3.6.)

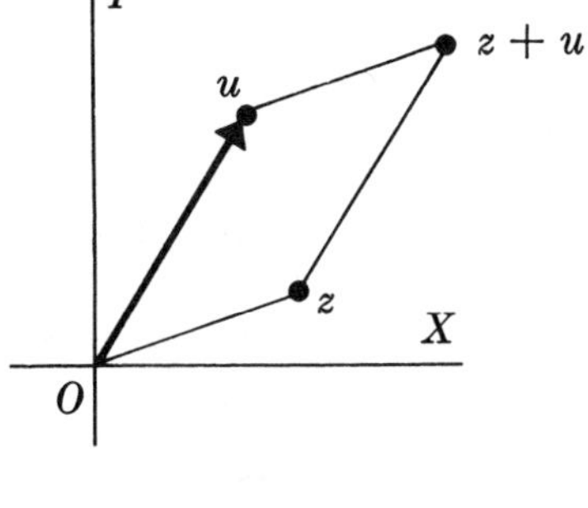

Figure 2.3.5

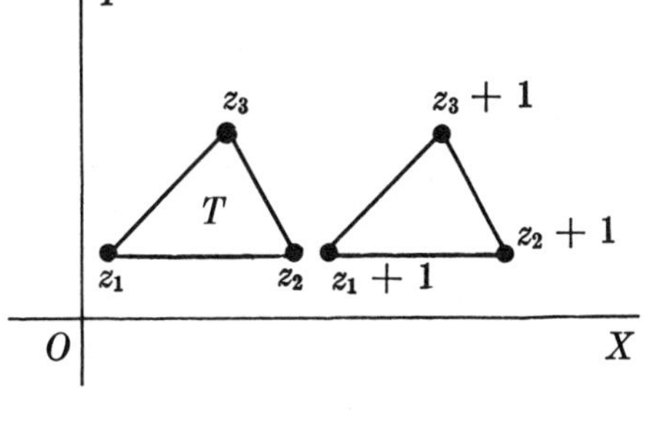

Figure 2.3.6

EXERCISES 2.3

I

1. Add, subtract, and multiply the following complex numbers:
(a) $2 + 3i$, $i - 1$. (b) i, $1 + i$. (c) $3 + i$, $4i - 2$.

2. Find the complex number z if
(a) $(2 + i)z = 3 - i$;
(b) $1 + i + iz = -1$;
(c) $z^2 = -1$;
(d) $z^2 = 1$.
In each case find all possible solutions, and present your solutions in the form $x + yi$ with real x, y.

3. Solve for the complex number z if $z^2 - (2 + 2i)z + 2i - 1 = 0$.

4. Prove or disprove the following conjecture: If z_1 and z_2 are complex numbers such that z_1z_2 is real, then both z_1 and z_2 are real.

5. Prove or disprove the following conjecture: If complex numbers z_1, z_2 satisfy $z_1 + z_2i = 0$, then $z_1 = z_2 = 0$.

6. Find a square with vertex at $2 + 3i$ congruent to the square with vertices at 0, 1, i, $1 + i$.

7. Let $z_1 = 1 + i$, $z_2 = 2 - 3i$, $z_3 = 4 + 2i$. Find a number z_4 such that z_1, z_2, z_3, z_4 are vertices of a parallelogram.

II

8. If z_1 and z_2 both lie on the circle with equation $x^2 + y^2 = 1$, does z_1z_2 also lie on this circle?

9. If z_1, z_2 both lie on the ellipse with equation $x^2 + 2y^2 = 1$, does z_1z_2 also lie on this ellipse?

10. If $z_1^2 + z_2^2 = 1$, where z_1, z_2 are complex numbers, do z_1, z_2 lie on the circle of radius 1 centered at O?

11. (a) Let z_1, z_2 lie on a straight line L through the origin O, with O not between z_1 and z_2. Let $z_3 = z_1 + z_2$. Verify that z_3 is on L as shown in the first figure (next page) and that distance Oz_3 = distance Oz_1 + distance Oz_2.
(b) Let z_1, z_2 again lie on a straight line L through O, with O between z_1 and z_2. Suppose z_1 is the closer to the origin of z_1, z_2. Let $z_3 = z_1 + z_2$. Show that z_3 lies on L, on the same side of O as z_2, and that distance Oz_3 = distance Oz_2 − distance Oz_1.

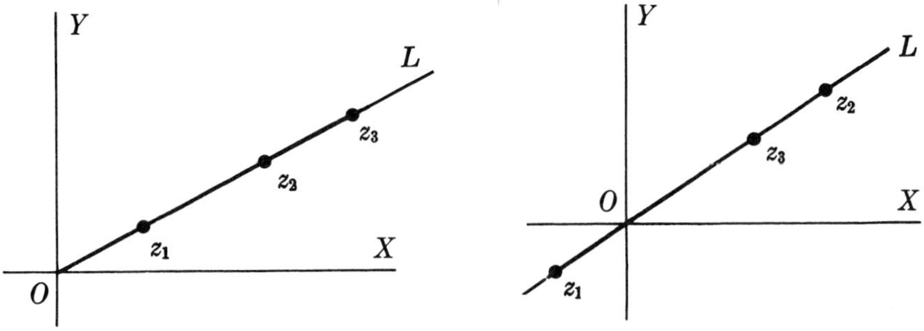

Figures for Exercise 11

12. Let z be a nonzero complex number. Show that O is the midpoint of the straight line segment joining $-z$ and z.

13. Show that it is impossible to find a nonempty subset $\mathcal{P}$ of **C** with these properties:

P1 For any complex number a, exactly one of the following is true: $a \in \mathcal{P}$; $a = 0$; $-a \in \mathcal{P}$.

P2 If $a \in \mathcal{P}$ and $b \in \mathcal{P}$, then $a + b \in \mathcal{P}$ and $ab \in \mathcal{P}$.
(In other words: *it is impossible to order the complex numbers.*) Hint: Consider $1^2 + i^2$.

14. Show, for any integer n, that $i^n = 1$, or i, or -1, or $-i$, according as $n = 4m$, $n = 4m + 1$, $n = 4m + 2$, $n = 4m + 3$. Here m is an integer.

III

15. Define a new multiplication $\times'$ on points by $(x_1, y_1) \times' (x_2, y_2) = (x_1x_2 + 2y_1y_2, x_1y_2 + x_2y_1)$. Show that, under addition of points as described in the text and multiplication $\times'$, the points of the plane form a field. Repeat with the multiplication $\times''$ defined by $(x_1, y_1) \times'' (x_2, y_2) = (x_1x_2 - 2y_1y_2, x_1y_2 + x_2y_1)$.

16. Let $z = x + yi$, and suppose $z^2 = a + bi$, where x, y, a, b are real. Set up a pair of simultaneous equations in x, y, a, b, and by solving them prove that

$$x = \pm\left(\frac{a + (a^2 + b^2)^{1/2}}{2}\right)^{1/2}, \qquad y = \pm\left(\frac{-a + (a^2 + b^2)^{1/2}}{2}\right)^{1/2},$$

where $(a^2 + b^2)^{1/2}$ denotes the nonnegative square root of $a^2 + b^2$. Discuss which choices of the $\pm$ signs should be used.

17. Let $z_1 = (x_1, y_1)$, $z_2 = (x_2, y_2)$ be points in the complex plane, and let $\vec{v_1} = \overrightarrow{Oz_1}$, $\vec{v_2} = \overrightarrow{Oz_2}$ be the vectors from O to z_1, z_2, respectively. The inner, or dot product $\vec{v_1} \cdot \vec{v_2}$ is the real number $x_1x_2 + y_1y_2$, and the cross product $\vec{v_1} \times \vec{v_2}$ is the vector perpendicular to the XY plane

with perpendicular component $x_1y_2 - x_2y_1$. Show that the real and imaginary parts of the complex number product $\bar{z}_1z_2$ are, respectively, the dot product $\vec{v}_1 \cdot \vec{v}_2$ and the perpendicular component of the cross product $\vec{v}_1 \times \vec{v}_2$. Here $\bar{z}_1 = (x_1, -y_1)$ is the *complex conjugate* of z_1, to be explained in the next section.

18. Endow the points of the Cartesian plane with addition as in the text.
    (a) When endowed with addition and the multiplication rule

$$(x_1, y_1)(x_2, y_2) = (x_1x_2 + y_1y_2, x_1y_2 + x_2y_1),$$

the points of the Cartesian plane are called the *split complex numbers.*

    (b) When endowed with addition and with multiplication given by

$$(x_1, y_1)(x_2, y_2) = (x_1x_2, x_1y_2 + x_2y_1),$$

the points of the Cartesian plane are called the *dual numbers.* Show that the split complex numbers and the dual numbers are rings containing the real numbers as a subring. Are they fields? If not, find some zero divisors. Can you find a connection between the split complex numbers and the ring in Example 1.6.6? Use the fact that $(\frac{1}{2}, \frac{1}{2})^2 = (\frac{1}{2}, \frac{1}{2})$, $(\frac{1}{2}, -\frac{1}{2})^2 = (\frac{1}{2}, -\frac{1}{2})$, $(\frac{1}{2}, \frac{1}{2})(\frac{1}{2}, -\frac{1}{2}) = 0$ in the split complex numbers. Can the split complex numbers or the dual numbers be ordered?

2.4 FURTHER PROPERTIES OF COMPLEX NUMBERS

Let $z = x + yi$ be a nonzero complex number, with real x and y. Let r be the distance between z and O and θ be the angle at O between the positive X axis and the line Oz. Remember the convention from trigonometry and calculus that an angle is positive if it corresponds to a rotation in the counterclockwise sense, and negative if it corresponds to a rotation in the clockwise sense. The angle θ indicated in Figure 2.4.1 is positive.

DEFINITION 2.4.1
The distance between O and z is called the modulus, *or* norm, *or* absolute value *of z. It is denoted by $|z|$.*

Thus $|z| = r$. From Pythagoras' theorem we have $r^2 = x^2 + y^2$. Hence:

$$\textit{if } z = x + yi \textit{ with } x, y \textit{ real, then } |z| = (x^2 + y^2)^{1/2}. \quad (1)$$

Notice that, when z is real, the absolute value just described reduces to the absolute value for real numbers given in §2.2.

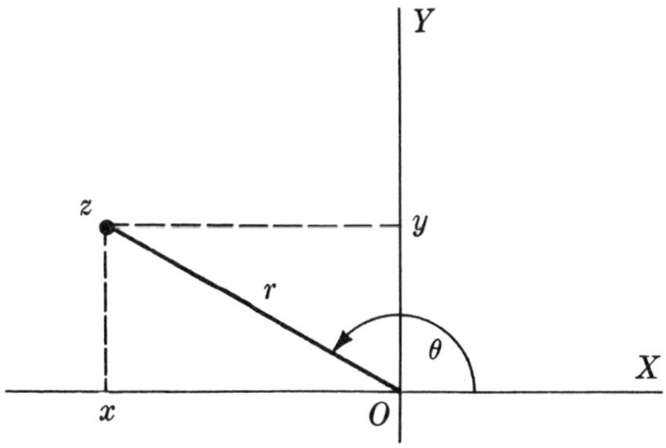

Figure 2.4.1

DEFINITION 2.4.2
The angle, *or* amplitude, *of z is the angle of rotation from the positive X axis to the line Oz.*

The angle θ shown in Figure 2.4.1 will do as angle z.

REMARK The angle of z is not uniquely determined. If a complete revolution is added to or subtracted from the angle θ in Figure 2.4.1 (that is, if 2π is added to or subtracted from θ), we obtain another perfectly acceptable angle for z. Thus the quantity angle z is determined only to within addition of integral multiples of 2π.

From elementary trigonometry (see Figure 2.4.1) we get the formulas

$$x = r \cos \theta, \qquad y = r \sin \theta.$$

Inserting these expressions into $z = x + yi$, we obtain the following representation, called the *polar form* of the nonzero complex number z:

$$z = |z| (\cos \theta + i \sin \theta) = r(\cos \theta + i \sin \theta). \tag{2}$$

Formula (2) is of fundamental importance. It presents the complex number z in terms of its distance from the origin $|z|$ and its angle θ. Moreover, this factorization of z is unique in the following sense: if also $z = r'(\cos \theta' + i \sin \theta')$ where $r' \geq 0$ and θ is an angle, then in fact $r' = |z|$ and $\theta' = \theta + 2\pi k$ for some integer k. This may be seen as follows. From the two forms for z we see, on equating real and imaginary parts, that $r \cos \theta = r' \cos \theta'$ and $r \sin \theta = r' \sin \theta'$. Squaring these equations, then adding, gives $r^2 = r'^2$, hence $r = r'$. Using this, we get $\cos \theta = \cos \theta'$ and $\sin \theta = \sin \theta'$, and so from elementary trigonometry we know that θ and θ' must agree to within an integral multiple of 2π.

We summarize these results in Theorem 2.4.3.

THEOREM 2.4.3
The nonzero complex number z may be written in polar form,

$$z = r(\cos\theta + i\sin\theta), \qquad r > 0.$$

Here r is unique and in fact $r = |z|$. The angle θ is uniquely determined to within addition of integral multiples of 2π.

When $z = 0$, we still speak of $|z|$ as the distance between z and O (in this case $|z| = 0$ and (1) still holds). However, no longer can we speak of the *angle of z*.

The next theorem (called DeMoivre's theorem) gives the behaviour of $r = |z|$ and $\theta =$ angle z under complex number multiplication.

THEOREM 2.4.4
Let z_1 and z_2 be complex numbers. Then:

$$|z_1 z_2| = |z_1|\,|z_2|, \tag{3}$$

$$\left|\frac{z_1}{z_2}\right| = \frac{|z_1|}{|z_2|}, \qquad \text{if } z_2 \neq 0. \tag{4}$$

If both z_1 and z_2 are nonzero, then:

$$\text{angle } z_1 z_2 = \text{angle } z_1 + \text{angle } z_2, \tag{5}$$

$$\text{angle } \frac{z_1}{z_2} = \text{angle } z_1 - \text{angle } z_2. \tag{6}$$

[Since angles of complex numbers are determined only to within addition of integral multiples of 2π, the meaning of (5) is that any angle for z_1 + any angle for z_2 yields an angle for $z_1 z_2$. A similar explanation goes with (6).]

PROOF Formula (3) is obviously correct if z_1 or z_2 is zero since each side then equals 0. Similarly, (4) holds when $z_1 = 0$. In the remainder of the proof we may assume both z_1 and z_2 are nonzero. Write z_1 and z_2 in polar form:

$$z_1 = |z_1|\,(\cos\theta_1 + i\sin\theta_1), \quad \text{where } \theta_1 = \text{angle } z_1, \tag{7}$$

$$z_2 = |z_2|\,(\cos\theta_2 + i\sin\theta_2), \quad \text{where } \theta_2 = \text{angle } z_2. \tag{8}$$

Recall the following trigonometric formulas:

$$\cos(\alpha + \beta) = \cos\alpha\cos\beta - \sin\alpha\sin\beta,$$

$$\sin(\alpha + \beta) = \sin\alpha\cos\beta + \cos\alpha\sin\beta.$$

Using these formulas with $\alpha = \theta_1$ and $\beta = \theta_2$, we see that

$$\begin{aligned}(\cos\theta_1 + i\sin\theta_1)(\cos\theta_2 + i\sin\theta_2) \\ &= \cos\theta_1\cos\theta_2 - \sin\theta_1\sin\theta_2 + i(\cos\theta_1\sin\theta_2 + \sin\theta_1\cos\theta_2) \\ &= \cos(\theta_1 + \theta_2) + i\sin(\theta_1 + \theta_2). \qquad (9)\end{aligned}$$

Multiply equations (7) and (8). Using (9), we get

$$z_1z_2 = |z_1|\,|z_2|\,(\cos(\theta_1 + \theta_2) + i \sin(\theta_1 + \theta_2)). \tag{10}$$

Since $|z_1|\,|z_2|$ is a positive real number, the uniqueness theorem (2.4.3) implies that $|z_1z_2| = |z_1|\,|z_2|$ and that $\theta_1 + \theta_2$ is an angle for z_1z_2. Thus (3) and (5) are proved.

If we apply (3) or (5) with z_1/z_2 in place of z_1, we get

$$|z_1| = \left|\frac{z_1}{z_2}\right| |z_2| \qquad \text{and} \qquad \text{angle } z_1 = \text{angle } \frac{z_1}{z_2} + \text{angle } z_2.$$

This gives (4) and (6). □

An immediate consequence is that if (7) and (8) are the polar forms of z_1 and z_2 respectively, then (10) is the polar form of z_1z_2, and the polar form of z_1/z_2 is

$$\frac{z_1}{z_2} = \frac{|z_1|}{|z_2|}(\cos(\theta_1 - \theta_2) + i \sin(\theta_1 - \theta_2)).$$

By DeMoivre's theorem, we may multiply complex numbers by multiplying their moduli and adding their angles. This is the geometric interpretation of multiplication promised earlier. To divide, we divide the moduli and subtract the angles. If z_1 and z_2 happen to lie on a circle centered at O and having radius 1 (thus $|z_1| = |z_2| = 1$) then multiplication is accomplished simply by adding the angles, and division is accomplished by subtracting the angles. This is illustrated in Figure 2.4.2.

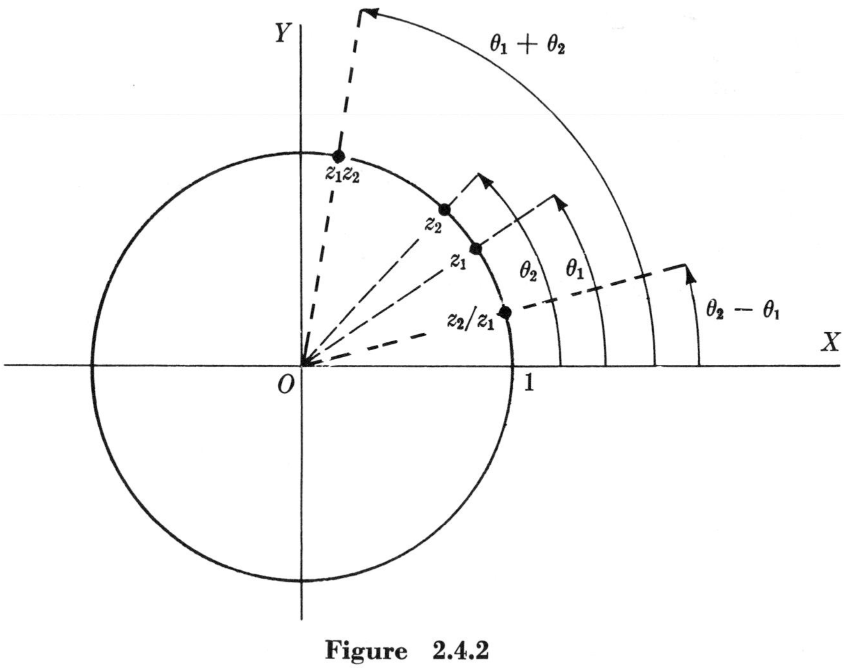

Figure 2.4.2

Theorem 2.4.4 may be used to derive interesting trigonometric identities. Here is an example. Theorem 2.4.4 gives

$$\cos 3\theta + i \sin 3\theta = (\cos \theta + i \sin \theta)^3.$$

Multiplying out the right-hand side, then comparing the real and imaginary parts on the two sides of the resulting equation yields

$$\cos 3\theta = \cos^3 \theta - 3 \cos \theta \sin^2 \theta,$$

$$\sin 3\theta = 3 \cos^2 \theta \sin \theta - \sin^3 \theta.$$

Using the absolute value concept, we may give a geometric interpretation of subtraction of complex numbers. Let z_1 and z_2 be complex numbers, and set $w = z_2 - z_1$. Then $z_1 + w = z_2$ and so, from the geometric interpretation of complex number addition in §2.3, we know that z_2 is the fourth vertex of a parallelogram whose other three vertices are O, z_1, w.

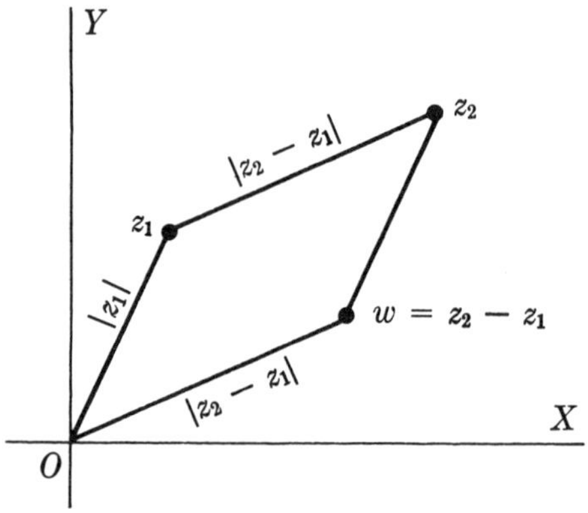

Figure 2.4.3

(See Figure 2.4.3.) In this figure, $|z_1|$ is the length of side Oz_1, and $|w| = |z_2 - z_1|$ is the length of side Ow. Because opposite sides of a parallelogram have equal length, we see that $|z_2 - z_1|$ *is the distance between* z_1 *and* z_2. Thus an inequality of the form $|z_2 - z_1| \leq R$ means that z_2 lies on or inside a circle C with center at z_1 and radius R. (See Figure 2.4.4.)

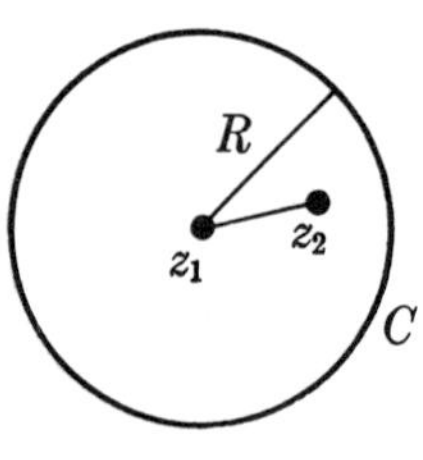

Figure 2.4.4

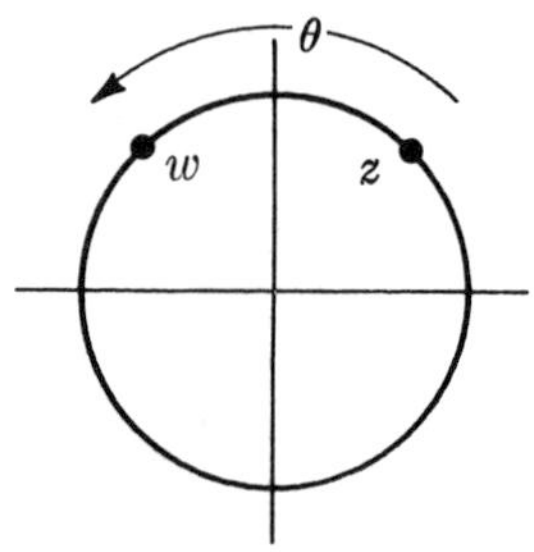

Figure 2.4.5

Complex number multiplication may be used to carry out rotations of the complex plane. For example, let z be a nonzero complex number and set $w = (\cos\theta + i\sin\theta)z$. Then $|w| = |z|$ and angle $w = \theta +$ angle z. Multiplying z by $\cos\theta + i\sin\theta$ thus has the effect of rotating z around the origin through the angle θ. (See Figure 2.4.5.) As an illustration of this idea, if z_1, z_2, z_3 are vertices of a triangle T, then iz_1, iz_2, iz_3 will be vertices of a triangle congruent to T, as shown in Figure 2.4.6.

The triangle inequality

We turn next to a fundamental inequality comparing $|z_1+z_2|$ with $|z_1|$ and $|z_2|$. We have already seen this inequality for integers in §1.3 and for rational and real numbers in §2.2.

THEOREM 2.4.5
If z_1, z_2 are complex numbers,

$$|z_1 + z_2| \leq |z_1| + |z_2|. \tag{11}$$

PROOF We base the proof on the geometric interpretation of addition that was given in §2.3. Let $z_3 = z_1 + z_2$. Then O, z_1, z_2, z_3 are vertices of a parallelogram, as shown in Figure 2.4.7. The length of the side Oz_1 is $|z_1|$. Because opposite sides of a parallelogram have equal length, the length of the side connecting z_2 and z_3 is also $|z_1|$. The length of the side Oz_2 is $|z_2|$ and the length of the diagonal Oz_3 is $|z_3|$. Consider now the triangle Oz_2z_3, for which the edges have the lengths $|z_2|, |z_1|, |z_3|$. It is an intuitive geometric fact that in a triangle the sum of the lengths of two edges is always at least the length of the third edge. Thus $|z_1| + |z_2| \geq |z_3|$. Since $z_3 = z_1 + z_2$, we obtain (11). □

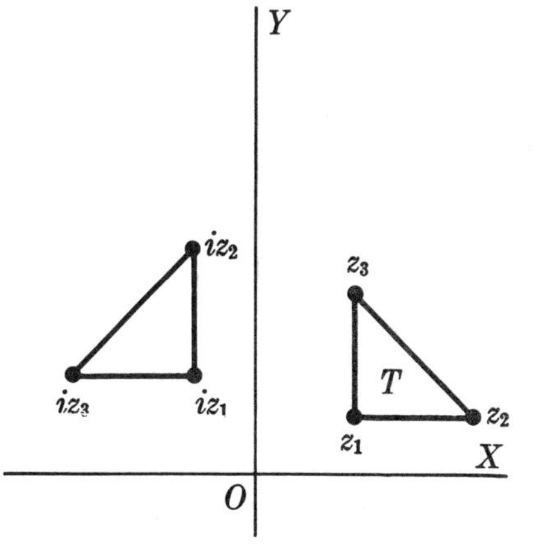

Figure 2.4.6

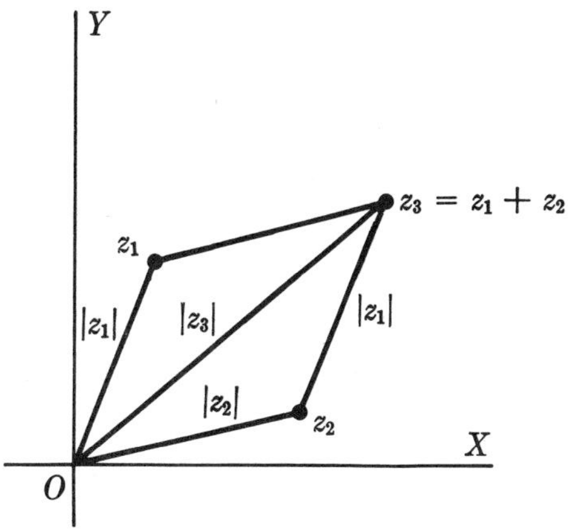

Figure 2.4.7

Because of the nature of its proof, the inequality (11) is usually called the triangle inequality. In Exercise 9 below we outline a proof of it using purely algebraic techniques.

Notice that the absolute value for complex numbers obeys the same rule V1–V4 that appeared in §1.3 and §2.2, namely:

V1 *For any* a, $|a|$ *is a nonnegative real number;*

V2 $|a| = 0$ *if and only if* $a = 0$;

V3 $|ab| = |a|\,|b|$ *for all* a, b;

V4 $|a + b| \leq |a| + |b|$ *for all* a, b.

Here a, b are any complex numbers.

Complex conjugation

This useful device is introduced in the next definition.

DEFINITION 2.4.6
Let $z = x + yi$ *be a complex number* (x, y *are real*). *Then the* complex conjugate *of* z, *denoted by* $\bar{z}$, *is defined by* $\bar{z} = x - yi$.

Thus $\bar{z}$ is obtained (algebraically) by changing the sign of the imaginary part of z. The next theorem gives the more important properties of complex conjugation.

***THEOREM* 2.4.7**
Let z, z_1 *and* z_2 *be complex numbers. Then:*

(a) $\overline{z_1 + z_2} = \bar{z}_1 + \bar{z}_2$;

(b) $\overline{z_1 z_2} = \bar{z}_1\,\bar{z}_2$;

(c) $\bar{\bar{z}} = z$;

(d) $|z|^2 = \bar{z}z = z\bar{z}$;

(e) $|\bar{z}| = |z|$;

(f) z *is real if and only if* $z = \bar{z}$.

(g) *If* (2) *is the polar form of* z, *the polar form of* $\bar{z}$ *is*

$$\bar{z} = |z|\,(\cos(-\theta) + i\sin(-\theta)) = |z|\,(\cos\theta - i\sin\theta).$$

PROOF (a) If $z_1 = x_1 + y_1 i$, $z_2 = x_2 + y_2 i$, with x_1, x_2, y_1, y_2 all real, then $z_1 + z_2 = x_1 + x_2 + (y_1 + y_2)i$. Thus $\overline{z_1 + z_2} = x_1 + x_2 - (y_1 + y_2)i = (x_1 - y_1 i) + (x_2 - y_2 i) = \bar{z}_1 + \bar{z}_2$.

(b) With z_1 and z_2 as in (a), we have $z_1z_2 = (x_1x_2 - y_1y_2) + (x_1y_2 + y_1x_2)i$. Thus

$$\begin{aligned}\overline{z_1z_2} &= (x_1x_2 - y_1y_2) - (x_1y_2 + y_1x_2)i \\ &= (x_1 - y_1i)(x_2 - y_2i) = \bar{z}_1\,\bar{z}_2.\end{aligned}$$

(c) $\bar{\bar{z}} = \overline{x - yi} = x + yi = z.$

(d) $\bar{z}z = z\bar{z} = (x + yi)(x - yi) = x^2 + y^2 = |z|^2.$

(e) $|\bar{z}| = (x^2 + (-y)^2)^{1/2} = (x^2 + y^2)^{1/2} = |z|.$

(f) z is real if and only if $y = 0$, and this is equivalent to $x + yi = x - yi$. (Why?)

(g) $\bar{z} = \overline{|z|\,(\cos\theta + i\sin\theta)} = |z|\,(\cos\theta - i\sin\theta)$

$= |z|\,(\cos(-\theta) + i\sin(-\theta)).$ □

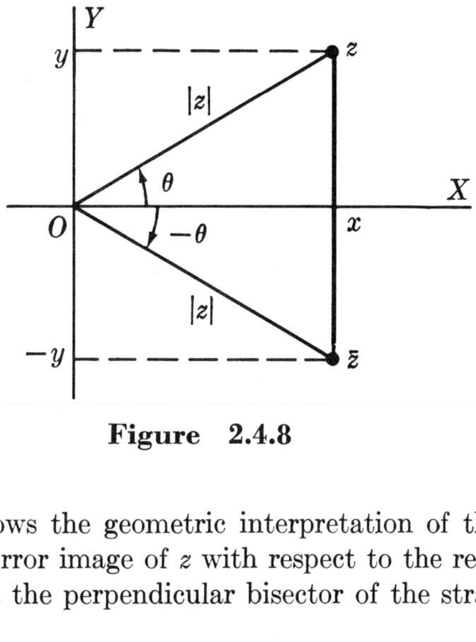

Figure 2.4.8

Figure 2.4.8 shows the geometric interpretation of the conjugation operation: $\bar{z}$ is the mirror image of z with respect to the real axis. This is because the X axis is the perpendicular bisector of the straight line joining z to $\bar{z}$.

We may use complex conjugation to solve equations like $z(1 + i) = 2 + i$ and also to compute reciprocals of complex numbers. To solve this equation, multiply each side by the complex conjugate of $1 + i$. This gives $2z = (2 + i)(1 - i) = 3 - i$, hence $z = \frac{3}{2} - \frac{1}{2}i$. To compute $1/(1 + i)$, multiply numerator and denominator by the conjugate of the denominator, thereby obtaining

$$\frac{1}{1+i} = \frac{1}{1+i}\frac{1-i}{1-i} = \frac{1-i}{2} = \frac{1}{2} - \frac{1}{2}i.$$

Complex exponentials

In more advanced courses the following formula is developed:

$$e^{i\theta} = \cos\theta + i\sin\theta. \tag{12}$$

Here $e = 2.71828\ldots$ is the natural logarithm base encountered in calculus, and θ is an arbitrary real number. We shall take (12) as the *definition* of $e^{i\theta}$. The full justification of this definition is to be found in textbooks in the *theory of a complex variable;* it is beyond the scope of this book to give it here. However, we are able to use the fact that angles add when complex numbers multiply to establish the following laws of exponents: *if θ_1 and θ_2 are real numbers, then*

$$e^{i(\theta_1+\theta_2)} = e^{i\theta_1}e^{i\theta_2}, \tag{13}$$

$$e^{i(\theta_1-\theta_2)} = e^{i\theta_1}/e^{i\theta_2}. \tag{14}$$

PROOF Applying the definition (12) of $e^{i\theta}$ with θ_1, θ_2, and $\theta_1+\theta_2$ in place of θ, and using (9), we get

$$\begin{aligned} e^{i(\theta_1+\theta_2)} &= (\cos(\theta_1+\theta_2) + i\sin(\theta_1+\theta_2)) \\ &= (\cos\theta_1 + i\sin\theta_1)(\cos\theta_2 + i\sin\theta_2) \\ &= e^{i\theta_1}e^{i\theta_2}. \end{aligned}$$

This proves (13). Applying (13) with $\theta_1 - \theta_2$ in place of θ_1, we get

$$e^{i\theta_1} = e^{i(\theta_1-\theta_2)}e^{i\theta_2},$$

from which (14) follows. [Note that $e^{i\theta}$ is not zero, for any real θ, since (12) yields $|e^{i\theta}| = (\cos^2\theta + \sin^2\theta)^{1/2} = 1$.] □

From (12) we see that

$$\overline{e^{i\theta}} = \cos\theta - i\sin\theta = \cos(-\theta) + i\sin(-\theta).$$

Thus

$$\overline{e^{i\theta}} = e^{-i\theta}.$$

Using complex exponentials, we may write the polar representation (2) of a complex number is the following simple form.

THEOREM 2.4.8
If z is a nonzero complex number, then

$$z = re^{i\theta}$$

where $r = |z|$, $\theta =$ any angle for z.

PROOF By (2), $z = |z|(\cos\theta + i\sin\theta) = |z|\,e^{i\theta} = re^{i\theta}$. □

Exercise 24 gives the definition and properties of e^z when z is any complex number.

Multiplicative groups of complex numbers

There are interesting multiplicative groups consisting of complex numbers lying on the circle of radius 1 centered at the origin. (This circle is often called the *unit circle.*) For example, the four complex numbers

$$e^0 = 1, \quad e^{i\pi/2} = i, \quad e^{i\pi} = -1, \quad e^{-i\pi/2} = -i$$

form a multiplicative group, as the reader may show for himself. Another multiplicative group consists of the six complex numbers

$$e^{ik\pi/3}, \quad \text{for } k = 0, 1, 2, 3, 4, 5.$$

A class of groups containing these examples as special cases will be seen in §5.5. These groups consist of complex numbers equally spaced around the unit circle. (See Figure 2.4.9.)

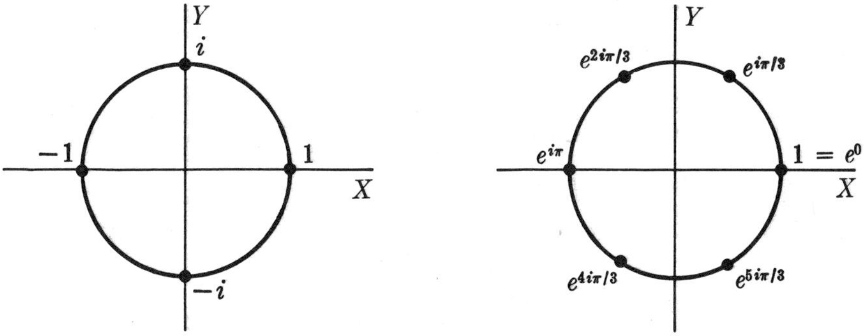

Figure 2.4.9

EXERCISES 2.4

In these exercises, z, z_1, z_2 denote complex numbers.

I

1. Convert the following complex numbers to polar form: $2i$, -7, $1 + i$, $1 - \sqrt{3}i$.

2. Convert the following complex numbers to cartesian form:

$$3\left(\cos\frac{\pi}{2} + i\sin\frac{\pi}{2}\right); \quad 2\left(\cos\left(\frac{-\pi}{4}\right) + i\sin\left(\frac{-\pi}{4}\right)\right);$$

$$6(\cos\pi + i\sin\pi).$$

3. Find the fourth powers of each of the numbers in Exercises 1 and 2.

4. Solve $(1 - i)z = 1 + i$ for z.

5. Find $\bar{z}$ when $z = i\cdot(2 - i)\cdot(-3) + i - 1$.

6. If $z\bar{z} = 0$, what can you say about z?

7. If $z + \bar{z} = 0$, what can you say about z?

8. Use DeMoivre's theorem to solve $z^8 = 1$ for z. Find all solutions.

II

9. Let z_1 and z_2 be complex numbers. Set $z_1\bar{z}_2 = x + yi$, with real x and y. By justifying each of the following steps, give an algebraic proof of the triangle inequality: $|z_1 + z_2|^2 = (z_1 + z_2)\overline{(z_1 + z_2)} = (z_1 + z_2)(\bar{z}_1 + \bar{z}_2) = z_1\bar{z}_1 + z_1\bar{z}_2 + \bar{z}_1 z_2 + z_2\bar{z}_2 = |z_1|^2 + 2x + |z_2|^2 \leq |z_1|^2 + 2\sqrt{x^2 + y^2} + |z_2|^2 = |z_1|^2 + 2\,|z_1 z_2| + |z_1|^2 = (|z_1| + |z_2|)^2$.

10. Show that a complex number z satisfies a monic quadratic equation with real coefficients, that is, an equation of the form $z^2 + az + b = 0$ with real a, b. Show that z and $\bar{z}$ satisfy the same equation. If z is not real, prove that it does not satisfy any equation $z + a = 0$ with a real.

11. Let $\mathcal{G}$ consist of all complex numbers z with $|z| = 1$, endowed with the operation of multiplication of complex numbers. Show that $\mathcal{G}$ is a commutative group.

12. Let n be a fixed integer. Show that the set $\mathcal{G}$ of all complex numbers z satisfying $z^n = 1$ is a commutative group under the operation of multiplication of complex numbers.

13. Prove that the complex numbers of the form $x + yi$ with $x, y \in \mathbf{Z}$ form an integral domain. (The complex numbers of this form are called *Gaussian integers*.) Show also that the complex numbers of the form $x + yi$ with $x, y \in \mathbf{Q}$ form a field, called the field of *Gaussian rationals*.

14. Show that the numbers 1 and $(-1 \pm \sqrt{-3})/2$ form the vertices of an equilateral triangle inscribed in the circle $|z| = 1$. Do these three numbers form a multiplicative group?

15. Let z be a complex number. Prove that there always is a complex number w such that $w^2 = z$.

16. Derive trigonometric identities expressing $\cos 4\theta$ and $\sin 4\theta$ in the terms of $\cos\theta$ and $\sin\theta$.

III

17. Determine under what conditions the equality sign holds in the triangle inequality.

18. Solve the following equations for z: $z^4 = 1$, $z^3 = -1$, $z^3 = 1$. Obtain all solutions and prove that you have done so.

19. Prove that $|z_1 - z_2| \geq | \, |z_1| - |z_2| \, |$.

20. Show that $|z_1 - z_2|^2 + |z_1 + z_2|^2 = 2\,|z_1|^2 + 2\,|z_2|^2$. Interpret this identity in terms of the lengths of the diagonals of the parallelogram with vertices O, z_1, z_2, $z_1 + z_2$.

21. Express, in terms of z_1 and z_2, the area of the parallelogram with vertices O, z_1, z_2, $z_1 + z_2$.

22. Describe the curve traversed by all complex numbers z satisfying $|z - a| + |z - b| = c$. Here a, b are fixed complex numbers, and c is a fixed positive real number. Then describe the region consisting of all z for which $|z - a| + |z - b| < c$.

23. Repeat the first part of 22 for $|z - a| - |z - b| = \pm c$.

24. If $z = x + yi$ is a complex number (real x and y), the *definition* of e^z is $e^z = e^x e^{iy}$. Prove: If z and w are complex numbers, then $e^{z+w} = e^z e^w$ and $e^{z-w} = e^z/e^w$. Prove also: $\overline{e^z} = e^{\bar{z}}$.

25. This exercise is addressed to readers who know the Maclaurin series expansions for e^x, $\cos x$, $\sin x$, and is intended to furnish additional justification for the definition (12) of $e^{i\theta}$. In the Maclaurin series $e^x = 1 + x/1! + x^2/2! + x^3/3! + \cdots$, substitute $i\theta$ for x. Without giving justification, rearrange the terms to get

$$\begin{aligned} e^{i\theta} &= (1 - \theta^2/2! + \theta^4/4! - \cdots) + i(\theta - \theta^3/3! + \theta^5/5! - \cdots) \\ &= \cos\theta + i\sin\theta. \end{aligned}$$

26. Let A, B, C be a triangle, with D, E the midpoints of AC and BC respectively. Use properties of complex numbers to prove that length DE is $\frac{1}{2}$(length AB).

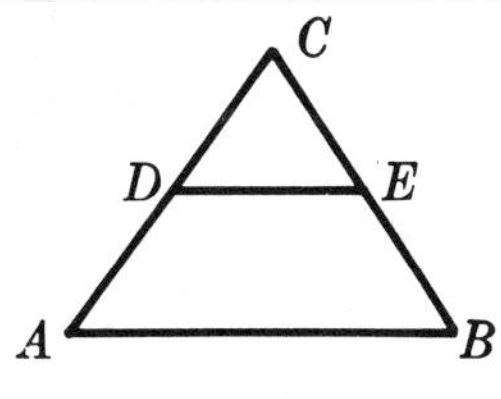

27. Show that $\frac{1}{2}(z_1 + z_2)$ is the midpoint of the straight line segment connecting complex numbers z_1, z_2. Use this and other properties of complex numbers to prove that the diagonals of a parallelogram bisect each other.

28. Let $\mathfrak{R}$ be the ring of split complex numbers (see Exercise 18, §2.3). Define the *norm* of a split complex number $z = (x, y)$ by $N(z) = |x^2 - y^2|^{1/2}$. If z_1, z_2 are split complex numbers, verify that $N(z_1 z_2) = N(z_1)N(z_2)$. Show that $N(z) = 0$ with $z \neq 0$ if and only if z is a zero divisor in $\mathfrak{R}$.

2.5 GROUPS OF MOTIONS OF THE COMPLEX PLANE (OPTIONAL)

In calculus one encounters functions of a real number. Recall that a function f of a real number is defined for all real numbers if, to any given real number x, there corresponds a real number $f(x)$. Often $f(x)$ is given by a formula, for example, $f(x) = x^2 + \sin x$, but this need not always be the case. This somewhat imprecise description of a function can be made fully precise by casting it into a more abstract setting; see Exercise 14. (We shall discuss this important point in greater detail in §7.4.) Notice that one is often interested in a function f rather than a particular functional value $f(x)$. The distinction here is that the function f is described by the totality of its functional values $f(x)$.

We shall now extend the concept of function to complex numbers. We say that f *is a function defined on the complex plane if, to each complex number* z, *there corresponds a complex number denoted by* $f(z)$. As in the real case, the letter f denotes the function itself, and $f(z)$ denotes the quantity obtained when f is evaluated at z.

We next define equality of functions. (For certain functions of real numbers this has already been done in Example 1.4.8.)

DEFINITION 2.5.1
Functions f and g on the complex plane are equal *when, and only when,* $f(z) = g(z)$ *for every complex number* z.

Let $\mathcal{S}$ be the set of all functions on the complex plane. Recall that in §1.2 we observed the necessity, when studying a set S, of having a precise rule specifying when elements of S are equal. For our set $\mathcal{S}$ of functions Definition 2.5.1 gives this precise rule. Later in this section we shall be interested in groups consisting of certain elements of $\mathcal{S}$. We prepare the way for this by introducing addition and multiplication operations for functions.

DEFINITION 2.5.2
Let f and g be functions on the complex plane. The sum s *and the* product p *of f and g are the functions defined by these rules:*

$$\left.\begin{aligned} s(z) &= f(z) + g(z), \\ p(z) &= f(g(z)), \end{aligned}\right\} \quad \text{for every complex number } z.$$

The addition of functions just introduced is called *pointwise addition.* Only rarely are other types of function addition used. However, several types of function multiplication occur in mathematics; the type introduced in Definition 2.5.2 is called *composition.* Two other types, called *pointwise multiplication* and *convolution,* will be seen later in this book, in §§7.4

and 8.6. For clarity we shall speak here of function addition and composition of functions (instead of function addition and function multiplication). Our primary object in this section is to study several types of functions on the complex plane, under the composition operation. However, because function addition will be encountered later in the book, we give first the following theorem, the proof of which introduces the *zero function* and the *additive inverse* (or *negative*) of a function.

THEOREM 2.5.3

Under pointwise addition, the set $\mathcal{S}$ of functions on the complex plane is a commutative group.

PROOF We must verify the axioms G1–G5 of Chapter 1, the operation $\circ$ in these axioms being function addition. The sum $f + g$ of functions f, g on the complex plane is (by Definition 2.5.2) again a function on the complex plane. Closure therefore holds. Let h be another function on the complex plane. The associative law, $(f + g) + h = f + (g + h)$ is verified by evaluating each side at an arbitrary complex number z and observing that equal results are obtained:

$$((f + g) + h)(z) = (f + g)(z) + h(z) = f(z) + g(z) + h(z),$$

$$(f + (g + h))(z) = f(z) + (g + h)(z) = f(z) + g(z) + h(z).$$

The equality of the functions $(f + g) + h$ and $f + (g + h)$ now follows from Definition 2.5.1. The verification of the commutative law, $f + g = g + f$, is similar. The identity for function addition and the additive inverse of f are, respectively, the functions $\mathcal{O}$ and F defined by

$$\mathcal{O}(z) = 0 \quad \text{and} \quad F(z) = -f(z), \quad \text{for all complex numbers } z.$$

It is an easy exercise (left to the reader) to prove that $\mathcal{O} + f = f + \mathcal{O} = f$ and $F + f = f + F = \mathcal{O}$. This completes the proof. (Note: The function $\mathcal{O}$ is called the *zero function* and is usually denoted simply by 0. The function F is usually denoted by $-f$; often it is called the negative of f.) □

Our discussion of composition will focus on the functions described in the next definition.

DEFINITION 2.5.4

A function f on the complex plane is a bijection *if it satisfies these two conditions:*

(a) *Whenever z_1 and z_2 are complex numbers such that $f(z_1) = f(z_2)$, then $z_1 = z_2$;*

(b) *If w is any complex number, a complex number z (dependent on w) exists such that $f(z) = w$.*

Example 2.5.5
Let u be a fixed complex number. The function t_u defined by

$$t_u(z) = z + u \qquad \text{for all complex numbers } z,$$

is called a *translation*. It is a bijection.

Proof: If z_1 and z_2 are complex numbers such that $t_u(z_1) = t_u(z_2)$, then $z_1 + u = z_2 + u$; hence $z_1 = z_2$. Thus (a) holds. If w is given and $z = w - u$, then $t_u(z) = w$. Thus (b) holds.

The reason for the name "translation" is that t_u acts on a complex number z by translating it in the direction and amount indicated by the arrow from the origin to u. (See Figure 2.5.1.) □

Example 2.5.6
Let θ be a fixed *real* number. The function r_θ defined by

$$r_\theta(z) = e^{i\theta}z \qquad \text{for all complex numbers } z,$$

is called a *rotation about the origin*. It is a bijection.

Proof: If $r_\theta(z_1) = r_\theta(z_2)$, then $e^{i\theta}z_1 = e^{i\theta}z_2$; hence $z_1 = z_2$ since $e^{i\theta} \neq 0$. This verifies (a). If w is given and $z = e^{-i\theta}w$, then $r_\theta(z) = w$; thus (b) also holds.

The reason for the name "rotation" is that r_θ acts on a complex number z by rotating it about the origin through the angle θ. (See Figure 2.5.2.) This is because $e^{i\theta}z$ is obtained from z simply by adding θ to the angle of z. □

Example 2.5.7
Complex conjugation is a bijection. That is, the function c defined by

$$c(z) = \bar{z} \qquad \text{for all complex numbers } z,$$

is a bijection. The very easy proof is left as an exercise. □

The subset of $\mathcal{S}$ consisting of all bijections on the complex plane will be denoted by $\mathcal{B}$. The three Examples above show that $\mathcal{B}$ contains all translations t_u, all rotations r_θ, and c.

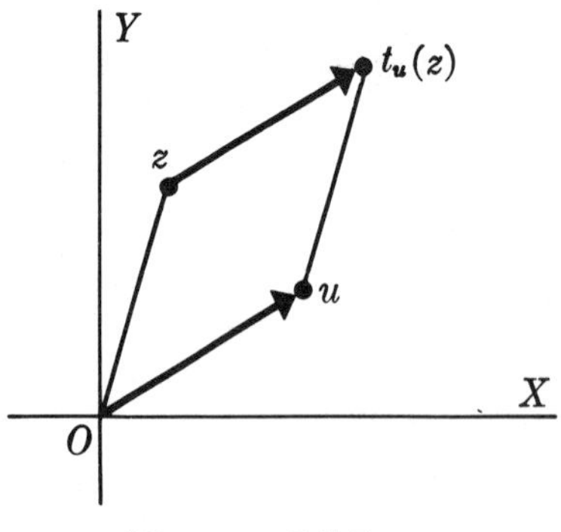

Figure 2.5.1

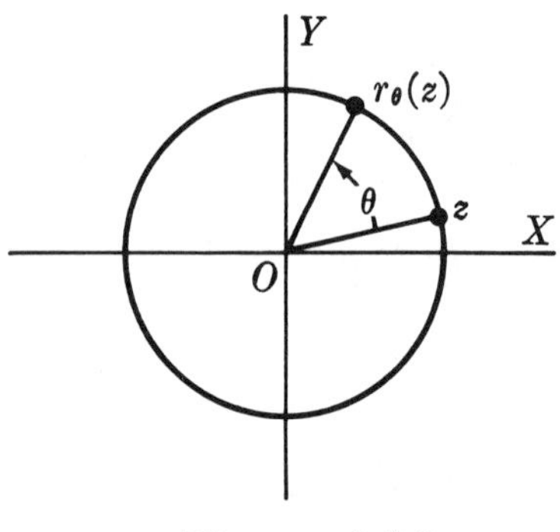

Figure 2.5.2

THEOREM 2.5.8
Under composition, the set $\mathfrak{B}$ of bijections on the complex plane is a noncommutative group.

PROOF Let f, g, h be bijections on the complex plane. We shall check, one by one, the group axioms.

Closure (G1) Let $p = fg$. We must show that p is a bijection: (a) If z_1 and z_2 are complex numbers such that $p(z_1) = p(z_2)$, then

$$f(g(z_1)) = f(g(z_2)).$$

This equation together with property (a) [applied to f] yields $g(z_1) = g(z_2)$, and then property (a) [applied to g] yields $z_1 = z_2$. This establishes (a) for p. Let w be given. Applying (b) to f, we obtain $w = f(z_1)$ for some z_1. Applying (b) to g, we obtain $z_1 = g(z)$ for some z. But then $p(z) = w$. This establishes (b) for p.

Associativity (G2) To prove that $(fg)h = f(gh)$, we evaluate each side at an arbitrary z, obtaining:

$$((fg)h)(z) = (fg)(h(z)) = f(g(h(z))),$$

$$(f(gh))(z) = f((gh)(z)) = f(g(h(z))).$$

The functions $(fg)h$ and $f(gh)$ therefore yield equal results when evaluated at an arbitrary z. This implies, by Definition 2.5.1, that $(fg)h = f(gh)$.

Existence of identity (G3) Define a function E by

$$E(z) = z \qquad \text{for all complex numbers } z.$$

It is a simple matter to check that E is a bijection and that $Ef = fE = f$ for any function f on the complex plane. We call E the *identity function.*

Existence of inverses (G4) Let f be a bijection on the complex plane. If z is a complex number, there is [by (a) and (b)] precisely *one* complex number z_1 such that $f(z_1) = z$. We define a function F on the complex plane by specifying that $F(z) = z_1$. Thus

$$F(z) = z_1 \qquad \text{if} \qquad f(z_1) = z. \tag{1}$$

We will show that F is the required inverse of f. First notice that (1) implies

$$f(F(z)) = z \qquad \text{for any complex number } z. \tag{2}$$

Furthermore, if z_1 is any complex number, by setting $z = f(z_1)$ and using (1), we also get

$$F(f(z_1)) = z_1 \qquad \text{for any complex number } z_1. \tag{3}$$

We contend that F is a bijection.

Proof: (a) Suppose z, z' are complex numbers with $F(z) = F(z')$. We are to show that $z = z'$. Applying (2) to z and z', we get

$$z = f(F(z)) = f(F(z')) = z'.$$

This verifies part (a) of Definition 2.5.4. To verify (b), let z_1 be a complex number. Then (3) shows that a solution z of $F(z) = z_1$ is $z = f(z_1)$. This verifies (b), and completes the proof that F is a bijection.

From (2) we get $(fF)(z) = E(z)$ for every z, so that $fF = E$. From (3) we get $(Ff)(z_1) = E(z_1)$ for every z_1, so that $Ff = E$. The equations $fF = E$ and $Ff = E$ show that F is indeed the inverse of f. This bijection F is usually denoted by f^{-1} and called the *inverse* of f. Notice that f^{-1}, when evaluated at z, is usually *not* $1/f(z)$. See the examples a few lines below.

This completes the proof that $\mathcal{B}$ is a group under the composition operation. To see that it is noncommutative, observe that

$$r_{\pi/2}t_1 \neq t_1 r_{\pi/2}. \tag{4}$$

To prove (4), evaluate each side at z. The left-hand side yields $i(z + 1)$, the right-hand side $iz + 1$, and these are clearly not equal. □

What is the inverse of a translation t_u, a rotation r_θ, or the conjugation c? The inverses are $t_u^{-1} = t_{-u}$, $r_\theta^{-1} = r_{-\theta}$, and $c^{-1} = c$, since

$$t_{-u}t_u = E = t_u t_{-u}, \qquad r_{-\theta}r_\theta = E = r_\theta r_{-\theta}, \qquad cc = E.$$

(*Proof:* Evaluate all parts of these equations at an arbitrary z.)

We now focus our attention on several subgroups of $\mathcal{B}$. By Theorem 1.6.10, to show that a nonempty subset of $\mathcal{B}$ is a subgroup, it is sufficient to show that it contains products and inverses of its elements.

THEOREM 2.5.9
Under composition, the set T of all translations of the complex plane is a commutative subgroup of $\mathcal{B}$.

PROOF Let u, u_1, u_2 be complex numbers. By evaluating at an arbitrary z, we deduce $t_{u_1}t_{u_2} = t_{u_1+u_2}$. As shown above, $t_u^{-1} = t_{-u}$. Products and inverses of translations are therefore translations. Hence T is a subgroup of $\mathcal{B}$. It is commutative because

$$t_{u_1}t_{u_2} = t_{u_1+u_2} = t_{u_2+u_1} = t_{u_2}t_{u_1}.$$

For a geometric representation of this commutativity, see Figure 2.5.3. This figure shows O, z, u_1, u_2, and the parallelograms used in computing $t_{u_1}(z)$, $t_{u_2}(t_{u_1}(z))$, $t_{u_2}(z)$, $t_{u_1}(t_{u_2}(z))$, $u_1 + u_2$, $t_{u_1+u_2}(z)$. □

QUESTION For which u is $E = t_u$?

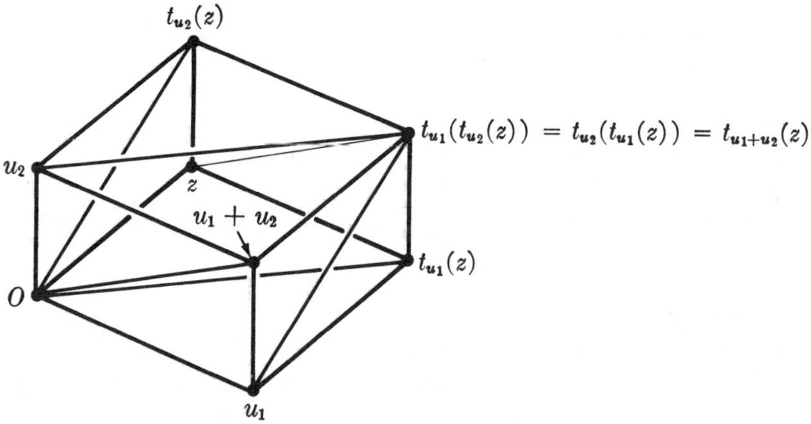

Figure 2.5.3

THEOREM 2.5.10
Under composition, the set R of all rotations of the complex plane about the origin is a commutative subgroup of $\mathcal{B}$.

PROOF Let θ, θ_1, θ_2 be real numbers. It is easy to see that $r_{\theta_1}r_{\theta_2} = r_{\theta_1+\theta_2}$. We already know that $r_\theta^{-1} = r_{-\theta}$. Products and inverses of rotations about the origin are therefore rotations about the origin; hence R is a subgroup of $\mathcal{B}$. It is commutative because

$$r_{\theta_1}r_{\theta_2} = r_{\theta_1+\theta_2} = r_{\theta_2+\theta_1} = r_{\theta_2}r_{\theta_1}. \qquad \square$$

QUESTION For which real numbers θ is $E = r_\theta$?

THEOREM 2.5.11
The two-element subset C of $\mathcal{B}$ consisting of E and c forms, under composition, a commutative subgroup of $\mathcal{B}$.

PROOF Observe that $cc = E$ and $c^{-1} = c$. The remaining details are left for the reader. $\square$

The element c is sometimes called a *reflection*. See Exercise 15 below for other two-element reflection groups.

We next consider functions m on $\mathbf{C}$ with either of the forms

$$\left.\begin{aligned} m(z) &= az + b, \\ m(z) &= a\bar{z} + b, \end{aligned}\right\} \begin{cases}\text{for all } z \in \mathbf{C}, \text{ where } a, b \text{ are} \\ \text{complex constants with } |a| = 1.\end{cases} \qquad (5)$$

The constants a, b and the particular form depend on m, but not on z.

The reader will observe that a translation, rotation about the origin, or conjugation always has one of the forms (5). We claim that any function m of either of these forms is a bijection on **C**.

Proof: (a) Let z_1, z_2 be complex numbers with $m(z_1) = m(z_2)$. Then $az_1 + b = az_2 + b$ or $a\bar{z}_1 + b = a\bar{z}_2 + b$, yielding $z_1 = z_2$ or $\bar{z}_1 = \bar{z}_2$, and therefore $z_1 = z_2$ in either case. (b) The equation $m(z) = w$ is satisfied by either $z = a^{-1}(w - b)$ or by $z = \bar{a}^{-1}(\bar{w} - \bar{b})$.

Notice that any function m of either of the forms (5) will satisfy

$$|m(z_1) - m(z_2)| = |z_1 - z_2| \tag{6}$$

for all complex numbers z_1, z_2. Thus m, when applied to any pair of points, preserves the distance between these points. Not every bijection has this distance preserving property; an example is provided by the bijection f defined by $f(x + yi) = x + 2yi$, for all real x, y. For this f,

$$|f(z_1) - f(z_2)| \neq |z_1 - z_2|$$

when $z_1 = i$ and $z_2 = 0$. This leads to the following definition.

DEFINITION 2.5.12
A rigid motion *of the complex plane is a bijection m satisfying* (6) *for all complex numbers z_1, z_2.*

The above remarks show that any bijection m of either of the forms (5) is a rigid motion; therefore translations, rotations about the origin, and complex conjugation are all rigid motions. Conceivably, there are other bijections m, besides those given by (5), which satisfy (6).

THEOREM 2.5.13
Under composition, the set M of all rigid motions of the complex plane is a noncommutative subgroup of $\mathcal{B}$ containing T, R, C.

PROOF By definition of rigid motion, each element of M is a bijection, so that M is a subset of $\mathcal{B}$. To prove that M is a subgroup, we need only verify that products and inverses of elements of M are again elements of M; i.e., are distance preserving. Suppose that $m_1, m_2 \in M$, so that

$$\begin{aligned} |m_1(z_1) - m_1(z_2)| &= |z_1 - z_2|, \\ |m_2(z_1) - m_2(z_2)| &= |z_1 - z_2|, \end{aligned} \tag{7}$$

for all complex numbers z_1, z_2. Then

$$\begin{aligned} |(m_1m_2)(z_1) - (m_1m_2)(z_2)| &= |m_1(m_2(z_1)) - m_1(m_2(z_2))| \\ &= |m_2(z_1) - m_2(z_2)| \\ &= |z_1 - z_2|. \end{aligned}$$

(We used here (7) with $m_2(z_1)$, $m_2(z_2)$ standing in place of z_1, z_2.) Therefore m_1m_2 is distance-preserving. Now let $m \in M$, take z_1, z_2 to be any complex numbers, and set $w_1 = m^{-1}(z_1)$, $w_2 = m^{-1}(z_2)$. Then $m(w_1) = z_1$, $m(w_2) = z_2$, and we have:

$$|m^{-1}(z_1) - m^{-1}(z_2)| = |w_1 - w_2| = |m(w_1) - m(w_2)| = |z_1 - z_2|.$$

Consequently, m^{-1} is also distance preserving. Thus M is a subgroup of $\mathcal{B}$. Plainly, by remarks above, M contains T, R, C. That M is noncommutative is shown by (4). □

Are there elements of M other than the bijections (5) above? No, for we shall now prove the following remarkable fact.

***THEOREM* 2.5.14**
Any rigid motion m of the complex plane has one of the forms (5).

PROOF Let m be a rigid motion of the complex plane, and set $w_0 = m(0)$, $w_1 = m(1)$, $w_2 = m(i)$. Notice that

$$|w_1 - w_0| = |m(1) - m(0)| = |1 - 0| = 1.$$

Therefore $w_1 - w_0 = e^{i\theta}$ for some real number θ. Put

$$\mu = r_{-\theta}t_{-w_0}m.$$

Then μ, as a product of elements $r_{-\theta}$, t_{-w_0}, m of M, is a rigid motion. Observe that

$$\mu(0) = r_{-\theta}(t_{-w_0}(m(0))) = r_{-\theta}(t_{-w_0}(w_0)) = r_{-\theta}(w_0 - w_0) = r_{-\theta}(0) = 0,$$
$$\mu(1) = r_{-\theta}(t_{-w_0}(m(1))) = r_{-\theta}(t_{-w_0}(w_1)) = r_{-\theta}(w_1 - w_0) = e^{-i\theta}e^{i\theta} = 1.$$

We claim that $\mu(i)$ is either i or $-i$. For let $w = \mu(i)$. Using $\mu(0) = 0$, $\mu(1) = 1$, we have:

$$|w - 0| = |\mu(i) - \mu(0)| = |i - 0| = 1,$$

$$|w - 1| = |\mu(i) - \mu(1)| = |i - 1| = \sqrt{2}.$$

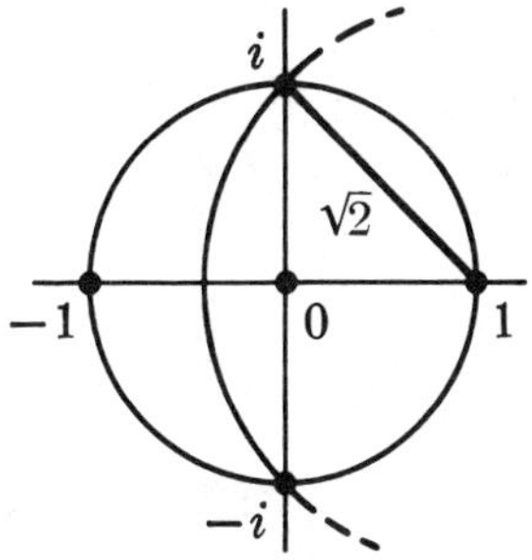

Figure 2.5.4

Thus $|w| = 1$ and $|w - 1| = \sqrt{2}$, implying that w is on a circle of radius 1 centered at 0, and also on a circle of radius $\sqrt{2}$ centered at 1. (Figure 2.5.4.)

Since the only complex numbers on both of these circles are i and $-i$, we see that $\mu(i) = w = i$ or $-i$.

We claim, if $\mu(i) = i$, that $\mu(z) = z$ for every complex number z. For let $z = x + yi$, $\mu(z) = a + bi$, where x, y, a, b are real. Using $\mu(0) = 0$, $\mu(1) = 1$, $\mu(i) = i$, we obtain:

$$a^2 + b^2 = |\mu(z)|^2 = |\mu(z) - \mu(0)|^2 = |z - 0|^2 = x^2 + y^2;$$

$$(a - 1)^2 + b^2 = |a + bi - 1|^2 = |\mu(z) - \mu(1)|^2 = |z - 1|^2 = (x - 1)^2 + y^2;$$

$$a^2 + (b - 1)^2 = |a + bi - i|^2 = |\mu(z) - \mu(i)|^2 = |z - i|^2 = x^2 + (y - 1)^2.$$

The equations $a^2 + b^2 = x^2 + y^2$, $(a - 1)^2 + b^2 = (x - 1)^2 + y^2$ together yield $a = x$, and $a^2 + b^2 = x^2 + y^2$, $a^2 + (b - 1)^2 = x^2 + (y - 1)^2$ together yield $b = y$. Thus $\mu(z) = a + bi = x + yi = z$, and this holds for any z.

Similarly, if $\mu(i) = -i$, then $a^2 + b^2 = x^2 + y^2$, $(a - 1)^2 + b^2 = (x - 1)^2 + y^2$, $a^2 + (b + 1)^2 = x^2 + (y - 1)^2$, yielding $a = x$, $b = -y$, and hence $\mu(z) = \bar{z}$ for any z.

Thus we either have $\mu(z) = z$, for all z, or $\mu(z) = \bar{z}$ for all z. In the first case, μ is E, the identity function, and in the second case, μ is the complex conjugation function c. From $\mu = r_{-\theta}t_{-w_0}m$, we obtain $m = t_{w_0}r_\theta\mu$, and hence $m = t_{w_0}r_\theta$ or $m = t_{w_0}r_\theta c$. Evaluating at z, we find that

$$m(z) = e^{i\theta}z + w_0 \quad \text{for all } z, \quad \text{or} \quad m(z) = e^{i\theta}\bar{z} + w_0 \quad \text{for all } z.$$

Thus m has one of the forms (5), with $a = e^{i\theta}$ and $b = w_0$. □

***COROLLARY* 2.5.15**

Any rigid motion m has one of the forms

$$m = t_w r_\theta \qquad \text{or} \qquad m = t_w r_\theta c, \tag{8}$$

with $w \in \mathbf{C}$ *and* $\theta \in \mathbf{R}$, *and therefore is the product of a rotation about the origin and a translation, or the product of complex conjugation with a rotation about the origin and a translation.*

Note that m has the first form in (5) precisely when it has the first form in (8).

EXERCISES 2.5

I

1. Define a function f on the complex plane by $f(z) = |z|$ for all z. Is f a bijection?
2. Define a function f on the complex plane by $f(z) = z^2$. Is f a bijection?

3. Find all rigid motions m of the complex plane such that m carries the straight line segment joining 1 and i to the straight line segment connecting -1 and i.

4. Find all rigid motions m of the complex plane which carry 1 to i, i to -1, and -1 to $-i$.

5. Find all rigid motions m of the complex plane which carry 1 to 1 and i to $-i$.

6. Let m be the rigid motion $r_\pi t_{1+i} c t_{-i} r_{\pi/2}$. What complex numbers are $m(1 + i)$, $m^{-1}(1 + i)$? Express m in the form given in Corollary 2.5.15. Repeat, with $t_i r_\pi c t_{-1} c r_\pi$.

II

7. Find all rigid motions f such that $f(r_1 z_1 + r_2 z_2) = r_1 f(z_1) + r_2 f(z_2)$ for all real numbers r_1, r_2 and all complex numbers z_1, z_2. Remark: A function f satisfying this equation for all real r_1, r_2 and all complex z_1, z_2 is called a *real linear transformation*. Prove that any product of rotations and conjugations is a real linear transformation, but that a translation is generally not a real linear transformation.

8. Show that, if f is a rigid motion of the complex plane which maps real numbers to real numbers, then f, when evaluated at a real number x, must have the form $f(x) = ax + b$ with real a, b.

9. If $w \in \mathbf{C}$, $\theta \in \mathbf{R}$, verify that the rigid motion m for which $m(z) = e^{i\theta}(z - w) + w$ is a rotation of the complex plane about w through angle θ.

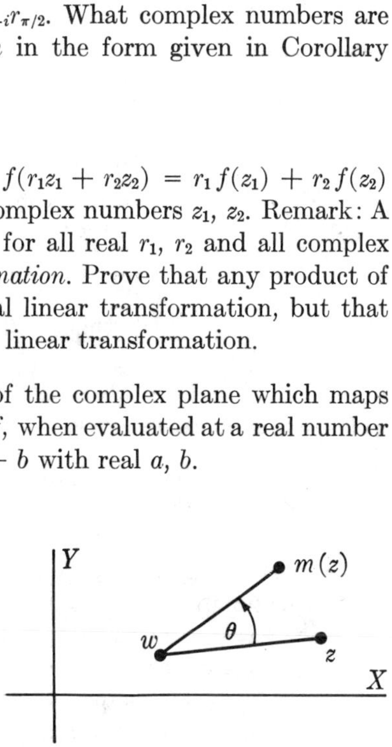

10. Let w be a fixed nonzero complex number. Define a function f on the complex plane by $f(z) = wz$, for all z. Is this function a bijection? Is it a real linear transformation as defined in Exercise 7? For which complex numbers w is it a rigid motion?

11. Show that the associative law $(fg)h = f(gh)$ for functions on the complex plane is valid when the functions are not necessarily bijections.

12. Let C_n be the set consisting of the rotations $r_{j2\pi/n}$ for $j = 0, 1, 2, \ldots, n - 1$. Here n is a fixed positive integer.
   (a) Show that C_n, under composition, is a subgroup of the group R of rotations.

(b) Show that $r_{j2\pi/n} = (r_{2\pi/n})^j$, i.e., each element of C_n is a power of one of its elements, namely a power of the rotation $r_{2\pi/n}$. Remark: A group, the elements of which are powers of one of its elements, is said to be *cyclic*. The group C_n therefore is a cyclic group consisting of exactly n elements.

13. Any product m of finitely many rotations about the origin, translations, and complex conjugations, must have the first form displayed in (8) if there are an even number of conjugations, and the second form if there are an odd number of conjugations. Prove this.

III

14. If S is an arbitrary nonempty set, let $S \times S$ be a set consisting of all pairs (s_1, s_2) in which s_1, s_2 are elements of S. Two pairs, (s_1, s_2) and (s_1', s_2') are, by definition, equal if and only if $s_1 = s_1'$ and $s_2 = s_2'$. We call $S \times S$ the *Cartesian product* of S with itself.

A function f *defined on* S *and taking values in* S is, by definition, a subset F of $S \times S$ such that, for an element s of S, the subset F contains precisely one pair (s_1, s_2) in which $s_1 = s$. If $(s_1, s_2) \in F$ and $s_1 = s$, we call s_2 the *functional value of* f *at* s, and we denote it by $s_2 = f(s)$.

Explain how this definition of a function incorporates the explanation of function given in the text and in your calculus course.

15. Let L be a fixed straight line in the complex plane. Define a function f on the complex plane in this manner: If z is a complex number, let z' be the complex number such that L is the perpendicular bisecter of the straight line segment joining z and z'. Then define $f(z)$ to be z'. This function f is called a *reflection with respect to the line* L.

(a) If the number u and angle θ are as indicated in the figure, prove that $f(z) = (t_u r_\theta c r_{-\theta} t_{-u})(z)$ for each complex number z, and hence that $f = t_u r_\theta c r_{-\theta} t_{-u}$.

(b) If $\theta = 0$ and $u = 0$, show that f is complex conjugation.

(c) Show that f is a rigid motion.

(d) Show that the rigid motions E, f form a two-element group under composition.

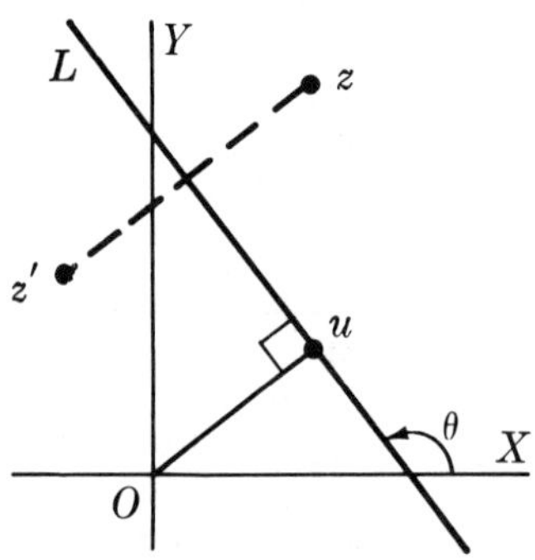

16. Let $\mathcal{G}$ be a group with operation $\circ$, and $\mathcal{H}$ a subgroup of $\mathcal{G}$. We call $\mathcal{H}$ a *normal* subgroup of $\mathcal{G}$ if $g^{-1} \circ h \circ g$ is an element of $\mathcal{H}$ for every h from $\mathcal{H}$ and every g from $\mathcal{G}$. Are T, R, or C normal subgroups of M?

17. Prove the Cartan-Dieudonné theorem: *A rotation of the complex plane is the product of two reflections* (with respect to lines).

18. (a) Define a function f by $f(z) = -z$, for all z. (This function is called a reflection through the origin.) Prove that f is a rigid motion.
    (b) Let u be a fixed complex number. Define a function f by $f(z) = z'$, for all z, where z' is such that u is the bisector of the line segment joining z and z'. (This function is called a reflection in the point u.) Prove that f is a rigid motion.

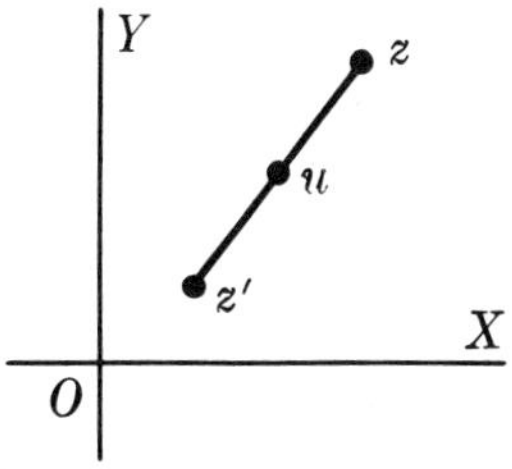

19. Is the factorization (8) of a rigid motion unique?

2.6 REVIEW

In this chapter we endowed the points in the XY-coordinate plane with operations of addition and multiplication, thereby obtaining a field called the field **C** of complex numbers. The complex numbers on the X axis constitute the real number field **R**, and it is a subfield of **C**. The field **R** contains the rational number field **Q** as a subfield. We derived several properties of complex numbers useful in theoretical and computational situations; these included the polar form, the multiplicative rule $|z_1z_2| = |z_1|\,|z_2|$ for the moduli of complex numbers z_1, z_2, the additive rule angle $z_1z_2 =$ angle $z_1 +$ angle z_2 for their angles, and the triangle inequality $|z_1 + z_2| \leq |z_1| + |z_2|$. Complex conjugation was also discussed. In the final section, rigid motions of the complex plane were studied; these consisted of products of rotations, reflections, and translations.

REVIEW EXERCISES 2.6

1. Show that the subset of **C** consisting of all z satisfying $z = \bar{z}$ is a subfield of **C**. What field is it?

2. How many complex numbers z are there satisfying $z^2 = -1$? Find them all. Do the same for $z^4 = -1$.

3. Show that a quadratic equation $z^2 + az + b = 0$ (where a, b may be real or complex) can always be satisfied by at least one complex number z.

4. Let $z = x + yi$ be a complex number, with x, y real. We say z has rational coordinates if x and y are rational numbers. Show that arbitrarily near a complex number z is a complex number z_0 with rational coordinates. (That is, find z_0 such that $|z - z_0|$ is small.) Show that the complex numbers with rational coordinates form a subfield of **C**.

5. Find the area of the triangle whose vertices are the complex numbers z_1, z_2, z_3.

6. Use complex numbers to find the equation of the curve consisting of all points P for which the distance of P from the Y axis equals the distance of P from 1.

3

INTEGERS AND POLYNOMIALS

3.1 INTRODUCTION

Two rings occupying a central place in elementary algebra are the familiar ring of integers and the ring of polynomials. These rings will be studied in some detail in this chapter, beginning with a thorough treatment of polynomial algebra, and going on to the concepts of greatest common divisor, least common multiple, prime, and factorization, for both integers and polynomials. Many readers will have some knowledge of these topics, probably acquired in high school. Nevertheless, this chapter should be read carefully, as it provides the theoretical foundation for all subsequent chapters.

3.2 POLYNOMIALS

Let $\mathcal{R}$ be a commutative ring with identity. Thus $\mathcal{R}$ may be **Z**, **Q**, **R**, or **C**. Apart from Chapter 7, the polynomial theory to be developed in this section will be used only when $\mathcal{R}$ is one of these four rings. Thus the reader may, if he wishes, assume now that $\mathcal{R}$ is **Z**, **Q**, **R**, or **C**, and defer until later the general case when $\mathcal{R}$ is an abstract ring.

Throughout this section the symbol 1 will denote the identity element of $\mathcal{R}$.

The discussion we are about to begin is a formal introduction to familiar concepts of polynomial algebra. Beginning with $\mathcal{R}$, we shall construct in detail the *ring of polynomials with coefficients in* $\mathcal{R}$. We introduce first the concept of an *infinite sequence* with entries in $\mathcal{R}$. This is simply a symbol of the form

$$(r_0, r_1, r_2, r_3, \ldots). \tag{1}$$

The quantities r_0, r_1, r_2, . . . are called the *components* or *coordinates* or *entries* of (1), and they are elements of $\mathcal{R}$. There are infinitely many entries in (1), which, however, need not be distinct. For example, all may be equal, possibly all equal to 0. The three dots in (1) before the right-hand parenthesis are to imply that there is no last component. We say that r_0 is the component in position zero, r_1 is the component in position one, r_2 is the component in position two, etc., and in general r_k is the component in position k, for $k = 0, 1, 2, \ldots$.

We define equality of infinite sequences in this way: Infinite sequences

$$(a_0, a_1, a_2, \ldots) \qquad \text{and} \qquad (b_0, b_1, b_2, \ldots)$$

are equal if and only if $a_0 = b_0$, $a_1 = b_1$, $a_2 = b_2$, . . ., that is, if and only if $a_k = b_k$ *for each nonnegative integer* k. Thus sequences are equal if and only if components in all corresponding positions are equal.

We say that a sequence $a = (a_0, a_1, a_2, \ldots)$ has *only finitely many nonzero entries* if, for some subscript m (which depends on a), we have $a_{m+1} = a_{m+2} = a_{m+3} = \cdots = 0$, that is, $a_k = 0$ for all $k > m$. Another way of expressing this is by saying that almost all entries of a are zero, *almost all* being a technical term meaning "all, apart from a finite number of exceptions." The finite number of nonzero entries may be different for different infinite sequences, and may even be zero; this happens for the "zero infinite sequence"

$$(0, 0, 0, \ldots) \tag{2}$$

in which every entry is zero.

We now introduce a set $\mathcal{S}$ that will eventually become the polynomial ring. This set $\mathcal{S}$ is to consist of those infinite sequences in which almost all entries are zero. Thus the zero infinite sequence is in $\mathcal{S}$, as are $(1, 0, 0, \ldots)$, $(0, 1, 0, 0, \ldots)$, $(0, 0, 1, 0, 0, \ldots)$. [These sequences have all entries zero, except for a single one at the indicated position.] We shall make $\mathcal{S}$ into a ring by introducing operations of addition and multiplication.

We first introduce addition for elements of $\mathcal{S}$. Let

$$a = (a_0, a_1, a_2, \ldots), \qquad b = (b_0, b_1, b_2, \ldots)$$

be in $\mathcal{S}$. If $s = (s_0, s_1, s_2, \ldots)$ is also an infinite sequence, we define addition by specifying that $s = a + b$ if and only if $s_0 = a_0 + b_0$, $s_1 = a_1 + b_1$, $s_2 = a_2 + b_2, \ldots$, and in general $s_k = a_k + b_k$ for all nonnegative integers k. This means: infinite sequences are added by adding entries in corresponding positions. For example, if $\mathcal{R}$ is $\mathbf{Z}$, we have

$$(0, 1, 2, 3, 1, 0, 0, \ldots) + (1, 1, 0, 1, 0, 0, 0, \ldots)$$
$$= (1, 2, 2, 4, 1, 0, 0, \ldots).$$

Obviously, the sum of infinite sequences is again an infinite sequence. Is it in $\mathcal{S}$? To be in $\mathcal{S}$, it must have almost all entries zero. This is evident: Let $a_k = 0$ for all $k > m$ and $b_k = 0$ for all $k > n$; then $s_k = a_k + b_k = 0 + 0 = 0$ for any subscript k exceeding both m and n. Thus almost all entries of s are zero and hence s is in $\mathcal{S}$.

It is now a simple matter to verify that $\mathcal{S}$, under this addition operation, is a commutative group. We must verify Axioms G1–G5. The verification of closure is done since we just showed that the sum of elements of $\mathcal{S}$ is in $\mathcal{S}$. To verify associativity, let a and b be as above, and let

$$c = (c_0, c_1, c_2, \ldots)$$

also be in $\mathcal{S}$. Using the definition of addition and the fact the entries of a, b, c satisfy the associative law (because they are elements of $\mathcal{R}$), we have:

$$\begin{aligned}
(a + b) + c &= ((a_0, a_1, \ldots) + (b_0, b_1, \ldots)) + c \\
&= (a_0 + b_0, a_1 + b_1, \ldots) + (c_0, c_1, \ldots) \\
&= ((a_0 + b_0) + c_0, (a_1 + b_1) + c_1, \ldots) \\
&= (a_0 + (b_0 + c_0), a_1 + (b_1 + c_1), \ldots) \\
&= (a_0, a_1, \ldots) + (b_0 + c_0, b_1 + c_1, \ldots) \\
&= (a_0, a_1, \ldots) + ((b_0, b_1, \ldots) + (c_0, c_1, \ldots)) \\
&= a + (b + c).
\end{aligned}$$

Thus associativity holds. Let 0 be the zero infinite sequence $(0, 0, 0, \ldots)$ in which each entry is the zero element of $\mathcal{R}$. (Notice that the symbol 0 is now being used in two meanings: as the zero element of $\mathcal{R}$, and as the infinite sequence in which each entry is the zero element of $\mathcal{R}$. The reader will not experience confusion from this double use of 0 if he understands the meaning of each symbol 0 as he encounters it.) It is straightforward to check that the infinite sequence 0 satisfies $0 + a = a + 0 = a$ for every $a \in \mathcal{S}$. Thus $\mathcal{S}$ has an additive identity. To verify the existence of additive inverses, let $h = (-a_0, -a_1, -a_2, \ldots)$ be obtained by taking the negative of each entry of a. Then $h + a = a + h = 0$ and $h \in \mathcal{S}$. (Why?)

Thus, under addition, $\mathcal{S}$ is a group. It is commutative because $a + b = (a_0 + b_0, a_1 + b_1, \ldots) = (b_0 + a_0, b_1 + a_1, \ldots) = b + a$.

Next we introduce sequence multiplication. Let $a = (a_0, a_1, a_2, \ldots)$ and $b = (b_0, b_1, b_2, \ldots)$ be as above, and set $p = (p_0, p_1, p_2, \ldots)$. The definition of sequence multiplication is that $p = ab$ if and only if

$$p_k = \sum_{i=0}^{k} a_i b_{k-i}, \qquad \text{for } k = 0, 1, 2, \ldots.$$

Thus

$$p_0 = a_0 b_0, \tag{3}$$

$$p_1 = a_0 b_1 + a_1 b_0,$$

$$p_2 = a_0 b_2 + a_1 b_1 + a_2 b_0,$$

$$\cdots\cdots\cdots\cdots\cdots\cdots$$

$$p_k = a_0 b_k + a_1 b_{k-1} + \cdots + a_k b_0, \tag{4}$$

$$\cdots\cdots\cdots\cdots\cdots\cdots$$

and in general p_k is the sum of all products of the form $a_i b_j$ (i.e., a component of a times a component of b) for which $i + j = k$. This rather strange method of combining the entries of a and b to yield those of p is called *convolution*. Later we shall see the motivation behind it. (A similar, though not identical, process also called "convolution" will be seen in §8.6.)

We shall now prove that $\mathcal{S}$, under addition and multiplication as defined above, is a ring. We have already verified Axiom R1, i.e., that $\mathcal{S}$ is an additive commutative group. To verify the closure law for multiplication, observe first that a product $p = ab$ of infinite sequences a, b from $\mathcal{S}$ is certainly an infinite sequence. To be in $\mathcal{S}$, it must have almost all entries zero. As before, let $a_k = 0$ for all $k > m$ and $b_k = 0$ for all $k > n$. Let $k > m + n$. Then

$$p_k = (a_0 b_k + a_1 b_{k-1} + \cdots + a_m b_{k-m}) \\ + (a_{m+1} b_{k-m-1} + a_{m+2} b_{k-m-2} + \cdots + a_k b_0).$$

Each term $a_i b_j$ in the first parentheses is zero since $b_j = 0$ (because the subscript j satisfies $j \geq k - m > (m + n) - m = n$.) Each term $a_i b_j$ in the second parentheses is also zero since $a_i = 0$ (because $i > m$). Thus $p_k = 0$ when $k > m + n$ since it is a sum of zero terms. This proves that almost all entries in p are zero and therefore $p \in \mathcal{S}$. Thus R2 is established. In fact we have proved this:

If $a_k = 0$ *for* $k > m$ *and* $b_k = 0$ *for* $k > n$, *then* $p_k = 0$ *for* $k > m + n$. (5)

In particular, if $m = n = 0$, we obtain:

If all entries in a and b after the initial entries are zero, then each entry in ab after the initial entry is also zero. (6)

Turning now to the verification of associativity of multiplication, let a, b, c be the sequences used previously. Then $ab = (p_0, p_1, \ldots)$ where p_h is the sum of all terms a_ib_j with $i + j = h$. Let $q = (ab)c$, and suppose $q = (q_0, q_1, q_2, \ldots)$. Then $q_l = p_0c_l + p_1c_{l-1} + \cdots + p_lc_0$ is the sum of all terms of the type p_hc_k with $h + k = l$. For fixed h we have

$$\begin{aligned} p_hc_k &= (a_0b_h + a_1b_{h-1} + \cdots + a_hb_0)c_k \\ &= (a_0b_h)c_k + (a_1b_{h-1})c_k + \cdots + (a_hb_0)c_k. \end{aligned}$$

Thus p_hc_k is the sum of all terms of the type $(a_ib_j)c_k$ with $i + j + k = l$ and $i + j = h$. Adding these sums for all possible h, we see that:

q_l is the sum of all terms of the type $(a_ib_j)c_k$ with $i + j + k = l$.

Now let $a(bc) = r = (r_0, r_1, r_2, \ldots)$. By a similar calculation we find that

r_l is the sum of all terms of the type $a_i(b_jc_k)$ in which $i + j + k = l$.

Since $(a_ib_j)c_k = a_i(b_jc_k)$ [because a_i, b_j, c_k are elements of $\mathcal{R}$ and associativity is valid in $\mathcal{R}$], we find that these descriptions of q_l and r_l are the same. Consequently $q_l = r_l$, and this holds for each $l = 0, 1, 2, \ldots$. Hence $q = r$. Thus $(ab)c = a(bc)$ and associativity of multiplication is established.

Finally, we verify the first formula in the distributive law R4; the second is verified similarly and is left as an exercise. We have $b + c = (b_0 + c_0, b_1 + c_1, b_2 + c_2, \ldots)$ and hence the entry of $a(b + c)$ in position k is

$$\begin{aligned} &a_0(b_k + c_k) + a_1(b_{k-1} + c_{k-1}) + \cdots + a_k(b_0 + c_0) \\ &\qquad = (a_0b_k + a_1b_{k-1} + \cdots + a_kb_0) + (a_0c_k + a_1c_{k-1} + \cdots + a_kc_0). \end{aligned}$$

Here we used the distributive and generalized commutative laws for elements of $\mathcal{R}$ to multiply out parentheses and then rearrange the order of the terms, and now this expression is clearly the sum of the entries of ab and ac in position k. Since this holds for every k, we see that $a(b + c) = ab + ac$. Distributivity is therefore verified.

We now know that $\mathcal{S}$ is a ring. That it is commutative follows from (4): by (4) the position-k entry of ab is $a_0b_k + a_1b_{k-1} + \cdots + a_{k-1}b_1 + a_kb_0$, whereas that of ba is $b_0a_k + b_1a_{k-1} + \cdots + b_{k-1}a_1 + b_ka_0$. These two expressions are clearly equal, since $\mathcal{R}$ is commutative, and this is true for each $k = 0, 1, 2, \ldots$. Thus $ab = ba$, and so $\mathcal{S}$ is a commutative ring.

We next consider the subset $\mathcal{S}_0$ of $\mathcal{S}$ consisting of the special infinite sequences of the form

$$(a_0, 0, 0, \ldots)$$

in which each entry after the initial entry is zero. Consider two such special sequences, $(a_0, 0, 0, \ldots)$ and $(b_0, 0, 0, \ldots)$. The rules for equality, addition, and multiplication of sequences show that:

$$(a_0, 0, 0, \ldots) = (b_0, 0, 0, \ldots) \quad \text{if and only if} \quad a_0 = b_0,$$

$$(a_0, 0, 0, \ldots) + (b_0, 0, 0, \ldots) = (a_0 + b_0, 0, 0, \ldots),$$

$$(a_0, 0, 0, \ldots)(b_0, 0, 0, \ldots) = (a_0b_0, 0, 0, \ldots).$$

(For the last part use (3) and (6).) From these formulas it is clear that special sequences behave under equality, addition, multiplication exactly as their initial entries behave. Because of this, we now make the notational change in which we agree to write the special sequence $(a_0, 0, 0, \ldots)$ more concisely as a_0. This means that we may regard the ring $\mathcal{R}$ as being embedded in $\mathcal{S}$, in the sense that the special sequences are nothing more than the elements of $\mathcal{R}$ in somewhat different notation. Henceforth we shall, whenever convenient, use the notations r and $(r, 0, 0, \ldots)$ interchangeably, for $r \in \mathcal{R}$. In particular, under this agreement the symbol 0 denotes both the zero element of $\mathcal{R}$ and the sequence in which each entry is zero, that is, the zero element of $\mathcal{S}$.

Consider the product

$$p = (p_0, p_1, p_2, \ldots) = (a_0, 0, 0, \ldots)b$$

of a special sequence and a sequence $b = (b_0, b_1, b_2, \ldots)$ of $\mathcal{S}$. Formula (4) gives $p_k = a_0b_k$ (since $a_1 = a_2 = \cdots = 0$). That is, $(a_0, 0, 0, \ldots) \times (b_0, b_1, b_2, \ldots) = (a_0b_0, a_0b_1, a_0b_2, \ldots)$. If we replace $(a_0, 0, 0, \ldots)$ by a_0, as we may under our agreement concerning notation, we obtain this rule:

$$a_0(b_0, b_1, b_2, \ldots) = (a_0b_0, a_0b_1, a_0b_2, \ldots).$$

In other words: *to multiply an element b of $\mathcal{S}$ by an element a_0 of $\mathcal{R}$, multiply each entry of b by a_0.* Since $\mathcal{S}$ is a commutative ring this rule is valid both for a_0b and ba_0. In particular, if we let $a_0 = 1$ be the identity element of $\mathcal{R}$, we see that $1b = b1 = b$ for every $b \in \mathcal{S}$. Thus *$\mathcal{S}$ is a commutative ring with identity.*

Now suppose that $\mathcal{R}$ is an integral domain. We shall show that $\mathcal{S}$ is also an integral domain. For this we must verify I1 and I2 of Theorem 1.6.4, i.e., that the multiplicative identity is nonzero and that there are no zero divisors. As for I1, if $\mathcal{R}$ is an integral domain, its identity element 1 is different from 0, hence the identity infinite sequence $(1, 0, 0, 0, 0, \ldots)$ is different from the zero infinite sequence $(0, 0, 0, 0, \ldots)$. Thus I1 holds for $\mathcal{S}$. To verify I2 we must show that the product of nonzero elements a, b of $\mathcal{S}$ is nonzero. Let

$$a = (a_0, a_1, \ldots, a_m, 0, 0, \ldots), \qquad b = (b_0, b_1, \ldots, b_n, 0, 0, \ldots),$$

where $a_m \neq 0$ and $b_n \neq 0$. Set $p = ab = (p_0, p_1, \ldots)$. By (5) we know that $p_k = 0$ for $k > m + n$. Let us compute p_{m+n}. Using (4) for $k = m + n$, we get

$$p_{m+n} = (a_0 b_{m+n} + a_1 b_{m+n-1} + \cdots + a_{m-1} b_{n+1}) + a_m b_n + (a_{m+1} b_{n-1} + a_{m+2} b_{n-2} + \cdots + a_{m+n} b_0).$$

Each term in the first parentheses is zero because each $b_j = 0$ since $j > n$. Each term in the second parentheses is also zero because each $a_i = 0$ since $i > m$. Therefore

$$p_{m+n} = a_m b_n, \qquad \text{and} \qquad p_k = 0 \quad \text{for all} \quad k > m + n. \tag{7}$$

Since the nonzero elements a_m, b_n belong to an integral domain $\mathcal{R}$, their product $a_m b_n$ is also nonzero. Thus $p_{m+n} \neq 0$ and therefore $p \neq 0$.

We have completed the proof that if *the ring* $\mathcal{R}$ *is an integral domain, the ring* $\mathcal{S}$ *of sequences is also an integral domain.*

We introduce the name λ (Greek lambda) for the sequence

$$(0, 1, 0, 0, 0, \ldots) \tag{8}$$

in which each entry is zero except for a 1 occupying position one. We prove by induction that, for each integer $i \geq 1$,

$$\lambda^i = (\underbrace{0, 0, 0, \ldots, 0}_{i \text{ zeros}}, 1, 0, 0, \ldots) \tag{9}$$

is a sequence in which all entries are zero except for a single 1 in position i. For $i = 1$ this is correct by definition of λ. This completes the first part of the proof by induction. For the second part, take as induction hypothesis the validity of (9) when i is l. Letting $a = \lambda^l$, we therefore have

$$a = \lambda^l = (a_0, a_1, a_2, \ldots)$$

where $a_0 = a_1 = \cdots = a_{l-1} = 0$, $a_l = 1$, $a_{l+1} = a_{l+2} = \cdots = 0$. Letting b be λ, so that $\lambda = (b_0, b_1, b_2, \ldots)$, where $b_0 = 0, b_1 = 1, b_2 = b_3 = \cdots = 0$, we have $\lambda^{l+1} = ab = p = (p_0, p_1, p_2, \ldots)$. By (3), $p_0 = 0$. For $k \geq 1$ we have

$$p_k = a_0 b_k + a_1 b_{k-1} + \cdots + a_{k-1} b_1 + a_k b_0 = a_{k-1} b_1 = a_{k-1}$$

since only b_1 is nonzero, and $b_1 = 1$. But a_{k-1} is nonzero only if $k - 1 = l$; so that the only nonzero entry of p is $p_{l+1} = a_l = 1$. Thus $\lambda^{l+1} = ab$ has all entries zero except for a single one in position $l + 1$. The verification needed for the second part of the induction proof is now complete and (9) is established. The laws of exponents (which are valid in integral domains—see Exercise 9, §1.5) require that a nonzero quantity raised to the power 0 be the identity element of the ring. Since the identity of $\mathcal{S}$ is the sequence $(1, 0, 0, \ldots)$, we see that (9) also holds for $i = 0$.

We may now express a sequence $a = (a_0, a_1, a_2, \ldots, a_m, 0, 0, \ldots)$ in terms of its entries and λ in the following way:

$$\begin{aligned} a &= a_0(1, 0, 0, \ldots) + a_1(0, 1, 0, 0, \ldots) + a_2(0, 0, 1, 0, 0, \ldots) \\ &\qquad + \cdots + a_m(0, 0, 0, \ldots, 0, 1, 0, \ldots) \\ &= a_0 + a_1\lambda + a_2\lambda^2 + \cdots + a_m\lambda^m. \end{aligned}$$

Thus: a is a *polynomial in* λ. We have, therefore, indeed been discussing polynomials. More precisely, we introduce the definition that a *polynomial is an element of the ring* $\mathcal{S}$, and a *polynomial in the indeterminate* λ is an element of $\mathcal{S}$ with λ denoting the particular element (8). There is nothing at all mysterious about λ; it is simply the name given to a certain element of $\mathcal{S}$. Any other letter, say x or y, could have been used in place of λ.

If we write $a = (a_0, a_1, \ldots)$ with all a_k for $k > m$ included (even though they are zero), we have

$$a = a_0 + a_1\lambda + a_2\lambda^2 + \cdots + a_m\lambda^m + a_{m+1}\lambda^{m+1} + \cdots = \sum_{k=0}^{\infty} a_k\lambda^k.$$

We call $a_0, a_1, a_2, \ldots$ the *coefficients* of a—they are elements of $\mathcal{R}$—and we call $a_k\lambda^k$ a *term* in the polynomial. The above formula represents a polynomial as an infinite sum of terms involving ascending powers of λ with coefficients from $\mathcal{R}$. It really isn't an infinite sum since almost all terms are zero, because almost all coefficients are zero. Nevertheless, it is convenient to call this representation an *infinite series*. Let

$$b = (b_0, b_1, b_2, \ldots) = \sum_{k=0}^{\infty} b_k\lambda^k$$

be another polynomial. According to the definition of equality of sequences, we have $a = b$ if and only if $a_k = b_k$ for every $k = 0, 1, 2, \ldots$. Thus: *polynomials a and b are equal if and only if the coefficients a_k, b_k of λ^k are equal, $a_k = b_k$, for every $k = 0, 1, 2, \ldots$.* Because sequences add by addition of corresponding entries, it follows that *polynomials add by combining terms involving the same power of* λ. That is:

$$\begin{aligned} &(a_0 + a_1\lambda + a_2\lambda^2 + \cdots) + (b_0 + b_1\lambda + b_2\lambda^2 + \cdots) \\ &\qquad = (a_0 + b_0) + (a_1 + b_1)\lambda + (a_2 + b_2)\lambda^2 + \cdots. \end{aligned}$$

Consider $p = ab$. By the generalized distributive law (R9), the product

$$\begin{aligned} p = ab &= (a_0 + a_1\lambda + a_2\lambda^2 + \cdots)\cdot(b_0 + b_1\lambda + b_2\lambda^2 + \cdots) \\ &= p_0 + p_1\lambda + p_2\lambda^2 + \cdots \end{aligned} \tag{10}$$

is obtained by multiplying each term in the first parentheses onto each term in the second parentheses. The product of the typical terms $a_i\lambda^i$ from a and $b_j\lambda^j$ from b is $a_ib_j\lambda^{i+j}$. If we combine all terms involving λ^k, that is, add all terms $a_ib_j\lambda^{i+j}$ for which $i + j = k$, we see that $p_k\lambda^k = (a_0b_k + a_1b_{k-1} + \cdots + a_kb_0)\lambda^k$. Thus p_k is the sum of all products of the form

a_ib_j for which $i + j = k$. But this is just the rule used to define multiplication of sequences. Thus the motivation behind the multiplication rule for sequences is that it is exactly the rule allowing sequence multiplication to agree with the natural laws for polynomial multiplication.

We say that the elements of $\mathcal{R}$ are *constants*. (This should not be interpreted to mean that λ is a variable. It isn't: λ is simply the name of a particular sequence.) The rule for multiplying a special sequence $(a_0, 0, 0, \ldots)$ and a general sequence $(b_0, b_1, b_2, \ldots)$ from $\mathcal{S}$ may now be stated as

$$a_0(b_0 + b_1\lambda + b_2\lambda^2 + \cdots) = a_0b_0 + (a_0b_1)\lambda + (a_0b_2)\lambda^2 + \cdots.$$

Thus: *to multiply a polynomial by a constant a_0, multiply each term by a_0.*

If

$$a = a_0 + a_1\lambda + \cdots + a_m\lambda^m$$

is a nonzero polynomial, with $a_m \neq 0$ (and where $a_0, a_1, \ldots, a_m$ are constants) we say that *the degree of a is m*. Let

$$\begin{aligned} b &= (b_0, b_1, \ldots, b_n, 0, 0, \ldots) \\ &= b_0 + b_1\lambda + \cdots + b_n\lambda^n, \qquad b_n \neq 0, \end{aligned}$$

also be a nonzero polynomial, with degree $b = n$. Set $p = ab$. By (7) we have

$$p = p_0 + p_1\lambda + \cdots + p_{m+n}\lambda^{m+n}, \qquad p_{m+n} = a_mb_n \neq 0.$$

The degree of $p = ab$ is therefore $m + n$. In other words: *when $\mathcal{R}$ is an integral domain,*

$$\textit{degree } ab = \textit{degree } a + \textit{degree } b \tag{11}$$

if $a \neq 0$, $b \neq 0$. This very important formula will be used many times. Notice that the definition of degree just given is applicable only to nonzero polynomials. To complete the definition, we *assign a degree to the zero polynomial by specifying its degree to be $-\infty$* (minus infinity). We assign a degree to the zero polynomial in order to gain conciseness and avoid the necessity of having to give additional explanation whenever a polynomial happens to be zero. However, we have to specify rules for working with the symbol ∞. The rules needed are these:

$$(-\infty) + (-\infty) = -\infty, \qquad -(-\infty) = \infty, \qquad \infty + \infty = \infty,$$

and when m is any integer,

$$-\infty \neq m, \qquad m \neq \infty, \qquad -\infty < m < \infty,$$

$$-\infty + m = m - \infty = -\infty, \quad m + \infty = \infty + m = \infty.$$

We shall also need this rule: if ρ is a number with $0 < \rho < 1$, then

$$\rho^\infty = 0.$$

The quantities that can be the degree of a polynomial are thus $-\infty$, 0, 1, 2, . . ., and only the zero polynomial has degree $-\infty$. We shall often abbreviate degree a as deg a.

With these agreements concerning the use of the symbol ∞, it is now easy to see that the degree equation (11) is valid when a, b, or both are 0.

We summarize our results in a theorem.

THEOREM 3.2.1

If $\mathcal{R}$ is an integral domain, the ring of polynomials with coefficients in $\mathcal{R}$ is also an integral domain. Furthermore, the degree formula (11) *is valid, as is the degree formula*

$$\deg (a + b) \leq \max (\deg a, \deg b). \tag{12}$$

When $\deg a \neq \deg b$, *the inequality* (12) *becomes the equality*

$$\deg (a + b) = \max (\deg a, \deg b). \tag{13}$$

[Max (deg a, deg b) denotes the larger of the quantities deg a, deg b.]

All is proved except (12) and (13). Both clearly hold if a or b is zero. If a and b are nonzero, inequality (12) follows from the observation that addition of polynomials may cause some powers of λ to cancel but cannot create new higher powers of λ. And if $\deg a > \deg b$, adding b to a cannot cancel the highest power of λ in a; hence $\deg (a + b) = \deg a$. This proves (13). □

There is some additional notation and terminology. Often a polynomial

$$a = a_0 + a_1\lambda + a_2\lambda^2 + \cdots + a_m\lambda^m$$

is denoted by $a(\lambda)$, and the ring $\mathcal{S}$ of polynomials is usually denoted by $\mathcal{R}[\lambda]$. (Notice: square brackets.) The symbol $\mathcal{R}[\lambda]$ therefore means the ring of polynomials with coefficients in $\mathcal{R}$, λ denoting the indeterminate (i.e., (8)). Referring to the polynomial a above, we call a_0 the *constant term*; and if $a_m \neq 0$, we call $a_m\lambda^m$ the *leading term* and a_m the *leading coefficient*. We say a is *monic* if $a_m = 1$. We call a a *monomial* if it has a single nonzero term. In general, we omit zero terms when writing out the terms of a. Finally, we say that a is a *constant* if $a_k = 0$ for all $k > 0$.

We showed, in (7), that if a_m is the leading coefficient of a, and b_n is the leading coefficient of b, then a_mb_n is the leading coefficient of $p = ab$. In particular, when $a_m = b_n = 1$, we see that

the product of monic polynomials is monic.

Notice that the ring $\mathbf{Z}[\lambda]$ is a subring of $\mathbf{Q}[\lambda]$, that $\mathbf{Q}[\lambda]$ is a subring of $\mathbf{R}[\lambda]$, and that $\mathbf{R}[\lambda]$ is a subring of $\mathbf{C}[\lambda]$. This is evident since polynomials with coefficients in one of the rings $\mathbf{Z}$, $\mathbf{Q}$, $\mathbf{R}$, $\mathbf{C}$ certainly have their coefficients in the next.

Substitution

The reader perhaps knows that in a polynomial or an equation involving polynomials one sometimes replaces the indeterminate λ with some other expression, i.e., one substitutes for λ. He has perhaps also seen differentiation of polynomials. We now give a careful examination of these two ideas.

We continue to assume that $\mathcal{R}$ is an integral domain. Suppose that

$$a(\lambda) = a_0 + a_1\lambda + \cdots + a_m\lambda^m,$$

where $a_0, a_1, \ldots, a_m \in \mathcal{R}$, and that α is an element of $\mathcal{R}[\lambda]$, that is, another polynomial. We define the *value of polynomial a at α* by this rule:

$$a(\alpha) = a_0 + a_1\alpha + a_2\alpha^2 + \cdots + a_m\alpha^m.$$

That is: $a(\alpha)$ is the quantity obtained when α is substituted for the indeterminate λ. We often say that $a(\alpha)$ is obtained by *evaluating* $a(\lambda)$ at α.

If we write $a(\lambda)$ as an infinite series,

$$a(\lambda) = \sum_{j=0}^{\infty} a_j\lambda^j, \qquad \text{then} \qquad a(\alpha) = \sum_{j=0}^{\infty} a_j\alpha^j.$$

This is valid even if α is zero, provided we agree that $0^0 = 1$. (See the discussion of 0^0 in §1.6.) Of course these infinite series are only finite sums since almost all coefficients a_j are zero.

There are several particularly useful rules concerning substitution. Let also $b(\lambda) \in \mathcal{R}[\lambda]$. Take $s = a + b$, $d = a - b$, $p = ab$ to be the sum, difference, and product of a and b. Then

$$s(\alpha) = a(\alpha) + b(\alpha), \qquad d(\alpha) = a(\alpha) - b(\alpha),$$
$$p(\alpha) = a(\alpha)b(\alpha).$$

In other words: evaluation at α is compatible with polynomial addition, subtraction, and multiplication. The proofs are easy. Write a, b, s, p as infinite series in which almost all coefficients are zero:

$$a(\lambda) = \sum_{k=0}^{\infty} a_k\lambda^k, \qquad b(\lambda) = \sum_{k=0}^{\infty} b_k\lambda^k,$$

$$s(\lambda) = \sum_{k=0}^{\infty} s_k\lambda^k, \qquad p(\lambda) = \sum_{k=0}^{\infty} p_k\lambda^k,$$

where $a_k, b_k, s_k, p_k \in \mathcal{R}$ for each $k = 0, 1, 2, \ldots$. We know that

$$s_k = a_k + b_k. \tag{14}$$

Multiply each side of (14) by α^k for each $k = 0, 1, 2, \ldots$, then add the resulting equations. We obtain $s(\alpha) = a(\alpha) + b(\alpha)$. The formula for differences is proved similarly. To prove the formula for products, consider

$$a(\alpha)b(\alpha) = (a_0 + a_1\alpha + a_2\alpha^2 + \cdots)(b_0 + b_1\alpha + b_2\alpha^2 + \cdots).$$

By repeating the discussion below equation (10) but replacing λ everywhere with α, we see that if we collect all terms in α^k after multiplying out the parentheses in the right-hand side, we get $(a_0b_k + a_1b_{k-1} + \cdots + a_kb_0)\alpha^k = p_k\alpha^k$. Hence $a(\alpha)b(\alpha) = p_0 + p_1\alpha + p_2\alpha^2 + \cdots = p(\alpha)$. □

> REMARK We shall reexamine these substitution rules from a more abstract point of view in §7.5, where we shall see that they provide an example of a type of mapping function called a *homomorphism*.

Differentiation

We continue to assume that $\mathcal{R}$ is an integral domain.

Very early in calculus the student meets the operation of differentiation, and he learns how to differentiate numerous types of functions. Here we are interested only in differentiation of polynomials, and we give a discussion that assumes no previous knowledge of differentiation. We introduce a differentiation operator, denoted by D, which is to act on a polynomial to yield another polynomial. Specifically, if

$$a(\lambda) = \sum_{k=0}^{\infty} a_k\lambda^k \tag{15}$$

where the a_k are constants, almost all zero, we define $D(a)$ by

$$D(a) = \sum_{k=1}^{\infty} ka_k\lambda^{k-1}. \tag{16}$$

Here ka_k means $a_k + \cdots + a_k$ (k terms). (See Exercise 12 of §1.4.) The properties of D that we shall need are these: if $a(\lambda)$ and $b(\lambda)$ are polynomials, then

(i) $D(a) = 0$ if a is a constant;

(ii) $D(a + b) = D(a) + D(b)$;

(iii) $D(ab) = aD(b)$ if a is a constant;

(iv) $D(ab) = aD(b) + D(a)b$;

(v) $D(a^n) = na^{n-1}D(a)$ if n is a positive integer.

These formulas are encountered in calculus, usually only when the coefficients are real numbers. However these polynomial differentiation formulas continue to hold when the coefficients are in any commutative ring with identity, and so, for example, are valid when the coefficients are complex numbers. The proofs we shall now give are perhaps different from those the reader sees in calculus.

Proof of (i): This is obvious from (15) and (16) when $a_1 = a_2 = a_3 = \cdots = 0$. □

Proof of (ii): With

$$b(\lambda) = \sum_{k=0}^{\infty} b_k\lambda^k,$$

where the b_k are constants, almost all zero, we have

$$D(a + b) = D(\sum_{k=0}^{\infty} (a_k + b_k)\lambda^k) = \sum_{k=1}^{\infty} k(a_k + b_k)\lambda^{k-1}$$

$$= \sum_{k=1}^{\infty} ka_k\lambda^{k-1} + \sum_{k=1}^{\infty} kb_k\lambda^{k-1} = D(a) + D(b). \qquad \square$$

Proof of (iii): If a is a constant,

$$D(ab) = D(\sum_{k=0}^{\infty} ab_k\lambda^k) = \sum_{k=1}^{\infty} kab_k\lambda^{k-1} = a\sum_{k=1}^{\infty} kb_k\lambda^{k-1} = aD(b). \qquad \square$$

Proof of (iv): When a is a constant, $D(ab) = aD(b) = aD(b) + D(a)b$ since $D(a) = 0$. Thus (iv) holds if a is a constant. Similarly it holds if b is a constant. We next verify (iv) when $a(\lambda) = \lambda^l$ and $b(\lambda) = \lambda^k$, for fixed exponents l and k. We have just done this if l or k is zero (since then a or b is a constant). Let $l > 0$, $k > 0$. Then

$$D(\lambda^l\cdot\lambda^k) = D(\lambda^{l+k}) = (l + k)\lambda^{l+k-1} = l\lambda^{l-1}\lambda^k + \lambda^l(k\lambda^{k-1})$$

$$= D(a)b + aD(b).$$

Thus

$$D(\lambda^l\cdot\lambda^k) = D(\lambda^l)\lambda^k + \lambda^l D(\lambda^k) \tag{17}$$

holds for all integer exponents $l \geq 0$, $k \geq 0$. Multiplying each side of (17) by the constant b_k and then summing $k = 0, 1, 2, \ldots$, by (ii) and (iii) we obtain

$$D(\lambda^l b(\lambda)) = D(\lambda^l)b(\lambda) + \lambda^l D(b(\lambda)). \tag{18}$$

Multiplying each side of (18) by the constant a_l and then summing $l = 0, 1, 2, \ldots$, by (ii) and (iii) we get

$$D(a(\lambda)b(\lambda)) = D(a(\lambda))b(\lambda) + a(\lambda)D(b(\lambda)). \qquad \square$$

Proof of (v): We use induction on n, taking the statement $P(n)$ to be: $D(a^n) = na^{n-1}D(a)$. Clearly $P(1)$ is correct, since $a^0 = 1$. This completes the first step in the proof by induction. For the second step, suppose $k - 1$ is a positive integer and that $D(a^{k-1}) = (k - 1)a^{k-2}D(a)$. This is the induction hypothesis. We wish to prove that $D(a^k) = ka^{k-1}D(a)$. Using the product rule (iv) with $b = a^{k-1}$ and then the induction hypothesis,

$$D(a^k) = D(a\cdot a^{k-1}) = aD(a^{k-1}) + D(a)a^{k-1}$$

$$= a(k - 1)a^{k-2}D(a) + D(a)a^{k-1}$$

$$= \{(k - 1)a^{k-1} + a^{k-1}\}D(a) = ka^{k-1}D(a).$$

Both steps needed for proof by induction are verified. By the first principle of induction it now follows that (v) is proved. □

In §5.6 we shall require the following generalization of (iv): if $a_1, a_2, \ldots, a_t$ are polynomials, then

$$D(a_1a_2\cdots a_t) = \sum_{j=1}^{t} a_1\cdots a_{j-1}D(a_j)a_{j+1}\cdots a_t. \tag{19}$$

In other words: to differentiate a product involving several factors, differentiate one factor of the product (leaving undifferentiated the other factors) in all possible ways, then add the products so obtained. The proof is by induction on t and is left as an exercise for the reader.

NOTATION Often the notation a' or $a^{(1)}$ is used to denote $D(a)$. Then $D(D(a))$ is denoted by $D^2(a)$, a'', or $a^{(2)}$; $D(D^2(a))$ is denoted by $D^3(a)$, a''', or $a^{(3)}$; etc. Because differentiation occurs relatively infrequently in this book, it will be explicitly stated whenever the notation a' is to indicate differentiation. Otherwise, in discussions involving several polynomials, the symbols a, a', a'', etc. will denote arbitrary polynomials.

Differentiation will be examined from another point of view in §8.6. There we shall see that if we introduce a multiplication for polynomials different from that discussed above, a mathematical framework will be created within which differentiation may be interpreted as a fraction.

The ring of formal power series (Optional)

If we repeat the above discussion, everywhere omitting the condition that almost all entries in the sequences are zero, we obtain a ring called the *ring of formal power series* with coefficients from $\mathcal{R}$, and denoted by $\mathcal{R}[[\lambda]]$. The typical element of $\mathcal{R}[[\lambda]]$ has the form

$$\sum_{j=0}^{\infty} a_j\lambda^j, \qquad a_j \in \mathcal{R} \quad \text{for all} \quad j \geq 0,$$

thus is an infinite series in powers of λ with coefficients a_j which are elements of $\mathcal{R}$, and with possibly infinitely many nonzero coefficients. An element of $\mathcal{R}[[\lambda]]$ is called a *formal power series,* and sometimes also called a *generating function.* Let

$$\sum_{j=0}^{\infty} b_j\lambda^j, \qquad b_j \in \mathcal{R} \quad \text{for all} \quad j \geq 0,$$

be another element of $\mathcal{R}[[\lambda]]$. The rules for working with these series are as follows:

$$\sum_{j=0}^{\infty} a_j\lambda^j = \sum_{j=0}^{\infty} b_j\lambda^j \qquad \text{if and only if } a_j = b_j \text{ for all } j \geq 0;$$

$$\sum_{j=0}^{\infty} a_j\lambda^j + \sum_{j=0}^{\infty} b_j\lambda^j = \sum_{j=0}^{\infty} (a_j + b_j)\lambda^j,$$

$$\left(\sum_{j=0}^{\infty} a_j\lambda^j\right) \cdot \left(\sum_{j=0}^{\infty} b_j\lambda^j\right) = \sum_{j=0}^{\infty} p_j\lambda^j,$$

with $p_j = a_0b_j + a_1b_{j-1} + \cdots + a_{j-1}b_1 + a_jb_0$ for all $j \geq 0$. The proofs given earlier in this section that $\mathcal{R}[\lambda]$ is a commutative ring with identity may be applied to formal power series without essential change to verify that $\mathcal{R}[[\lambda]]$ is a commutative ring with identity, and that it contains $\mathcal{R}[\lambda]$ as a subring. Although an element of $\mathcal{R}[[\lambda]]$ is a sum of infinitely many terms, no conceptual difficulty arises because of the precise rules just given for working with these sums. However, evaluation of an element of $\mathcal{R}[[\lambda]]$ at an element of $\mathcal{R}$ does create a conceptual difficulty because we have given no meaning to sums of infinitely many elements of $\mathcal{R}$. We shall discuss this in §§6.5 and 8.4. Furthermore, because there need not be a highest power of λ with nonzero coefficient in an infinite series, the degree concept does not apply, and another procedure is need for measuring the size of a formal power series. One method that may be adopted is to use the lowest power of λ. (See exercise 20.) However, we do have the fact that $\mathcal{R}[[\lambda]]$ is an integral domain when $\mathcal{R}$ is an integral domain.

EXERCISES 3.2

If no explicit statement is made, the polynomials in the following exercises are to have coefficients in an integral domain.

I

1. Add, subtract, and multiply the following polynomials $a(\lambda)$, $b(\lambda)$ in $\mathbf{Z}[\lambda]$.
   (a) $a = 1 + 2\lambda + \lambda^3$, $b = 2 + \lambda^2 + \lambda^3$.
   (b) $a = 3 + \lambda$, $b = 2\lambda^3 - \lambda$.
   (c) $a = -\lambda^4 - \lambda^3 + \lambda^2$, $b = \lambda^2 + \lambda + \lambda^4$.
2. Find all monic polynomials of degree zero.
3. What is the leading coefficient and what is the constant term in the following product of polynomials in $\mathbf{Z}[\lambda]$: $(2 - \lambda^3 + \lambda^2) \cdot (3\lambda - 4 + \lambda^3)$. Do not multiply out in full.
4. Suppose a, b, p are polynomials and that $p = ab$. What is a if $p \neq 0$ and $\deg p = \deg b$? If also b and p are monic, what can you say about a?

5. Find all polynomials a for which

$$\lambda^3 + 2\lambda^2 + 3\lambda - 1 = a(\lambda)(\lambda + 1) + \lambda^2 + 2\lambda - 2.$$

6. Find $D((\lambda + 3)^{10})$, $D^2((\lambda + 3)^{10})$, $D^3((\lambda + 3)^{10})$, etc.

7. If $\deg a = 3$, $\deg b = 5$, what are $\deg (a + 2b)$, $\deg (a - b)$, $\deg (ab)$, $\deg (a^3 + b^4)$, $\deg (a^2b + a^3b^2)$? Can you draw any conclusion about $\deg (a^5 - 3b^3)$?

8. The product of monic polynomials is monic. Is this true for sums?

9. Is the ring of polynomials a field? Is it a group under addition? Is it a group under multiplication?

10. Can it happen that $D(a(\lambda)) = 0$ even though $a(\lambda)$ is not a constant?

11. If $\alpha_0 + \alpha_1\lambda + \cdots + \alpha_m\lambda^m = 0$, where $\alpha_0, \alpha_1, \ldots, \alpha_m$ are elements of $\mathcal{R}[\lambda]$, does it follow that $\alpha_0 = \alpha_1 = \cdots = \alpha_m = 0$? What if $\alpha_0, \alpha_1, \ldots, \alpha_m$ are in $\mathcal{R}$?

II

12. Let $\mathcal{R}$ be an integral domain. Consider polynomials a, b, c in $\mathcal{R}[\lambda]$. Prove this cancellation rule: If $ab = ac$ and $a \neq 0$, then $b = c$.

13. Let $a(\lambda)$ be a polynomial of degree m. Find the degree of the polynomial $a(\lambda)^n$.

14. Let a be a nonconstant polynomial in $\mathcal{R}[\lambda]$, where $\mathcal{R}$ is an integral domain. Let $c_0, c_1, \ldots, c_t$ be constants. What are $c_0, c_1, \ldots, c_t$ if $c_0 + c_1a + c_2a^2 + \cdots + c_ta^t = 0$? Give complete details in your proof. Use your result to answer this question: Can a nonconstant polynomial a in $\mathcal{R}[\lambda]$ satisfy an equation with coefficients in $\mathcal{R}$? Does this conclusion still hold if a is a constant polynomial?

15. Let $\mathcal{F}$ be a field and $a(\lambda) = a_0 + a_1\lambda + \cdots + a_m\lambda^m \in \mathcal{F}[\lambda]$, where $a_0, a_1, \ldots, a_m$ are constants with $a_m \neq 0$. Show that $a_m^{-1}a(\lambda)$ is monic.

16. Write out the proof of generalization (19).

17. Let $a(\lambda) = a_0 + a_1\lambda + \cdots + a_m\lambda^m$ be a polynomial, where $a_0, a_1, \ldots, a_m$ are constants. Let $\alpha(\lambda)$ be another polynomial. Set $A(\lambda) = a(\alpha(\lambda))$. Prove the *chain law for differentiation*,

$$DA(\lambda) = a'(\alpha)\alpha'(\lambda),$$

as follows: by applying (v) to $\alpha(\lambda)^k$, show that $D(\alpha(\lambda)^k) = k\alpha(\lambda)^{k-1}D(\alpha(\lambda))$. Multiply this equation by a_k, and sum over k.

18. Let $a(\lambda)$ and $b(\lambda)$ be polynomials with integral coefficients, and suppose every coefficient of $a(\lambda)$ is an even integer. Is it true that every coefficient of ab is an even integer?

19. Let $\mathfrak{F}$ be a field. Which elements in $\mathfrak{F}[\lambda]$ have inverses under multiplication?

III

20. Let $\mathfrak{R}$ be a commutative ring identity.

(i) Imitating the polynomial discussion in the text, verify that $\mathfrak{R}[[\lambda]]$ is a commutative ring with identity, containing $\mathfrak{R}[\lambda]$ as a subring.

(ii) If $a(\lambda)$, $b(\lambda)$ are formal power series with

$$a(\lambda) = a_m\lambda^m + \text{higher powers of } \lambda, \qquad a_m \in \mathfrak{R},$$
$$b(\lambda) = b_n\lambda^n + \text{higher powers of } \lambda, \qquad b_n \in \mathfrak{R},$$

verify that

$$a(\lambda)b(\lambda) = a_m b_n \lambda^{m+n} + \text{higher powers of } \lambda.$$

If $\mathfrak{R}$ is an integral domain, conclude that $\mathfrak{R}[[\lambda]]$ is also an integral domain.

(iii) Let $\mathfrak{R}$ be an integral domain, and take ρ to be a fixed real number with $0 < \rho < 1$. Define the *size* $|a(\lambda)|$ of the formal powers series $a(\lambda)$ by

$$|a(\lambda)| = 0 \text{ if } a(\lambda) = 0,$$
$$|a(\lambda)| = \rho^m \text{ if } a(\lambda) = a_m\lambda^m + \text{higher powers of } \lambda,$$
$$a_m \neq 0, \qquad a_m \in \mathfrak{R}.$$

Verify that this measuring function $|\ |$ satisfies the properties V1–V4 listed in §§1.3 and 2.2. Applying it to $\mathfrak{R}[\lambda]$, conclude that it provides a measurement of the size of a polynomial in which attention is focused on the lowest degree term (and not on the highest degree term).

21. Let $a(\lambda)$ be a polynomial in $\mathbf{C}[\lambda]$, and let ρ be a real number with $0 < \rho < 1$. Define a function V by $V(a) = \rho^{-\deg a}$. If a, b are polynomials in $\mathbf{C}[\lambda]$, prove that $V(a + b) \leq V(a) + V(b)$.

22. Let $\alpha_1, \alpha_2, \ldots, \alpha_m$ be constants, and set $a(\lambda) = (\lambda + \alpha_1)(\lambda + \alpha_2)\cdots(\lambda + \alpha_m)$. Show that $\deg a(\lambda) = m$ and that if $a(\lambda) = a_0 + a_1\lambda + \cdots + a_m\lambda^m$ has constant coefficients $a_0, a_1, \ldots, a_m$, then

$$a_m = 1, \qquad a_{m-1} = \alpha_1 + \alpha_2 + \cdots + \alpha_m, \qquad a_{m-2} = \sum_{i<j} \alpha_i\alpha_j,$$

$$a_0 = \alpha_1\alpha_2\cdots\alpha_m.$$

Use induction on m. Here $\sum_{i<j}$ means a sum extending over all pairs of integers i and j with $1 \leq i < j \leq m$.

23. [Extension of 22] Show that

$$a_{m-t} = \sum \alpha_{i_1}\alpha_{i_2}\cdots\alpha_{i_t}, \qquad t = 1, 2, \ldots, m,$$

where the sum extends over all collections of integers $i_1, i_2, \ldots, i_t$ satisfying $1 \le i_1 < i_2 < i_3 < \cdots < i_t \le m$. Use induction on m.

24. Let $\mathcal{R}$ be an integral domain, and let $\mathcal{R}[\mu]$ denote the ring of polynomials in indeterminate μ. Using $\mathcal{R}[\mu]$ in place of $\mathcal{R}$ in the discussion of §3.2, form the ring of $\mathcal{R}[\mu][\lambda]$. This ring is usually denoted by $\mathcal{R}[\mu, \lambda]$ and called the *polynomial ring over* $\mathcal{R}$ *in two indeterminates* μ *and* λ. Show that the typical element of $\mathcal{R}[\mu, \lambda]$ has the form

$$\sum_{i=0}^{m}\sum_{j=0}^{n} a_{i,j}\mu^i\lambda^j$$

where $a_{i,j}$ are elements of $\mathcal{R}$ (constants). If $b_{i,j}$ are also constants, show that

$$\sum_{i=0}^{m}\sum_{j=0}^{n} a_{i,j}\mu^i\lambda^j = \sum_{i=0}^{m}\sum_{j=0}^{n} b_{i,j}\mu^i\lambda^j$$

if and only if $a_{i,j} = b_{i,j}$ for all $i = 0, \ldots, m$ and $j = 0, \ldots, n$.

25. Let f be a function defined on the polynomial ring $\mathbf{C}[\lambda]$ and taking functional values in $\mathbf{C}[\lambda]$, satisfying these rules: for arbitrary polynomials a, b in $\mathbf{C}[\lambda]$,

(i) $f(a + b) = f(a) + f(b)$,

(ii) $f(ca) = cf(a)$ if c is a constant,

(iii) $f(ab) = af(b) + f(a)b$.

(a) Prove: There is a fixed polynomial F such that $f(a) = FD(a)$ for every polynomial a. In other words: f must be fixed polynomial times differentiation. Conversely, show that this f satisfies (i), (ii), (iii).

(b) If f also satisfies the chain law, namely if $f(a(\alpha(\lambda)))$ equals $f(a(\lambda))$ evaluated at α, times $f(\alpha)$, determine f.

26. Is the degree formula (11) for products of polynomials necessarily valid when $\mathcal{R}$ is a commutative ring with identity but not an integral domain? Answer this also for the degree formulas (12), (13) for sums of polynomials.

27. Find every operator D on $\mathbf{Q}$ to itself such that $D(a + b) = D(a) + D(b)$ and $D(ab) = D(a)b + aD(b)$, for all $a, b \in \mathbf{Q}$.

Generating functions may be used to solve certain types of equations involving differences of unknowns. The following example is chosen to illustrate the technique.

28. Find numbers $y_0, y_1, y_2, \ldots$ such that $y_j - y_{j-1} = j$ for all $j \geq 1$, and $y_0 = 0$. Outline of solution:

    (i) Set $a(\lambda) = \sum_{j=0}^{\infty} y_j\lambda^j$. Using $y_0 = 0$ and $y_j - y_{j-1} = j$, verify that $(1 - \lambda)a(\lambda) = \sum_{j=0}^{\infty} j\lambda^j$.

    (ii) By multiplying out the left-hand side, verify that $(1 - \lambda)(\sum_{j=0}^{\infty} \lambda^j) = 1$, and hence that $(1 - \lambda)^{-1} = \sum_{j=0}^{\infty} \lambda^j$.

    (iii) Multiplying the equation in (i) by $(1 - \lambda)^{-1}$, obtain $a(\lambda) = (\sum_{j=0}^{\infty} \lambda^j)(\sum_{j=0}^{\infty} j\lambda^j) = \sum_{j=0}^{\infty} (0 + 1 + \cdots + j)\lambda^j = \sum_{j=0}^{\infty} \frac{1}{2}j(j + 1)\lambda^j$. Conclude that $y_j = \frac{1}{2}j(j + 1)$, for all $j \geq 0$.

3.3 THE DIVISION ALGORITHM IN $\mathbf{Z}$ AND $\mathfrak{F}[\lambda]$

Though simple to state and easy to prove, the theorems about division in this section will yield many consequences.

THEOREM 3.3.1 Division Algorithm for Integers
If a and b are integers, with $b \neq 0$, unique integers q and r exist such that

$$a = qb + r \qquad \text{and} \qquad 0 \leq r < |b|. \tag{1}$$

PROOF The theorem contains an existence assertion and a uniqueness assertion. Each requires proof. We give the existence proof first.

Let x be a variable integer. Consider the set S consisting of all integers of the form $a - xb$ as x ranges over $\mathbf{Z}$. This set S is a subset of $\mathbf{Z}$. We claim that it contains at least one nonnegative integer. To show this, let u be either $+1$ or -1 such that $a = u\,|a|$. Then $u + b^2 \geq 0$ since b^2, being a positive integer, is no smaller than 1. Setting $x = -b\,|a|$, we get:

$$a - xb = u\,|a| + b^2\,|a| = (u + b^2)\,|a| \geq 0.$$

Thus S contains a nonnegative integer. Therefore, by the well-ordering principle, it contains a smallest nonnegative integer. Letting r be this minimal nonnegative integer in S, we have $0 \leq r$ and $r = a - qb$ for some integer q. We claim that $r < |b|$. Suppose this is not so. Then $r \geq |b|$. Choose $\epsilon = +1$ or -1 such that $b = \epsilon\,|b|$. Then

$$0 \leq r - |b| = r - \epsilon b = a - qb - \epsilon b = a - (q + \epsilon)b.$$

Thus the nonnegative integer $r - |b|$ has the form $a - xb$ for $x = q + \epsilon$, and hence is in S. This is a contradiction since r is the smallest nonnegative integer in S. Consequently $r \geq |b|$ is impossible. We therefore have $a = qb + r$ (since $r = a - qb$) and $0 \leq r < |b|$. The existence proof is now complete.

Now we give the uniqueness proof. Suppose we also had $a = q'b + r'$ where $q', r' \in \mathbf{Z}$ and $0 \leq r' < |b|$. We wish to show that $q = q'$ and $r = r'$. One of the integers r, r' is the larger, say $r' \geq r$. From $a = qb + r = q'b + r'$, we get $r' - r = (q - q')b$. If $q - q' \neq 0$, we would have $|q - q'| \geq 1$ and hence:

$$|b| \leq |q - q'|\,|b| = |(q - q')b| = |r' - r| = r' - r \leq r' < |b|.$$

But this yields $|b| < |b|$, obviously a contradiction. Hence $q - q' = 0$, thus $r' - r = 0$, whence we have $r = r'$ and $q = q'$, as desired. □

What is the meaning of (1)? It shows that subtraction of a suitable multiple of b (namely qb) from a produces a result r between 0 and $|b| - 1$. This is exactly what is achieved by the procedures for division taught in grade school. Thus Theorem 1 asserts that division is possible. We therefore call

q the quotient and r the remainder

when a is divided by b.

Sometimes the following alternative form of the division algorithm is useful.

THEOREM 3.3.2

If a and b are integers, with $b \neq 0$, unique integers q' and r' exist such that

$$a = q'b + r' \qquad \text{and} \qquad \frac{-|b|}{2} < r' \leq \frac{|b|}{2}. \tag{2}$$

PROOF Let $a = qb + r$ where the integers q, r satisfy (1). If it happens that $r \leq |b|/2$, take $q' = q$, $r' = r$; and then (2) is satisfied. If it happens that $|b|/2 < r < |b|$, let $\epsilon = \pm 1$ be such that $b = \epsilon\,|b|$, and take $q' = q + \epsilon$ and $r' = r - \epsilon b = r - |b|$. Then r' satisfies $-|b|/2 < r' < 0$. This establishes the existence of a representation of a in the form (2). To show uniqueness, suppose we also had $a = q''b + r''$ with $-|b|/2 < r'' \leq |b|/2$. Then

$$-|b|/2 < r' \leq |b|/2,$$
$$-|b|/2 \leq -r'' < |b|/2.$$

Adding these inequalities gives

$$-|b| < r' - r'' < |b|.$$

Hence $|r' - r''| < |b|$. From $a = q'b + r' = q''b + r''$ we get $r' - r'' = (q'' - q')b$. If $q'' - q'$ is not zero, then $|q'' - q'| \geq 1$, and then:

$$|b| > |r' - r''| = |(q'' - q')b| = |q'' - q'|\,|b| \geq |b|.$$

Thus $|b| > |b|$, a contradiction. So $q'' - q' = 0$, and hence $r' - r'' = 0$. Therefore $q' = q''$ and $r' = r''$. □

NOTE $|b|/2$ is not necessarily an integer.

REMARK The integer r' in Theorem 3.3.2 is called the *numerically least remainder* obtained on dividing a by b, whereas the remainder r in Theorem 3.3.1 is called the *least nonnegative remainder.*

In Chapter 4 we shall study the arithmetical operations, including division, in some detail.

We turn now to the corresponding facts for polynomials. Let $\mathfrak{F}$ be a field. Then $\mathfrak{F}$ is automatically an integral domain, and hence (by §3.2) so is the polynomial ring $\mathfrak{F}[\lambda]$. The reader who elected in §3.2 to take the ring $\mathfrak{R}$ to be **Z**, **Q**, **R**, or **C**, should now regard $\mathfrak{F}$ as one of **Q**, **R**, or **C**.

THEOREM 3.3.3 Division Algorithm for Polynomials

If a and b are polynomials in $\mathfrak{F}[\lambda]$, with $b \neq 0$, unique polynomials q, r exist in $\mathfrak{F}[\lambda]$ such that

$$a = qb + r \qquad \text{and} \qquad \deg r < \deg b. \tag{3}$$

PROOF The proof imitates that of Theorem 3.3.1. We give the existence proof first. Let x be a variable polynomial in $\mathfrak{F}[\lambda]$, and consider the set S consisting of all polynomials of the form $a - xb$, as x ranges over $\mathfrak{F}[\lambda]$. Either S contains the zero polynomial or it does not. If it does, let r be the zero polynomial. If it does not, among all polynomials in S let r be one with lowest degree. There is such a polynomial because the degrees of the various polynomials in S constitute a nonempty set of nonnegative integers which, by the well-ordering principle, must contain a smallest nonnegative integer. In either case, the polynomial r is in S and $\deg r \leq \deg s$ for every polynomial s in S. Since r is in S, there is a polynomial q in $\mathfrak{F}[\lambda]$ such that $a - qb = r$. We wish to prove that $\deg r < \deg b$. If this is not so, let $k = \deg r$ and $n = \deg b$. Then $k \geq n$. Set

$$b(\lambda) = b_0 + b_1\lambda + \cdots + b_n\lambda^n, \qquad r(\lambda) = r_0 + r_1\lambda + \cdots + r_k\lambda^k,$$

where $b_0, b_1, \ldots, b_n, r_0, r_1, \ldots, r_k$ are constants with $b_n \neq 0$, $r_k \neq 0$. Let $r' = a - q'b$, where $q'(\lambda)$ is the polynomial $q(\lambda) + r_k b_n^{-1}\lambda^{k-n}$. Then r' belongs to S, and we have:

$$\begin{aligned} r' &= a - q'b = a - (q + r_k b_n^{-1}\lambda^{k-n})b \\ &= a - qb - (r_k\lambda^k + \text{terms in } \lambda^{k-1}, \lambda^{k-2}, \ldots) \\ &= r - r_k\lambda^k - \text{terms in } \lambda^{k-1}, \lambda^{k-2}, \ldots \\ &= \text{a polynomial involving only } \lambda^0, \lambda^1, \ldots, \lambda^{k-1}. \end{aligned} \tag{4}$$

It follows from (4) that $\deg r' < k$. Thus $r' \in S$ and $\deg r' < \deg r$, contradicting the fact that r has the lowest degree among all polynomials in S. Hence $\deg r \geq \deg b$ is impossible, and therefore we have (3).

The uniqueness proof is easy. Suppose $Q(\lambda)$ and $R(\lambda)$ are polynomials in $\mathfrak{F}[\lambda]$ with $a = Qb + R$ and $\deg R < \deg b$. Then $a = qb + r = Qb + R$; hence $R - r = (q - Q)b$. If $q - Q \neq 0$, then $R - r \neq 0$; by Theorem 3.2.1 $\deg (R - r) \leq \max (\deg R, \deg r) < \deg b$. This yields

$$\deg b > \deg (R - r) = \deg (q - Q)b = \deg (q - Q) + \deg b \geq \deg b.$$

Thus $\deg b > \deg b$, a contradiction. Therefore $q - Q = 0$ and $r - R = 0$, completing the proof. □

As with integers, we call the polynomials $q(\lambda)$, $r(\lambda)$ in (3) the quotient and remainder, respectively, when $a(\lambda)$ is divided by $b(\lambda)$.

We now apply the discussion of evaluation of polynomials in §3.2 to derive a basic result called the *remainder theorem*. It gives the form of the remainder when a polynomial is divided by a monic first degree polynomial.

THEOREM 3.3.4

Let $a(\lambda) \in \mathfrak{F}[\lambda]$ and $\alpha \in \mathfrak{F}$. Then a polynomial $q(\lambda) \in \mathfrak{F}[\lambda]$ and a constant $r \in \mathfrak{F}$ exist such that

$$a(\lambda) = q(\lambda)(\lambda - \alpha) + r.$$

In fact, $r = a(\alpha)$.

PROOF Apply Theorem 3.3.3 with $b(\lambda) = \lambda - \alpha$. The remainder $r(\lambda)$ satisfies $\deg r(\lambda) < \deg (\lambda - \alpha) = 1$. Therefore it is a constant, so that $a(\lambda) = q(\lambda)b(\lambda) + r$ with $q(\lambda)$ a polynomial and r a constant. Evaluating at α, we get $a(\alpha) = q(\alpha)(\alpha - \alpha) + r = r$. Thus $r = a(\alpha)$. □

Let $a(\lambda)$ and $b(\lambda)$ be polynomials with integral coefficients. Suppose $b(\lambda)$ is monic. Taking the field $\mathfrak{F}$ to be $\mathbf{Q}$ and applying Theorem 3.3.3, we see that (3) holds where the quotient $q(\lambda)$ and remainder $r(\lambda)$ have coefficients in $\mathbf{Q}$. However, because $b(\lambda)$ is monic, we are able to assert that $q(\lambda)$ and $r(\lambda)$ have *integral* coefficients.

THEOREM 3.3.5

Let $a(\lambda)$ and $b(\lambda)$ be polynomials with integral coefficients, with $b(\lambda)$ monic. Then the quotient $q(\lambda)$ and remainder $r(\lambda)$ in Theorem 3.3.3 have integral coefficients.

PROOF Let $\deg b(\lambda) = n$, and suppose $q(\lambda) = q_l\lambda^l + \cdots + q_0$, where $q_l, \ldots, q_0$ are rational numbers, some possibly zero. Because $\deg r < \deg b$ and b is monic, the only term involving λ^{l+n} in $qb + r$ is $q_l\lambda^{l+n}$. This must be the only term involving λ^{l+n} in the equal polynomial $a(\lambda)$. Hence q_l is an integer. From this case we deduce that all of $q_l, \ldots, q_0$ are integers, as follows. If we have proved that $q_l, \ldots, q_{j+1}$ are all integers, rewrite $a = qb + r$ as

$$a(\lambda) - (q_l\lambda^l + \cdots + q_{j+1}\lambda^{j+1})b(\lambda) = (q_j\lambda^j + \cdots + q_0)b(\lambda) + r(\lambda).$$

Since the polynomial on the left-hand side has integral coefficients, we may apply the case proved above to deduce that q_j is an integer. Repeating this, we eventually see that all coefficients of $q(\lambda)$ are integers. Since $r = a - qb$, it then follows that r also has integral coefficients. □

EXERCISES 3.3

I

1. Compute the quotient q and remainder r when the integer a is divided by the integer b, if (i) $a = 125$, $b = 7$; (ii) $a = -125$, $b = 7$; (iii) $a = 125$, $b = -7$; (iv) $a = -125$, $b = -7$, (v) $a = 7$, $b = 125$; (vi) $a = -7$, $b = 125$. In each case first require that r be the least nonnegative remainder, then require that it be the numerically least remainder.

2. Repeat all parts of Exercise 1, beginning with $a = 67$, $b = 9$.

3. Compute the quotient q and the remainder r when the polynomial a is divided by the polynomial b, if: (i) $a = \lambda^4 + \lambda^3 + \lambda^2 + \lambda - 2$, $b = \lambda + 1$, (ii) $a = \lambda^2 + \lambda$, $b = \lambda^3 + \lambda$; (iii) $a = i\lambda^2 - \lambda + i$, $b = (1 + i)\lambda$; (iv) $a = \lambda^4 + 1$, $b = \lambda - 1$.

4. If $a(\lambda) = 7\lambda^3 + 3\lambda^2 + 2\lambda - 4$, evaluate $a(2)$ by two methods:
   (a) direct substitution of 2 in place of λ;
   (b) dividing $a(\lambda)$ by $\lambda - 2$.

5. When $\lambda^4 + \lambda^3 + \lambda + x$ is divided by $\lambda - 1$, the remainder is 4. Find the number x.

6. If a, b, q, r, q_1, r_1 are integers ($b \neq 0$) with $a = qb + r = q_1b + r_1$, does it follow that $r = r_1$? Give either a proof or an example in which this is not so.

7. Is the conclusion of Theorem 3.3.5 still correct if $b(\lambda)$ is not monic?

II

8. Let $a(\lambda)$ and $b(\lambda)$ be polynomials in which the coefficients of the odd powers of λ are all zero, that is, only even powers of λ actually

appear. Show that this is also true for the quotient $q(\lambda)$ and remainder $r(\lambda)$ obtained on dividing a by b. Can an analogous conclusion be drawn if only odd powers of λ are present in a and b?

9. Let α be a rational number. The *greatest integer in* α is by definition the largest integer m for which $m \leq \alpha$. For example, $m = 1, -1, 5$ respectively when $\alpha = 7/4, -3/4$, or 5. Suppose a and b are integers, and let $a = qb + r$, where q, r are integers with $0 \leq r < |b|$. Show that q is the greatest integer in a/b. If we require instead that $-|b|/2 < r \leq |b|/2$, show that q is the integer *nearest* a/b. (There may be two integers nearest a/b. Which one is q?)

10. If a and b are even integers, are the quotient q and remainder r obtained on dividing a by b necessarily even? How are your answers changed if we substitute "odd" for "even" in this question?

11. Let a, b, q, r be integers with $a = qb + r$. Does it follow that $|r| \leq |b|$?

12. Let a, b, q, r be positive integers with $a = qb + r$ and $0 \leq r < b$ Show that $0 \leq q \leq a/b$.

13. Show that the well-ordering principle is valid in the set $\{-\infty, 0, 1, 2, \ldots\} = \{-\infty\} \cup \{\text{nonnegative integers}\}$ i.e., any nonempty subset has a least element. Using this, simplify the proof of Theorem 3.3.3, avoiding separate discussions of the cases in which S contains, or does not contain, the zero polynomial.

III

14. Let a and b be polynomials with coefficients in a commutative ring $\mathcal{R}$ with identity. Assume that b is monic. Show that there is a quotient q and remainder r with coefficients in $\mathcal{R}$ such that $a = qb + r$ and $\deg r < \deg b$. Does this conclusion continue to hold if b is not monic?

15. Let

$$a(\lambda) = a_m\lambda^m + a_{m-1}\lambda^{m-1} + \cdots + a_1\lambda + a_0$$

where $a_0, a_1, \ldots, a_m$ are constants. Show that

$$a(\lambda) = (\cdots(((a_m\lambda + a_{m-1})\lambda + a_{m-2})\lambda + a_{m-3})\lambda + \cdots + a_1)\lambda + a_0.$$

Use this to establish the validity of the following simple scheme for evaluating $a(\lambda)$ at α: Let $d_0 = a_m$, then determine $d_1, d_2, \ldots$, in succession by this rule: $d_1 = d_0\alpha + a_{m-1}$, $d_2 = d_1\alpha + a_{m-2}$, $d_3 = d_2\alpha + a_{m-3}, \ldots, d_{m-1} = d_{m-2}\alpha + a_1$, $d_m = d_{m-1}\alpha + a_0$. Prove that $d_m = a(\alpha)$. Use this procedure to compute $a(3)$ when $a(\lambda) = \lambda^4 + \lambda^3 + 2\lambda^2 + \lambda + 5$.

16. [Continuation of 15] Let $a(\lambda)$ be as in Exercise 15, and assume that α is a constant. Let $q(\lambda)$ be the quotient and r the remainder when $a(\lambda)$ is divided by $\lambda - \alpha$. Suppose $q(\lambda) = q_{m-1}\lambda^{m-1} + \cdots + q_0$, where $q_{m-1}, \ldots, q_0$ are constants. By comparing coefficients on the

two sides of the equation $a(\lambda) = q(\lambda)(\lambda - \alpha) + r$, prove that $q_{m-1} = a_m$, $q_{m-2} = a_{m-1} + \alpha q_{m-1}$, $q_{m-3} = a_{m-2} + \alpha q_{m-2}$, . . ., $q_0 = a_1 + \alpha q_1$, $r = a_0 + \alpha q_0$. Using Exercise 15, deduce that $q_{m-1} = d_0$, $q_{m-2} = d_1$, . . ., $q_0 = d_{m-1}$. Hence show that the d's in Exercise 15 yield the coefficients of $q(\lambda)$. Specifically: $q(\lambda) = d_0\lambda^{m-1} + \cdots + d_{m-1}$ and $r = d_m$.

3.4 DIVISIBILITY

Throughout the rest of this chapter we shall discuss both the integer ring **Z** and the polynomial ring $\mathfrak{F}[\lambda]$, where $\mathfrak{F}$ is a field. To achieve economy of language, henceforth in this chapter we let the symbol $\mathfrak{R}$ denote either of these rings. As noted in the previous section, on first reading the student may assume that $\mathfrak{F}$ is one of **Q**, **R**, or **C**.

DEFINITION 3.4.1
Let a and b be elements of $\mathfrak{R}$. If $\mathfrak{R}$ contains an element q with $a = qb$, we say that "b divides a" and write $b \mid a$. If $\mathfrak{R}$ contains no such element q, we say "b does not divide a" and write $b \nmid a$.

Some alternative terminology is: "b is a divisor of a," or "b is a factor of a," or "a is a multiple of b," or "a is divisible by b," all in place of "b divides a."

When $b = 0$, it is clear that $b \mid a$ if and only if $a = 0$.

When b is nonzero, $b \mid a$ if and only if the remainder r in Theorems 3.3.1, 3.3.2, *or* 3.3.3 *is zero.* PROOF: If the quantity r in these theorems is zero, then $a = qb$ and hence $b \mid a$. Conversely, if $b \mid a$, then $a = qb$ for some q in $\mathfrak{R}$, that is, $a = qb + 0$ so that (by the uniqueness of the remainder) $r = 0$. □

Suppose $b \neq 0$. It is worth noting the various uses of the words "divide" and "quotient." When speaking of a fraction a/b with integral numerator a and denominator b, we often call a/b the quotient of a by b, or the fraction obtained upon dividing a by b. This was the language used when introducing rational numbers in Example 1.5.3. If we have integers a, b, q, r with $a = qb + r$, as for example in Theorem 3.3.1, we also call q the quotient and r the remainder on dividing a by b. These uses of "quotient" do not mean the same thing since usually $a/b \neq q$; in fact $a/b = q + r/b$. When b divides a, as in Definition 3.4.1, then $r = 0$ and $a/b = q$. Otherwise $a/b \neq q$. If we say "divide b into a" or "divide a by b," we usually mean "find q and r in $a = qb + r$, as in §3.3," but on infrequent occasions we mean "form the fraction a/b." Context will always indicate the meaning intended.

The next two lemmas give useful facts about divisibility.

LEMMA 3.4.2
Let $a_1, a_2, \ldots, a_n, b$, be elements of $\mathfrak{R}$, and suppose $b \mid a_1, b \mid a_2, \ldots, b \mid a_n$. Then $b \mid (x_1a_1 + \cdots + x_na_n)$ for any elements $x_1, \ldots, x_n$ of $\mathfrak{R}$.

PROOF Since $b \mid a_i$ there is an element q_i in $\mathfrak{R}$ such that $a_i = q_ib$; $i = 1, \ldots, n$. Then

$$x_1a_1 + \cdots + x_na_n = (x_1q_1 + \cdots + x_nq_n)b.$$

But this implies that $b \mid (x_1a_1 + \cdots + x_na_n)$. □

DEFINITION 3.4.3
An element u in $\mathfrak{R}$ is a unit *if and only if $u \mid 1$.*

Thus u is a unit in $\mathfrak{R}$ if and only if $uq = 1$ for some element q of $\mathfrak{R}$. When $\mathfrak{R}$ is $\mathbf{Z}$, the integer u satisfies $uq = 1$ for some integer q only when $u = \pm 1$. When $\mathfrak{R}$ is $\mathfrak{F}[\lambda]$, the polynomial u satisfies $uq = 1$ for some polynomial q only when u is a nonzero constant. Thus *the only units in* $\mathbf{Z}$ *are* 1 and -1, and *the only units in* $\mathfrak{F}[\lambda]$ *are the nonzero constants.* In either case, the units in $\mathfrak{R}$ are precisely the elements u for which $\mathfrak{R}$ contains a multiplicative inverse $q = u^{-1}$.

LEMMA 3.4.4
Let a and b be elements of $\mathfrak{R}$. If $b \mid a$ and u is a unit, then $(ub) \mid a$.

PROOF If $a = qb$, where q is in $\mathfrak{R}$, then $a = (qu^{-1})(ub)$. Since u^{-1} is in $\mathfrak{R}$, so is qu^{-1}. Hence $(ub) \mid a$. □

The greatest common divisor

This is the central concept in this section.

DEFINITION 3.4.5
Let $a_1, a_2, \ldots, a_n$ be elements of $\mathfrak{R}$. An element d in $\mathfrak{R}$ is called the greatest common divisor of $a_1, a_2, \ldots, a_n$ if it satisfies these three conditions:
(i) $d \mid a_1, d \mid a_2, \ldots, d \mid a_n$;
(ii) *if e is an element of $\mathfrak{R}$ such that $e \mid a_1, e \mid a_2, \ldots, e \mid a_n$, then $e \mid d$;*
(iii) *d, if nonzero, is positive when $\mathfrak{R}$ is $\mathbf{Z}$, and monic when $\mathfrak{R}$ is $\mathfrak{F}[\lambda]$.*

When $a_1, \ldots, a_n$ are all zero, conditions (i), (ii), (iii) are satisfied by taking $d = 0$. Suppose $a_1, \ldots, a_n$ are not all zero. Then, by (i), d cannot be zero, and we may explain the term "the greatest common divisor" in this way: Property (i) shows that d is a *common* divisor since it divides all of $a_1, \ldots, a_n$. Property (ii) shows that d is the *greatest* among common divisors since it is a multiple of any other common divisor. (See Exercise 9.) Property (iii) provides a uniqueness condition so that we can speak of *the* greatest common divisor. (This will become evident upon reading the discussion below. See Exercise 16 also.)

Conditions (i), (ii), (iii) are rather stringent, and it is not obvious that $\mathcal{R}$ always contains an element d satisfying these conditions.

THEOREM 3.4.6

If $a_1, \ldots, a_n$ are elements of $\mathcal{R}$, their greatest common divisor exists and is unique.

PROOF First suppose that $a_1, \ldots, a_n$ are all zero. Then $d = 0$ satisfies (i), (ii), (iii), and by taking e in (ii) to be 0, we see that only $d = 0$ satisfies these three conditions.

In the remainder of the proof we suppose $a_1, \ldots, a_n$ are not all zero. Consider the subset S of $\mathcal{R}$ consisting of all elements of the form $x_1a_1 + x_2a_2 + \cdots + x_na_n$, as $x_1, x_2, \ldots, x_n$ each vary over all elements of $\mathcal{R}$. By taking $x_1 = 1$, $x_2 = x_3 = \cdots = x_n = 0$ we see that $a_1 \in S$, and similarly $a_2, a_3, \ldots, a_n$, are all in S. Thus S contains some nonzero elements (since $a_1, \ldots, a_n$ are not all zero). Let d_1 be an element of S with the following property:

$$\left.\begin{array}{rl} \multicolumn{2}{l}{\text{Among all nonzero elements of } S,} \\ |d_1| & \text{is least, when } \mathcal{R} \text{ is } \mathbf{Z}, \\ \deg d_1 & \text{is least, when } \mathcal{R} \text{ is } \mathfrak{F}[\lambda]. \end{array}\right\} \quad (1)$$

The existence of d_1 is guaranteed by the well-ordering principle if we apply it (for integers) to the set of all $|s|$ or (for polynomials) to the set of all $\deg s$, for s an arbitrary nonzero element of S.

We now show that d_1 is a divisor of any element s of S. Since $a_1, \ldots, a_n$ are all elements of S, this will imply that d_1 is a common divisor of $a_1, \ldots, a_n$. Let $s \in S$. Then

$$s = x_1'a_1 + \cdots + x_n'a_n \quad (2)$$

for certain $x_1', \ldots, x_n' \in \mathcal{R}$. Since $d_1 \in S$, we also have

$$d_1 = x_1a_1 + \cdots + x_na_n \quad (3)$$

for certain $x_1, \ldots, x_n \in \mathcal{R}$. Let $s = qd_1 + r$ where $q, r \in \mathcal{R}$ and

$$\begin{array}{rl} 0 \le r < |d_1|, & \text{when } \mathcal{R} \text{ is } \mathbf{Z}, \\ \deg r < \deg d_1, & \text{when } \mathcal{R} \text{ is } \mathfrak{F}[\lambda]. \end{array} \quad (4)$$

Theorems 3.3.1 and 3.3.3 (the division algorithms) guarantee the existence of $q, r \in \mathcal{R}$ with these properties. Multiply (3) by q and subtract the result from (2). We get

$$r = (x_1' - qx_1)a_1 + \cdots + (x_n' - qx_n)a_n.$$

This equation implies that $r \in S$ since the right-hand side is a sum of multiples of the a_i. If r is nonzero, it must therefore satisfy

$$|r| \geq |d_1| \text{ when } \mathcal{R} = \mathbf{Z}, \qquad \deg r \geq \deg d_1 \text{ when } \mathcal{R} = \mathfrak{F}[\lambda]$$

(by the minimality property (1) of d_1). Owing to (4), r cannot satisfy these conditions. Consequently $r = 0$, yielding $s = qd_1$, and hence $d_1 \mid s$. Thus d_1 is indeed a common divisor of $a_1, \ldots, a_n$.

We next show that any common divisor e of $a_1, \ldots, a_n$ must divide d_1. For if $e \mid a_1, \ldots, e \mid a_n$, then (by Lemma 3.4.2) e divides $x_1a_1 + \cdots + x_na_n = d_1$. Thus the common divisor d_1 of $a_1, \ldots, a_n$ is divisible by any common divisor e of $a_1, \ldots, a_n$.

Now let d_1 be any element of $\mathcal{R}$ (possibly, but not necessarily, the element found above) which is a common divisor of $a_1, \ldots, a_n$ and which is divisible by any common divisor e of $a_1, \ldots, a_n$. There is a unit u in $\mathcal{R}$ such that ud_1 is positive (when $\mathcal{R}$ is $\mathbf{Z}$) or monic (when $\mathcal{R}$ is $\mathfrak{F}[\lambda]$) since d_1, as a divisor of a nonzero element (one of $a_1, \ldots, a_n$) cannot be zero. (See Exercise 15 of §3.2.) Let $d = ud_1$. Then d satisfies (iii). By Lemma 3.4.4, ud_1 and therefore d is a common divisor of $a_1, \ldots, a_n$. Hence d satisfies (i). If e is a common divisor of $a_1, \ldots, a_n$, then $e \mid d_1$, and by Lemma 3.4.2 $e \mid ud_1$, that is, $e \mid d$. Hence d satisfies (ii). We have completed the proof that d exists.

Now we give the uniqueness part of the proof. Suppose that d' satisfies the same three conditions that d satisfies, namely,

(i′) d' is a common divisor of $a_1, \ldots, a_n$;

(ii′) if e is a common divisor of $a_1, \ldots, a_n$, then $e \mid d'$;

(iii′) d' is positive (when $\mathcal{R}$ is $\mathbf{Z}$) or monic (when $\mathcal{R}$ is $\mathfrak{F}[\lambda]$).

We wish to prove that $d' = d$. Combining properties (i′) and (ii) [take $e = d'$ in (ii)], we see that $d' \mid d$. Combining properties (i) and (ii′) [take $e = d$ in (ii′)], we see that $d \mid d'$. Since $d' \mid d$ and $d \mid d'$, we have $d = qd'$ and $d' = q'd$ for suitable q, q' in $\mathcal{R}$. Hence $d = qq'd$. Because $\mathcal{R}$ is an integral domain, we may cancel the nonzero element d, to obtain $1 = qq'$. Thus q is a unit, hence is ± 1 (when $\mathcal{R}$ is $\mathbf{Z}$) or a nonzero constant (when $\mathcal{R}$ is $\mathfrak{F}[\lambda]$). From $d = qd'$ and the fact that d, d' are both positive (for $\mathcal{R} = \mathbf{Z}$) or monic (for $\mathcal{R} = \mathfrak{F}[\lambda]$), we find that $q = 1$. Hence $d = d'$. □

NOTATION The symbol $(a_1, \ldots, a_n)$ will denote the greatest common divisor d of $a_1, \ldots, a_n$. This notation enables us to avoid introducing a new symbol each time we have a new greatest common divisor to discuss.

COROLLARY 3.4.7

Elements $y_1, \ldots, y_n$ exist in $\mathcal{R}$ such that

$$(a_1, \ldots, a_n) = y_1a_1 + \cdots + y_na_n. \tag{5}$$

PROOF Any $y_1, \ldots, y_n \in \mathcal{R}$ will satisfy (5) when $a_1, \ldots, a_n$ are all zero. When $a_1, \ldots, a_n$ are not all zero, we found above that $d = ud_1$, where d_1 is given by (3). Multiply (3) by u, and set $y_1 = ux_1, \ldots, y_n = ux_n$. □

In the course of the proof above, we established this fact:

If $a_1, \ldots, a_n$ are not all zero, $(a_1, \ldots, a_n) = ud_1$ if d_1 is a common divisor of $a_1, \ldots, a_n$ divisible by every common divisor, and u is a unit such that ud_1 is positive or monic. (6)

Let a and b be elements of $\mathcal{R}$, with $b \neq 0$. If $b \mid a$, it is easy to compute (a, b). For $b \mid a$ (by hypothesis), and any common divisor e of a and b certainly divides b. Thus b has the properties of d_1 in (6), and therefore $(a, b) = ub$ for a suitably chosen unit u. We now want to show how to compute (a, b) in the more difficult case when $b \nmid a$.

LEMMA 3.4.8

Let a and b be elements of $\mathcal{R}$, with b nonzero. Suppose that $r_0 = a$, $r_1 = b$, and that $r_2, \ldots, r_m, q_1, \ldots, q_m$ are further elements of $\mathcal{R}$, such that these equations hold:

$$\begin{aligned} r_0 &= q_1r_1 + r_2 \\ r_1 &= q_2r_2 + r_3 \\ &\vdots \\ r_j &= q_{j+1}r_{j+1} + r_{j+2} \\ &\vdots \\ r_{m-2} &= q_{m-1}r_{m-1} + r_m \\ r_{m-1} &= q_mr_m. \end{aligned} \tag{7}$$

Then r_m is a common divisor of a, b divisible by every common divisor.

PROOF The last equation of (7) implies that $r_m \mid r_{m-1}$. Clearly $r_m \mid r_m$. Thus there is a value of j (namely $j = m - 2$) such that $r_m \mid r_{j+1}$ and $r_m \mid r_{j+2}$. If we have a value of j such that $r_m \mid r_{j+1}$ and $r_m \mid r_{j+2}$, then $r_m \mid (q_{j+1}r_{j+1} + r_{j+2})$ [by Lemma 3.4.2], and hence $r_m \mid r_j$ [by (7)]. Thus $r_m \mid r_j$ and $r_m \mid r_{j+1}$. Therefore we may decrease j; eventually we get $r_m \mid r_0$ and $r_m \mid r_1$. That is, $r_m \mid a$ and $r_m \mid b$, so that r_m is a common divisor of a and b. Suppose now that e is a common divisor of a and b. Then $e \mid r_0$ and $e \mid r_1$. Thus there is a value of j (namely, $j = 0$) such that $e \mid r_j$ and $e \mid r_{j+1}$. If we have a value of j such that $e \mid r_j$ and $e \mid r_{j+1}$, then $e \mid (r_j - q_{j+1}r_{j+1})$

[by Lemma 3.4.2] and hence $e \mid r_{j+2}$ [by (7)]. Thus $e \mid r_{j+1}$ and $e \mid r_{j+2}$. Therefore we may increase j. Eventually we get $e \mid r_{m-1}$ and $e \mid r_m$. That is: $e \mid r_m$, as required.

The significance of Lemma 3.4.8 is this: from (6) we know $(a, b) = ur_m$ for a unit u chosen to make ur_m positive or monic.

We now give a procedure, called *Euclid's algorithm*, for constructing a system of equations of type (7). By the division algorithms (Theorems 3.3.1, 3.3.2, or 3.3.3), applied to r_0 and r_1, we obtain elements q_1 and r_2 in $\mathcal{R}$ with $r_0 = q_1r_1 + r_2$ and (for $\mathcal{R} = \mathbf{Z}$) $|r_2| < |r_1|$ or (for $\mathcal{R} = \mathfrak{F}[\lambda]$) $\deg r_2 < \deg r_1$. If $r_2 \neq 0$, by applying the division algorithms to r_1 and r_2, we obtain q_2 and r_3 in $\mathcal{R}$ with $r_1 = q_2r_2 + r_3$ and (for $\mathcal{R} = \mathbf{Z}$) $|r_3| < |r_2|$ or (for $\mathcal{R} = \mathfrak{F}[\lambda]$) $\deg r_3 < \deg r_2$. If $r_3 \neq 0$, by applying the division algorithms to r_2 and r_3, we obtain elements q_3 and r_4 in $\mathcal{R}$ with $r_2 = q_3r_3 + r_4$ and (for $\mathcal{R} = \mathbf{Z}$) $|r_4| < |r_3|$ or (for $\mathcal{R} = \mathfrak{F}[\lambda]$) $\deg r_4 < \deg r_3$. We continue in this way, producing at each step a new quotient q_j and remainder r_{j+1} until we arrive at a step at which the new remainder r_{j+1} is zero. Let r_{m+1} be the first remainder which is zero. We then have the system (7). Must there always be a first remainder which is zero? If not, for $\mathcal{R} = \mathbf{Z}$ we have $|r_1| > |r_2| > |r_3| > \cdots$ so that (because the $|r_j|$ are positive integers) we would have a nonterminating decreasing sequence of positive integers. This is impossible by the well-ordering principle. When $\mathcal{R} = \mathfrak{F}[\lambda]$ we have $\deg r_1 > \deg r_2 > \deg r_3 \cdots$ so that (because the quantities $\deg r_j$ are nonnegative integers) we would have a nonterminating decreasing sequence of nonnegative integers, and this also is impossible, by the well-ordering principle. Hence $r_{m+1} = 0$ for some integer m; and then we have (7).

We know that $(a, b) = y_1a + y_2b$ for certain elements $y_1, y_2 \in \mathcal{R}$. When $b \mid a$ and $(a, b) = ub$ with u a unit, we may take $y_1 = 0$, $y_2 = u$. When $b \nmid a$, can we obtain y_1, y_2 from (7)? We now show how to do this. Suppose, for a certain value of j we have found ring elements A_{j+1} and B_{j+1} such that $r_m = A_{j+1}r_{j+1} + B_{j+1}r_{j+2}$. [We can certainly do this for $j = m - 3$ since the next-to-last equation (7) gives $r_m = A_{m-2}r_{m-2} + B_{m-2}r_{m-1}$ with $A_{m-2} = 1$, $B_{m-2} = -q_{m-1}$.] From (7) we have $r_{j+2} = r_j - q_{j+1}r_{j+1}$. Substituting into the expression above for r_m, we get $r_m = A_{j+1}r_{j+1} + B_{j+1}(r_j - q_{j+1}r_{j+1}) = A_jr_j + B_jr_{j+1}$, where $A_j = B_{j+1}$, $B_j = A_{j+1} - q_{j+1}B_{j+1}$. That is: we can lower j. Eventually we get $r_m = A_0r_0 + B_0r_1 = A_0a + B_0b$. Multiplying by u yields $(a, b) = y_1a + y_2b$ where $y_1 = uA_0$, $y_2 = uB_0$.

The name *Euclid's algorithm* is most often applied to the procedure outlined above for computing the greatest common divisor of integers a, b if throughout we use least nonnegative remainders. This is because the procedure is given in the writings of the ancient Greek geometer Euclid. One may however use the name Euclid's algorithm to describe any calculation of (a, b) which is based on a chain of equations of the form (7).

Here is a numerical example. Let $a = 1096$, $b = 336$. Taking $r_0 = a$, $r_1 = b$, and using numerically least remainders, we find

$$\begin{aligned} r_0 &= 3r_1 + r_2\,, & r_2 &= 88,\\ r_1 &= 4r_2 + r_3\,, & r_3 &= -16,\\ r_2 &= (-5)r_3 + r_4\,, & r_4 &= 8,\\ r_3 &= (-2)r_4 + 0, & r_5 &= 0. \end{aligned}$$

Thus $(a, b) = 8$. Then working backwards, we obtain $8 = r_4 = r_2 + 5r_3 = r_2 + 5(r_1 - 4r_2) = 5r_1 - 19r_2 = 5r_1 - 19(r_0 - 3r_1) = -19r_0 + 62r_1$. Thus $(a, b) = -19a + 62b$.

How do we compute $(a_1, a_2, \ldots, a_n)$ if $n > 2$? The procedure is given by this lemma.

LEMMA 3.4.9

Let $a_1, \ldots, a_n$ be elements of $\mathcal{R}$, with $a_1, \ldots, a_{n-1}$ not all zero. Let $\delta = (a_1, \ldots, a_{n-1})$, $d = (a_1, \ldots, a_n)$. Then $d = (\delta, a_n)$.

PROOF Let $d' = (\delta, a_n)$. We wish to prove that $d' = d$. Certainly $d' \mid \delta$ and $d' \mid a_n$. Because $d' \mid \delta$ and δ is a common divisor of $a_1, \ldots, a_{n-1}$, it follows that d' is a common divisor of $a_1, \ldots, a_{n-1}$, and therefore d' is a common divisor of $a_1, \ldots, a_n$. Consequently $d' \mid d$. On the other hand, d is a common divisor of $a_1, \ldots, a_n$, hence a common divisor of $a_1, \ldots, a_{n-1}$, so that $d \mid \delta$. Because $d \mid \delta$, $d \mid a_n$ we get $d \mid d'$. Thus $d \mid d'$ and $d' \mid d$ and therefore (since d', d are both positive or monic) we get $d = d'$. □

This lemma shows that repeated application of the procedure for computing the greatest common divisor of two quantities will yield $(a_1, \ldots, a_n)$ when $n > 2$. Moreover, we can find $y_1, \ldots, y_n$ to express $(a_1, \ldots, a_n)$ as $y_1a_1 + \cdots + y_na_n$ by repeatedly applying the $n = 2$ procedure given above, in this way: If we have found elements $x_1, \ldots, x_{n-1}$ of $\mathcal{R}$ such that $\delta = (a_1, \ldots, a_{n-1}) = x_1a_1 + \cdots + x_{n-1}a_{n-1}$, and also found elements Y_1, Y_2 such that $(\delta, a_n) = Y_1\delta + Y_2a_n$, then $d = (\delta, a_n) = Y_1(x_1a_1 + \cdots + x_{n-1}a_{n-1}) + Y_2a_n = y_1a_1 + \cdots + y_na_n$, for $y_1 = Y_1x_1$, $\ldots, y_{n-1} = Y_1x_{n-1}$, $y_n = Y_2$.

As a continuation of the above example, let us compute (a, b, c) when $a = 1096$, $b = 336$, $c = 3$. We already know $(a, b) = 8 = -19a + 62b$. Compute next $(8, c)$:

$$8 = 3c - 1, \qquad c = 3(-1) + 0.$$

Thus $(8, c) = 1$. Hence $(a, b, c) = 1$. And we also have $1 = 3c - 8 = 3c - (-19a + 62b) = 19a - 62b + 3c$.

Elements a, b in $\mathcal{R}$ for which $(a, b) = 1$ occur so frequently that it is useful to have a name for this property: we say a and b are *relatively prime* if $(a, b) = 1$. The following simple fact is very useful: *If a, b are elements of $\mathcal{R}$, not both zero, then a/d and b/d are relatively prime, where* $d = (a, b)$. To prove this, notice that from $d = y_1a + y_2b$ (for certain y_1, y_2 in $\mathcal{R}$) we get

$$1 = y_1 \frac{a}{d} + y_2 \frac{b}{d};$$

hence any common divisor of a/d and b/d must divide 1. Thus

$$\left(\frac{a}{d}, \frac{b}{d}\right) = 1.$$

More generally, if $a_1, a_2, \ldots, a_n$ are elements of $\mathcal{R}$, not all zero, and $d = (a_1, a_2, \ldots, a_n)$, then $(a_1/d, a_2/d, \ldots, a_n/d) = 1$. The proof is the same, using the equation $y_1a_1 + y_2a_2 + \cdots + y_na_n = d$.

Having discussed the common divisors of $a_1, \ldots, a_n$, we shall now discuss the common multiples. First we give a lemma needed in this discussion and in §3.5.

LEMMA 3.4.10
Let a, b, c be elements of $\mathcal{R}$, with c nonzero. If $c \mid (ab)$ and a, c are relatively prime, then $c \mid b$.

PROOF Since $(c, a) = 1$, elements $x, y \in \mathcal{R}$ exist such that $1 = xc + ya$. Multiplying by b, we get $b = (xb)c + y(ab)$. Now $c \mid c$ and $c \mid ab$, hence $c \mid (xbc + yab)$, and hence $c \mid b$. □

The least common multiple (Optional)

DEFINITION 3.4.11
If $a_1, \ldots, a_n$ are nonzero elements of $\mathcal{R}$, their least common multiple is an element m of $\mathcal{R}$ with these properties:

(i) $a_1 \mid m, a_2 \mid m, \ldots, a_n \mid m$;

(ii) *if e is an element of $\mathcal{R}$ such that $a_1 \mid e, a_2 \mid e, \ldots, a_n \mid e$, then $m \mid e$;*

(iii) *m is positive (when $\mathcal{R} = \mathbf{Z}$) or monic (when $\mathcal{R} = \mathcal{F}[\lambda]$).*

Property (i) asserts that m is a *common* multiple of $a_1, \ldots, a_n$, property (ii) that m is *least* among the common multiples. It requires a theorem to establish that m exists and is unique. As with greatest common divisors, (iii) guarantees the uniqueness of m.

THEOREM 3.4.12

The least common multiple of nonzero elements $a_1, \ldots, a_n$ of $\mathcal{R}$ exists and is unique.

NOTATION We denote the unique quantity m satisfying (i), (ii), (iii) by $\langle a_1, a_2, \ldots, a_n\rangle$.

PROOF We give the uniqueness part of the proof first. Let (i)′, (ii)′, (iii)′ be the statements obtained by everywhere substituting m' for m in (i), (ii), (iii). We prove that if m satisfies (i), (ii), (iii) and m' satisfies (i)′, (ii)′, (iii)′ then $m = m'$. By (i) and (ii)′ we see that $m' \mid m$. By (i)′ and (ii) we see that $m \mid m'$. Thus $m \mid m'$ and $m' \mid m$ and so, by (iii) and (iii)′ we deduce (as in the last steps of the proof of Theorem 3.4.6) that $m = m'$.

We establish the existence of m by induction on n, starting with $n = 2$. If m satisfies (i) and (ii) but not necessarily (iii), then um will also satisfy (i) and (ii), for any unit u, and we may choose u so that um is positive or monic. Thus, for $n = 2$, it is sufficient to find an element m in $\mathcal{R}$ such that (i), $a_1 \mid m$, $a_2 \mid m$ and (ii), $m \mid e$ whenever $a_1 \mid e$ and $a_2 \mid e$. Take $m = a_1a_2/(a_1, a_2)$. We have $m = a_1b$ where $b = a_2/(a_1, a_2)$. Since $b \in \mathcal{R}$ [because $(a_1, a_2) \mid a_2$], we see that $a_1 \mid m$; and similarly $a_2 \mid m$. Thus (i) holds. We now prove (ii). Let $d = (a_1, a_2)$. Then $d \mid a_1$, $d \mid a_2$ and also $(a_1/d, a_2/d) = 1$. Since $a_2 \mid e$ we have $e = ta_2$ for a suitable $t \in \mathcal{R}$. Thus $e/d = t(a_2/d)$. Since $a_1 \mid e$, we have $(a_1/d) \mid (e/d)$. So a_1/d divides $t(a_2/d)$ and a_1/d, a_2/d are relatively prime. It follows from Lemma 3.4.10 that $(a_1/d) \mid t$. Hence $t = \alpha(a_1/d)$ for some $\alpha \in \mathcal{R}$, and thus $e/d = \alpha(a_1/d)(a_2/d)$. Therefore $e = \alpha(a_1a_2/d) = \alpha m$. That is: $m \mid e$.

The existence of the least common multiple of any pair of nonzero elements of $\mathcal{R}$ is now known. We use this in the next part of the proof.

Let $P(n)$ be the statement that $\langle a_1, \ldots, a_n\rangle$ exists. Since $P(2)$ is true, the first step in a proof by induction is completed, and we now give the second step. Let k be an integer with $k - 1 \geq 2$, and suppose (as induction hypothesis) that $\langle a_1, \ldots, a_{k-1}\rangle$ exists. We will demonstrate that the quantity μ defined by $\mu = \langle\langle a_1, \ldots, a_{k-1}\rangle, a_k\rangle$ satisfies the three properties required of a least common multiple of $a_1, \ldots, a_k$. This, when proved, will imply that $\langle a_1, \ldots, a_k\rangle$ exists (and, in fact, equals μ), completing the proof by induction. To show that μ satisfies the three properties, let $M = \langle a_1, \ldots, a_{k-1}\rangle$. Then $\mu = \langle M, a_k\rangle$. By the first property (applied to M), each of $a_1, \ldots, a_{k-1}$ divides M, and by the first property (applied to $\langle M, a_k\rangle$), both M and a_k divide μ. Therefore μ has the first property required of a least common multiple of $a_1, \ldots, a_k$, namely, it is divisible by all of $a_1, \ldots, a_k$. If e is divisible by all of $a_1, \ldots, a_k$, then it is divisible by all of $a_1, \ldots, a_{k-1}$, so that $M \mid e$ (by the second property applied to M). Thus $M \mid e$ and $a_k \mid e$, yielding $\langle M, a_k\rangle \mid e$, by applying the second property to $\langle M, a_k\rangle$. Consequently, μ has the second property required of a least common multiple of $a_1, \ldots, a_k$. Since it plainly is positive or monic, it has the third property. □

The above proof gives the method to be used in numerical examples for calculating the least common multiple of two quantities, and also shows how to calculate the least common multiple of more than two quantities by repeatedly applying the procedure for two quantities. Specifically,

$$\langle a, b\rangle = uab/(a, b), \quad \text{for a suitable unit } u,$$

and $\langle a_1, \ldots, a_k\rangle$ is obtained from $\langle a_1, \ldots, a_{k-1}\rangle$ and a_k by

$$\langle a_1, \ldots, a_k\rangle = \langle\langle a_1, \ldots, a_{k-1}\rangle, a_k\rangle.$$

Forward methods for calculating y_1, y_2 in $y_1a + y_2b = (a, b)$ (Optional)

We showed above not only how to compute the greatest common divisor (a, b) of elements a, b of $\mathcal{R}$, but also how to calculate elements y_1, y_2 of $\mathcal{R}$ satisfying $(a, b) = y_1a + y_2b$. The procedure was, first, to compute (a, b) by setting $r_0 = a$, $r_1 = b$, then calculating further quantities r_2, r_3, $\ldots$, r_m by successive divisions, continuing until a zero remainder r_{m+1} is reached. The last nonzero remainder, r_m, is then (a, b), apart from a unit factor. The second part of the procedure is to use equations (7) produced by these divisions in backwards order, that is, beginning with the next to last, to calculate y_1, y_2. Because the equations are used in reverse order, it is necessary that they be recorded as they are calculated. This causes no difficulty in the simple examples likely to be encountered in text books. However, modern applications of algebra (for example, in coding theory—see the Berlekamp book cited on page 455) require that these calculations be made on electronic computing equipment for polynomials of very high degree, and then the necessity for recording the numerous coefficients of the many polynomials $q_1, \ldots, q_{m-1}, r_0, \ldots, r_m$ appearing in the equations becomes an unwelcome complication. However, there is an alternative "forward" procedure avoiding this difficulty, which we now describe. This procedure is valid both for integers and polynomials.

The procedure is to calculate, in recursive fashion, three quantities: r_j, N_j, D_j, for $j = 0, 1, 2, \ldots$. The r_j are the same as before. We begin by setting

$$\begin{aligned} r_0 &= a, & r_1 &= b, \\ N_0 &= 0, & N_1 &= 1, \\ D_0 &= 1, & D_1 &= 0. \end{aligned}$$

If, for some $j \geq 0$, we have calculated

$$r_j, \quad r_{j+1}, \quad N_j, \quad N_{j+1}, \quad D_j, \quad D_{j+1},$$

then we obtain r_{j+2}, N_{j+2}, D_{j+2} as follows:

$$\begin{aligned} r_j &= q_{j+1}r_{j+1} + r_{j+2}\,, \\ N_{j+2} &= q_{j+1}N_{j+1} + N_j, \\ D_{j+2} &= q_{j+1}D_{j+1} + D_j. \end{aligned} \tag{8}$$

That is, r_{j+2} is obtained from r_j and r_{j+1} by a division, exactly as before, then the quotient q_{j+1} from the division and N_j, N_{j+1}, D_j, D_{j+1} are used to calculate N_{j+2}, D_{j+2}. At each stage of this process, we record only the last two r's, the last two N's, and the last two D's.

Observe that

$$\begin{aligned} N_{j+1}r_{j+2} + N_{j+2}r_{j+1} &= N_jr_{j+1} + N_{j+1}r_j, \\ D_{j+1}r_{j+2} + D_{j+2}r_{j+1} &= D_jr_{j+1} + D_{j+1}r_j, \\ N_{j+2}D_{j+1} - D_{j+2}N_{j+1} &= -(N_{j+1}D_j - D_{j+1}N_j). \end{aligned}$$

To verify these formulas, simply substitute $r_{j+2} = r_j - q_{j+1}r_{j+1}$, $N_{j+2} = q_{j+1}N_{j+1} + N_j$, $D_{j+2} = q_{j+1}D_{j+1} + D_j$ into the left-hand sides. Repeated use of these formulas, for steadily decreasing j, leads to

$$\begin{aligned} N_{j+1}r_{j+2} + N_{j+2}r_{j+1} &= N_0r_1 + N_1r_0 = a, \\ D_{j+1}r_{j+2} + D_{j+2}r_{j+1} &= D_0r_1 + D_1r_0 = b, \\ N_{j+2}D_{j+1} - D_{j+2}N_{j+1} &= (-1)^{j+1}(N_1D_0 - D_1N_0) = (-1)^{j+1}. \end{aligned} \tag{9}$$

Setting $j = m - 1$ in (9), and using $r_{m+1} = 0$, we obtain

$$\begin{aligned} N_{m+1}r_m = a, \qquad D_{m+1}r_m = b, \\ N_{m+1}D_m - D_{m+1}N_m = (-1)^m. \end{aligned} \tag{10}$$

Multiplying the last equation in (10) by r_m, then using the other two equations (10), we get

$$\begin{aligned} (r_mN_{m+1})D_m - (r_mD_{m+1})N_m &= (-1)^mD_m, \\ aD_m - bN_m &= (-1)^mr_m, \end{aligned}$$

or

$$r_m = ax_1 - bx_2 \quad \text{with} \quad x_1 = (-1)^mD_m, \quad x_2 = (-1)^mN_m.$$

Multiplying by a unit factor to convert r_m to (a, b), we then obtain values y_1, y_2 satisfying $(a, b) = y_1a + y_2b$.

For hand computing, the process may be arranged in a tabular form, as in the following example, in which the greatest common divisor of $a = 177$, $b = 78$ is calculated using least nonnegative remainders.

q		2	3	1	2	2	
r	177	78	21	15	6	3	0
N	0	1	2	7	9	25	59
D	1	0	1	3	4	11	26

q		2	3	1	2	2	
r	177	78	21	15	6	3	0
N	0	1	2	7	9	25	59
D	1	0	1	3	4	11	26

The row labelled r begins with $r_0 = a$, $r_1 = b$, the N row with $N_0 = 0$, $N_1 = 1$, and the D row with $D_0 = 1$, $D_1 = 0$. Each new entry in the r row is obtained by division of the two most recent entries, the quotient being placed in the q row above the most recent entry, and the remainder being the new entry. This quotient times the entry in the N (or D) row under it, plus the previous entry in the N (or D) row, gives the new entry in the N (or D) row. The last nonzero entry in the r row is $r_m = (a, b)$, and beneath it are integers N_m, D_m satisfying $aD_m - bN_m = (-1)^m(a, b)$. The number of the column in which r_m appears is m, if the initial column is called column 0; e.g., $r_m = 3$ appears in column 5 above, so $m = 5$. As a computational check, one more entry in the N and D rows may be computed; these entries times r_m must equal a, b respectively, by (10).

Numerically least remainders may be used in this computational arrangement, and it may also be used with polynomials.

A closely related discussion will be found in Chapter 8, in the section on continued fractions, and the interested reader may look ahead to it.

EXERCISES 3.4

I

1. Find $(a_1, \ldots, a_n)$ and (optional: $\langle a_1, \ldots, a_n \rangle$) when:

   (a) $n = 2$, $a_1 = 154$, $a_2 = 112$.
   (b) $n = 3$, $a_1 = 154$, $a_2 = 112$, $a_3 = 175$.
   (c) $n = 4$, $a_1 = 154$, $a_2 = 112$, $a_3 = 175$, $a_4 = 363$.

   In each case find integers $x_1, \ldots, x_n$ such that $(a_1, \ldots, a_n) = x_1a_1 + \cdots + x_na_n$.

2. Repeat Exercise 1 with $a_1 = 100$, $a_2 = 370$, $a_3 = 265$, $a_4 = 77$.

3. Repeat parts (a) and (b) of Exercise 1, this time with polynomials, taking $a_1 = \lambda^2 - 1$, $a_2 = \lambda^3 - \lambda^2 + \lambda - 1$, $a_3 = \lambda^3 + \lambda^2 + \lambda + 1$.

4. Show, by a numerical example, that $c \mid ab$ does not imply $c \mid a$ or $c \mid b$.

5. Find all integers x and y such that $3x + 4y = 1$. Do the same for the equations $5x + 6y = 3$ and $9x + 12y = 7$.

6. Find the greatest common divisor (and *optional*: the least common multiple) of the following pairs of integers.

   (a) $a = 217, \quad b = 42;$
   (b) $a = 72, \quad b = 88;$
   (c) $a = 91, \quad b = 143;$
   (d) $a = 165, \quad b = 102.$

II

7. Show: if $a \mid b$ and $b \mid c$, then $a \mid c$. Produce an example to demonstrate that $a \mid b$ and $a \mid c$ do not imply either $b \mid c$ or $c \mid b$.

8. What conclusion can you draw if $a \mid b$, $b \mid c$, and $c \mid a$?

9. If $e \mid d$ with e and d nonzero show that $|d| \geq |e|$ (when $\mathcal{R}$ is $\mathbf{Z}$) or $\deg d \geq \deg e$ (when $\mathcal{R}$ is $\mathfrak{F}[\lambda]$).

10. (Optional) Let a, b be nonzero elements of $\mathbf{Z}$. Show that $(a, b)\langle a, b\rangle = |ab|$. What is the corresponding fact for polynomials?

11. Let $a(\lambda), b(\lambda) \in \mathbf{Q}[\lambda]$. Then $a, b \in \mathbf{R}[\lambda]$ and $\mathbf{C}[\lambda]$. Show that the greatest common divisor of a, b within $\mathbf{Q}[\lambda]$ equals that within $\mathbf{R}[\lambda]$ and that within $\mathbf{C}[\lambda]$. (Optional: Do the same for the least common multiple.)

12. Let a_1 and a_2 be nonzero polynomials in $\mathbf{C}[\lambda]$. Suppose both a_1 and a_2 have rational numbers as coefficients. Prove that (a_1, a_2) and (optional) $\langle a_1, a_2\rangle$ also have rational numbers as coefficients.

13. Let a, b, c be nonzero elements of $\mathbf{Z}$. Show that $(ca, cb) = |c|\,(a, b)$. [*Hint:* Show that each side of this equation divides the other, or multiply every equation in (7) by c.] Formulate and prove the corresponding fact for polynomials.

14. Let a and b be in $\mathbf{Z}$. What are the possibilities for $(a + b, a - b)$?

15. Let $a(\lambda), b(\lambda)$ be monic polynomials with coefficients in $\mathbf{Z}$. It will follow from §7.6 that $d(\lambda) = (a(\lambda), b(\lambda))$ will also be a monic polynomial with integral coefficients. Thus, for each integer α, $a(\alpha)$, $b(\alpha)$, $d(\alpha)$ are integers. Is it true, for every integer α, that $d(\alpha) = (a(\alpha), b(\alpha))$?

16. Let a and b be nonzero integers, and let $d = (a, b)$. Show that both d and $-d$ will satisfy (i) and (ii) of Definition 3.4.5.

III

17. If $a \mid c$, $b \mid c$, and a, b are relatively prime, show that $ab \mid c$. (*Hint:* Use Lemma 3.4.10.)

18. If a, b, c are nonzero elements of $\mathcal{R}$, with $(a, b) = 1$, show that $(ab, c) = (a, c)(b, c)$. (*Hint:* Show that each side of this equation

divides the other. Let $d_1 = (a, c)$, $d_2 = (b, c)$, $d = (ab, c)$. Use Exercise 17 to show that $d_1d_2 \mid ab$ and $d_1d_2 \mid c$, hence that $d_1d_2 \mid d$. Write $d_1 = xa + yc$, $d_2 = x'b + y'c$, and multiply, to show $d \mid d_1d_2$.) Show also that $\langle ab, c\rangle = \langle a, c\rangle\langle b, c\rangle/c$ when c is positive or monic.

19. Let $a(\lambda) = \lambda^3 + 1$, $b(\lambda) = \lambda^3 - 1$. Show that $(a(\lambda), b(\lambda)) = 1$, and that for an integer t, $(t^3 + 1, t^3 - 1)$ is 1 or 2. When is it 1 and when 2?

20. Let a and b be nonzero integers. Suppose the lowest three digits in both a and b are zero. Show that this is also true of (a, b) and $\langle a, b\rangle$.

21. Let $a(\lambda) = (\lambda - \alpha)(\lambda - \beta)$ be a polynomial in $\mathbf{C}[\lambda]$, where $\alpha, \beta \in \mathbf{C}$. What is (a, a') [where a' denotes the derivative of a]?

22. In equations (7) we did not require that all r_j are nonzero.

    (i) Prove that two consecutive r_j are never both zero and that $r_m \neq 0$.
    (ii) Prove that $(r_0, r_1) = (r_1, r_2) = (r_2, r_3) = \cdots = (r_{m-1}, r_m)$.

23. Prove that the greatest common divisor of the polynomials $\lambda^m - 1$, $\lambda^n - 1$ is $\lambda^d - 1$, where d is the greatest common divisor of the integers m, n.

3.5 THE FUNDAMENTAL THEOREM OF ARITHMETIC

A fact often recognized quite early in life is that an integer may be factored into a product of primes. In this section we shall study this theorem for both $\mathbf{Z}$ and $\mathfrak{F}[\lambda]$. We continue to let $\mathcal{R}$ denote either of these rings.

Recall (Definition 3.4.3) that a *unit* is a divisor of 1, so that an element u is a unit in $\mathcal{R}$ if and only if it has a multiplicative inverse u^{-1} in $\mathcal{R}$. The units in $\mathbf{Z}$ are ± 1 and the units in $\mathfrak{F}[\lambda]$ are the nonzero constants. Notice that products and inverses of units are units.

Plainly, an element u of $\mathbf{Z}$ or $\mathfrak{F}[\lambda]$ is a unit if and only if

$$|u| = 1 \quad (\text{when } \mathcal{R} = \mathbf{Z}), \qquad \deg u = 0 \quad (\text{when } \mathcal{R} = \mathfrak{F}[\lambda]).$$

A nonzero element a in $\mathcal{R}$ which is *not* a unit satisfies

$$|a| > 1 \quad (\text{when } \mathcal{R} = \mathbf{Z}), \qquad \deg a > 0 \quad (\text{when } \mathcal{R} = \mathfrak{F}[\lambda]).$$

DEFINITION 3.5.1
A nonzero element p of $\mathcal{R}$ is a prime if it is not a unit and if a factorization $p = ab$ (where a; b are in $\mathcal{R}$) implies that a or b is a unit.

REMARK 1 If u is a unit, we may always factor p as $p = ub$ where $b = u^{-1}p$. We call this type of factorization a *trivial* factorization.

(A nontrivial factorization would be one in which neither factor is a unit.) Definition 3.5.1 states that only trivial factorizations are possible for primes.

REMARK 2 If, in $p = ab$, both a and b were units, then p, as a product of units, would then be a unit. Thus exactly one of the factors a, b is a unit when p is a prime.

REMARK 3 An element of $\mathcal{R}$ which is not 0, not a prime, and not a unit, is said to be *composite*. A composite element therefore has a factorization in which neither factor is a unit. Evidently the prime elements, the composite elements, 0, and the units are a partition of $\mathcal{R}$ with four components.

Example 3.5.2

In $\mathbf{Z}$, 2, -3, 3, 5, -7 are all primes, whereas 4 and -6 are composite. In $\mathbf{Q}[\lambda]$, $\lambda - 1$, $\lambda + 1$, and $\lambda^2 + 1$ are primes, whereas $\lambda^2 - 1$ is composite. (Can you supply the proof that $\lambda^2 + 1$ is a prime in $\mathbf{Q}[\lambda]$?) In $\mathbf{C}[\lambda]$, $\lambda + 1$ is a prime but $\lambda^2 + 1 = (\lambda + i)(\lambda - i)$ is composite. The integer 2 is a prime in $\mathbf{Z}$ but a unit in $\mathbf{Q}[\lambda]$. An important though simple fact is that *in* $\mathfrak{F}[\lambda]$, *any first degree polynomial is a prime.* For let $\deg a(\lambda) = 1$, and suppose $a = bc$ is a factorization of a into a product of polynomials. Then $1 = \deg a = \deg b + \deg c$; hence either $\deg b = 0$ or $\deg c = 0$, so that either b or c is a nonzero constant, i.e., a unit. Thus a is a prime since it is not a unit.

It is very common to call a prime polynomial an *irreducible polynomial.*

Notice that a product of primes is not a unit. In $\mathbf{Z}$ this is clear since $|p| > 1$ if p is a prime, so that a product $p_1 \cdots p_r$ of primes satisfies $|p_1 \cdots p_r| > 1$. In $\mathfrak{F}[\lambda]$, $\deg p > 0$ if p is a prime, so that a product $p_1 \cdots p_r$ of primes satisfies $\deg(p_1 \cdots p_r) > 0$.

It is straightforward to see that $q = up$ is a prime if p is a prime and u a unit. We say that $q = up$ is an *associate* of p, or that p and $q = up$ are *associates*. For example, in $\mathbf{Z}$, 3 and -3 are associates, and in $\mathbf{Q}[\lambda]$, $\lambda^2 + 1$ and $2\lambda^2 + 2$ are associates. The following simple properties are easily proved.

(i) A prime p is always an associate of itself.

(ii) If p, q are primes, with p an associate of q, then q is an associate of p.

(iii) If p_1, p_2, p_3 are primes, with p_1 an associate of p_2, and p_2 an associate of p_3, then p_1 is an associate of p_3.

We now come to a basic theorem called the *fundamental theorem of arithmetic* or the *unique factorization theorem*. It presents a familiar fact, which however requires proof. See Example 3.5.5 and Exercise 16 for mathematical situations in which this fact, however familiar it may be, is *not* true. See also Exercise 11 in the review section at the end of the chapter.

THEOREM 3.5.3 Unique Factorization Theorem
A nonzero element a of $\mathcal{R}$ may be factored as

$$a = up_1p_2\cdots p_r, \tag{1}$$

where u is a unit and $p_1, p_2, \ldots, p_r$ are primes in $\mathcal{R}$, with $p_1, \ldots, p_r$ positive (for $\mathcal{R} = \mathbf{Z}$) or monic (for $\mathcal{R} = \mathfrak{F}[\lambda]$). If

$$a = vq_1\cdots q_s \tag{2}$$

is another factorization of a, where v is a unit and $q_1, \ldots, q_s$ are positive or monic primes in $\mathcal{R}$, then $s = r$ and $q_1, \ldots, q_s$ are $p_1, \ldots, p_r$ in some order.

We may have $r = 0$ in (1). This happens if and only if a is a unit. Note that $p_1, \ldots, p_r$ are not claimed to be distinct.

PROOF The theorem contains an existence part and a uniqueness part. Each part requires proof. We give the existence proof first.

We first call attention to a fact needed in the proof. Suppose A is a nonzero element of $\mathcal{R}$, but not a prime. Let $A = A_1A_2$ be a factorization of A in which A_1, A_2 are in $\mathcal{R}$ and neither is a unit. We then have

$$|A_1| > 1, \qquad |A_2| > 1, \quad \text{when } \mathcal{R} = \mathbf{Z},$$
$$\deg A_1 > 0, \qquad \deg A_2 > 0, \quad \text{when } \mathcal{R} = \mathfrak{F}[\lambda],$$

yielding

$$|A| > |A_1|, \qquad |A| > |A_2|, \quad \text{when } \mathcal{R} = \mathbf{Z},$$
$$\deg A > \deg A_1, \qquad \deg A > \deg A_2, \quad \text{when } \mathcal{R} = \mathfrak{F}[\lambda].$$

Thus the factors in a nontrivial factorization of A must have smaller absolute value or degree than A.

Starting with a, factor it, if possible, in a nontrivial way, say as $a = A_1A_2$. Factor A_1 or A_2, if possible, in nontrivial ways, then factor their factors nontrivially, if possible. Continue until we get $a = a_1\cdots a_r$, where none of $a_1, \ldots, a_r$ can be factored further except in trivial ways. This must eventually happen since each nontrivial factorization can only introduce new factors b in which the positive integer $|b|$ or the nonnegative integer $\deg b$ is reduced in size; and by the well-ordering principle these quantities cannot indefinitely grow smaller. Then, because $a_1, \ldots, a_r$ are capable only of trivial factorizations, each must be a prime. Let $u_1, \ldots, u_r$ be units such that $p_1 = u_1a_1, \ldots, p_r = u_ra_r$ are positive or monic. Set $u = u_1^{-1}\cdots u_r^{-1}$. Then $a = up_1\cdots p_r$, with u a unit and $p_1, \ldots p_r$ positive or monic primes.

Before proving uniqueness, we give a lemma needed in the proof.

LEMMA 3.5.4
Let p, a, b be elements of $\mathcal{R}$, with p a prime. If $p \mid ab$ then $p \mid a$, or $p \mid b$, or both.

PROOF If $p \mid a$, there is nothing to prove. So assume $p \nmid a$. Let $d = (a, p)$. Then $d \mid p$, so that d, as a factor of a prime, must either be an associate of p or a unit. If d is an associate of p, then (because $d \mid a$) we would have $p \mid a$. This is not so; hence d is a unit. Because the only positive or monic unit is 1, we see that $d = 1$. Thus $p \mid ab$ and $(p, a) = 1$. By Lemma 3.4.10, it follows that $p \mid b$. □

We now prove uniqueness. From (1) and (2) we get

$$up_1 \cdots p_r = vq_1 \cdots q_s. \tag{3}$$

In this equation p_1 divides the left-hand side; hence it divides the right-hand side, and by Lemma 3.5.4, it must divide at least one of $v, q_1, \ldots, q_s$. It cannot divide v since v is a unit. So p_1 divides at least one of $q_1, \ldots, q_s$. (In particular we must have $s \geq 1$.) Renumbering the q's, if necessary, we may assume $p_1 \mid q_1$. But q_1 is a prime: its only divisors are units and associates of itself. Since p_1 is not a unit, it is an associate of q_1, say $p_1 = wq_1$ where w is a unit. Because p_1 and q_1 are both positive or monic, the unit $w = 1$. Hence $p_1 = q_1$. Cancelling p_1 from (3), we get

$$up_2 \cdots p_r = vq_2 \cdots q_s. \tag{4}$$

Repeating this argument, but using (4) and p_2 (instead of p_1), we see that p_2 equals one of $q_2, \ldots, q_s$ (in particular, we must have $s \geq 2$). Changing the numbering of the q's, we may assume $p_2 = q_2$. Cancelling p_2 gives

$$up_3 \cdots p_r = vq_3 \cdots q_s.$$

We then repeat the argument, applying it to p_3. We may plainly continue until we have proved (after renumbering the q's) that $p_i = q_i$ for $i = 1, \ldots, r$. In particular $s \geq r$. If $s > r$, we would have

$$u = vq_{r+1} \cdots q_s.$$

But this implies that $q_{r+1} \cdots q_s$ equals the unit uv^{-1}, a contradiction since a product of primes cannot be a unit. Thus $r = s$, hence $u = v$, and the proof is complete. □

Example 3.5.5
Let S consist of all integers s of the form $s = 4t + 1$ where t is a variable positive integer. Endow S with the operation of multiplication. Then S is closed under multiplication. To see this, let $s_1 = 4t_1 + 1$, $s_2 = 4t_2 + 1$

where t_1, t_2 are positive integers. Then $s_1s_2 = 4T + 1$ where $T = 4t_1t_2 + t_1 + t_2 > 0$. Hence $s_1s_2 \in S$. We define an element s of S to be a *prime for* S if it is impossible to factor s as $s = s_1s_2$ where s_1 and s_2 are in S. Thus 5, 9, 13, 17, 21 are primes in S, but 25 is not. Imitating the existence proof for Theorem 3.5.3, we find that any element s of S may be factored as a product of primes for S. This representation is not unique. For example, $441 = 9 \cdot 49 = 21 \cdot 21$ and 9, 49, 21 are all primes for S. Thus the unique factorization theorem does not hold in S.

See Exercise 16 for a *ring* within which unique factorization fails.

REMARK When the nonzero element a of $\mathfrak{R}$ has been factored into a product of primes and a unit, say as $a = u\pi_1 \cdots \pi_r$, where u is a unit and $\pi_1, \ldots, \pi_r$ are positive or monic primes, the primes $\pi_1, \ldots, \pi_r$ need not be distinct. Let $p_1, \ldots, p_k$ be the distinct primes among $\pi_1, \ldots, \pi_r$, and suppose p_1 occurs e_1 times, p_2 occurs e_2 times, . . ., p_k occurs e_k times. Then *a is uniquely representable, apart from the order of the factors, as*

$$a = up_1^{e_1}p_2^{e_2} \cdots p_k^{e_k} \tag{5}$$

where $p_1, \ldots, p_k$ are distinct positive or monic primes, and $e_1, \ldots, e_k$ are positive integers. We shall frequently use this form of the fundamental theorem of arithmetic. An element of $\mathfrak{R}$ of the form p^e is said to be *primary* if p is a prime and e a positive integer. Thus (5) displays a as a product of a unit u and primary components $p_1^{e_1}, \ldots, p_k^{e_k}$ belonging to distinct positive or monic primes $p_1, \ldots, p_k$. The exponents $e_1, \ldots, e_k$ are called the *multiplicities* of the prime factors $p_1, \ldots, p_k$, respectively.

Notice that the uniqueness part of the unique factorization theorem is applicable only to primary factors. Distinct factorizations are possible when nonprimary factors are allowed, as is shown by $36 = 4 \cdot 9 = 6 \cdot 6$.

Example 3.5.6

The unique factorization theorem may be used to give rapid proofs of the irrationality of many numbers. For example, to prove that $\sqrt{3}$ is irrational, begin with $\sqrt{3} = m/n$, m, n being integral. Squaring and rearranging gives $3n^2 = m^2$. Since the prime 3 must be a factor of any square with even multiplicity, its multiplicity in $3n^2$ is odd, but even in m^2. This contradicts the uniqueness part of Theorem 3.5.3. To prove that $\sqrt{2} + \sqrt{3}$ is irrational, assume that it is rational, then square, to deduce that $\sqrt{6}$ is rational, say $\sqrt{6} = m/n$, with integral m, n. Squaring again yields $6n^2 = m^2$, a contradiction because 2 appears with odd multiplicity on one side of this equation and even multiplicity on the other.

EXERCISES 3.5

I

1. Factor each of the polynomials in Exercise 3 of §3.4 into a product of monic primes in $\mathbf{Q}[\lambda]$. Do the same in $\mathbf{R}[\lambda]$ and $\mathbf{C}[\lambda]$.

2. Prove that 101 is a prime in $\mathbf{Z}$. Is 1001 a prime in $\mathbf{Z}$? If not, factor it. What about 10001? (See Exercises 3 and 7 below.)

3. Make a table of all positive primes ≤ 100 in $\mathbf{Z}$ by this method: List on a piece of paper the integers 2, 3, 4, 5, . . ., 100. The first integer in this list, 2, is a prime. Put a star $*$ beside 2, then cross it out and all multiples of it (i.e., every second number). The first number not crossed out must be a prime (prove this!), put a $*$ beside it, then cross out it and all its multiples (i.e., every third number). The first number not crossed out must be a prime. Put a star beside it, then cross it out and all its multiples. Continue in this way until all numbers are crossed out. The starred numbers are then the primes ≤ 100. (This method is called the *sieve of Eratosthenes*.)

4. Is 441 the smallest integer in the set S of Example 3.5.5 that can be written as a product of primes for S in two different ways?

5. Change the set S of Example 3.5.5 by considering instead all integers of the form $5t + 1$ for integral $t > 1$. For this new set S find an element of S that can be written as a product of primes for S in two different ways. See Exercise 19, §1.3.

II

6. Show that there are an infinite number of primes in $\mathcal{R}$. (Hint: Suppose $p_1, \ldots, p_r$ are all the primes in $\mathcal{R}$. Consider $m = 1 + p_1 \cdots p_r$. Show that none of $p_1, \ldots, p_r$ can be a factor of m. By Theorem 3.5.3, there is at least one prime, call it p, that is a factor of m. Prove that p is not any of $p_1, \ldots, p_r$.)

7. Let n be a positive integer, not a prime or a unit. Show that n has at least one positive prime factor p with $p \leq \sqrt{n}$. (Hint: What would happen if every prime factor p of n satisfied $p > \sqrt{n}$?)

8. Show, using Exercise 7, if we reach a point in Exercise 3 where we star and cross out all multiples of a number $\geq \sqrt{100}$, every remaining not crossed out number on the list is a prime.

9. If $\alpha_1, \alpha_2 \in \mathfrak{F}$ and $a(\lambda) = \lambda^2 + \alpha_1\lambda + \alpha_2$ is not a prime in $\mathfrak{F}[\lambda]$, there is an element γ in $\mathfrak{F}$ such that $a(\gamma) = 0$. Prove this.

10. Repeat Exercise 9 for the cubic polynomial $a(\lambda) = \lambda^3 + \alpha_1\lambda^2 + \alpha_2\lambda + \alpha_3$, where $\alpha_1, \alpha_2, \alpha_3 \in \mathfrak{F}$. Show that this conclusion does not hold for fourth degree polynomials $a(\lambda)$.

11. Show that the units in $\mathcal{R}$ form a group under multiplication.

III

12. Let p_1, p_2, p_3, p_4 be four distinct positive prime integers, each of the form $4t - 1$. Prove that $p_1p_2p_3p_4$ belongs to the set S of Example 3.5.5, and moreover, that it can be written as a product of primes for S in two distinct ways.

13. If x, y are real numbers, let min (x, y) denote the smaller of the numbers x, y, and max (x, y) the larger. Let a and b be nonzero elements of $\mathcal{R}$, and suppose

$$a = up_1^{e_1}p_2^{e_2}\cdots p_r^{e_r}$$

$$b = vp_1^{f_1}p_2^{f_2}\cdots p_r^{f_r}$$

are factorizations of a and b into products of distinct positive or monic primes $p_1, \ldots, p_r$. Here $e_1, \ldots, e_r, f_1, \ldots, f_r$ are nonnegative integers and u, v are units. (For example, if $p_1 \mid b$ but $p_1 \nmid a$, we insert p_1 into the factorization of a by taking exponent $e_1 = 0$.) Prove that

$$(a, b) = p_1^{\min(e_1,f_1)}p_2^{\min(e_2,f_2)}\cdots p_r^{\min(e_r,f_r)},$$

$$\langle a, b\rangle = p_1^{\max(e_1,f_1)}p_2^{\max(e_2,f_2)}\cdots p_r^{\max(e_r,f_r)}.$$

14. Rework Exercise 18 of §3.4, using Exercise 13 above.

15. Using Exercise 13 above, show that $(a, b)\langle a, b\rangle = |ab|$ when a, b are integers.

16. Let $\mathcal{R}$ consist of all complex numbers of the form $x + y\sqrt{5}i$, for integral x, y. Endow $\mathcal{R}$ with the usual operations of complex number addition and multiplication.

    (a) Prove that $\mathcal{R}$ is a ring. Define "divides," "unit," "prime" for this ring $\mathcal{R}$ exactly as in Definitions 3.4.1, 3.4.3, 3.5.1.
    (b) Prove that ± 1 are the only units in this ring $\mathcal{R}$.
    (c) Prove that any nonzero element of $\mathcal{R}$ is a product of primes. [Hint: Imitate the existence proof in the unique factorization theorem 3.5.3, using the integer $|a|^2$ in place of $|a|$ (which, for $a \in \mathcal{R}$, is not usually an integer).]
    (d) Show that 3, 7, $4 + \sqrt{5}i$, $4 - \sqrt{5}i$ are all primes in $\mathcal{R}$. By observing that $21 = 3\cdot 7 = (4 + \sqrt{5}i)(4 - \sqrt{5}i)$, show that unique factorization fails for $\mathcal{R}$.

3.6 CONGRUENCES

Congruences form one of the most useful tools in algebraic investigations. We shall use them in Chapters 4 and 7. In this section they will be defined and their properties developed.

Throughout this section we continue to let $\mathcal{R}$ be either **Z** or $\mathcal{F}[\lambda]$.

DEFINITION 3.6.1
Let m be an element of $\mathcal{R}$. Elements a, b in $\mathcal{R}$ are said to be congruent modulo m *if $m \mid (a - b)$. This relation is written as*

$$a \equiv b \pmod{m}.$$

We read this formula as "a is congruent to *b* mod *m." If a is not congruent to $b \pmod{m}$, we write $a \not\equiv b \pmod{m}$.*

The most basic properties of congruences are expressed in these three facts: If a, b, c, $m \in \mathcal{R}$, then

(i) $a \equiv a \pmod{m}$;

(ii) *if* $a \equiv b \pmod{m}$, *then* $b \equiv a \pmod{m}$;

(iii) *if* $a \equiv b \pmod{m}$, *and* $b \equiv c \pmod{m}$, *then* $a \equiv c \pmod{m}$.

Proof of (i): $a \equiv a \pmod{m}$ since $a - a = 0$ and $m \mid 0$. □

Proof of (ii): If $a \equiv b \pmod{m}$, then $m \mid (a - b)$. Hence $m \mid -(a - b)$, that is, $m \mid (b - a)$, whence $b \equiv a \pmod{m}$. □

Proof of (iii): From $m \mid (a - b)$ and $m \mid (b - c)$ it follows that $m \mid (a - b + b - c)$, that is, $m \mid (a - c)$. But this implies $a \equiv c \pmod{m}$. □

Some basic algebraic properties of congruences are given next.

(iv) *Let m, a, b, c, d be elements of $\mathcal{R}$ such that*

$$a \equiv b \pmod{m}, \tag{1}$$

$$c \equiv d \pmod{m}. \tag{2}$$

Then

$$a + c \equiv b + d \pmod{m}, \tag{3}$$

$$a - c \equiv b - d \pmod{m}, \tag{4}$$

$$ac \equiv bd \pmod{m}. \tag{5}$$

That is, congruences (mod m) may be added, subtracted or multiplied. For proof, note that (1), (2) give $m \mid (a - b)$ and $m \mid (c - d)$. Hence $m \mid ((a - b) + (c - d))$, that is, $m \mid ((a + c) - (b + d))$, and this proves (3). Also $m \mid ((a - b) - (c - d))$, that is, $m \mid ((a - c) - (b - d))$, and this proves (4). Furthermore, $m \mid (c(a - b) + b(c - d))$, hence $m \mid (ac - bd)$, and this proves (5). □

Another fact:

(v) *If* $a \equiv b \pmod{m}$ *and if* $n \mid m$, *then* $a \equiv b \pmod{n}$.

For if $m \mid (a - b)$ and $n \mid m$, then surely $n \mid (a - b)$. □

Taking $d = c$ in (iv), we get from (5): $ca \equiv cb \pmod m$ follows from $a \equiv b \pmod m$. Thus a congruence may be multiplied on both sides by an element c of $\mathcal{R}$. A stronger statement can be made however:

(vi) *From $a \equiv b \pmod m$ follows $ca \equiv cb \pmod{cm}$.*

For if $m \mid (a - b)$ then $cm \mid c(a - b)$. □

As for cancelling a common factor from a congruence, we have:

(vii) *Let $m \neq 0$. If $ca \equiv cb \pmod m$, then $a \equiv b \pmod{m/(m, c)}$.*

PROOF From $ca \equiv cb \pmod m$, we get $m \mid c(a - b)$, and hence

$$\frac{m}{(m, c)} \,\Big|\, \frac{c}{(m, c)}\,(a - b). \tag{6}$$

From remarks above Lemma 3.4.10, we obtain

$$\left(\frac{m}{(m, c)}, \frac{c}{(m, c)}\right) = 1. \tag{7}$$

By Lemma 3.4.10, (6) and (7) imply that

$$\frac{m}{(m, c)} \,\Big|\, (a - b).$$

This gives $a \equiv b \pmod{m/(m, c)}$, as desired. □

Examples

(1) Let $\mathcal{R}$ be $\mathbf{Z}$. Show that $3^4 \equiv 1 \pmod 4$.
SOLUTION Since $3 \equiv -1 \pmod 4$, $3^4 \equiv (-1)^4 \pmod 4 \equiv 1 \pmod 4$.

(2) Let $\mathcal{R}$ be $\mathbf{Q}[\lambda]$. Show that $\lambda^4 \equiv 1 \pmod{\lambda^2 + 1}$.
SOLUTION $(\lambda^2 + 1) \mid (\lambda^4 - 1)$ since $\lambda^4 - 1 = (\lambda^2 - 1)(\lambda^2 + 1)$.

The following often useful fact determines when a congruence involving an unknown x has a solution.

(viii) *Let $a, b, m \in \mathcal{R}$. An element x in $\mathcal{R}$ exists such that*

$$ax \equiv b \pmod m \tag{8}$$

if and only if

$$(a, m) \mid b. \tag{9}$$

PROOF First suppose (8) has a solution α. Then $a\alpha \equiv b \pmod m$, hence $m \mid (a\alpha - b)$, and hence $a\alpha - b = tm$ for some $t \in \mathcal{R}$. That is, $b = a\alpha - tm$. Since $(a, m) \mid a$, and $(a, m) \mid m$, we get $(a, m) \mid (a\alpha - tm)$ and therefore $(a, m) \mid b$. Thus (9) is satisfied if (8) has a solution.

Now suppose (9) is satisfied. We know that $(a, m) = ay_1 + my_2$ for suitable $y_1, y_2 \in \mathcal{R}$. Since $(a, m) \mid b$, we have $b = s(a, m)$ for some $s \in \mathcal{R}$. Then $b = s(a, m) = a(sy_1) + m(sy_2)$; hence $a(sy_1) - b$ is a multiple of m. Letting $x = sy_1$, we have $m \mid (xa - b)$ and hence (8) is satisfied. □

In Chapter 7, when we study *finite fields*, we shall see that solving an equation $ax = b$ for x in one of these fields amounts to solving a congruence of the form $ax \equiv b \pmod{m}$ for x, where a, b, x, m are integers or polynomials. The second part of the proof just given provides a method for efficiently finding a solution: First, compute elements y_1, y_2 in $\mathcal{R}$ satisfying $ay_1 + my_2 = (a, m)$, then multiply y_1 by $s = b/(a, m)$ to obtain $x = y_1 s$. The calculation of y_1, y_2 (only y_1 is actually needed) is most efficiently carried through using the forward procedure described at the end of §3.4. An example in §7.3, pertaining to the theory of codes, will illustrate one way in which solutions of these equations and congruences are used in applications of algebra.

Congruences for real numbers

The concepts of divisibility and congruence can be extended to real numbers. If a, m are real numbers, i.e., not necessarily integers, we say that m divides a if $a = tm$ for an *integer* t. We write this as before: $m \mid a$. If b is also a real number, we then define congruence modulo m exactly as above:

$$a \equiv b \pmod{m} \quad \textit{if and only if} \quad m \mid (a - b).$$

The reader may check that properties (i)–(vi) above hold for this extended notion of congruence, where a, b, c, d, m are real numbers. The proofs are the same. However, (vii), (viii) are meaningful as stated only for integers (and polynomials).

This extended notion of congruence is useful when considering the various angles possible for a nonzero complex number z. If θ is an angle for z, another real number θ' can serve as angle z only when it differs from θ by a full revolution; i.e.,

$$\theta' = \theta + 2\pi t \quad \text{where } t \in \mathbf{Z}.$$

But this is equivalent to

$$\theta' \equiv \theta \pmod{2\pi}. \tag{10}$$

THEOREM 3.6.2

Let z, z_1, z_2 be nonzero complex numbers. If θ is an angle for z, then θ' is also an angle for z if and only if θ, θ' satisfy (10). *If* "angle z" *denotes an arbitrary angle for z, then* angle z_1z_2, angle z_1/z_2 *are related to* angle z_1, angle z_2 *by*

$$\begin{aligned} \text{angle } z_1z_2 &\equiv \text{angle } z_1 + \text{angle } z_2 \pmod{2\pi}, \\ \text{angle } z_1/z_2 &\equiv \text{angle } z_1 - \text{angle } z_2 \pmod{2\pi}. \end{aligned} \tag{11}$$

PROOF Equation (5) of §2.4 shows that angle z_1 + angle z_2 is an acceptable value for angle z_1z_2. If we replace this value by any other acceptable value, we must by (10) replace angle z_1 + angle z_2 by a number congruent to it (mod 2π). This proves the first congruence (11), and the second is proved similarly, using equation (6) of §2.4. □

The Chinese remainder theorem (Optional)

The next result is called the *Chinese remainder theorem* because it was known to the ancient Chinese. In recent years it has become important in applied mathematics and in particular it has been employed to develop numerical procedures for solving systems of equations using electronic computing machines. (See page 416 of §8.4 and the article by R. T. Gregory cited on page 455.)

THEOREM 3.6.3
Let $a_1, a_2, \ldots, a_n$ be elements of $\mathcal{R}$, and suppose $m_1, m_2, \ldots, m_n$ are nonzero pairwise relatively prime elements of $\mathcal{R}$, i.e.,

$$(m_i, m_j) = 1 \qquad \text{whenever } i \neq j, \text{ for } 1 \leq i, j \leq n. \tag{12}$$

Then the system of simultaneous congruences in unknown x,

$$x \equiv a_i \pmod{m_i}, \qquad i = 1, 2, \ldots, n, \tag{13}$$

has a solution x in $\mathcal{R}$. The solution is uniquely determined modulo M, where $M = m_1m_2\cdots m_n$.

PROOF We first show that an element x exists satisfying all congruences (13). For each fixed i with $1 \leq i \leq n$, let

$$M_i = m_1\cdots m_{i-1}m_{i+1}\cdots m_n = M/m_i.$$

We claim that $(m_i, M_i) = 1$. If this were not so, there would be a prime, call it p, such that $p \mid m_i$ and $p \mid M_i$. By Lemma 3.5.4, the prime p, as a divisor of M_i, would have to be a divisor of at least one of $m_1, \ldots, m_{i-1}, m_{i+1}, \ldots, m_n$, say $p \mid m_j$ for some $j \neq i$. But this contradicts the hypothesis that $(m_i, m_j) = 1$. Therefore $(m_i, M_i) = 1$; hence [by (viii)] there is an element y_i of $\mathcal{R}$ such that

$$y_iM_i \equiv 1 \pmod{m_i}. \tag{14}$$

Let

$$x = \sum_{j=1}^{n} y_jM_ja_j. \tag{15}$$

For fixed i, each of $M_1, \ldots, M_{i-1}, M_{i+1}, \ldots, M_n$ contains m_i as a factor, hence $M_j \equiv 0 \pmod{m_i}$ if $j \neq i$. This fact, (14) and (15) show that

$$x \equiv y_iM_ia_i \equiv a_i \pmod{m_i}.$$

Thus x, given by (15), is indeed a solution of all congruences (13).

Let X be another solution of all congruences (13). Then $X \equiv a_i \pmod{m_i}$ and $x \equiv a_i \pmod{m_i}$, hence $X \equiv x \pmod{m_i}$, and therefore $m_i \mid (X - x)$. Because no two of $m_1, \ldots, m_n$ have a common factor, we must have $(m_1 \cdots m_n) \mid (X - x)$. Therefore $X \equiv x \pmod{M}$, and thus the solution x is uniquely determined modulo M. Furthermore, if x satisfies each congruence (13) and $X \equiv x \pmod{M}$, then X will indeed satisfy each congruence (13). For we have $X \equiv x \pmod{m_i} \equiv a_i \pmod{m_i}$. □

In Chapter 6 we shall encounter a very general concept of relation on a set called an *equivalence relation*. It will turn out that congruence is an example of an equivalence relation.

EXERCISES 3.6

I

1. Let $\mathcal{R} = \mathbf{Z}$. Which of the following congruences are valid?

(i) $2 \equiv 7 \pmod{5}$;
(ii) $3 \equiv 4 \pmod{-1}$;
(iii) $11^2 \equiv 7 \pmod{3}$;
(iv) $11^5 + 11 \equiv 7 \pmod{3}$;
(v) $9^{1000} \equiv 1 \pmod{5}$;
(vi) $30 \cdot 15 \equiv 0 \pmod{9}$;
(vii) $121^4 \equiv 2 \pmod{7}$;
(viii) $13^{1000000} \equiv -1 \pmod{14}$.

2. Let $\mathcal{R}$ be $\mathbf{Q}[\lambda]$. Which of the following congruences are valid?

(i) $\lambda^2 + 1 \equiv 2\lambda \pmod{(\lambda - 1)}$;
(ii) $\lambda^4 + 1 \equiv 0 \pmod{(\lambda^2 + 1)}$;
(iii) $(\lambda^8 - 2)^{10} \equiv 1 \pmod{(\lambda - 1)}$;
(iv) $\lambda^{999} \equiv -1 \pmod{(\lambda + 1)}$.

3. Construct a numerical example demonstrating that $ca \equiv cb \pmod{m}$ does not imply $a \equiv b \pmod{m}$.

4. Let $\mathcal{R}$ be $\mathbf{Z}$. Which of the following congruences in unknown integer x have a solution? When solutions exist, find one.

(i) $16x \equiv 20 \pmod{36}$;
(ii) $16x \equiv 21 \pmod{36}$;
(iii) $9x + 12 \equiv 6 \pmod{15}$;
(iv) $3x^2 + 2x + 4 \equiv 0 \pmod{5}$;
(v) $3x^2 + 2x + 3 \equiv 0 \pmod{5}$.

5. Let $\mathcal{R}$ be $\mathbf{Q}[\lambda]$. Which of the following congruences in unknown polynomial x have a solution? When solutions exist find one.

(i) $16x \equiv 21 \pmod{36}$;
(ii) $(\lambda^2 + 1)x \equiv 1 \pmod{(\lambda + 1)^2}$;
(iii) $(\lambda^2 - 1)x \equiv 2\lambda + 2 \pmod{(\lambda + 1)^2}$.

II

6. (i) Let $m \neq 0$. Prove: If $ax \equiv 0 \pmod{m}$ then x has the form $x = tm/(a, m)$, where $t \in \mathcal{R}$. Furthermore, any element x of this form satisfies $ax \equiv 0 \pmod{m}$.
   (ii) If x_1 and x_2 satisfy $ax_1 \equiv b \pmod{m}$ and $ax_2 \equiv b \pmod{m}$, show that $a(x_2 - x_1) \equiv 0 \pmod{m}$. Deduce that $x_2 = x_1 + tm/(a, m)$ for $t \in \mathcal{R}$.
   (iii) Let x_1 be a fixed solution of $ax \equiv b \pmod{m}$. Show that every solution of this congruence is given by the formula $x = x_1 + tm/(a, m)$ as t ranges over $\mathcal{R}$.
   (iv) Let $\mathcal{R}$ be $\mathbf{Z}$. Find a formula giving all solutions of the congruence $6x \equiv 21 \pmod{15}$.
   (v) Using your formula in (iv), determine all simultaneous solutions of $6x \equiv 21 \pmod{15}$ and $11x \equiv 7 \pmod{13}$.

7. Show: if $m \neq 0$ and $(a, m) = 1$, then $ax \equiv 1 \pmod{m}$ for a suitable $x \in \mathcal{R}$. Is this conclusion correct without the hypothesis $(a, m) = 1$?

8. Observe that $2 \cdot 3 \equiv 0 \pmod{6}$ even though neither 2 nor 3 is congruent to zero (mod 6). Show that this is typical of any composite integer m: If $m > 1$ is not a prime, positive integers m_1, m_2 exist such that $m_1 m_2 \equiv 0 \pmod{m}$, but $m_1 \not\equiv 0 \pmod{m}$ and $m_2 \not\equiv 0 \pmod{m}$.

9. Show that $s \equiv 1 \pmod{4}$ for each s in the set S of Example 3.5.5.

10. If a is divided by b to get $a = qb + r$, where q is the quotient and r the remainder, show that $a \equiv r \pmod{b}$.

11. Let m be a positive integer, a an integer. Show that (modulo m), a is congruent to exactly one of $0, 1, 2, \ldots, m - 1$.

12. If $a \equiv b \pmod{m}$, show that $(a, m) = (b, m)$. (*Optional:* Is it true that $\langle a, m \rangle = \langle b, m \rangle$?)

13. If $m = 0$, prove that $a \equiv b \pmod{m}$ if and only if $a = b$.

III

14. If a is an odd integer, show that

$$a^{2^n} \equiv 1 \pmod{2^{n+2}}$$

for each positive integer n. (Hint: $a^{2^{n+1}} = (a^{2^n})^2$.)

15. Let m_1, m_2 be relatively prime elements of $\mathcal{R}$. If $a \equiv b \pmod{m_1}$ and $a \equiv b \pmod{m_2}$, show that $a \equiv b \pmod{m_1 m_2}$. (See Exercise 17 of §3.4.)

16. Show that $3^n \equiv 1 \pmod{5}$ if and only if the nonnegative integer n satisfies $n \equiv 0 \pmod{4}$.

17. Let $\mathcal{R} = \mathbf{Z}$. (i) If $p_1 \equiv 3 \pmod 4$ and $p_2 \equiv 3 \pmod 4$, prove that $p_1 p_2 \equiv 1 \pmod 4$.
(ii) Prove that there are infinitely many positive prime integers p with $p \equiv 3 \pmod 4$. [Hint: Suppose $p_1, \ldots, p_r$ are all the positive primes p for which $p \equiv 3 \pmod 4$. Let

$$M = p_1 \cdots p_r + 2 \qquad \text{if } r \text{ is even,}$$

$$M = p_1^2 p_2 p_3 \cdots p_r + 2 \quad \text{if } r \text{ is odd.}$$

Show that $M \equiv 3 \pmod 4$. Then show that $q \equiv 1 \pmod 4$ cannot hold for every prime factor q of M. Show also that none of $p_1, \ldots, p_r$ divides M. Deduce your result from these facts. (Compare with Exercise 6 of §3.5.)]

18. Modify the scheme in Exercise 17 to prove the existence of infinitely many primes p with $p \equiv 5 \pmod 6$.

3.7 REVIEW

A central theme in this book is a discussion of the many common properties possessed by integers and polynomials. This chapter began the discussion by introducing divisibility and its allied notions: the greatest common divisor, the unique factorization theorem, and congruences. Of considerable importance from the point of numerical computations is Euclid's algorithm for calculating a greatest common divisor. Congruences are employed as a working tool, principally in Chapter 7, and the unique factorization theorem (for the ring $\mathfrak{F}[\lambda]$) plays an important role in the development of the theory of roots of polynomials in Chapter 5.

REVIEW EXERCISES 3.7

1. Let $a = -1214$, $b = 1111$. Find (a, b) when a, b are regarded as elements of: (i) $\mathbf{Z}$; (ii) $\mathbf{Q}[\lambda]$; (iii) $\mathbf{C}[\lambda]$. (*Optional:* Find $\langle a, b\rangle$ in each of these cases.)

2. (*Optional:* Let $\mathcal{R} = \mathbf{Z}$. If $\langle a, b\rangle = |ab|$, what is (a, b)? Can you determine when $\langle a, b, c\rangle = |abc|$?)

3. Let $a \in \mathbf{C}[\lambda]$. If $a^5 + a + 1 = 0$, must $a \in \mathbf{C}$?

4. Let $a \in \mathbf{C}[\lambda]$ and $\alpha \in \mathbf{C}$. Find the unique constant c such that $a \equiv c \pmod{(\lambda - \alpha)}$.

5. Determine the smallest possible value of $|m|$ if the integer m satisfies $27 \equiv 7m \pmod{10}$?

6. Let a, b be nonzero elements of $\mathcal{R}$. Show that $\mathcal{R}$ contains an element x such that $ax \equiv (a, b) \pmod{b}$.

7. Let S be the set of all positive integers s for which $s \equiv 1 \pmod{7}$. Show that S is closed under multiplication. To what extent is the unique factorization theorem valid in S?

8. What is the remainder obtained upon dividing $\lambda^5 + 3\lambda^4 + \lambda^2 + \lambda + 2$ by $\lambda - 2$? (Do not do the division). If we divide instead by $2\lambda - 6$, what is the remainder now?

9. Find $D^2((\lambda^5 + 1)(\lambda^2 + 2\lambda + 4))$.

10. If integers a and b are relatively prime to m, is ab also relatively prime to m? If $b \mid a$, is a/b relatively prime to m? Are $a + b$ or $a - b$ relatively prime to m?

11. Consider the set of all complex numbers of the form

$$x + y\sqrt{14}i, \qquad x, y \in \mathbf{Z}.$$

(a) Under the usual addition and multiplication of complex numbers, prove that this set is a ring. Denote it by $\mathcal{R}$.
(b) Define "divides," "unit," "prime" for $\mathcal{R}$ exactly as in Definitions 3.4.1, 3.4.3, 3.5.1. Show that ± 1 are the only units in $\mathcal{R}$.
(c) By imitating the proof in the unique factorization theorem 3.5.3, but using $|a|^2$ in place of $|a|$, show that each nonzero element of $\mathcal{R}$ is factorable as a product of primes and a unit.
(d) Show that 2, 3, $2 + \sqrt{14}i$, $2 - \sqrt{14}i$ are primes in $\mathcal{R}$. Example: If $2 + \sqrt{14}i = (x + y\sqrt{14}i)(x' + y'\sqrt{14}i)$, taking absolute values gives $18 = (x^2 + 14y^2)(x'^2 + 14y'^2)$. This is an equation in integers and all solutions are easily found.
(e) Observe that $18 = 2 \cdot 3 \cdot 3 = (2 + \sqrt{14}i)(2 - \sqrt{14}i)$ are distinct factorizations of 18 as products of primes. Conclude that unique factorization fails for $\mathcal{R}$, and even that the number of prime factors of an element of $\mathcal{R}$ is not unique.

4

INTEGRAL ARITHMETIC

4.1 INTRODUCTION

In his first few years of schooling, the student encounters the natural numbers. He learns the method for representing natural numbers using decimal digits and the procedures for addition, subtraction, multiplication, and division. In this chapter we shall give the theory underlying this notation and these procedures, for integers and, as well, for polynomials. Besides proving that the familiar procedures for performing arithmetic of integers are indeed valid, we shall develop some alternative methods perhaps new to the reader and, for polynomials, we shall obtain a theorem often encountered in calculus called Taylor's theorem.

Using notation from Chapter 3, throughout Chapter 4 we shall let the symbol $\mathfrak{R}$ denote either the integer ring $\mathbf{Z}$ or the polynomial ring $\mathfrak{F}[\lambda]$, where $\mathfrak{F}$ is a field. The reader may assume, if he wishes, that $\mathfrak{F}$ is one of $\mathbf{Q}$, $\mathbf{R}$, or $\mathbf{C}$.

4.2 REPRESENTATION IN BASE b

Let b be a nonzero element of $\mathcal{R}$. When $\mathcal{R}$ is $\mathbf{Z}$, we assume $b > 1$; when $\mathcal{R}$ is $\mathfrak{F}[\lambda]$, we assume $\deg b \geq 1$, so that b is not a constant.

Our first results may be regarded as generalizations of the division algorithms.

THEOREM 4.2.1

Let b be as described above, and take a to be an element of $\mathcal{R}$, nonnegative when $\mathcal{R}$ is $\mathbf{Z}$. Then an integer l and elements $d_0, d_1, \ldots, d_l$ of $\mathcal{R}$ exist such that

$$a = d_l b^l + d_{l-1} b^{l-1} + \cdots + d_1 b + d_0, \tag{1}$$

with $d_l \neq 0$ when a is nonzero, and

$$\text{for } \mathcal{R} = \mathbf{Z}: \quad 0 \leq d_j < b, \qquad j = 0, \ldots, l; \tag{2.1}$$

$$\text{for } \mathcal{R} = \mathfrak{F}[\lambda]: \quad \deg d_j < \deg b, \qquad j = 0, \ldots, l; \tag{2.2}$$

We give first a pair of facts needed in the proof.

(i) *Let $\mathcal{R} = \mathbf{Z}$. If b, K, k, d are integers with $b > 1$, $K > 0$ and*

$$K = kb + d, \qquad 0 \leq d < b, \tag{3.1}$$

then

$$0 \leq k < K; \tag{4.1}$$

(ii) *Let $\mathcal{R} = \mathfrak{F}[\lambda]$. If b, K, k, d are polynomials with $\deg b > 0$, $K \neq 0$, and*

$$K = kb + d, \qquad \deg d < \deg b, \tag{3.2}$$

then

$$\deg k < \deg K. \tag{4.2}$$

In case (i) we have: $K = kb + d \geq kb$, so that $k \leq K/b < K$ (because $b > 1$). Also $K = kb + d < kb + b = (k+1)b$; hence $k + 1 > K/b > 0$. Thus $k > -1$, implying that $k \geq 0$ (because k is an integer). In case (ii), if $k \neq 0$, we have $\deg kb = \deg k + \deg b \geq \deg b$. Since $\deg d < \deg b$, we thus have $\deg (kb + d) = \deg kb$, and hence: $\deg K = \deg (kb + d) = \deg kb = \deg k + \deg b > \deg k$. If $k = 0$, it is obvious that $\deg k < \deg K$.

Proof of Theorem 4.2.1 If $a = 0$, we take $d_0 = d_1 = \cdots = d_l = 0$, and then (1) holds (for any l). Assume now that $a \neq 0$. Letting $q_0 = a$, we begin a recursive calculation.

Applying the division algorithms (Theorems 3.3.1, 3.3.3) to q_0 and b, we obtain elements q_1 and d_0 in $\mathcal{R}$ satisfying

$$q_0 = a = q_1 b + d_0\,, \quad \begin{cases} 0 \le d_0 < b & \text{if } \mathcal{R} = \mathbf{Z}, \\ \deg d_0 < \deg b & \text{if } \mathcal{R} = \mathfrak{F}[\lambda]. \end{cases} \tag{5}$$

Applying the division algorithms to q_1 and b, we obtain elements q_2 and d_1 in $\mathcal{R}$ satisfying

$$q_1 = q_2 b + d_1\,, \quad \begin{cases} 0 \le d_1 < b & \text{if } \mathcal{R} = \mathbf{Z}, \\ \deg d_1 < \deg b & \text{if } \mathcal{R} = \mathfrak{F}[\lambda]. \end{cases}$$

Substituting for q_1 in (5), we get

$$a = q_2 b^2 + d_1 b + d_0\,. \tag{6}$$

Applying the division algorithms to q_2 and b, we obtain elements q_3 and d_2 in $\mathcal{R}$ satisfying

$$q_2 = q_3 b + d_2\,, \quad \begin{cases} 0 \le d_2 < b & \text{if } \mathcal{R} = \mathbf{Z}, \\ \deg d_2 < \deg b & \text{if } \mathcal{R} = \mathfrak{F}[\lambda]. \end{cases}$$

Substituting for q_2 in (6), we get

$$a = q_3 b^3 + d_2 b^2 + d_1 b + d_0\,.$$

We continue in this fashion.

Suppose we have obtained elements $d_0, \ldots, d_j$ (all satisfying (2)) and $q_0, \ldots, q_{j+1}$ in $\mathcal{R}$, such that

$$a = q_{j+1} b^{j+1} + d_j b^j + \cdots + d_1 b + d_0\,, \tag{7}$$

$$q_0 = q_1 b + d_0\,, \qquad q_1 = q_2 b + d_1\,, \ldots, \qquad q_j = q_{j+1} b + d_{j+1}\,. \tag{8}$$

Applying the division algorithms to q_{j+1} and b, we obtain elements q_{j+2} and d_{j+1} in $\mathcal{R}$ satisfying

$$q_{j+1} = q_{j+2} b + d_{j+1}\,, \quad \begin{cases} 0 \le d_{j+1} < b & \text{if } \mathcal{R} = \mathbf{Z}, \\ \deg d_{j+1} < \deg b & \text{if } \mathcal{R} = \mathfrak{F}[\lambda]. \end{cases}$$

Substituting for q_{j+1} in (7), we get

$$a = q_{j+2} b^{j+2} + d_{j+1} b^{j+1} + \cdots + d_1 b + d_0\,,$$

and also

$$q_0 = q_1 b + d_0\,, \ldots, \qquad q_j = q_{j+1} b + d_j\,, \qquad q_{j+1} = q_{j+2} b + d_{j+1}\,.$$

Thus we may progressively increase j in (7) and (8).

We know q_0 is positive (when $\mathcal{R} = \mathbf{Z}$) or nonzero (when $\mathcal{R} = \mathfrak{F}[\lambda]$). Applying (4) to $q_0 = q_1 b + d_0$, we see that

$$q_0 > q_1 \ge 0 \qquad \text{or} \qquad \deg q_0 > \deg q_1\,.$$

If $q_1 \neq 0$, applying (4) to $q_1 = q_2 b + d_1$, we get

$$q_1 > q_2 \geq 0 \qquad \text{or} \qquad \deg q_1 > \deg q_2 .$$

If $q_2 \neq 0$, applying (4) to $q_2 = q_3 b + d_2$, we get

$$q_2 > q_3 \geq 0 \qquad \text{or} \qquad \deg q_2 > \deg q_3 .$$

We continue applying (4) in this manner, as long as we do not get a zero q. If $q_0, \ldots, q_j$ are nonzero, we therefore have

$$q_0 > q_1 > \cdots > q_j > 0 \quad \text{or} \quad \deg q_0 > \deg q_1 > \cdots > \deg q_j \geq 0. \tag{9}$$

However, (9) cannot hold for indefinitely increasing j since $q_0, \ldots, q_j$ or $\deg q_0, \ldots, \deg q_j$ are integers. Consequently, there will be a subscript l such that

$$q_0, \ldots, q_l \text{ all are nonzero,} \qquad \text{and } q_{l+1} = 0.$$

Taking j in (7) to be l, we obtain (1), and d_l is nonzero since $d_l = q_l - q_{l+1}b = q_l$. □

By putting $d_j = 0$ for $j > l$, we may write (1) as an infinite series

$$a = \sum_{j=0}^{\infty} d_j b^j \tag{10}$$

in which

$$\text{for } \quad \mathcal{R} = \mathbf{Z}: \qquad 0 \leq d_j < b, \qquad\qquad j = 0, 1, 2, \ldots, \tag{11.1}$$

$$\text{for } \quad \mathcal{R} = \mathfrak{F}[\lambda]: \quad -\infty \leq \deg d_j < \deg b, \qquad j = 0, 1, 2, \ldots, \tag{11.2}$$

and almost all d_j are zero. (Recall that "almost all" means "all, apart from a finite number of exceptions.") Of course, the infinite series (10) is really a finite series since only a finite number of its terms are nonzero.

We now establish the uniqueness of this representation of a.

THEOREM **4.2.2**
The d_j in (10) satisfying (11) are uniquely determined by a.

PROOF Suppose we also had

$$a = \sum_{j=0}^{\infty} d_j' b^j \tag{10'}$$

with $0 \leq d_j' < b$ when $\mathcal{R} = \mathbf{Z}$, or $-\infty \leq \deg d_j' < \deg b$ when $\mathcal{R} = \mathfrak{F}[\lambda]$, and where almost all d_j' are zero. We wish to prove that $d_j = d_j'$ for every j. We may write (10) and (10′) as

$$a = qb + d_0 = q'b + d_0',$$

where

$$q = \sum_{j=1}^{\infty} d_j b^{j-1}, \qquad q' = \sum_{j=1}^{\infty} d_j' b^{j-1}$$

and

$$0 \leq d_0 < b, \qquad 0 \leq d_0' < b, \qquad \text{when } \mathfrak{R} = \mathbf{Z},$$

$$-\infty \leq d_0 < \deg b, \qquad -\infty \leq \deg d_0' < \deg b, \qquad \text{when } \mathfrak{R} = \mathfrak{F}[\lambda].$$

That is, q and q' are quotients and d_0 and d_0' are remainders on dividing a by b. It follows from the uniqueness of the remainder that $d_0 = d_0'$. Applying this result to $(a - d_0)/b = d_1 + d_2 b + \cdots = d_1' + d_2' b + \cdots$, we get $d_1 = d_1'$. Applying it again to $(a - d_0 - d_1 b)/b^2 = d_2 + d_3 b + \cdots = d_2' + d_3' b + \cdots$, we get $d_2 = d_2'$. Plainly this can be continued; hence $d_j = d_j'$ for every j. □

DEFINITION 4.2.3.
The unique quantities $d_0, d_1, d_2, \ldots$ *in* (10) *satisfying* (11) *are called the* digits of a in the base b. *The expression* (10) *in which these digits appear is called the* b-adic expansion of a.

Base b representations for nonnegative integers

For the moment let us confine our attention to the case $\mathfrak{R} = \mathbf{Z}$. It is customary to write (1) in the following shorthand form:

$$a = d_l d_{l-1} d_{l-2} \ldots d_3 d_2 d_1 d_0 . \tag{12}$$

Thus formula (12) stands in place of (1), and the digits d_j in it satisfy $0 \leq d_j < b$ for each j. One may introduce names (i.e., symbols) for these digits; for example, when b is ten, the symbols used are 0, 1, 2, 3, 4, 5, 6, 7, 8, 9. When b is thirteen, we need to have symbols for thirteen digits; these could be 0, 1, 2, 3, 4, 5, 6, 7, 8, 9, A, B, C where A denotes ten, B eleven, C twelve. Only two symbols are needed for digits when b is two; we may use 0 and 1. The base sixteen is used in some computer programming centers, where it is customary to let A be ten, B eleven, C twelve, D thirteen, E fourteen, F fifteen. We shall adhere to this convention.

In numerical examples we need to be able to indicate which base b is being used. We do this by attaching a subscript giving the base; e.g., (12) becomes

$$a = d_l d_{l-1} \ldots d_1 d_{0_b} .$$

Let us agree that the subscript giving the base is always written in the familiar base ten notation.

Example 4.2.4 Convert 12114_{10} to base five.

SOLUTION The procedure, according to the proof of Theorem 4.2.1, is to divide 12114 by five, obtain the remainder (the initial digit d_0) and a quotient q_1. Then divide q_1 by five. Obtain a remainder (the next digit d_1) and a quotient q_2. Next divide q_2 by five; etc. Continue the process until a zero quotient is obtained. Arithmetic is easiest when done in base ten, and so we give the calculations in this base. We have

$$\begin{aligned}
12114 &= 5\cdot 2422 + 4, & q_1 &= 2422, & d_0 &= 4,\\
2422 &= 5\cdot 484 + 2, & q_2 &= 484, & d_1 &= 2,\\
484 &= 5\cdot 96 + 4, & q_3 &= 96 & d_2 &= 4,\\
96 &= 5\cdot 19 + 1, & q_4 &= 19, & d_3 &= 1,\\
19 &= 5\cdot 3 + 4, & q_5 &= 3, & d_4 &= 4,\\
3 &= 0\cdot 3 + 3, & q_6 &= 0, & d_5 &= 3.
\end{aligned}$$

Thus $12114_{10} = 341424_5$, and this is a shorthand for

$$12114_{10} = 3\cdot 5^5 + 4\cdot 5^4 + 1\cdot 5^3 + 4\cdot 5^2 + 2\cdot 5 + 4.$$

Example 4.2.5 Express 22294_{10} in base thirteen.

SOLUTION We have

$$\begin{aligned}
22294 &= 13\cdot 1714 + 12, & q_1 &= 1714, & d_0 &= 12,\\
1714 &= 13\cdot 131 + 11, & q_2 &= 131, & d_1 &= 11,\\
131 &= 13\cdot 10 + 1, & q_3 &= 10, & d_2 &= 1,\\
10 &= 13\cdot 0 + 10, & q_4 &= 0, & d_3 &= 10.
\end{aligned}$$

Thus (using A = ten, B = eleven, C = twelve):

$$22294_{10} = \mathrm{A1BC}_{13} = 10\cdot 13^3 + 1\cdot 13^2 + 11\cdot 13 + 12.$$

Example 4.2.6 Express $\mathrm{BA928}_{12}$ in base ten.

SOLUTION We have (see Exercise 15 of §3.3):

$$\begin{aligned}
\mathrm{BA928}_{12} &= 11\cdot 12^4 + 10\cdot 12^3 + 9\cdot 12^2 + 2\cdot 12 + 8\\
&= (((11\cdot 12 + 10)12 + 9)12 + 2)12 + 8\\
&= (((142)12 + 9)12 + 2)12 + 8\\
&= ((1713)12 + 2)12 + 8\\
&= (20558)12 + 8\\
&= 246704_{10}.
\end{aligned}$$

Base b representations for polynomials

In the next part of the discussion we take $\mathcal{R}$ to be $\mathfrak{F}[\lambda]$. Here $a = a(\lambda)$ is a polynomial and the polynomials $d_0, \ldots, d_l$ in (1) satisfy (2.2). It is not customary with polynomials to rewrite (1) in the form (12), although there is no reason why this could not be done. Moreover, the d_j are not usually called the digits of a in the base b, although again there is no reason why this language could not be used. Formula (1) [with condition (2.2)] is very often encountered in the special case where $b = \lambda - \alpha$ with $\alpha \in \mathfrak{F}$. What are the d_j? In this case they satisfy: $\deg d_j < \deg b = 1$. Thus $\deg d_j$ equals 0 or $-\infty$, so that the d_j are *constants*. We have completed part of the proof of the next theorem.

THEOREM 4.2.7 Taylor's Theorem
If $a(\lambda)$ is a polynomial of degree n in $\mathfrak{F}[\lambda]$, and α is a constant, constants $d_0, d_1, \ldots, d_n$ exist such that

$$a(\lambda) = d_0 + d_1(\lambda - \alpha) + d_2(\lambda - \alpha)^2 + \cdots + d_n(\lambda - \alpha)^n, \quad d_n \neq 0. \tag{13}$$

This formula is often called the *Taylor series expansion of* $a(\lambda)$ *about* α. It may also be called *the* $(\lambda - \alpha)$-adic *expansion of* $a(\lambda)$.

PROOF Formula (13) is just (1) rewritten; but we have to show that the integer l in (1) is n. If d_j is a nonzero constant, the degree of $d_jb^j = d_j(\lambda - \alpha)^j$ is j. Therefore the degree of the term d_lb^l in (1) is l, and the degrees of the other terms $d_{l-1}b^{l-1}, \ldots, d_1b, d_0$ are at most $l - 1$. Thus the right-hand side of (1) has degree l, whereas the left-hand side has degree n; hence $l = n$. □

It is possible, when $\mathfrak{F}$ is **C** or a subfield of **C**, to evaluate explicitly the digits $d_0, d_1, \ldots$ of $a(\lambda)$ in the base $(\lambda - \alpha)$. We do this now. Evaluating both sides of (13) at α, we obtain $d_0 = a(\alpha)$. Differentiating each side of (13) with respect to λ, we get

$$a'(\lambda) = d_1 + 2d_2(\lambda - \alpha) + 3d_3(\lambda - \alpha)^2 + \cdots$$

where $a'(\lambda)$ denotes the derivative of a. Evaluating at α gives $d_1 = a'(\alpha)$. Differentiating once more, we get

$$a''(\lambda) = 2d_2 + 3\cdot 2d_3(\lambda - \alpha) + 4\cdot 3d_4(\lambda - \alpha)^2 + \cdots.$$

and evaluating at α gives $d_2 = \frac{1}{2}a''(\alpha)$. Continuing in this way, we see that *the digits of* $a(\lambda)$ *in the base* $\lambda - \alpha$ *are*

$$a(\alpha),\quad a'(\alpha)/1!,\quad a''(\alpha)/2!,\quad a'''(\alpha)/3!, \ldots.$$

As before, we may rewrite (13) as an infinite series

$$a(\lambda) = \sum_{j=0}^{\infty} d_j(\lambda - \alpha)^j$$

in which the d_j are constants, almost all zero. There are many functions besides polynomials representable as an infinite series of this form, except that no longer is it true that almost all d_j are zero. The reader will encounter this in calculus. For example, for each *real number* λ,

$$e^{\lambda} = \sum_{j=0}^{\infty} \frac{1}{j!} \lambda^j.$$

(Near the end of §2.2 we used the special case of this formula when λ is 1.) Thus we may say that the digits of the function e^{λ} in the base λ are the numbers $1/0!$, $1/1!$, $1/2!$, $1/3!$, ...

Consequences of Taylor's theorem for polynomial with real coefficients

Taylor's theorem may be used to prove an intuitively evident fact about derivatives of polynomials with real coefficients. Suppose $a(\lambda) \in \mathbf{R}[\lambda]$, with $\deg a = n > 0$. Further suppose α, β are real numbers, with $\alpha < \beta$. *If*

$$a(x) \geq a(\alpha) \quad \textit{for all } x \textit{ satisfying} \quad \alpha \leq x < \beta,$$

$$\textit{then } a'(\alpha) \geq 0.$$

In other words: if $a(x)$ does not decrease as x is increased from α, then $a'(\alpha) \geq 0$. (See Figure 4.2.1.)

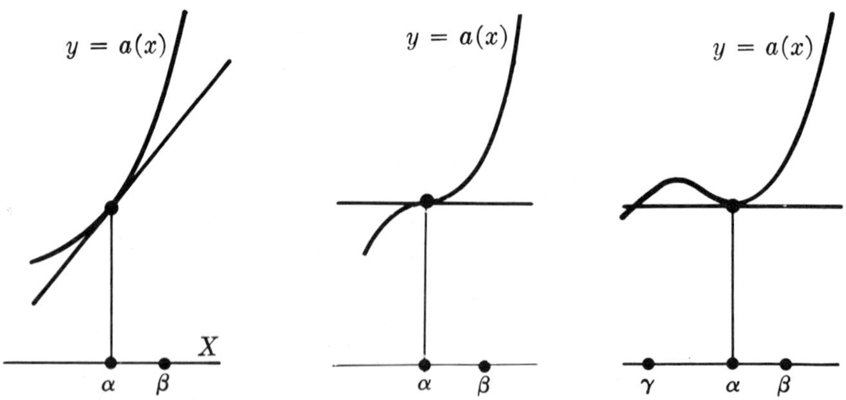

Figure 4.2.1

PROOF Writing $a(\lambda)$ in base $\lambda - \alpha$, as in (13), then evaluating at x and transposing terms, we get

$$a(x) - a(\alpha) = (x - \alpha)(d_1 + d_2(x - \alpha) + \cdots + d_n(x - \alpha)^{n-1}).$$

Here $d_1 = a'(\alpha)$. Denying the desired conclusion, suppose $d_1 = a'(\alpha) < 0$. If x is near α, all terms $d_2(x - \alpha), \ldots, d_n(x - \alpha)^{n-1}$ are small. We may, by taking x sufficiently* near α, satisfy both $\alpha < x < \beta$ and

$$d_2(x - \alpha) + \cdots + d_n(x - \alpha)^{n-1} < -\tfrac{1}{2}d_1.$$

(Note: $-\frac{1}{2}d_1$ is positive.) For this x,

$$a(x) - a(\alpha) < (x - \alpha)(d_1 - \tfrac{1}{2}d_1) = (x - \alpha)\tfrac{1}{2}d_1 < 0.$$

This contradicts the hypothesis $a(x) \geq a(\alpha)$. Therefore $a'(\alpha) < 0$ is impossible. Hence $a'(\alpha) \geq 0$. □

The reader may prove similarly that *if* $\gamma < \alpha$ *and*

$$a(x) \geq a(\alpha) \quad \textit{for all } x \textit{ satisfying} \quad \gamma < x \leq \alpha,$$

$$\textit{then } a'(\alpha) \leq 0.$$

We say that α gives a *local minimum* for $a(\lambda)$ if there are numbers γ, β with $\gamma < \alpha < \beta$ such that $a(x) \geq a(\alpha)$ for all x satisfying $\gamma < x < \beta$. (See the third graph in Figure 4.2.1.) If instead, $a(x) \leq a(\alpha)$ for all x with $\gamma < x < \beta$, we say that α gives a *local maximum*. Suppose α gives a local minimum. From the last two paragraphs we get $a'(\alpha) \geq 0$ and $a'(\alpha) \leq 0$, so that $a'(\alpha) = 0$. If α gives a local maximum, a similar discussion shows that again $a'(\alpha) = 0$. Thus: $a'(\alpha) = 0$ *when* α *gives a local minimum or maximum for the polynomial* $a(\lambda)$. Consequently, Taylor's theorem leads (for polynomials) to the procedure taught in calculus for finding local minima and maxima: Set $a'(x) = 0$, treating this as an equation in unknown x, and solve it to find α.

The *second derivative test,* also usually taught in calculus, gives a criterion for distinguishing between a local minimum and a local maximum: α gives a local minimum if $a''(\alpha) > 0$ and a local maximum if $a''(\alpha) < 0$. A proof, for polynomials, can be obtained using Taylor's theorem. (See Exercise 24.)

The Binomial Theorem

We describe now a special case of Taylor's theorem of frequent occurrence. Let β be a complex number, and take the polynomial $a(\lambda)$ to be $(\lambda + \beta)^n$. Setting α equal to zero, from Taylor's theorem we obtain

$$(\lambda + \beta)^n = d_0 + d_1\lambda + d_2\lambda^2 + \cdots + d_n\lambda^n.$$

---

* For a precise estimate, see Exercise 23.

As above, $d_0 = a(0)$, $d_1 = a'(0)$, etc. In general, $d_j = a^{(j)}(0)/j!$. By induction on j one may prove that

$$a^{(j)}(\lambda) = D^j(a(\lambda)) = n(n-1)\cdots(n-(j-1))(\lambda+\beta)^{n-j}.$$

We leave this proof as an exercise for the reader. Evaluating at 0 gives

$$a^{(j)}(0) = n(n-1)\cdots(n-(j-1))\beta^{n-j} = \frac{n!}{(n-j)!}\beta^{n-j}.$$

Thus we obtain the formula

$$(\lambda+\beta)^n = \sum_{j=0}^{n} \frac{n!}{j!(n-j)!}\beta^{n-j}\lambda^j. \tag{14}$$

The assertion that the two sides of (14) are equal is called the *binomial theorem*. It states that the digits of $(\lambda+\beta)^n$ in the base λ are the quantities

$$\frac{n!}{j!(n-j)!}\beta^{n-j}, \qquad j = 0, 1, 2, \ldots, n.$$

A proof of the binomial theorem that does not use differentiation will be found in Exercises 16, 17, 18 below.

It is customary to denote the quantity

$$\frac{n!}{j!(n-j)!} \quad \text{by} \quad \binom{n}{j} \quad \text{or by} \quad C_j^n, \qquad j = 0, 1, \ldots, n,$$

and to call these numbers the *binomial coefficients*. Using this notation, (14) becomes

$$(\lambda+\beta)^n = \sum_{j=0}^{n} \binom{n}{j}\beta^{n-j}\lambda^j.$$

Note that the binomial coefficients are integers. (They seem to be fractions because of the presence of denominator terms.) To see this, take $\beta = 1$. Then $(\lambda+1)^n$, as a product of polynomials $\lambda+1, \ldots, \lambda+1$ with integral coefficients, has integral coefficients. The coefficients are the numbers $\binom{n}{j}$ for $j = 0, 1, \ldots, n$. These numbers are therefore integers.

Digital representations based on numerically least remainders (Optional)

We now return to the case $\mathcal{R} = \mathbf{Z}$. When establishing the b-adic representation (1) in this case, we used the least nonnegative remainder algorithm, Theorem 3.3.1. We could equally well have used the numerically least remainder algorithm, Theorem 3.3.2. If we use numerically least remainders, we get a slightly different result, that has the virtue of being valid both for positive and negative integers. We state it as Theorem 4.2.8.

THEOREM 4.2.8

Let b be an integer, with $|b| > 2$. If a is an integer, perhaps negative, integers d_0, d_1, . . . exist such that

$$a = \sum_{j=0}^{\infty} d_j b^j \tag{15}$$

where

$$-|b|/2 < d_j \leq |b|/2 \qquad \text{for } j = 0, 1, 2, \ldots \tag{16}$$

and almost all d_j are zero. Moreover, this representation is unique.

PROOF We use this fact: If K, k, d are integers with $K \neq 0$, $K = kb + d$, and $-|b|/2 < d \leq |b|/2$, then $|k| < |K|$. To show this, calculate as follows, using first $k = (K - d)/b$, then the triangle inequality, $|d| \leq |b|/2$, and $|b| \geq 3$, to get: $|k| = |K - d|/|b| \leq (|K| + |d|)/|b| = |K|/|b| + |d|/|b| \leq |K|/3 + 1/2 < |K|$; the last step follows from $|K| \geq 1$.

The existence part of the proof of Theorem 4.2.8 is parallel to the existence part of the proof of Theorem 4.2.1, with this difference: in place of (9) we get $|a| > |q_1| > |q_2| > \cdots > |q_j| > 0$. Thus eventually $q_{l+1} = 0$ for some l, and then we have (10) (i.e., (15)) with (16) satisfied.

The proof of uniqueness of the representation (15) is parallel to the proof of uniqueness of the representation (10) and relies on the uniqueness of the numerically least remainder. □

Taking b = ten, we see that an integer a is uniquely representable in the form $d_0 + d_1 10 + d_2 10^2 + \cdots$, where each digit d_j is one of the numbers -4, -3, -2, -1, 0, 1, 2, 3, 4, 5, and this is possible not only for nonnegative a but also for negative a.

The requirement that $|b| > 2$ in Theorem 4.2.8 is essential. If $b = 2$, Theorem 4.2.8 applies only to nonnegative numbers.

When, for integers, we speak of the representation in base b, we shall (unless we specifically state otherwise) mean the representation given by Theorem 4.2.1, and not that given by Theorem 4.2.8.

Complements (Optional)

Besides that given by Theorem 4.2.8, there is another way of obtaining a base b representation for negative integers. To develop this, we ask: What happens if we apply to the negative integer $-a$ the procedure that led to the base b representation of the positive integer a? The procedure for positive integer a was to obtain digits d_0, d_1, d_2, . . . satisfying $0 \leq d_j < b$

for each j, and integers $q_1, q_2, q_3, \ldots$, recursively from these equations:

$$a = q_1 b + d_0, \qquad q_1 = q_2 b + d_1, \qquad q_2 = q_3 b + d_2, \ldots,$$

$$q_j = q_{j+1} b + b_j, \ldots, \qquad q_l = q_{l+1} b + d_l, \tag{17}$$

the process continuing until we arrive at the first q, call it q_{l+1}, which is zero. (Then $d_l \neq 0$ because $q_{l+1} = 0$ and $q_l \neq 0$.) By taking $d_{l+1} = d_{l+2} = d_{l+3} = \cdots = 0$, and $q_{l+1} = q_{l+1} = q_{l+2} = \cdots = 0$, and setting $q_0 = a$, we may write (17) as the following infinite system of equations:

$$q_0 = a,$$

$$q_j = q_{j+1} b + d_j, \qquad 0 \le d_j < b; j = 0, 1, 2, \ldots. \tag{18}$$

We may regard (18) as equations giving the q's and d's recursively. That is, having computed q_j, we compute both d_j and the next q, namely q_{j+1}, as specified by (18). We start by knowing q_0 $(= a)$ and thus obtain in succession all the q's and d's. Of course, almost all d's and almost all q's are zero.

In order to explain what happens when we apply this procedure to $-a$, we use $d_0, d_1, d_2, \ldots$ to define recursively new integers $\delta_0, \delta_1, \delta_2, \ldots, c_0, c_1, c_2, \ldots$ according to these rules:

$$c_0 = 0,$$

$$\delta_j = c_{j+1} b - c_j - d_j, \quad c_{j+1} = 0 \text{ or } 1, \qquad 0 \le \delta_j < b; j = 0, 1, 2, \ldots. \tag{19}$$

That is, if we know c_j, equal to zero or 1, we find c_{j+1}, also equal to zero or 1, such that $0 \le c_{j+1} b - c_j - d_j < b$, then set $\delta_j = c_{j+1} b - c_j - d_j$. Is there a value of c_{j+1}, equal to zero or 1, such that $0 \le c_{j+1} b - c_j - d_j < b$? Because $0 \le d_j < b$ and $c_j = 0$ or 1, we have $0 \le c_j + d_j \le b$, so we take $c_{j+1} = 0$ if $c_j = d_j = 0$ and otherwise we take $c_{j+1} = 1$. Thus (19) enables us to begin with c_0 and compute in succession all the c's and δ's.

For $j = 0, 1, 2, \ldots$, define integer Q_j by $Q_j = -q_j - c_j$. If we use (19) to express d_j in terms of δ_j, c_j, and c_{j+1}, and substitute the result into (18), we get:

$$Q_0 = -a,$$

$$Q_j = Q_{j+1} b + \delta_j, \qquad 0 \le \delta_j < b; j = 0, 1, 2, \ldots. \tag{20}$$

Formulas (20) are parallel to (18), except that we begin with $-a$. That is, we put $Q_0 = -a$, and then Q_{j+1} is the quotient and δ_j the remainder when Q_j is divided by b. Applying to $-a$ the procedure used to obtain the base b representation of a leads therefore to digits $\delta_0, \delta_1, \delta_2, \ldots, \delta_l, \delta_{l+1}, \delta_{l+2}, \ldots$.

In order to investigate further the digits $\delta_0, \delta_1, \ldots$ encountered here, let us notice these facts:

$$c_{l+1} = c_{l+2} = c_{l+3} = \cdots = 1, \tag{21.1}$$

$$\delta_{l+1} = \delta_{l+2} = \delta_{l+3} = \cdots = b - 1, \tag{21.2}$$

$$Q_{l+1} = Q_{l+2} = Q_{l+3} = \cdots = -1. \tag{21.3}$$

We prove (21.1) first. Because $d_l \neq 0$, a remark some lines above shows that $c_{l+1} = 1$. If we know, for some subscript $j \geq l + 1$, that $c_j = 1$, then by this same remark we get $c_{j+1} = 1$. Thus, by induction, (21.1) is proved. For $j \geq l + 1$ we now have $d_j = 0$, $c_j = 1$, $c_{j+1} = 1$, hence (19) gives $\delta_j = b - 1$. Thus (21.2) is proved. For $j \geq l + 1$, we see (using $q_j = 0$) that $Q_j = -q_j - c_j = -1$. Thus (21.3) is established.

Formula (21.2) implies that *all digits δ_j for $j > l$ are equal to $b - 1$.* In other words, the repetitive process for calculating the digits of $-a$ does *not* terminate. Instead, it yields a nonending sequence of digits, almost all of which equal $b - 1$. Parallel to (7), for any $t \geq 0$ we obtain

$$-a = Q_{t+1}b^{t+1} + \delta_t b^t + \delta_{t-1}b^{t-1} + \cdots + \delta_1 b + \delta_0 .$$

If $t \geq l$ this becomes [by (21.3)],

$$-a = -b^{t+1} + \delta_t b^t + \delta_{t-1}b^{t-1} + \cdots + \delta_1 b + \delta_0 . \tag{22}$$

It would be nice to ignore the term $-b^{t+1}$ in (22) and simply say that the base b representation of $-a$ is the infinite series

$$\cdots + \delta_t b^t + \delta_{t-1}b^{t-1} + \cdots + \delta_1 b + \delta_0 = \sum_{s=0}^{\infty} \delta_s b^s. \tag{23}$$

Unfortunately, this series is *divergent** since (22) gives

$$\delta_0 + \delta_1 b + \cdots + \delta_t b^t = b^{t+1} - a, \tag{24}$$

so that $\lim_{t\to\infty} (\delta_0 + \delta_1 b + \cdots + \delta_t b^t) = \lim_{t\to\infty} (b^{t+1} - a) = +\infty$.

Divergent though it is, the series (23) is nevertheless useful to manufacturers of computing machinery. To see why, we now describe briefly some of its properties. Let us call the series (23) *the complement of a in base b*, and write it as

$$-a \sim \cdots + \delta_t b^t + \cdots \cdot \delta_l b^l + \cdots + \delta_0$$

or, more concisely, as an infinite string of digits

$$-a \sim \ldots \delta_t \delta_{t-1} \ldots \delta_l \ldots \delta_0 . \tag{25}$$

(Read $\sim$ as "corresponds to.") This infinite string of digits is also called the *complement of a in base b*, and it has this interesting property:

* The notion of *convergence* of series has been discussed in a preliminary manner in the remark following Corollary 2.2.4. We give here a general definition. A series $\alpha_1 + \alpha_2 + \alpha_3 + \cdots$ is *convergent* if $\lim_{k\to\infty} (\alpha_1 + \alpha_2 + \cdots + \alpha_k)$ exists and is a finite number α (i.e. not ∞ or $-\infty$). An equivalent definition is: whenever a tolerance ϵ (a positive real number) is given, then $|\alpha - (\alpha_1 + \alpha_2 + \cdots + \alpha_k)| < \epsilon$ for all k exceeding an integer K (which may depend on ϵ.) Convergence therefore means that a finite number of terms of the series provide an approximation of α within tolerance ϵ, with the additional property that approximation of α within this tolerance ϵ is retained if a larger number of terms is used. The reason K depends on ϵ is that closer approximations of α (i.e., smaller ϵ) will generally require a larger number of terms (bigger K).

If we add it to a and use the normal rules for calculating the digits of the sum, the result of the addition is what it should be, namely, zero. This we shall show in §4.3. Suitably interpreted, the complement also behaves properly under multiplication—see §8.4. Thus the complement has desirable arithmetical properties.

From (24) we see that the lowest digits $\delta_0, \ldots, \delta_t$ of the base b complement of a are (for $t \geq l$) the base b digits of $b^{t+1} - a$. These facts make complements useful to manufacturers of computing machinery because they enable a negative integer $-a$ to be represented in the computer logic by a positive integer $b^{t+1} - a$. Some remarks amplifying this point are in Exercise 16 of §4.3.

Further discussion of the series (23) will be found in §8.4, where we shall see that if it is interpreted from another point of view,* it becomes convergent.

Example 4.2.9

Let b be ten. Take $a = 1$. Then $d_0 = 1$, $q_1 = 0$, $l = 0$; hence by (19), $\delta_0 = 9$. Since $l = 0$, (21.2) yields $\delta_1 = \delta_2 = \delta_3 = \cdots = 9$. Thus the complement of 1 (in base ten) is

$$-1 \sim \ldots 999 \ldots 999 \tag{26}$$

in which there are infinitely many digits, every one a 9.

(*Exercise:* Add (26) to 1, using the rules for carrying of digits taught in elementary school.) The complement for 6235 is

$$\ldots 99993765,$$

all digits except the right-hand four being 9's. If it is added to 6235, the result is 0.

(*Exercise:* If a desk calculating machine employing complements† is available for your use, perform on it this operation: Subtract 1 from 0. The result will be a string of 9's, as in (26). Then add this string of 9's to 1. The result will be 0. Perform also the operation 0 minus 6235, then verify the calculation by adding the result to 6235.)

---

* "Another point of view" means a different method for measuring the precision of an approximation.

† Many calculating machines are built for commercial use. These often perform differently from calculators designed for scientific use. A desk calculator for commercial use may, if a calculation yields a negative result, show the corresponding positive number in red type or with a symbol, such as CR, denoting "Credit." Examples of desk calculators using complements are the Marchant, Model AB10FA and the Friden, Model SRQ. These machines may sometimes be found in campus computing centers. Recently,

EXERCISES 4.2

I

1. Convert 20747_{10} to base two, to base five, to base twelve, and to base sixteen.

2. Convert $BEFOD_{16}$ to base ten. Do the same with 51704_8 and 51704_9.

3. Convert $ABBA_{16}$ to base five. Convert 1234321_5 to base twelve.

4. Let $a(\lambda) = \lambda^3 - \lambda^2 + 2\lambda + 1$ be in $\mathbf{Q}[\lambda]$. Find the digits of $a(\lambda)$ in base λ. Generalize.

5. Find the digits of $a(\lambda) = 3\lambda^5 - 2\lambda^4 + 8\lambda^3 - 5\lambda^2 + 6\lambda - 2$ in base $\lambda^2 + 1$.

6. Find the digits of $(\lambda + 2)^4$ in base λ.

7. Find the digits of the function $a(\lambda) = 1/(1 - \lambda)$ in base λ.

8. *(Optional)* Find the representation of the integer $a = 12345_{10}$ in base five using digits $-2, -1, 0, 1, 2$. Do the same in base six, using digits $-2, -1, 0, 1, 2, 3$. Find the base ten number c (using digits $0, 1, 2, \ldots, 9$) for which the representation of c in base ten (using digits $-4, -3, \ldots, 4, 5$) is $5(-1)4(-2)3$. Do the same for $d = (-1)2(-3)10$.

9. *(Optional)* Find the base ten complement of 2345_{10}. Find the base seven complement of 2345. Find the base ten digits of $10^7 - 6$.

II

10. Show that a positive integer is divisible by 16 if and only if the number formed from its four lowest base ten digits is divisible by sixteen. Generalize, putting 2^n in place of 16.

11. Use the fact that $10 \equiv 1 \pmod 9$ to prove: A base ten number $a = d_l d_{l-1} \ldots d_1 d_0$ satisfies $a \equiv d_l + d_{l-1} + \cdots + d_1 + d_0 \pmod 9$. Similarly use the fact that $10 \equiv -1 \pmod{11}$ to show that $a \equiv d_0 - d_1 + d_2 - d_3 + d_4 - \cdots \pmod{11}$.

12. Using repeatedly the first fact in Exercise 11, show that (base ten) $146354 \equiv 23 \equiv 5 \pmod 9$.

---

many desk calculating machines have come on the market embodying transistor-based electronic circuitry rather than the gears and turning wheels in older machines. None (to the author's knowledge) use complements to present a negative result of a calculation. Complements may be used in the design of the internal circuitry, however. For large-scale computing machines, complements are not visible if the programming of the machine is done in an interpretation language such as FORTRAN, but if the programming is done in the machine's own language, complements may then be used and sometimes are used.

13. Let $\gamma < \alpha < \beta$ be real numbers. Let $a(\lambda) \in \mathbf{R}[\lambda]$. If $\gamma < x < \beta$ and $a(x)$ is maximal for these x when $x = \alpha$, imitate the discussion in the text and prove that $a'(\alpha) = 0$.

14. Let b be an integer, at least 4. Show that 121_b is a perfect square, and 1331_b is a perfect cube.

15. Show that, when weighing using a balance, weights of 1, 2, 2^2, 2^3, . . ., 2^{n-1} will weigh any integral weight not exceeding $2^n - 1$.

III

16. If n is a nonnegative integer and j an integer with $0 \leq j \leq n$, define C_j^n as in the text by

$$C_j^n = \frac{n!}{j!(n-j)!}.$$

Prove: $C_0^n = 1$ for all n, $C_n^n = 1$ for all n; and if $0 < j < n$, prove that $C_j^n = C_j^{n-1} + C_{j-1}^{n-1}$.

17. Construct a table (called *Pascal's triangle*) in this manner:

1					
1	1				
1	2	1			
1	3	3	1		
1	4	6	4	1	
1	5	10	10	5	1
.					.
.					. .
.					. . .

The topmost entry is a 1; each row begins with a 1, has one more entry than the previous row, ends with a 1, and an intermediate entry is obtained as the sum of the entry immediately above and the entry above and one position to the left (e.g., $10 = 4 + 6$ as indicated in the table). Using Exercise 16, show that the entries in Pascal's triangle are the numbers C_j^n. In particular, verify that each C_j^n is an integer.

18. Let a and b be elements in a commutative ring. Using induction on n, show that

$$(a+b)^n = \sum_{j=0}^{n} C_j^n a^j b^{n-j}.$$

(Hint: Write out this formula with $n-1$ in place of n. Multiply each side by $a+b$. Collect terms containing like powers of a and b, and use Exercise 16 to obtain the formula for $(a+b)^n$.)

19. Show that the complement in base b of the number 1 is

$$\ldots (b-1)(b-1)\ldots(b-1),$$

in which each digit is $b-1$. This symbol stands for

$$\sum_{j=0}^{\infty}(b-1)b^j = (b-1)\sum_{j=0}^{\infty}b^j.$$

Show that if we use Corollary 2.2.4 to sum the geometric series here, even though the condition $-1<b<1$ is *not* satisfied, the sum nevertheless is the correct value, namely -1.

20. Let $b>2$. Suppose that the representation of the number a in base b, as in Theorem 4.2.8, is $a = d_0 + d_1 b + \cdots + d_l b^l$, where $-b/2 < d_j \le b/2$ for each j and $d_l \neq 0$. Show that $a>0$ if $d_l>0$ and $a<0$ if $d_l<0$. Use the formula for the sum of a finite geometric series (Lemma 2.2.3.)

21. Using Exercise 20, show that any positive integer can be written uniquely in the form $3^l + \epsilon_{l-1}3^{l-1} + \cdots + \epsilon_1 3 + \epsilon_0$, where each ϵ_j is -1, 0, 1. Use this fact to show that with a balance, weights 1, 3, $3^2, \ldots, 3^l$ are capable of weighing any integral weight not exceeding $(3^{l+1}-1)/2$. (You may want to place weights on both pans of the balance.)

22. A certain number x has all of the following properties:
   (i) In base two, its three lowest digits are zero, but its fourth lowest digit is 1.
   (ii) In base seven its lowest digit is zero, and its second lowest digit is not zero.
   (iii) In base eleven its lowest digit is zero, and its second lowest digit is not zero.
   (iv) In every other base b, with twelve exceptions, the lowest digit is not zero.

   Find x and find the twelve exceptions.

23. This exercise refers to the footnote on page 157. If $n \ge 2$ and if

$$\alpha < x < \text{the least of } \{\beta,\ \alpha+1,\ \alpha - \tfrac{1}{2}d_1/(|d_2| + \cdots + |d_n|)\},$$

prove that $d_2(x-\alpha) + \cdots + d_n(x-\alpha)^{n-1} < -\tfrac{1}{2}d_1$.

24. Let $a(\lambda) \in \mathbf{R}[\lambda]$, with $\deg a(\lambda) = n \ge 2$. Suppose the derivative of $a(\lambda)$ is zero when evaluated at a real number α. Write $a(\lambda)$ in base $\lambda - \alpha$, then evaluate at a real number x, to get

$$a(x) - a(\alpha) = (x-\alpha)^2(d_2 + d_3(x-\alpha) + \cdots + d_n(x-\alpha)^{n-2}).$$

Use this formula for x near α to prove that α gives a local minimum of $a(\lambda)$ if $a''(\alpha) > 0$, and a local maximum if $a''(\alpha) < 0$. If $a''(\alpha) = 0$, show by examples that there may or may not be a local minimum or maximum. Develop a criterion to distinguish the cases that may arise. (Hint: Consider the first nonzero term in Taylor's series.)

4.3 BASIC ARITHMETIC

In this section we outline the theory underlying the basic procedures for addition, subtraction, multiplication, and division, and we shall as well describe some variations of these procedures. We shall work entirely in base b when presenting the theory; to simplify the appearance of the formulas, the subscript b will be omitted. The subscript giving the base will be retained, as needed, in the examples.

Addition

Let $a_1, \ldots, a_n, b$ be nonnegative integers, with $b > 1$. Express $a_1, \ldots, a_n$ in the base b, and let $d_l^{(i)}, \ldots, d_0^{(i)}$ be the digits of a_i; $i = 1, \ldots, n$. Thus:

$$\begin{aligned} a_1 &= d_l^{(1)}d_{l-1}^{(1)} \ldots d_1^{(1)}d_0^{(1)} = d_l^{(1)}b^l + \cdots + d_1^{(1)}b + d_0^{(1)}, \\ a_2 &= d_l^{(2)}d_{l-1}^{(2)} \ldots d_1^{(2)}d_0^{(2)} = d_l^{(2)}b^l + \cdots + d_1^{(2)}b + d_0^{(2)}, \qquad (1) \\ &\cdots\cdots\cdots\cdots\cdots\cdots\cdots\cdots \\ a_n &= d_l^{(n)}d_{l-1}^{(n)} \ldots d_1^{(n)}d_0^{(n)} = d_l^{(n)}b^l + \cdots + d_1^{(n)}b + d_0^{(n)}, \end{aligned}$$

Each digit $d_i^{(j)}$ satisfies $0 \leq d_i^{(j)} < b$. By including zero digits (e.g., $d_l^{(1)}$ may be zero), we may suppose all a_i have the same number of digits. Let $A = a_1 + \cdots + a_n$. We wish to find the representation of A in base b. Adding equations (1), we get

$$\begin{aligned} A &= (d_l^{(1)} + \cdots + d_l^{(n)})b^l + \cdots + (d_1^{(1)} + \cdots + d_1^{(n)})b \\ &\qquad + (d_0^{(1)} + \cdots + d_0^{(n)}) \\ &= D_lb^l + \cdots + D_1b + D_0, \qquad (2) \end{aligned}$$

where $D_i = d_i^{(1)} + \cdots + d_i^{(n)}$, $i = 0, 1, \ldots, l$. The difficulty with the expression (2) is that the coefficient D_i of b^i may well exceed $b - 1$. So after adding $a_1, \ldots, a_n$ to obtain the representation (2) of A, we *reduce and carry*: Divide D_0 by b, to get integers c_1 and r_0 such that

$$D_0 = c_1b + r_0 \quad \text{with } 0 \leq r_0 < b.$$

Substituting this into (2) gives

$$A = D_lb^l + \cdots + D_2b^2 + (D_1 + c_1)b + r_0. \qquad (3)$$

We call c_1 the *carry integer* from position zero into position one. Next, divide $D_1 + c_1$ by b, to get integers c_2 and r_1 such that $D_1 + c_1 = c_2b + r_1$, with $0 \le r_1 < b$. Substituting this into (3) gives

$$A = D_lb^l + \cdots + D_3b^3 + (D_2 + c_2)b^2 + r_1b + r_0 . \tag{4}$$

We call c_2 the *carry integer* from position one into position two. Next, divide $D_2 + c_2$ by b, to get $D_2 + c_2 = c_3b + r_2$, $0 \le r_2 < b$, and hence

$$A = D_lb^l + \cdots + D_4b^4 + (D_3 + c_3)b^3 + r_2b^2 + r_1b + r_0 .$$

Plainly we may continue in this way. When we reach position l we have

$$A = (D_l + c_l)b^l + r_{l-1}b^{l-1} + \cdots + r_1b + r_0 .$$

If $D_l + c_l$ exceeds $b - 1$, we reduce and carry, and so get c_{l+1} and r_l such that $D_l + c_l = c_{l+1}b + r_l$ with $0 \le r_l < b$. Hence

$$A = c_{l+1}b^{l+1} + r_lb^l + \cdots + r_1b + r_0 .$$

If c_{l+1} exceeds $b - 1$, we again reduce and carry, and so get c_{l+2} and r_{l+1} such that $c_{l+1} = c_{l+2}b + r_{l+1}$ with $0 \le r_{l+1} < b$; hence

$$A = c_{l+2}b^{l+2} + r_{l+1}b^{l+1} + r_lb^l + \cdots + r_0 .$$

If c_{l+2} still exceeds b, we have $c_{l+2} = c_{l+3}b + r_{l+2}$, then $A = c_{l+3}b^{l+3} + r_{l+2}b^{l+2} + \cdots + r_0$. Since $c_l > c_{l+1} > c_{l+2} > \cdots \ge 0$, eventually we get a carry integer $c_{L+1} = 0$ (for some $L \ge l$) and then the process terminates as

$$A = r_Lb^L + \cdots + r_lb^l + \cdots + r_0 .$$

The process may be compactly arranged as follows:

$$
\begin{array}{cccccccll}
 & & d_l^{(1)} & \cdots & d_2^{(1)} & d_1^{(1)} & d_0^{(1)} & & \\
 & & d_l^{(2)} & \cdots & d_2^{(2)} & d_1^{(2)} & d_0^{(2)} & & \\
 & & \cdot & \cdots & \cdot & \cdot & \cdot & & \\
 & & d_l^{(n)} & \cdots & d_2^{(n)} & d_1^{(n)} & d_0^{(n)} & & \text{First step:} \\
\cline{3-7}
 & & & & & & & & \text{add the digits} \\
 & & D_l & \cdots & D_2 & D_1 & D_0 & \leftarrow & \text{in each column} \\
\cdots c_{l+2} & c_{l+1} & c_l & \cdots & c_2 & c_1 & & \leftarrow & \text{Second step:} \\
\cline{1-7}
 & & & & & & & & \text{reduce and} \\
\cdots r_{l+2} & r_{l+1} & r_l & \cdots & r_2 & r_1 & r_0 & \leftarrow & \text{carry}
\end{array}
\tag{5}
$$

Of course, it is possible to add the column zero digits $d_0^{(1)}, \ldots, d_0^{(n)}$ to get D_0, then reduce D_0 to get the digit r_0 and the integer c_1 carried into column one, before adding the digits in column one. Then add all the integers $d_1^{(1)}, \ldots, d_1^{(n)}, c_1$ in column one, reduce the sum to get the digit r_1 and the integer c_2 carried into column two, before adding any column two digits, etc. Thus we may arrange the work as follows, placing the carry integers at the top of each column:

$$\begin{array}{ccccccc} \ldots c_{l+2} & c_{l+1} & c_l & \ldots & c_2 & c_1 & \\ & & d_l^{(1)} & \ldots & d_2^{(1)} & d_1^{(1)} & d_0^{(1)} \\ & & d_l^{(2)} & \ldots & d_2^{(2)} & d_1^{(2)} & d_0^{(2)} \\ & & \cdot & \cdot\ \cdot\ \cdot & \cdot & \cdot & \cdot \\ & & d_l^{(n)} & \ldots & d_2^{(n)} & d_1^{(n)} & d_0^{(n)} \\ \hline \ldots r_{l+2} & r_{l+1} & r_l & \ldots & r_2 & r_1 & r_0 \end{array}$$

Example Calculate the sum of 4564, 2605, and 6661. These numbers are in base seven.

SOLUTION

$$\begin{array}{r} 2211 \\ 4564 \\ 2605 \\ 6661 \\ \hline 20463 \end{array}$$

Here $4 + 5 + 1$, on being divided by seven, gives remainder $r_0 = 3$ and a carry of 1. Then $1 + 6 + 0 + 6$, on being divided by seven, gives a remainder of 6 and a carry of 1. Then $1 + 5 + 6 + 6$, on being divided by seven, gives a remainder of 4 and a carry of 2. Next, $2 + 4 + 2 + 6$, on being divided by seven, gives a remainder of 0 and a carry of 2. In base seven, the required sum therefore is 20463.

Complements (Optional)

In §4.2, we saw that if $a = d_l d_{l-1} \ldots d_1 d_0$ is the representation of positive integer a in base b, there is a representation of the negative integer $-a$ in base b given by

$$-a \sim \ldots (b-1)(b-1)(b-1)\delta_l\delta_{l-1} \ldots \delta_1\delta_0 \tag{6}$$

where $\delta_0, \ldots, \delta_l$ are related to $d_0, \ldots, d_l$ by

$$d_0 + \delta_0 = c_1 b, \quad d_1 + \delta_1 + c_1 = c_2 b, \ldots, \quad d_l + \delta_l + c_l = c_{l+1} b, \tag{7}$$

and each of $c_1, \ldots, c_l$ is 0 or 1, with $c_{l+1} = 1$. [See (19) and (21) of §4.2.] Let us add $-a$ to a according to the rules above, using (6). In the first stage we get

$$\begin{array}{cccccccc} \ldots\ 0 & 0 & 0 & d_l & \ldots & d_2 & d_1 & d_0 \\ \ldots (b-1) & (b-1) & (b-1) & \delta_l & \ldots & \delta_2 & \delta_1 & \delta_0 \\ \hline \ldots (b-1) & (b-1) & (b-1) & (d_l+\delta_l) & \ldots & (d_2+\delta_2) & (d_1+\delta_1) & (d_0+\delta_0) \end{array}$$

Formulas (7) show that if we reduce and carry, beginning at the right, (the c's in these formulas turn out to be the carry integers), we obtain $\ldots 0\,0\,0 \ldots 0$. Thus $(-a) + a = 0$ when we use the addition scheme described above and the representation (6) of $-a$.

Subtraction

Let a_1 and a_2 be positive integers, with $a_1 \geq a_2$, represented in base b as in the discussion of addition. We wish to obtain the base b digits of $a_1 - a_2$. We have

$$a_1 = d_l^{(1)}b^l + \cdots + d_2^{(1)}b^2 + d_1^{(1)}b + d_0^{(1)} = d_l^{(1)} \ldots d_1^{(1)}d_0^{(1)},$$

$$a_2 = d_l^{(2)}b^l + \cdots + d_2^{(2)}b^2 + d_1^{(2)}b + d_0^{(2)} = d_l^{(2)} \ldots d_1^{(2)}d_0^{(2)}. \qquad (8)$$

Let $S = a_1 - a_2$. Subtracting equations (8), we get

$$S = D_lb^l + \cdots + D_2b^2 + D_1b + D_0, \qquad (9)$$

where now

$$D_i = d_i^{(1)} - d_i^{(2)}, \qquad i = 0, \ldots, l. \qquad (10)$$

Certainly each of $D_0, D_1, \ldots, D_l$ is $<b$ [by (10)]. The trouble, though, is that perhaps some are negative. Certainly each D_i is $>(-b)$ [by (10)]. So we go through the following process which the reader will recognize as the grade school process of "borrowing." Divide D_0 by b (i.e., reduce and carry) to get $D_0 = c_1b + r_0$, $0 \leq r_0 < b$. Since $-b < D_0 < b$ we have $c_1 = 0$ or $c_1 = -1$. Substituting for D_0 in (9), we get

$$S = D_lb^l + \cdots + D_2b^2 + (D_1 + c_1)b + r_0. \qquad (11)$$

We call c_1 a "borrow digit" or "carry" digit. Divide $D_1 + c_1$ by b (i.e., reduce and carry) to get $D_1 + c_1 = c_2b + r_1$, $0 \leq r_1 < b$. Since $b > D_1 + c_1 \geq -b$ (because $b > D_1 > -b$ and $c_1 = 0$ or -1), we get $c_2 = 0$ or -1. Substituting into (11), we obtain

$$S = D_lb^l + \cdots + D_3b^3 + (D_2 + c_2)b^2 + r_1b + r_0.$$

We also call c_2 a "borrow digit." Then $b > D_2 + c_2 \geq -b$, hence $D_2 + c_2 = c_3b + r_2$, where $0 \leq r_2 < b$ and $c_3 = 0$ or -1. We may continue in this way, and eventually we will arrive at

$$S = (D_l + c_l)b^l + r_{l-1}b^{l-1} + \cdots + r_1b + r_0, \qquad (12)$$

where the digits $r_0, \ldots, r_{l-1}$ lie between 0 and $b - 1$. At this point, $D_l + c_l \geq 0$ and our subtraction process has come to an end. Why is $D_l + c_l \geq 0$? Each $r_j \leq b - 1$. Using this and the formula for the sum of a finite geometric series (Lemma 2.2.3), we obtain

$$r_{l-1}b^{l-1} + \cdots + r_1b + r_0 \leq (b-1)b^{l-1} + \cdots + (b-1)b + (b-1)$$

$$= (b-1)\frac{b^l - 1}{b - 1} = b^l - 1.$$

Hence, by (12), $S \leq (D_l + c_l)b^l + b^l - 1$. Since $D_l + c_l$ is an integer, if $D_l + c_l < 0$ then $D_l + c_l \leq -1$, and this yields $S \leq (-1)b^l + b^l - 1 = -1$. But this contradicts the fact that $S \geq 0$.

The work may be arranged this way:

$$\begin{array}{cccccll}
d_l^{(1)} & \ldots & d_2^{(1)} & d_1^{(1)} & d_0^{(1)} & & \\
d_l^{(2)} & \ldots & d_2^{(2)} & d_1^{(2)} & d_0^{(2)} & & \text{First step:} \\
\hline
 & & & & & & \text{subtract digits} \\
D_l & \ldots & D_2 & D_1 & D_0 & \leftarrow & \text{in each column} \\
c_l & \ldots & c_2 & c_1 & & \leftarrow & \text{Second step:} \\
\hline
 & & & & & & \text{reduce and} \\
r_l & \ldots & r_2 & r_1 & r_0 & \leftarrow & \text{carry}
\end{array}$$

As with addition, we may reduce and carry in each column before moving to the next column. And then the work is arranged this way:

$$\begin{array}{ccccc}
c_l & \ldots & c_2 & c_1 & \\
d_l^{(1)} & \ldots & d_2^{(1)} & d_1^{(1)} & d_0^{(1)} \\
d_l^{(2)} & \ldots & d_2^{(2)} & d_1^{(2)} & d_0^{(2)} \\
\hline
r_l & \ldots & r_2 & r_1 & r_0
\end{array}$$

Example In base five let $a_1 = 4222$, $a_2 = 3244$. Find $a_1 - a_2$.

SOLUTION

$$\begin{array}{rcccc}
 & {\scriptstyle -1} & {\scriptstyle -1} & {\scriptstyle -1} & \\
 & 4 & 2 & 2 & 2 \\
- & 3 & 2 & 4 & 4 \\
\hline
 & 0 & 4 & 2 & 3
\end{array}$$

Here $2 - 4$, on being divided by five, produces a remainder of 3 and a carry of -1. Then $-1 + 2 - 4$, on being divided by five, produces a remainder of 2 and a carry of -1. Then $-1 + 2 - 2$, on being divided by five, produces a remainder of 4 and a carry of -1. And finally $-1 + 4 - 3 = 0$. Thus $4222_5 - 3244_5 = 423_5$.

Complements and subtraction (Optional)

When calculating $a_1 - a_2$, we may replace $-a_2$ by its b-complement

$$\ldots (b-1)(b-1) \ldots (b-1)\delta_m\delta_{m-1} \ldots \delta_2\delta_1\delta_0$$

and add this to a_1. This means that a subtraction can be viewed as an addition. For example (in base ten), let us calculate $9247 - 6235$. We have (see the end of §4.2) to perform the following addition:

$$\begin{array}{r} \ldots 0\,0\,0\,9\,2\,4\,7 \\ \ldots 9\,9\,9\,3\,7\,6\,5 \\ \hline \ldots 0\,0\,0\,3\,0\,1\,2 \end{array}$$

The difference therefore is 3012.

Let a_1 and a_2 be positive integers with $a_2 > a_1$. If we subtract a_2 from a_1 using the rules for reducing and carrying (i.e., borrowing) as explained above, we find that it can be carried out without difficulty, but does not terminate. In fact what results is a complement. Thus, as an example (base ten) if we subtract 9247 from 6235 we get:

$$\begin{array}{r} 6\,2\,3\,5 \\ -9\,2\,4\,7 \\ \hline \ldots 9\,9\,9 \quad 6\,9\,8\,8 \end{array}$$

This answer, ... 9996988, is indeed -3012 because when we add it to 3012 we get zero.

Multiplication

Let the base b representation of the positive integer a be

$$a = d_l b^l + \cdots + d_1 b + d_0 = d_l \ldots d_0 ,$$

where the d_j are the digits. Let α be another nonnegative integer. How can we compute the base b representation of αa? We have

$$\begin{aligned} \alpha a &= (\alpha d_l) b^l + \cdots + (\alpha d_2) b^2 + (\alpha d_1) b + \alpha d_0 \\ &= D_l b^l + \cdots + D_1 b + D_0 , \end{aligned}$$

where $D_i = \alpha d_i$ for each i. The quantities $D_0, D_1, \ldots, D_l$ will usually be too large to be base b digits. But we can reduce and carry: $D_0 = c_1 b + r_0$, $0 \le r_0 < b$. Thus $\alpha a = D_l b^l + \cdots + D_2 b^2 + (D_1 + c_1) b + r_0$. Then $D_1 + c_1 = c_2 b + r_1$, $0 \le r_1 < b$, and hence

$$\alpha a = D_l b^l + \cdots + D_3 b^3 + (D_2 + c_2) b^2 + r_1 b + r_0 .$$

Continuing in this manner we eventually arrive at

$$\alpha a = (D_l + c_l) b^l + r_{l-1} b^{l-1} + \cdots + r_1 b + r_0 .$$

Reducing and carrying again, let $D_l + c_l = c_{l+1} b + r_l$, $0 \le r_l < b$. Then $\alpha a = c_{l+1} b^{l+1} + r_l b^l + \cdots + r_1 b + r_0$. Reducing and carrying once more, we have $c_{l+1} = c_{l+2} b + r_{l+1}$, $0 \le r_{l+1} < b$, and then $\alpha a = c_{l+2} b^{l+2} + r_{l+1} b^{l+1} + \cdots + r_1 b + r_0$. We may continue reducing and carrying this way. Since $c_l > c_{l+1} > c_{l+2} > \cdots \ge 0$, we eventually arrive at a carry of zero, and then the multiplication is complete.

In particular this process is valid when α is a single digit (i.e., $0 \le \alpha < b$), and in this case it is quite easy to carry out.

Example Compute 5×6243. The base is seven.

SOLUTION On division by seven, 5×3 produces a remainder of 1 and a carry of 2. Then $2 + 5 \times 4$, on division by seven, produces a remainder of 1 and a carry of 3. Then $3 + 5 \times 2$, on division by seven, produces a remainder of 6 and a carry of 1. Next, $1 + 5 \times 6$ produces a remainder of 3 and a carry of 4. Thus $5 \times 6243 = 43611$ when the base is seven.

Suppose now that the base b representation of α is $d_k'd_{k-1}' \ldots d_1'd_0'$; thus

$$\alpha = d_k'b^k + d_{k-1}'b^{k-1} + \cdots + d_1'b + d_0'.$$

Let $M = \alpha a$. Then

$$M = (d_k'a)b^k + (d_{k-1}'a)b^{k-1} + \cdots + (d_1'a)b + (d_0'a). \tag{13}$$

Use the procedure above to express the products $d_0'a, d_1'a, \ldots, d_k'a$ (each of which is of the type: a single digit $\times\ a$) in base b, say as

$$\begin{aligned} d_0'a &= \ldots d_2^{(0)}d_1^{(0)}d_0^{(0)}, \\ d_1'a &= \ldots d_2^{(1)}d_1^{(1)}d_0^{(1)}, \\ d_2'a &= \ldots d_2^{(2)}d_1^{(2)}d_0^{(2)}, \\ &\cdots\cdots \\ d_k'a &= \ldots d_2^{(k)}d_1^{(k)}d_0^{(k)}. \end{aligned}$$

Here the $d_i^{(j)}$, $i = 0, 1, 2, \ldots$ are the digits of $d_j'a$. Thus

$$d_j'a = \cdots + d_2^{(j)}b^2 + d_1^{(j)}b + d_0^{(j)}.$$

Substituting these expressions into (13), we get

$$\begin{aligned} M = \quad & \cdots + d_3^{(0)}b^3 + d_2^{(0)}b^2 + d_1^{(0)}b + d_0^{(0)} \\ + & \cdots + d_2^{(1)}b^3 + d_1^{(1)}b^2 + d_0^{(1)}b \\ + & \cdots + d_1^{(2)}b^3 + d_0^{(2)}b^2 \\ + & \cdots + d_0^{(3)}b^3 \\ + & \cdots \\ = \quad & \cdots + \Delta_3 b^3 + \Delta_2 b^2 + \Delta_1 b + \Delta_0 \end{aligned} \tag{14}$$

where $\Delta_0 = d_0^{(0)}$, $\Delta_1 = d_1^{(0)} + d_0^{(1)}$, $\Delta_2 = d_2^{(0)} + d_1^{(1)} + d_0^{(2)}$, etc. Applying reduction and carrying to the integers $\Delta_0, \Delta_1, \Delta_2, \ldots$ in (14), one then obtains the base b representation of M.

This is the usual multiplication process taught to the reader around the third or fourth grade. It may be presented in this compact form:

$$
\begin{array}{rccccl}
 & & & & a & \\
\times \ldots & d_3' & d_2' & d_1' & d_0' & \\
\hline
\ldots & d_3^{(0)} & d_2^{(0)} & d_1^{(0)} & d_0^{(0)} & \\
\ldots & d_2^{(1)} & d_1^{(1)} & d_0^{(1)} & & \\
\ldots & d_1^{(2)} & d_0^{(2)} & & & \\
\ldots & d_0^{(3)} & & & & \\
\cdot & & & & & \\
\cdot\cdot & & & & & \\
\cdot\cdot\cdot & & & & & \\
\hline
\end{array}
$$

These lines are obtained by multiplying the individual digits d_0', d_1', d_2', . . . onto a.

Obtain the answer here by adding the digits above, reducing and carrying.

Example Multiplying in base six, we have:

$$
\begin{array}{r}
4\,2\,3\,5 \\
\times\ 3\,4\,5 \\
\hline
3\,4\,1\,1\,1 \\
2\,5\,4\,3\,2 \\
2\,1\,1\,5\,3 \\
\hline
2\,4\,5\,2\,1\,3\,1
\end{array}
$$

Multiplication by halving and doubling (Optional)

There is another method of multiplication which is sometimes simpler. We illustrate the procedure with an example in base ten, but it can be done in any base. To multiply 103×41, put either factor, say 103, at the top of a column, and the other fact, 41, at the top of a second column. In the first column repeatedly half 103; whenever the division by 2 is not exact discard the remainder; continue until zero results. In the second column repeatedly double 41. Then strike out every number in the second column opposite an even number in the first column. The sum of the remaining numbers in the second column, 4223, is then 103×41:

103	41
51	82
25	164
12	~~328~~
6	~~656~~
3	1312
1	2624
	4223

Here is the proof that this method always works. Let a be written in base 2:

$$a = d_l d_{l-1} \ldots d_1 d_0 = d_l 2^l + \cdots + d_1 2 + d_0\,,$$

where each base two digit d_i is 0 or 1. We are to multiply $a \times \alpha$. Repeatedly halving in the first column and doubling in the second column, we get:

$$\begin{array}{r|c}
d_l 2^l + \quad \cdots \quad + d_1 2 + d_0 & \alpha \\
d_l 2^{l-1} + \quad \cdots \quad + d_2 2 + d_1 & 2\alpha \\
d_l 2^{l-2} + \cdots + d_3 2 + d_2 & 2^2\alpha \\
\cdots\cdots\cdots\cdots & \cdots \\
d_l 2^{l-i} + \cdots + d_i & 2^i\alpha \\
\cdots\cdots\cdots & \cdots \\
d_l & 2^l\alpha
\end{array}$$

The number $d_l 2^{l-i} + \cdots + d_{i+1} 2 + d_i$ is even if $d_i = 0$ and odd if $d_i = 1$. Thus $2^i\alpha$ is struck out if $d_i = 0$, and not struck out if $d_i = 1$. The sum of the not struck out numbers in the second column is therefore the sum of the $2^i\alpha$ for all values of i satisfying $d_i = 1$, that is

$$\text{the sum of all } 2^i\alpha d_i \text{ for those } i \text{ with } d_i = 1,$$

and this is the same as

$$\text{the sum of all } 2^i\alpha d_i\,,$$

since for the remaining values of i we have $2^i\alpha d_i = 0$ because $d_i = 0$. Thus the sum of the not struck out numbers in the second column is

$$\sum_{i=0}^{l} 2^i\alpha d_i = \left(\sum_{i=0}^{l} d_i 2^i\right)\alpha = a\alpha.$$

Division

Let A and B be integers, with $B \neq 0$. To divide A by B means to find integers Q, R with $A = QB + R$ and $0 \leq R < |B|$. But this says: from A subtract a multiple of B (namely, QB) to produce a result R in the range

$$0 \leq R < |B|. \tag{15}$$

Division is therefore nothing other than repeated subtraction. An example in base ten is shown in display (16). Let $A = 15221$, $B = 42$.

$$
\begin{array}{r|r}
42\overline{)\ 15221} & \\
4200 & 100 \\
\hline
11021 & \\
12600 & 300 \\
\hline
-1579 & \\
-1260 & -30 \\
\hline
-319 & \\
-420 & -10 \\
\hline
101 & \\
168 & 4 \\
\hline
-67 & \\
-84 & -2 \\
\hline
17 &
\end{array}
\tag{16}
$$

The explanation of (16) is this: We first made a rough guess of how many B's should be subtracted from A; we tried subtracting $100B$ from A. The result is 11021. We clearly have not subtracted nearly enough B's to get into the range (15), so we then proceeded to subtract some more B's, namely $300B$. This produced -1579, i.e., we overshot the mark. So then we subtracted $-30B$ from -1579. This produced -319, still negative, and so we subtracted a further $-10B$. This produced 101, too large. So next we subtracted $4B$, yielding -67, another overshot. Next we subtracted $-2B$ from -67, and this finally gives a result, 17, in the desired range. Thus we have $A - (100B + 300B + (-30)B + (-10)B + 4B + (-2)B) = 17$, that is,

$$A = 362B + 17.$$

The quotient 362 is the sum of the integers at the right-hand edge of the display (16), and the remainder is the number at the bottom.

This method of doing long division has the attractive feature that if your guess of the trial divisor is wide of the mark and you either overshoot or undershoot, there is no need to erase and guess again; simply carry on and continue to subtract more multiples of B, always aiming to get a result in the range (15).

If the numerically least remainder is desired instead of the least non-negative remainder, the procedure is the same except that we try to get the remainder in the range $-|B|/2 < R \leq |B|/2$.

Suppose A and B are both positive. The long division process taught in grade school is the special case of the above process in which an undershot or overshot is never allowed. In base b it goes this way [see (17)]:

Locate (by trial and error) the largest exponent l and the largest digit d_l (nonzero and $<b$) such that $d_l b^l \times B$ does not exceed A. Subtract to get $R_l = A - d_l b^l B$. Then find (again by trial and error) the largest digit d_{l-1} ($<b$ and possibly zero) such that $d_{l-1} b^{l-1} \times B$ does not exceed R_l. Subtract to get $R_{l-1} = R_l - d_{l-1} b^{l-1} B$. Then locate (once more by trial and error) the largest digit d_{l-2} ($<b$ and possibly zero) such that $d_{l-2} b^{l-2} \times B$ does not exceed R_{l-1}. Subtract to get $R_{l-2} = R_{l-1} - d_{l-2} b^{l-2} B$. Continue in this way. Using these equations we find

$$
\begin{aligned}
A &= d_l b^l B + R_l = (d_l b^l + d_{l-1} b^{l-1}) B + R_{l-1} \\
&= (d_l b^l + d_{l-1} b^{l-1} + d_{l-2} b^{l-2}) B + R_{l-2} \\
&= \cdots\cdots\cdots\cdots\cdots\cdots \\
&= (d_l b^l + d_{l-1} b^{l-1} + \cdots + d_0) B + R_0 .
\end{aligned}
$$

One can prove that R_0 satisfies $0 \le R_0 < B$. (See Exercise 18 below.) Thus the quotient is $Q = d_l b^l + d_{l-1} b^{l-1} + \cdots + d_0$ and the remainder R is R_0. The work may be arranged as shown in (17).

At the right-hand edge, digits $d_l, \ldots, d_0$ of the quotient appear explicitly. In grade school one is usually taught to write these digits at the top of the configuration, i.e., above A, in the correct base b position. This placing of the digits correctly as one proceeds through the calculation eliminates the necessity of writing down the powers b^l, b^{l-1}, etc.

When worked without undershooting or overshooting, the long division example above takes the form shown in (18). Thus $Q = 300 + 60 + 2 = 362$ and $R = 17$. If one places the digits at the top, the calculation takes the appearance shown in display (19).

Division of polynomials

The procedure for long division of polynomials is similar to that for integers. If we wish to divide $a(\lambda)$ by $b(\lambda)$, we simply subtract polynomial multiples of $b(\lambda)$ from $a(\lambda)$ until we arrive at a remainder having degree smaller than $\deg b(\lambda)$. The most convenient procedure is, at each stage, to subtract (a monomial) $\times\, b(\lambda)$ of the form

$$(\text{constant} \times \lambda^t) \times b(\lambda), \quad \text{for a suitable integer exponent } t, \tag{20}$$

the constant and the exponent t being chosen so that the subtraction cancels the highest present power of λ. For example, if $a(\lambda) = 2\lambda^5 - 3\lambda^4 + 4\lambda^3 + 2\lambda^2 + \lambda - 1$ and $b(\lambda) = \lambda^2 - 2\lambda + 2$, the details of the division are shown in display (21) on page 178.

$B\overline{)\ A}$		
$d_l b^l B$	$d_l b^l$	
R_l		
$d_{l-1} b^{l-1} B$	$d_{l-1} b^{l-1}$	
R_{l-1}		
$d_{l-2} b^{l-2} B$	$d_{l-2} b^{l-2}$	(17)
R_{l-2}		
⋮	⋮	
R_1		
$d_0 b^0 B$	$d_0 b^0$	
R_0		

(18)	$42\overline{)15221}$	
	12600	300
	2621	
	2520	60
	101	
	84	2
	17	

362	
$42\overline{)15221}$	
12600	
2621	
2520	(19)
101	
84	
17	

$$
\begin{array}{rrrrrrr|l}
\lambda^2 - 2\lambda + 2\,\big) & 2\lambda^5 & - 3\lambda^4 & + 4\lambda^3 & + 2\lambda^2 & + \lambda & - 1 & \\
 & 2\lambda^5 & - 4\lambda^4 & + 4\lambda^3 & & & & 2\lambda^3 \\
\cline{2-7}
 & & \lambda^4 & & + 2\lambda^2 & + \lambda & - 1 & \\
 & & \lambda^4 & - 2\lambda^3 & + 2\lambda^2 & & & \lambda^2 \\
\cline{3-7}
 & & & 2\lambda^3 & & + \lambda & - 1 & \\
 & & & 2\lambda^3 & - 4\lambda^2 & + 4\lambda & & 2\lambda \\
\cline{4-7}
 & & & & 4\lambda^2 & - 3\lambda & - 1 & \\
 & & & & 4\lambda^2 & - 8\lambda & + 8 & 4 \\
\cline{5-7}
 & & & & & 5\lambda & - 9 &
\end{array}
\tag{21}
$$

Here we first subtracted $2\lambda^3 \times b(\lambda)$, the monomial $2\lambda^3$ being chosen so that the subtraction causes the λ^5 term to cancel. Next we subtracted $\lambda^2 \times b(\lambda)$, the monomial λ^2 being chosen so that the subtraction causes the highest remaining power of λ (the λ^4 power) to cancel. Then we subtracted $2\lambda \times b(\lambda)$, the monomial 2λ being chosen so that the subtraction causes the highest remaining power of λ (the λ^3 power) to cancel. Finally the subtraction of $4 \times b(\lambda)$ causes the λ^2 term to cancel. Thus

$$a(\lambda) - \{2\lambda^3 b(\lambda) + \lambda^2 b(\lambda) + 2\lambda b(\lambda) + 4b(\lambda)\} = 5\lambda - 9;$$

hence

$$a(\lambda) = \{2\lambda^3 + \lambda^2 + 2\lambda + 4\}b(\lambda) + 5\lambda - 9.$$

The quotient is therefore $2\lambda^3 + \lambda^2 + 2\lambda + 4$, and the remainder is $5\lambda - 9$. The remainder appears in the last line of the display, and the quotient is the sum of the monomials at the right-hand edge.

Suppose $a(\lambda)$ and $b(\lambda)$ have integral coefficients and $b(\lambda)$ is monic. At a particular stage of the calculation, when we cancel a power of λ by subtracting (20), the constant will [because $b(\lambda)$ is monic] simply be the coefficient of this power of λ. If all powers of λ still present have integral coefficients, the constant will be an integer, and after (20) has been subtracted, the powers of λ that then remain will again all have integral coefficients. Thus each stage of the calculation will leave a polynomial with integral coefficients. The remainder will therefore have integral coefficients, and so will the quotient since it is the sum of all monomials of the type (constant $\times\ \lambda^t$) used in these subtractions. We have just given a second proof of Theorem 3.3.5.

EXERCISES 4.3

I

1. Let $a = 14132_b$, $c = 2430_b$. Find, and express in base b: $a + c$, $a - c$, ac, when b is five, nine, or thirteen. Also find and express in bases five, nine, or thirteen the integers q, r satisfying $a = qc + r$, $0 \leq r < c$.

2. Add: $13214_b + 4204_b + 3124_b + 200_b$. Express the result in base b, when b is five, nine, thirteen.

3. Let $a = 1784_9$, $c = 268_9$. Find $a + c$, $a - c$, ac, and express the results in base nine. Verify your calculations by converting a, c, $a + c$, $a - c$, ac to base ten.

4. In which base b is the base ten number 137 equal to 254_b?

5. The number a, when written in base eight, has 40 as lowest digits. What are the six lowest digits of a in base two? What are the three lowest digits in base four?

6. Find the quotient q and remainder r in base b when $a = qc + r$, in each of the following cases. (You are given a, c in base b and expected to do all your work in base b):
   (i) b is two, $a = 10111011$, $c = 1101$;
   (ii) b is four, $a = 2133201$, $c = 2311$;
   (iii) b is eight, $a = 4174043$, $c = 777$;
   (iv) b is sixteen, $a = \text{CAF008D}$, $c = \text{BAFF}$.
   In each case find q and r when r is the least nonnegative remainder and when r is the numerically least remainder. In each case, convert a, c, q, r to base ten and verify that $a = qc + r$.

7. (*Optional*) Let $a = 2(-3)4(-2)$ and $c = 1(-1)(-2)4$ be integers presented in the form described in Theorem 4.2.8, where the base b is eight. Find, for this b, the representation of $a + c$, $a - c$, ac in the form described in Theorem 4.2.8. Verify your results by converting to base ten.

8. (*Optional*) Add: $a_1 + a_2 + a_3$ where (in base ten) $a_1 = 42531$, $a_2 = \ldots 999213$, $a_3 = \ldots 994567$. (Here a_2, a_3 are negative integers in the complement form.) Find also a_1a_2 and a_2a_3. Find $a_1 - a_2$.

9. (*Optional*) Use the method of halving and doubling to multiply $a \times c$ when $a = 1234$, $c = 421$; $a = 2479$, $c = 1541$; $a = 411$, $c = 2429$. (These numbers are in base ten.)

II

10. Let a and c be positive integers. Suppose $a + c \equiv 0 \pmod 8$. Let a_0 and c_0 be the lowest digits in the base eight representations of a and c. Show that $a_0 + c_0 \equiv 0 \pmod 8$.

11. If a be a positive integer in base ten, let $s(a)$ be the sum of the digits of a, $s(s(a))$ be the sum of the digits of $s(a)$, etc. Form $s(a)$, $s(s(a))$, $s(s(s(a)))$, etc., until eventually a single digit is obtained. For example, if $a = 787796798$, then $s(a) = 68$, $s(s(a)) = 14$, $s(s(s(a))) = 5$. Call the digit resulting this way $S(a)$. Prove: for integers a and c, $S(a + c) = S(S(a) + S(c))$, $S(ac) = S(S(a)S(c))$.
[Hint: Use the fact that $10 \equiv 1 \pmod 9$, hence $a \equiv S(a) \pmod 9$. See Exercise 11, §4.2.] (This procedure is called "casting out nines.")

12. Show that the rules developed in the text for performing arithmetic operations apply equally well when adding, subtracting, multiplying, and dividing polynomials in base b. In particular, show that addition and subtraction for polynomials is simpler than for integers in the sense that there never will be a "carry." Show that this is also true for multiplication of polynomials if the base $b(\lambda)$ has degree 1.

13. Find the base b if $12_b \times 21_b = 302_b$.

14. Let a and c be positive integers. Suppose the lowest digit of a in base b is 3, and that of c is 2. If the lowest digit of $a + c$ is 1, find b. If instead, the lowest digit of ac is 1, find b.

15. (*Optional*) Construct a "taking thirds and tripling" multiplication method as follows. Let a and α be positive integers. From two columns, headed by a and α. Form new entries in these columns this way: Divide the current lowest number in the a column by 3; if the remainder is zero put a line through the opposite number in the α column; if the remainder is 2 write the number in the α column again. Put the quotient obtained in the division as the new lowest number in the a column and put a new lowest number in the α column by tripling the previous entry in the α column. Show that the sum of the numbers in the α column not having a line struck through them is $a\alpha$. Example: $a = 179$, $\alpha = 41$,

179	41	
	41	
59	123	
	123	
19	369	$a\alpha = 7339$
6	~~1107~~	
2	3321	
	3321	
	7339	

(Hint: Imitate the proof in the text for the halving and doubling method.)

III

16. Consider a base b scientific calculating machine. It can record a maximum of $l + 1$ digits, so that if an addition or subtraction results in a number $d_k d_{k-1} \ldots d_1 d_0$ in which $k > l$, the machine records only $d_l d_{l-1} \ldots d_1 d_0$, and loses the other digits. (This phenomenon is called *overflow*.) If one subtracts positive integer $a = d_l d_{l-1} \ldots d_1 d_0$ (d_l may be zero here) from zero on this machine, it attempts to yield the complement of a, but because it can carry only $l + 1$ digits, it yields instead the *adding machine complement* of a, namely $\delta_l \delta_{l-1} \ldots \delta_1 \delta_0 = b^{l+1} - a$. (Notation here is as in the discussion at the end of §4.2, except that we may have $d_l = 0$.)

(i) Show that if one adds $\delta_l \ldots \delta_0$ to $d_l \ldots d_0$ on this machine, the result will indeed be zero.

(ii) Suppose the machine has a double length register for multiplication, i.e., if two positive integers $a = d_l \ldots d_0$ and $c = d_l' \ldots d_0'$ are multiplied, all $2l + 2$ digits of the product ac appear in the multiplication register. (Many adding machines and electronic calculating machines designed for scientific usage have this feature.) The operator wishes to multiply two negative numbers, $-a$ and $-c$. So he enters their adding machine complements $b^{l+1} - a$ and $b^{l+1} - c$ into his machine, then has the machine multiply them. The machine multiplies $b^{l+1} - a$ and $b^{l+1} - c$, treating them as positive integers, and shows a $2l + 2$ digit result. Will the operator be able to recognize the product he wishes, namely ac, in the result showing on the machine? More generally, to what extent are the rules of the minus sign valid when multiplications involve adding machine complements?

If there is a suitable adding machine available for your use, make some experiments on it. An example of an adding machine with the features described above is the Marchant, model AB10FA. For this machine $l + 1 = 10$.

See §8.4 for more details.

17. Investigate how to modify the usual addition, subtraction, and multiplication schemes when using the representation of integers with positive and negative digits described in Theorem 4.2.8.

18. This exercise refers to the discussion of long division of integers given in the text, and uses the notation of the text.

(a) Prove that $R_l < b^l B \leq A$. (Hint: If not, one could increase either l or d_l.)

(b) Prove that $R_t < b^t B$, in succession for $t = l, l - 1, l - 2, \ldots, 0$. (Hint: The case $t = l$ is in part (a). Suppose we know $R_t < b^t B$. We have $R_t - d_{t-1} b^{t-1} B = R_{t-1}$. If $R_{t-1} \geq b^{t-1} B$, then $R_t \geq (d_{t-1} + 1) b^{t-1} B$, and this would imply either that d_{t-1} is not the largest possible digit or else that $R_t \geq b^t B$.) [*Cont.*]

(c) Conclude that $R_0 < B$.
(d) Deduce that R_0 is indeed the remainder obtained on division of A by B.

19. In the usual procedure for a multiplication $a \times \alpha$, show how one may replace a complicated multiplication of the type (a large digit of α) $\times\, a$ by two multiplications of the type (a small digit) $\times\, a$ combined with an extra addition. For example, if (in base ten) the multiplication $9 \times a$ is encountered, replace it with the multiplications $5 \times a$ and $4 \times a$, both of which are easier.

4.4 APPLICATIONS (OPTIONAL)

We shall give two applications of the theory of §4.2. The first application is to estimate the number of steps needed to calculate the greatest common divisor of a pair of integers by Euclid's algorithm. We make our estimate when least nonnegative remainders are used throughout. So let a and b be positive integers, with $a > b$. Put $r_0 = a$, $r_1 = b$, and consider this system of equations:

$$\begin{aligned}
r_0 &= q_1 r_1 + r_2\,, & 0 &< r_2 < r_1\,,\\
r_1 &= q_2 r_2 + r_3\,, & 0 &< r_3 < r_2\,,\\
&\cdots\cdots\cdots\cdots\cdots\cdots\\
r_j &= q_{j+1} r_{j+1} + r_{j+2}\,, & 0 &< r_{j+2} < r_{j+1}\,,\\
&\cdots\cdots\cdots\cdots\cdots\cdots\\
r_{m-2} &= q_{m-1} r_{m-1} + r_m\,, & 0 &< r_m < r_{m-1}\,,\\
r_{m-1} &= q_m r_m\,.
\end{aligned} \tag{1}$$

The number of divisions required in (1) is m, and the last nonzero remainder r_m is (a, b). We wish to compare m with b. Lamé's theorem, which we prove next, provides a rapid comparison.

THEOREM 4.4.1 Lamé's Theorem
If a, b are positive integers, with $b \leq a$, the number of divisions required in computing (a, b) by Euclid's algorithm based on least nonnegative remainders is at most five times the number of base ten digits in b.

PROOF The theorem is certainly true if $b \mid a$ since then only one division is required. So assume $b \nmid a$.

Observe that the quotients $q_1, \ldots, q_m$ in (1) are all at least 1 (since, for example, if $q_{j+1} = 0$ for $j < m - 1$ then $r_j = r_{j+2}$, contradicting

$r_{j+2} < r_{j+1} < r_j$) and q_m itself satisfies $q_m \geq 2$ (since $q_m = 1$ implies $r_m = r_{m-1}$). Using these estimates, from (1) we get $r_m \geq 1$, $r_{m-1} \geq 2r_m$, $r_{m-2} \geq r_{m-1} + r_m$, $r_{m-3} \geq r_{m-2} + r_{m-1}$, . . ., $r_2 \geq r_3 + r_4$, $r_1 \geq r_2 + r_3$.

Let $\xi = (1 + \sqrt{5})/2$. Then $\xi^2 = \xi + 1$. We claim that

$$r_m \geq \xi^0, \quad r_{m-1} \geq \xi^1, \quad r_{m-2} \geq \xi^2, \ldots, \quad r_{m-j} \geq \xi^j, \ldots, \quad r_1 \geq \xi^{m-1}.$$

Certainly $r_m \geq \xi^0$ since $r_m \geq 1$ and $\xi^0 = 1$. Furthermore, $r_{m-1} \geq 2r_m \geq 2$ and $\xi < 2$, so that $r_{m-1} \geq \xi$. Thus the inequalities

$$r_{m-(j-1)} \geq \xi^{j-1} \qquad \text{and} \qquad r_{m-j} \geq \xi^j \tag{2}$$

are both valid when $j = 1$. If these two inequalities are valid for a certain value of j, then

$$r_{m-(j+1)} \geq r_{m-j} + r_{m-(j-1)} \geq \xi^j + \xi^{j-1} = \xi^{j-1}(\xi + 1) = \xi^{j-1}\xi^2 = \xi^{j+1}.$$

Therefore both $r_{m-j} \geq \xi^j$ and $r_{m-(j+1)} \geq \xi^{j+1}$ are valid. That is, beginning with $j = 1$, we may progressively increase j in (2) until we reach $j = m - 1$. For $j = m - 1$, the second inequality (2) becomes

$$b \geq \xi^{m-1},$$

since $b = r_1$. We shall use this below.

Suppose the base ten representation of b is

$$b = d_0 + d_1 10 + d_2 10^2 + \cdots + d_{k-1}10^{k-1}, \qquad d_{k-1} \neq 0.$$

Here k is the number of base ten digits in b. Then

$$b \leq 9 + 9 \cdot 10 + \cdots + 9 \cdot 10^{k-1} = 9\,\frac{10^k - 1}{10 - 1} = 10^k - 1 < 10^k.$$

Combining this with $\xi^{m-1} \leq b$, we get $\xi^{m-1} < 10^k$. Taking logarithms (to base ten) yields $m - 1 < k/\log \xi$. From a table of logarithms we find that $\log \xi > 1/5$. Hence $m - 1 < 5k$. Since $m - 1$ and k are integers, we see that $m \leq 5k$. □

If numerically least remainders are employed at each step in Euclid's algorithm, r_2, r_3, . . . will satisfy these inequalities [in place of the inequalities in (1)]:

$$\begin{aligned} -|r_1|/2 &< r_2 \leq |r_1|/2, \\ -|r_2|/2 &< r_3 \leq |r_2|/2, \\ &\cdots\cdots \\ -|r_{m-1}|/2 &< r_m \leq |r_{m-1}|/2. \end{aligned} \tag{3}$$

Because of the presence of the factor 1/2, it is reasonable to think that the remainders in (3) will decrease in size very rapidly, perhaps more rapidly than the remainders in (1), and therefore fewer divisions will be required. If the reader tries numerical examples, he will find that the number of steps using numerically least remainders is never more (and often less) than the number of steps using other types of remainders. In fact. this is always true, and may be stated as a theorem.

THEOREM 4.4.2
The number of divisions required in the computation of the greatest common divisor of a pair of integers by Euclid's algorithm is smallest if numerically least remainders are used at each stage.

The proof is too long to give here and may be found in the book *Elementary Number Theory*, by J. V. Uspensky and M. A. Heaslot, New York, 1939. The number of divisions needed using numerically least remainders does not exceed $\frac{8}{3}(k + \frac{1}{2})$, where k is the number of base ten digits in b. In Exercise 5 below we outline a proof of this estimate.

As the second application of the theory, we describe the game of *Nim*, a game played by two people. There are three piles of chips or similar tokens. The players play alternately. A player, at his move, picks up as many chips as he wishes (but at least one chip) from any one of the three piles of chips. The player who picks up the last chip is the winner.

We can use the representation of integers in base two to devise a strategy such that a player will never lose, provided that he knows the strategy and is allowed to determine how many chips are in each pile at the beginning of the game.

Suppose the three piles have n, n', and n'' chips. Write the integers n, n', n'' in base two:

$$n = d_l d_{l-1} \ldots d_1 d_0 ,$$

$$n' = d_l' d_{l-1}' \ldots d_1' d_0',$$

$$n'' = d_l'' d_{l-1}'' \ldots d_1'' d_0''.$$

Here $d_0, \ldots, d_l, d_0', \ldots, d_l', d_0'', \ldots, d_l''$ are all 0 or 1. (We may assume l is the same for all three integers n, n', n'' by including additional zero digits, e.g., $17 = 0010001$.) We now say that the configuration is type I if

$$d_j + d_j' + d_j'' \text{ is even} \tag{4}$$

for each $j = 0, 1, \ldots, l$. If there is a value of j for which (4) does not hold, we say the configuration is type II.

If the configuration is type I, and a player makes any move whatsoever, a configuration of type II must result. For suppose the move is to remove some chips from the first pile. This move causes the integer n to change to some other value, say N. Let

$$N = D_l D_{l-1} \ldots D_1 D_0 .$$

Each D_i is 0 or 1. At least one digit of N is different from that of n, since $N \neq n$. Suppose $D_k \neq d_k$. Thus either $D_k = d_k + 1$ or $D_k = d_k - 1$. But then the sum (4) for N, n', n'' with $j = k$ is

$$D_k + d_k' + d_k'' = d_k + d_k' + d_k'' \pm 1 = \text{even} \pm 1 = \text{odd}.$$

Thus a type II configuration must result.

Suppose the configuration is type II. We claim there is at least one move that will create a type I configuration. Since the configuration is type II, there is at least one value of j for which (4) does not hold. Let the largest value of j for which (4) does not hold be J. Then at least one of d_J, d_J', d_J'' is a 1, since $d_J + d_J' + d_J''$ is not even. Suppose $d_J = 1$. The move will then be to change the number of chips in the first pile from n to N, where N is given in base 2 as $N = d_l \ldots d_{j+1}D_JD_{J-1} \ldots D_2D_1D_0$, the digits D_J, $D_{J-1}, \ldots, D_1, D_0$ being determined in this way: $D_J = 0$ and $D_{J-1}, D_{J-2}, \ldots$, D_0 are chosen to be 0 or 1 such that

$$D_j + d_j' + d_j'' \text{ is even}, \qquad \text{for } j = 0, 1, \ldots, J - 1.$$

This is always possible: if $d_j' + d_j''$ is even, take $D_j = 0$; if $d_j' + d_j''$ is odd, take $D_j = 1$. The move is then: remove $n - N$ chips from the first pile. By our choice of N this will result in a type I configuration. We must, however, show that the move is a valid move, that is, show that $N < n$. We have

$$n = d_l2^l + \cdots + d_J2^J + \cdots + d_0, \qquad d_J = 1,$$

$$N = d_l2^l + \cdots + D_J2^J + \cdots + D_0, \qquad D_J = 0,$$

hence

$$n - N = 2^J + (d_{J-1} - D_{J-1})2^{J-1} + \cdots + (d_0 - D_0).$$

We have $(d_{J-1} - D_{J-1}) \geq -1, \ldots, d_0 - D_0 \geq -1$; hence

$$\begin{aligned} n - N &\geq 2^J - 2^{J-1} - \cdots - 2 - 1 \\ &= 2^J - (2^J - 1) = 1. \end{aligned}$$

(Here we used Lemma 2.2.3 to sum the finite geometric series $1 + 2 + 2^2 + \cdots + 2^{J-1}$.)

Thus, if the configuration is type II, there is a move that will create a type I configuration.

Example Let $n = 31$, $n' = 27$, $n'' = 10$. Then

$$31 = 11111_2, \qquad 27 = 11011_2, \qquad 10 = 01010_2,$$

so that $J = 3$, and any one of the following moves will create a type I configuration: replace 11111_2 with 10001_2, or replace 11011_2 with 10101_2, or replace 01010_2 with 00100_2. We should therefore take 14 chips from the first pile, or 6 from either the second or the third pile.

Let players A and B play Nim. The configuration is type I, and A offers B the first move. So B must leave a type II configuration. Then A chooses his move to create a type I configuration. Again B must create a type II configuration, and then A chooses his move to once more create a type I configuration. The play continues in this fashion, A always choosing his move to create a type I configuration, and B therefore after his move always leaving a type II configuration. Since a type II configuration has $d_j + d_j' + d_j'' =$ odd for at least one j, at least one of d_j, d_j', d_j'' is 1, and hence at least one pile contains a chip. Thus B, after each of his moves, has left a chip, and therefore cannot be the winner. Consequently A wins.

EXERCISES 4.4

I

1. Play some games of Nim with your nonmathematical friends; and win.

2. Compute $(-265, 370)$, $(154, -112)$, $(112, 175)$, in each case using the least possible number of divisions.

3. In the game of Nim there are 27, 3, 7 chips in the three piles. Find the unique move which will create a type I configuration.

II

4. Read the sections of the book of Uspensky and Heaslot devoted to the least remainder algorithm and the game of Nim.

III

5. Let the remainders in equations (1) be the numerically least remainders, so that $|r_m| \le \frac{1}{2}|r_{m-1}|$, $|r_{m-1}| \le \frac{1}{2}|r_{m-2}|$, . . ., $|r_2| \le \frac{1}{2}|r_1|$, and $r_m \ne 0$.
   (i) Rewrite $r_1 = q_2 r_2 + r_3$ as $|r_1| = Q_2|r_2| + \epsilon|r_3|$, where $Q_2 = \pm q_2$ and $\epsilon = \pm 1$.
   (ii) Prove that $Q_2 \ge 2$. (Hint: If $Q_2 \le 1$, then $|r_1| \le |r_2| + |r_3| \le 3|r_1|/4$.)
   (iii) If $\epsilon = -1$, prove that $Q_2 \ge 3$. (Hint: If $Q_2 = 2$, then $|r_1| < |r_1|$.)
   (iv) Prove that $|r_1| \ge 2|r_2| + |r_3|$. (Hint: Consider separately the cases $\epsilon = 1$ and $\epsilon = -1$.)
   (v) Similarly show that

$$|r_1| \ge 2|r_2| + |r_3|, \qquad |r_2| \ge 2|r_3| + |r_4|, \ldots,$$
$$|r_{m-2}| \ge 2|r_{m-1}| + |r_m|,$$

   and that $|r_{m-1}| \ge 2$. Let $\eta = 1 + \sqrt{2}$.
   (vi) Using the inequalities in (v), prove that $|r_{m-j}| \ge .82\eta^j$ in succession for $j = 0, 1, \ldots, m-1$. Imitate the proof of Lamé's theorem.

(vii) Deduce that $|b| \geq .82\eta^{m-1}$.

(viii) Then prove that $10^k > .82\eta^{m-1}$, where k is as in the proof of Lamé's theorem. From this, using a table of common logarithms, prove that $m \leq \frac{8}{3}(k + \frac{1}{2})$.

4.5 REVIEW

The theme of the chapter has been the various ways of representing integers with respect to a base, and how these representations can be used to perform the fundamental arithmetic operations of addition, subtraction, multiplication, division. We also developed the representation theory of a polynomial with respect to a base, obtaining thereby Taylor's theorem for polynomials and, as a special case, the binomial theorem.

REVIEW EXERCISES 4.5

1. Let $a = d_l d_{l-1} \ldots d_1 d_0$ be a positive integer represented in base ten. Let $d(a) = d_0 - d_1 + d_2 - d_3 + d_4 - \cdots$ be the alternating sum of the digits of a. Apply $d(a)$ repeatedly, i.e., compute $d(d(a))$, $d(d(d(a)))$, . . ., until a single digit results. Call this digit $D(a)$. Using the fact that $10 \equiv -1 \pmod{11}$, show that $a \equiv d(a) \equiv d(d(a)) \equiv \cdots \equiv D(a) \pmod{11}$. Hence prove: If a and b are integers, then $D(D(a) + D(b)) \equiv D(a + b) \pmod{11}$, $D(D(a)D(b)) \equiv D(ab) \pmod{11}$.

2. In base ten an integer a is divisible by 5 if and only if its lowest digit is divisible by 5. In what other bases b is this true?

3. Estimate the number of steps needed to calculate the greatest common divisor of a pair of polynomials by Euclid's algorithm.

4. Let a be a positive integer. Suppose that the lowest digit of a in base three is 0, but the next lowest is not zero. Suppose this is also true in base seven. Suppose in every other base, with one exception, the lowest digit is not zero. Find a, and find the exception.

5

POLYNOMIALS AND THEIR ZEROS

5.1 INTRODUCTION

Very likely the reader has solved linear or quadratic equations of the form $ax + b = 0$ or $ax^2 + bx + c = 0$ for an unknown x, where a, b, c are known quantities, with a nonzero. When a, b, c are real numbers, he probably has found that the solution of the first of these equations is unique and is a real number, whereas usually there are two solutions of the second, but sometimes only one, and that the solutions may be complex numbers. This chapter gives a reasonably complete account of the solutions of equations of the form $a(x) = 0$, where $a(\lambda)$ is a polynomial with coefficients in a field $\mathfrak{F}$, which usually will be $\mathbf{Q}$, $\mathbf{R}$, or $\mathbf{C}$. The cornerstone of the theory will be the unique factorization theorem for the polynomial ring $\mathfrak{F}[\lambda]$.

5.2 ZEROS OF POLYNOMIALS—GENERAL THEORY

Let $\mathfrak{F}$ be a field. Since we shall need the theory of this section only when $\mathfrak{F}$ is $\mathbf{Q}$, $\mathbf{R}$, or $\mathbf{C}$, the reader may assume that $\mathfrak{F}$ is one of these fields. Throughout this section $f(\lambda)$ denotes a nonzero polynomial with coefficients in $\mathfrak{F}$.

In §3.2 we discussed the evaluation of a polynomial at another polynomial α (i.e., substitution of α for the indeterminate λ), and we showed that evaluation at α is compatible with polynomial addition and multiplication. In this chapter we restrict α to be an element of $\mathfrak{F}$, i.e., a constant.

DEFINITION 5.2.1.
An element α of $\mathfrak{F}$ is a root, *or* zero, *of $f(\lambda)$ if $f(\alpha) = 0$.*

For example, 1 and $-5/2$ are roots of $2\lambda^2 + 3\lambda - 5$.

The link between the theory in Chapter 3 and zeros of polynomials is furnished by the next lemma.

***LEMMA* 5.2.2**
An element α of $\mathfrak{F}$ is a root of a polynomial $f \in \mathfrak{F}[\lambda]$ if and only if $(\lambda - \alpha) \mid f(\lambda)$.

PROOF Suppose first that $(\lambda - \alpha) \mid f(\lambda)$. Then a polynomial $q(\lambda)$ in $\mathfrak{F}[\lambda]$ exists such that $f(\lambda) = (\lambda - \alpha)q(\lambda)$. Evaluating at α, we get $f(\alpha) = (\alpha - \alpha)q(\alpha) = 0$.

Conversely, suppose that $f(\alpha) = 0$. Divide $f(\lambda)$ by $\lambda - \alpha$ to get $f(\lambda) = q(\lambda)(\lambda - \alpha) + r$, where $q(\lambda) \in \mathfrak{F}[\lambda]$ and $r \in \mathfrak{F}$. The remainder theorem (3.3.4) gives $r = f(\alpha)$, hence $r = 0$, and therefore $(\lambda - \alpha) \mid f(\lambda)$. □

Applying the unique factorization theorem (3.5.3) to the element $f(\lambda)$ of $\mathfrak{F}[\lambda]$, we find that

$$f = up_1^{e_1}p_2^{e_2} \ldots p_m^{e_m} \tag{1}$$

where $p_1, p_2, \ldots, p_m$ are monic distinct prime elements of $\mathfrak{F}[\lambda]$ (i.e., irreducible polynomials) and $e_1, \ldots, e_m$ are positive integers. The quantity u is a nonzero constant. In general, the degrees of some of $p_1, \ldots, p_m$ will be 1, the rest having higher degrees. We may choose the notation so that $\deg p_1 = \cdots = \deg p_t = 1$, $\deg p_{t+1} > 1, \ldots, \deg p_m > 1$. Since $p_1, \ldots, p_m$ are distinct and monic,

$$p_1(\lambda) = \lambda - \alpha_1, \quad p_2(\lambda) = \lambda - \alpha_2, \ldots, \quad p_t(\lambda) = \lambda - \alpha_t,$$

where $\alpha_1, \ldots, \alpha_t$ are distinct elements of $\mathfrak{F}$. Thus the factorization (1) becomes

$$f(\lambda) = u(\lambda - \alpha_1)^{e_1} \ldots (\lambda - \alpha_t)^{e_t} p_{t+1}(\lambda)^{e_{t+1}} \ldots p_m(\lambda)^{e_m}, \tag{2}$$

where

$\alpha_1, \ldots, \alpha_t$ are distinct elements of $\mathfrak{F}$,

u is a nonzero element of $\mathfrak{F}$,

$p_{t+1}, \ldots, p_m$ are distinct monic primes in $\mathfrak{F}[\lambda]$, each of degree at least 2, (3)

$e_1, \ldots, e_m$ are positive integers.

(In special cases it may happen that $t = 0$ or $t = m$.)

***THEOREM* 5.2.3**
Let (2) and (3) be the factorization of $f(\lambda)$ into a product of powers of distinct monic primes in $\mathfrak{F}[\lambda]$. Then $\alpha_1, \ldots, \alpha_t$ are the only roots of f in $\mathfrak{F}$.

PROOF If α is a root of f in $\mathfrak{F}$, then (by Lemma 5.2.2) $(\lambda - \alpha) \mid f(\lambda)$. Since $\lambda - \alpha$ is a monic *prime* (see Example 3.5.2), by the *unique* factorization theorem, it must coincide with one of the monic prime factors in (2). Since $p_{t+1}, \ldots, p_m$ each have degree at least 2, it can coincide only with one of $\lambda - \alpha_1, \ldots, \lambda - \alpha_t$, implying that α is one of $\alpha_1, \ldots, \alpha_t$. On the other hand, evaluating f at any of $\alpha_1, \ldots, \alpha_t$ clearly gives 0, so that $\alpha_1, \ldots, \alpha_t$ are indeed roots of f. □

DEFINITION 5.2.4.
The root α_j belonging to the factor $(\lambda - \alpha_j)^{e_j}$ in (2) is called a root of f of multiplicity e_j, *and also called a* simple *root when $e_j = 1$ or a* multiple *root when $e_j > 1$; $j = 1, 2, \ldots, t$. We say that t is the number of distinct roots of f in $\mathfrak{F}$ and that $e_1 + \cdots + e_t$ is the number of roots counting multiplicities.*

Since $t \leq e_1 + \cdots + e_t$ (because each $e_j \geq 1$), the number of distinct roots does not exceed the number of roots counting multiplicities.

COROLLARY 5.2.5
The number of roots in $\mathfrak{F}$, counting multiplicities, of a nonzero polynomial $f \in \mathfrak{F}[\lambda]$ is at most $\deg f$.

PROOF Computing the degree of each side of (2), we get

$$\deg f = e_1 + \cdots + e_t + e_{t+1} \deg p_{t+1} + \cdots + e_m \deg p_m$$
$$\geq e_1 + \cdots + e_t . \qquad \square$$

Note that a polynomial with coefficients in $\mathfrak{F}$ need not have any roots in $\mathfrak{F}$. For example, the polynomial $\lambda^2 + 1$ with coefficients in $\mathbf{R}$ has no roots in $\mathbf{R}$.

There is an interesting relationship between the multiplicities of the roots of f and those of its derivative f'. This relationship is not valid for all fields $\mathfrak{F}$, but is valid when $\mathfrak{F}$ is $\mathbf{Q}$, $\mathbf{R}$, or $\mathbf{C}$.

THEOREM 5.2.6
Let the field $\mathfrak{F}$ be $\mathbf{Q}$, $\mathbf{R}$, or $\mathbf{C}$. If α is a root of multiplicity e of a nonconstant polynomial $f \in \mathfrak{F}[\lambda]$, then α is a root of multiplicity $e - 1$ of its derivative f'. In particular, a simple root of f is not a root of f'.

PROOF If α is a root of f of multiplicity e, taking α_1 and e_1 in the factorization (2) to be α and e, we get

$$f(\lambda) = (\lambda - \alpha)^e g(\lambda), \quad \text{where } g(\lambda) \text{ is a polynomial with } g(\alpha) \neq 0.$$

Differentiating, we obtain

$$f'(\lambda) = (\lambda - \alpha)^{e-1}(eg(\lambda) + (\lambda - \alpha)g'(\lambda)) = (\lambda - \alpha)^{e-1}G(\lambda),$$

where $G(\lambda) = eg(\lambda) + (\lambda - \alpha)g'(\lambda)$. Then $G(\alpha) = eg(\alpha) \neq 0$, so that $(\lambda - \alpha) \nmid G(\lambda)$. From $f' = (\lambda - \alpha)^{e-1}G$ and $(\lambda - \alpha) \nmid G$, it follows that the multiplicity of the prime factor $\lambda - \alpha$ of f' is $e - 1$. That is: α is a root of f' of multiplicity $e - 1$. □

COROLLARY 5.2.7

Take $\mathfrak{F}$ as in Theorem 5.2.6. Let $d = (f, f')$ be the greatest common divisor of f and f', and set $F(\lambda) = f(\lambda)/d(\lambda)$. Then the roots of F in $\mathfrak{F}$ are the distinct roots of f in $\mathfrak{F}$, and each is a simple root of F.

PROOF Continuing the notation used above, we have $f(\lambda) = (\lambda - \alpha)^e g(\lambda)$ and $f'(\lambda) = (\lambda - \alpha)^{e-1}G(\lambda)$, where $(\lambda - \alpha) \nmid g(\lambda)$ and $(\lambda - \alpha) \nmid G(\lambda)$. Since $(\lambda - \alpha)^{e-1}$ divides both f and f', it must also divide (f, f'). Thus $d(\lambda) = (\lambda - \alpha)^{e-1}h(\lambda)$, where $h(\lambda)$ is a polynomial. If $(\lambda - \alpha) \mid h(\lambda)$, then $(\lambda - \alpha)^e \mid d(\lambda)$, and (because $d \mid f'$) we would have $(\lambda - \alpha)^e \mid f'$; this we showed above to be impossible. Thus $(\lambda - \alpha) \nmid h(\lambda)$. From $f = (\lambda - \alpha)^e g$, $d = (\lambda - \alpha)^{e-1}h$ and $f = dF$, we get $(\lambda - \alpha)g(\lambda) = h(\lambda)F(\lambda)$. The prime $\lambda - \alpha$ appears on the left-hand side of this equation exactly once. It must therefore appear exactly once on the right-hand side. It is not a factor of $h(\lambda)$, so it must be a simple factor of $F(\lambda)$. This shows that a root α of $f(\lambda)$ is a simple root of $F(\lambda)$. From $f(\lambda) = d(\lambda)F(\lambda)$ it is clear that any root of $F(\lambda)$ is also a root of $f(\lambda)$. So $F(\lambda)$ can have no roots other than simple roots coinciding with the roots of f. □

EXERCISES 5.2

Throughout I and II, all polynomials are in $\mathbf{C}[\lambda]$, the symbols α, β denote constants, f and g denote polynomials, and f' denotes the derivative of f, unless stated otherwise.

I

1. Find all roots in $\mathbf{Q}$ of the following polynomials: $\lambda^2 + 1$; $\lambda^2 + 2\lambda + 3$; $\lambda - 1$; $2\lambda + 5$; $\lambda^2 - 2$; $\lambda^3 - 1$; $\lambda^3 + 1$; $\lambda^4 - 4\lambda^2 + 4$. Find also all roots of these polynomials in $\mathbf{R}$ and all roots in $\mathbf{C}$.

2. Find all multiple roots of the polynomial $\lambda^4 - 4\lambda^3 + 5\lambda^2 - 4\lambda + 4$. Then find all roots in $\mathbf{C}$ of this polynomial. Find a root of its derivative. Then find all roots of its derivative.

3. Construct an example demonstrating that the sum, difference, product, or quotient of roots of a polynomial f need not be roots of f.

4. Find all roots in $\mathbf{C}$ of the polynomial $\lambda^2 - i\sqrt{12} + 2$. Use the method in Exercise 16 of §2.3.

5. The number 2 is a root of $f(\lambda)$ of multiplicity three and $f'(\alpha) \neq 0$ for every other root α of f. What is (f, f')?

6. Can it happen that a simple root α of $f(\lambda)$ is a root of $D^2f = f''$? Can it be a multiple root of f''?

II

7. Let $f(\alpha) = 0, f(\beta) = 0$ where $\alpha \neq \beta$. Show that $(\lambda - \alpha)(\lambda - \beta) \mid f(\lambda)$.

8. If α is a root of f and if $f \mid g$, show that α is a root of g. Is the converse true?

9. Consider the polynomial $(\lambda - 1)^2$ in $\mathfrak{F}[\lambda]$, where $\mathfrak{F}$ is the two-element field of Exercise 3, §1 5. Show that Theorem 5.2.6 does not hold for this polynomial.

10. If f has only simple roots, show that polynomials a and b exist such that $1 = af + bf'$. Is this conclusion still valid if f has multiple roots?

11. If α is a simple root of f', is α a root of f of multiplicity two?

12. Suppose $a_n, a_{n-1}, \ldots, a_0 \in \mathbf{C}$ and $f(\lambda) = a_n\lambda^n + a_{n-1}\lambda^{n-1} + \cdots + a_0 \in \mathbf{C}[\lambda]$. If $a_n + a_{n-1} + \cdots + a_1 + a_0 = 0$, find a root of f.

III

13. Let $f, g \in \mathfrak{F}[\lambda]$. Suppose $\deg f \leq n$, $\deg g \leq n$, and that $f(\alpha_j) = g(\alpha_j)$, $j = 1, 2, \ldots, n + 1$, where $\alpha_1, \alpha_2, \ldots, \alpha_{n+1}$ are distinct elements of $\mathfrak{F}$. Deduce that $f = g$. (Hint: how many roots does $f - g$ have?)

14. Let $\alpha_1, \alpha_2, \ldots, \alpha_n$ be elements of a field $\mathfrak{F}$. By induction on n, show that, when multiplied out, $(\lambda - \alpha_1)(\lambda - \alpha_2) \ldots (\lambda - \alpha_n) = \lambda^n - a_1\lambda^{n-1} + a_2\lambda^{n-2} - \cdots + (-1)^n a_n$ where $a_1, \ldots, a_n \in \mathfrak{F}$ and

$$a_1 = \alpha_1 + \cdots + \alpha_n ,$$

$$a_n = \alpha_1 \ldots \alpha_n .$$

Generalize, and show that

$$a_j = \sum_{1 \leq i_1 < \cdots < i_j \leq n} \alpha_{i_1}\alpha_{i_2} \ldots \alpha_{i_j} .$$

(The sum extends over all sets of integers $i_1, \ldots, i_j$ with $1 \leq i_1 < \cdots < i_j \leq n$.)

15. If $f(\lambda) = (\lambda - \alpha_1) \cdots (\lambda - \alpha_n)$, where $\alpha_1, \ldots, \alpha_n$ are in $\mathfrak{F}$, show that $f'(\alpha_j) = (\alpha_j - \alpha_1) \cdots (\alpha_j - \alpha_{j-1})(\alpha_j - \alpha_{j+1}) \cdots (\alpha_j - \alpha_n)$.

16. Show that any root α of a nonzero polynomial $f(\lambda)$ in $\mathfrak{F}[\lambda]$ *must* be a constant. That is, if $\alpha(\lambda) \in \mathfrak{F}[\lambda]$ and $f(\alpha) = 0$, then $\alpha \in \mathfrak{F}$.

5.3 THE FUNDAMENTAL THEOREM OF ALGEBRA

The following theorem is often called the Fundamental Theorem of Algebra.

THEOREM 5.3.1
A nonconstant polynomial with complex coefficients always has a complex root.

When applied to a polynomial with real coefficients, Theorem 5.3.1 asserts that a complex root exists; in special cases there may be a real root. However, it is *not* true that a nonconstant polynomial with real coefficients always has a real root (consider $\lambda^2 + 1$).

The remainder of this section consists solely of two proofs of Theorem 5.3.1. The reader who is willing to accept the theorem without proof may immediately go on to the next section. However, the reader who wishes to understand *why* the theorem is true will find reading these proofs worthwhile. He will also gain insight into procedures to be described later for calculating the roots.

Let $f(\lambda)$ be a nonconstant polynomial with complex coefficients. We are to find a complex number α for which $f(\alpha) = 0$. Multiplying f by a nonzero constant, we may assume without loss of generality that f is monic. We may also assume that $f(0) \neq 0$ since otherwise 0 is a root of f and the theorem is proved.

First proof Let z denote a variable complex number and set

$$t = |f(z)|. \tag{1}$$

Suppose $z = x + yi$ where x, y are real. Form a three-dimensional coordinate system—see Figure 5.3.1—with axes X, Y, T. Here $z = (x, y) = x + yi$ is a point in the XY plane and (x, y, t) is a point in three-dimensional space. As (x, y) varies over all points in the XY plane, the point (x, y, t) describes a surface S in 3-space. Since $t = |f(z)| \geq 0$, S is always in or above the XY plane.

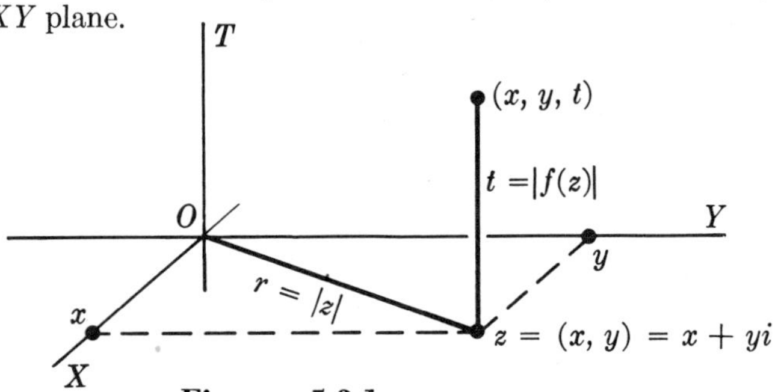

Figure 5.3.1

We wish to make a rough sketch of S. To do so, we first study S when $r = |z|$ is large, i.e., (x, y) is far from the origin. Let

$$f(\lambda) = \lambda^n + a_{n-1}\lambda^{n-1} + \cdots + a_1\lambda + a_0\,, \tag{2}$$

where $a_{n-1}, \ldots, a_0$ are complex numbers. Then

$$f(z) = z^n\left(1 + \frac{a_{n-1}}{z} + \cdots + \frac{a_1}{z^{n-1}} + \frac{a_0}{z^n}\right)$$

$$= z^n(1 + W) \tag{3}$$

where

$$W = a_{n-1}/z + \cdots + a_0/z^n. \tag{4}$$

By the triangle inequality,

$$|W| \le |a_{n-1}|/r + \cdots + |a_0|/r^n, \tag{5}$$

so that $|W|$ is small if r is large. Consequently, $1 + W$ is a complex number near 1 when r is large. Now

$$t = |f(z)| = |z|^n\,|1 + W| = r^n\,|1 + W|.$$

Since $|1 + W|$ is also near 1 when r is large, the nature of S for large r can be obtained by sketching the surface $t = r^n = (x^2 + y^2)^{n/2}$. This latter surface has a bowl-shaped structure, rising as r increases. (See Figure 5.3.2.) Thus S also has a bowl-shaped structure: when (x, y) is far from O, S is rising, but when (x, y) is near O, S may be quite different from the surface in Figure 5.3.2. Since the surface S has a bowl-shaped structure and is everywhere on or above the XY plane, it will have a lowest point.* Let P be a lowest point on S; if there are several equally low points, take P to be any one of these (See Figure 5.3.3.) Let $P = (x_0, y_0, t_0)$ and set $z_0 = (x_0, y_0) = x_0 + y_0 i$. Thus $t_0 = |f(z_0)|$. There are two possibilities: (i) P is in the XY plane; (ii) P is above the XY plane.

In case (i), $P = (x_0, y_0, 0)$ so that $|f(z_0)| = t_0 = 0$. Thus $f(z_0) = 0$ and the theorem is proved since z_0 is a root of f.

We dispose of case (ii) by proving that a point P on S above the XY plane can never be a lowest point on S, so that case (ii) never occurs.

LEMMA 5.3.2

If P is a point on the surface S, with P strictly above the XY plane, then S contains a point $\tilde{P}$ strictly lower than P and in the vicinity of P.

---

* At this point the proof relies on geometric intuition. By developing the properties of real numbers somewhat more deeply than is possible in this book, a completely satisfactory argument may be given to support the claim that S has a lowest point.

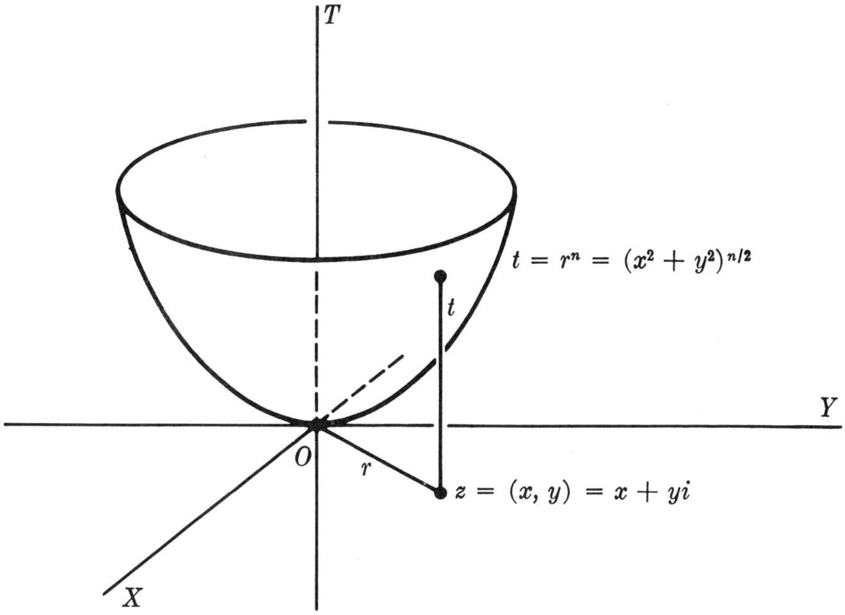

Figure 5.3.2

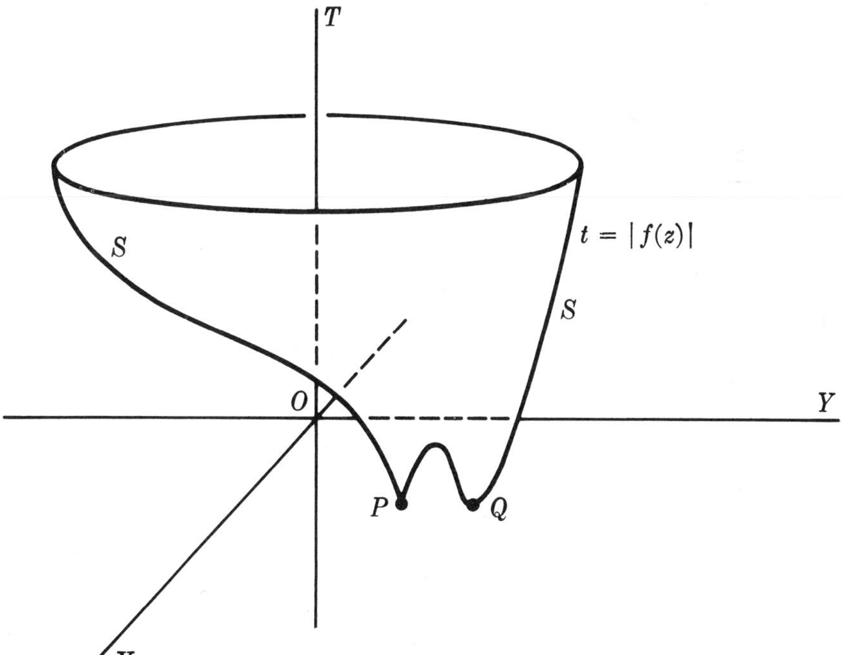

Figure 5.3.3

PROOF Let $P = (x_0, y_0, t_0)$. Then $t_0 = |f(z_0)|$, where $z_0 = x_0 + y_0 i$. We shall produce a complex number $z = x + yi$ such that $|f(z)| < |f(z_0)|$, with z near z_0. Setting $\tilde{P} = (x, y, t)$, where $t = |f(z)|$, we will then have $\tilde{P}$ on S, near P, and lower than P. The proof therefore amounts to finding a complex number z near z_0 such that

$$|f(z)| < |f(z_0)|. \tag{6}$$

Since P is not in the XY plane, $f(z_0) \neq 0$. Let $d_0 = f(z_0)$ and $d = f(z)$. Plot d_0 and d as points in the XY plane. Then d_0 is not the origin. The inequality (6) will be satisfied if we find a value of z such that $|d| < |d_0|$, i.e., such that d is within the circle centered at the origin and passing through d_0.

To find z, we use Taylor's theorem (4.2.7). Write $f(\lambda)$ in the base $\lambda - z_0$. Then

$$f(\lambda) = f(z_0) + d_1(\lambda - z_0) + d_2(\lambda - z_0)^2 + \cdots + d_n(\lambda - z_0)^n \tag{7}$$

where $d_1, d_2, \ldots, d_n$ are constants, and $d_0 = f(z_0)$. Evaluating at z, we get

$$d = f(z) = d_0 + d_1(z - z_0) + d_2(z - z_0)^2 + \cdots + d_n(z - z_0)^n.$$

Set $h = z - z_0$. Then

$$d = d_0 + d_1 h + \cdots + d_n h^n \tag{8}$$

and we are to choose h such that $|d| < |d_0|$, with $|h| = |z - z_0|$ small. Not all of $d_1, \ldots, d_n$ are zero since otherwise $f(\lambda)$ would be a constant, contrary to hypothesis. Let d_m be the first nonzero coefficient among $d_1, d_2, \ldots, d_n$. That is, $d_1 = 0, \ldots, d_{m-1} = 0$, $d_m \neq 0$. Then (8) becomes

$$d = d_0 + d_m h^m + d_{m+1} h^{m+1} + \cdots + d_n h^n, \qquad d_m \neq 0.$$

Choose a small positive real number ρ satisfying

$$\rho^m |d_m| < \frac{1}{2}|d_0|, \quad \rho\left(\left|\frac{d_{m+1}}{d_m}\right| + \cdots + \left|\frac{d_n}{d_m}\right|\right) < \frac{1}{2}, \quad \rho < 1. \tag{9}$$

These inequalities may all be satisfied if ρ is taken sufficiently small. Let $\theta = \text{angle } d_0$, $\varphi = \text{angle } d_m$, and set $c = \rho^m |d_m|$. Now take h to be this complex number:

$$h = \rho\left(\cos\frac{\theta - \varphi + \pi}{m} + i \sin\frac{\theta - \varphi + \pi}{m}\right).$$

Then $|h| = \rho$. Moreover, angle $d_m h^m = \pi + \theta = \pi +$ angle d_0, since angles add when complex numbers multiply (DeMoivre's theorem, 2.4.4). From angle $d_m h^m = \pi +$ angle d_0, it follows that the origin is between d_0 and $d_m h^m$. (See Figure 5.3.4.) Because $|d_m h^m| = |d_m| \rho^m = c < \frac{1}{2}|d_0|$, $d_m h^m$ is nearer the origin than d_0. The point $B = d_0 + d_m h^m$ will thus lie on the line segment joining O and d_0, a distance c towards the origin from d_0

(since $|B - d_0| = c$). We have $d = B + d_m h^{m+1} H$, where $H = (d_{m+1} + d_{m+2}h + \cdots + d_n h^{n-m-1})/d_m$. The triangle inequality and the second inequality (9) yield $\rho\,|H| < \frac{1}{2}$. Hence

$$|d - B| = |d_m h^{m+1} H| = c\rho\,|H| < \tfrac{1}{2}c.$$

Consequently, d lies in a circle centered on B with radius $\frac{1}{2}c$, and is therefore within the circle of radius $|d_0|$ centered at the origin. Thus $|d| < |d_0|$, as required. □

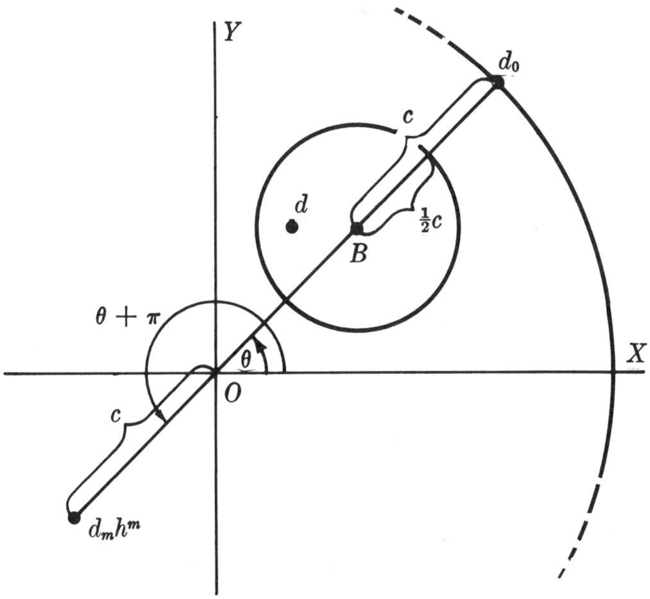

Figure 5.3.4

REMARK Let C be the smallest of the constants on the right-hand sides of the inequalities (9). The above proof shows that if ρ is any positive number satisfying $\rho < C$, then z can be found with $|z - z_0| = \rho$ and $|f(z)| < |f(z_0)|$. This gives a quantitive aspect to the statement "z near z_0," and by taking ρ very tiny, shows that z satisfying $|f(z)| < |f(z_0)|$ can be found arbitrarily near z_0.

A similar proof, which we omit as the result it establishes is not needed later, shows that *near a point P on S, there is a point $\tilde{P}$ on S strictly higher than P*. This assertion is sometimes called the *maximum modulus principle*; we may call Lemma 5.3.2 the *minimum modulus principle.* A consequence of these principles is that S cannot have any "local highest points P" or "local lowest points P," as indicated in Figure 5.3.5, apart from lowest points P with P in the XY plane.

Second proof Let $f(\lambda)$ be given by (2), and set $w = f(z)$. With W as in (4), from (3) we get

$$w = z^n(1 + W). \tag{10}$$

We consider two coordinate planes, one called the XY plane, its points denoted by $z = (x, y) = x + yi$, the other called the UV plane, with its points denoted by $w = (u, v) = u + vi$. (See Figure 5.3.6.)

For each z in the XY plane, equation (10) yields a point w in the UV plane. Consider a circle C_r of radius r centered at the origin in the XY plane. Let z be a variable point in the XY plane and let it travel once about the origin on the circumference of C_r, beginning at the point $(r, 0)$. The corresponding point w in the UV plane then travels over some curve, call it K_r. The curve K_r may be very complicated, but because z returns to its starting value, so does w, and hence K_r is a closed curve. We wish to determine the nature of K_r when r is very large. First observe by (5) that $|W|$ is small for large r; hence, for large r, the number $1 + W$ is close to 1. To be specific, choose r so large that $|W| \leq \frac{1}{2}$. Let $\mathcal{W} = 1 + W$. Then $|\mathcal{W} - 1| \leq \frac{1}{2}$ and hence $\mathcal{W}$ is inside a circle Δ of radius $\frac{1}{2}$ centered at 1. (See Figure 5.3.7.) The number $\mathcal{W}$ depends on z, and as z travels once about the origin on the large circle C_r, the quantity $\mathcal{W}$ travels over some curve contained within Δ.

Since $w = z^n\mathcal{W}$, angles for z, w, $\mathcal{W}$ may be found such that

$$\text{angle } w = n \text{ angle } z + \text{angle } \mathcal{W}. \tag{11}$$

As z travels once about the origin on the large circle C_r, angle z increases by 2π. Because $\mathcal{W}$ stays inside Δ, it cannot travel about the origin, and therefore angle $\mathcal{W}$ returns to its initial value. Consequently the net increase in angle $\mathcal{W}$ is zero. Thus, by (11), the net increase in angle w is $2\pi n$. This means: the curve K_r must loop the origin in the UV plane n times, and in particular, must enclose the origin. All this was for large r.

Let $r = 0$. Then K_r is the single point $f(0) \neq 0$. Thus for $r = 0$ the curve K_r does not enclose the origin.

We now use this fact: as r decreases in a continuous manner, the curve K_r shifts in a continuous fashion. This is because a polynomial in z is a continuous function of z. Let r decrease continuously from a large value to zero. The curve K_r begins by enclosing the origin, shifts continuously, and finishes by not enclosing the origin. For some intermediate value of r the curve K_r must pass through the origin of the UV plane. For this value of r there is a number α on C_r such that $f(\alpha) = w = 0$. Thus $f(\alpha) = 0$, and the proof is ended. We remark that a fully rigorous proof not relying on geometric intuition can be given to show that K_r does change continuously as r changes and that at some intermediate stage it does pass through the origin. It is beyond the scope of this book to develop the concepts necessary for such a proof. □

Figure 5.3.5

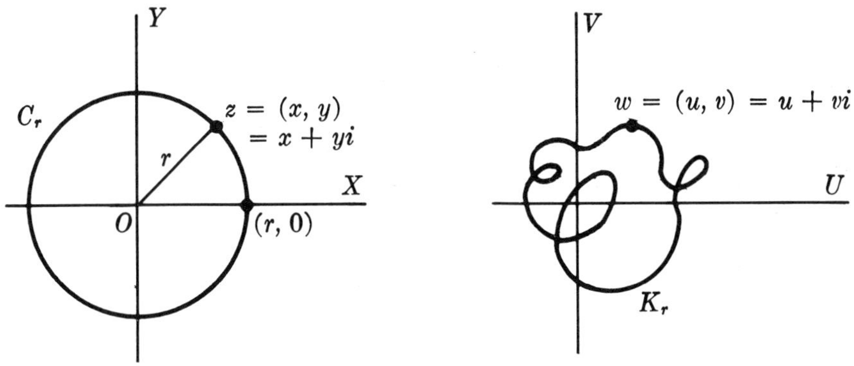

Figure 5.3.6

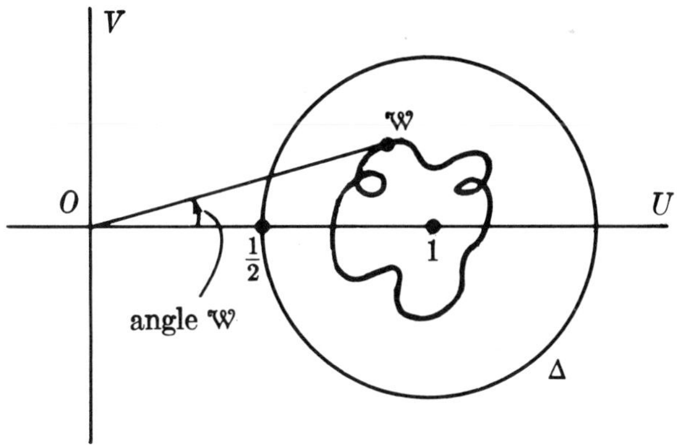

Figure 5.3.7

Other proofs of the fundamental theorem of algebra can be given that are more algebraic than the two above. See, for example, page 192 of the MacDuffie book cited on page 455. Every proof, however, contains some nonalgebraic ideas.

EXERCISES 5.3

I

1. Make a sketch of the surface S when (*i*) $f(\lambda) = \lambda$; (ii) $f(\lambda) = \lambda^2$, (iii) $f(\lambda) = \lambda^2 - \lambda$; (iv) $f(\lambda) = \lambda^2(\lambda - 1)$.

2. Let $r > 0$. Sketch the curve K_r when $f(\lambda) = \lambda^2 - 1$. Do this for $r = \frac{1}{2}$, 1, 2. It may be helpful to introduce the polar form $z = r(\cos\theta + i\sin\theta)$, and express the real and imaginary parts of w in terms of θ.

3. Let n be an even integer. Produce an example of a polynomial with real coefficients and degree n which has no real root.

II

4. Let α be a simple root of the polynomial $f(\lambda)$. Show that the surface S, which touches the XY plane at α, is *pointed* there; that is, touches the XY plane in the manner indicated for P in Figure 5.3.3. If, however, α is a multiple root of f, show that at α the surface S is *rounded*; that is, touches the XY plane in the manner indicated for Q in Figure 5.3.3.

5. Let $f(\lambda) = (\lambda - \alpha_1)(\lambda - \alpha_2)\cdots(\lambda - \alpha_n)$, where $\alpha_1, \alpha_2, \ldots, \alpha_n$ are complex constants with $|\alpha_1| < |\alpha_2| < \cdots < |\alpha_n|$. If $|\alpha_k| < r < |\alpha_{k+1}|$, show that the curve K_r loops the origin exactly k times. (Hint: Use angle w = angle $(z - \alpha_1)$ + angle $(z - \alpha_2)$ + $\cdots$ + angle $(z - \alpha_n)$, and consider the variation in angle $(z - \alpha_1), \ldots,$ angle $(z - \alpha_n)$ as z travels once around C_r. Drawing a figure may help.)

6. If f is a nonconstant polynomial and w is a complex number, prove that a complex number α exists such that $f(\alpha) = w$.

III

7. Formulate the concept of continuity for functions $f(z)$ of a complex variable z, then prove that $f(z)$ is a continuous function of z when f is a polynomial with complex coefficients. (Adapt the definition and proof for functions of a real number x given in your calculus book.)

8. If z_0 is a complex number such that the *tangent plane* to the surface S at z_0 is horizontal (that is, parallel to the XY plane), but above the XY plane, prove that $f'(z_0) = 0$. (The prime denotes derivative.)

   Outline: Let $\bar{f}(\lambda)$ be the polynomial obtained from $f(\lambda)$ by replacing each coefficient with its complex conjugate. Rewrite $t = |f(z)|$ as

$$t^2 = f(x + yi)\bar{f}(x - yi).$$

   Differentiate with respect to x, keeping y fixed, and with respect to y, keeping x fixed; then use the fact that the curves C_1, C_2 in the figure have horizontal tangent lines at z_0, to get

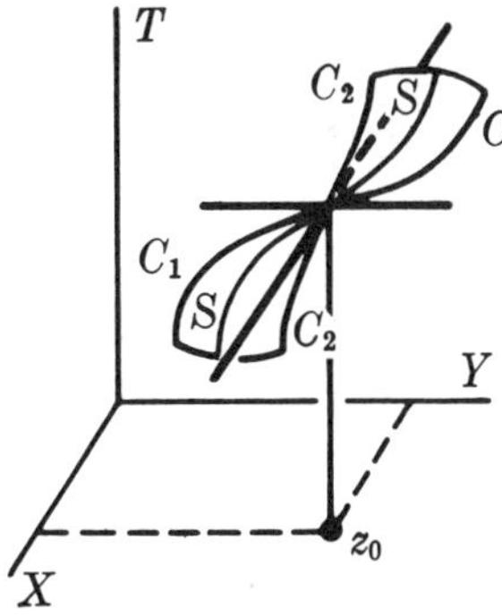

$$0 = f'(z_0)\bar{f}(\bar{z}_0) + f(z_0)\bar{f}'(\bar{z}_0),$$
$$0 = if'(z_0)\bar{f}(\bar{z}_0) - if(z_0)\bar{f}'(\bar{z}_0).$$

Deduce the result by solving these equations for $f'(z_0)$.

9. Let $f(\lambda) = 1 + \lambda^2$. Then $f'(\lambda)$ has a simple root at 0. Sketch the surface $t = |f(z)|$ for z near 0. Repeat with $f(\lambda) = 1 + \lambda^3$, for which $f'(\lambda)$ has a double root at 0. From these examples, draw conclusions about the nature of $t = |f(z)|$ when z is near a root of $f'(\lambda)$ of multiplicity m. Here $f'(\lambda)$ denotes the derivative of $f(\lambda)$. (It may be helpful to construct models of the surface using children's modelling clay.)

5.4 POLYNOMIALS WITH RATIONAL, REAL, AND COMPLEX COEFFICIENTS

Let $\mathfrak{F}$ be a field. From Example 3.5.2 we know that a polynomial in $\mathfrak{F}[\lambda]$ is a prime if its degree is 1. As a consequence of the Fundamental Theorem of Algebra, we shall now prove that every prime in $\mathbf{C}[\lambda]$ is a first degree polynomial. In passing, we remark that a corresponding complete description of the primes in $\mathbf{Z}$ is not known.

THEOREM 5.4.1

A polynomial $p(\lambda)$ in $\mathbf{C}[\lambda]$ is a prime if and only if $\deg p(\lambda) = 1$.

PROOF Let $p(\lambda)$ be a prime in $\mathbf{C}[\lambda]$. Then it is not a constant (see the definition of a prime, 3.5.1) and therefore, by the Fundamental Theorem of Algebra, 5.3.1, $p(\alpha) = 0$ for some complex number α. Hence $(\lambda - \alpha) \mid p(\lambda)$ (by Lemma 5.2.2), so that $p(\lambda) = (\lambda - \alpha)q(\lambda)$ for some polynomial $q(\lambda) \in \mathbf{C}[\lambda]$. Since $p(\lambda)$ is a prime, either $\lambda - \alpha$ or $q(\lambda)$ must be a unit. Thus q is a nonzero constant, and therefore $\deg p(\lambda) = 1$. This completes the proof since we already know that $p(\lambda)$ is a prime if $\deg p(\lambda) = 1$. □

Applying the unique factorization theorem (3.5.3) to $\mathbf{C}[\lambda]$, we now have:

THEOREM 5.4.2

If $f(\lambda)$ is a nonzero polynomial with complex coefficients and degree n, distinct complex numbers $\alpha_1, \ldots, \alpha_t$ and a complex number u exist such that

$$f(\lambda) = u(\lambda - \alpha_1)^{e_1}(\lambda - \alpha_2)^{e_2}\cdots(\lambda - \alpha_t)^{e_t}, \qquad u \neq 0. \qquad (1)$$

The exponents $e_1, \ldots, e_t$ are positive integers satisfying $e_1 + \cdots + e_t = n$. This factorization is unique, except for the order of the factors.

PROOF In (2) of §5.2 there can be no factors $p_{t+1}, \ldots, p_m$ since all primes in $\mathbf{C}[\lambda]$ are first degree polynomials. □

COROLLARY 5.4.3

If f is a nonzero polynomial of degree n with complex coefficients, then f has precisely n complex roots, counting multiplicities.

This follows immediately from the factorization (1). (See §5.2 for the term "counting multiplicities.") □

Turning now to polynomials with real coefficients, let us classify all primes in $\mathbf{R}[\lambda]$. The principal step, given in the next lemma, is the classification of the primes $p(\lambda)$ for which $\deg p > 1$.

LEMMA 5.4.4

A monic polynomial $p(\lambda)$ in $\mathbf{R}[\lambda]$, with $\deg p(\lambda) > 1$, is a prime in $\mathbf{R}[\lambda]$ if and only if

$$p(\lambda) = (\lambda - a)^2 + b^2,$$

where a, b are real numbers, with b nonzero.

PROOF Let $p(\lambda)$ be a prime in $\mathbf{R}[\lambda]$ with $\deg p > 1$. By the Fundamental Theorem of Algebra, $p(\alpha) = 0$ for some complex number α, since p is also a nonconstant polynomial with complex coefficients. Let $\alpha = a + bi$, where a, b are real numbers. If $b = 0$, then $\alpha = a$ is a real number with $p(\alpha) = 0$, implying (by Lemma 5.2.2) that the real polynomial $\lambda - \alpha$ divides $p(\lambda)$. This cannot be since $p(\lambda)$ is a prime in $\mathbf{R}[\lambda]$ with $\deg p > 1$ and so cannot have a first degree factor from $\mathbf{R}[\lambda]$. Therefore b is nonzero.
Within the ring $\mathbf{R}[\lambda]$, divide $p(\lambda)$ by $(\lambda - a)^2 + b^2$ to obtain a quotient $q(\lambda)$ and a remainder. The remainder will have degree at most 1, so that

$$p(\lambda) = ((\lambda - a)^2 + b^2)q(\lambda) + A\lambda + B \qquad (2)$$

where $q(\lambda)$ has real coefficients and A, B are real numbers. Evaluating (2) at α, (permissible because (2) is also an equation involving polynomials in $\mathbf{C}[\lambda]$), and using $p(\alpha) = 0$, we get

$$0 = ((a + bi - a)^2 + b^2)q(a + bi) + A(a + bi) + B$$
$$= 0 \cdot q(a + bi) + Aa + B + Abi = Aa + B + Abi.$$

Because $Aa + B$ and Ab are real numbers, from $Aa + B + Abi = 0$ we get $Aa + B = 0$ and $Ab = 0$. Since $b \neq 0$, we have $A = 0$; hence also $B = 0$. Therefore (2) becomes

$$p(\lambda) = ((\lambda - a)^2 + b^2)q(\lambda).$$

Since $p(\lambda)$ is a prime, one of the factors here must be a unit, implying that $q(\lambda)$ is a nonzero constant. When p is monic, $q = 1$. This completes half of the proof.

To finish the proof we show that

$$(\lambda - a)^2 + b^2 \tag{3}$$

is a prime in $\mathbf{R}[\lambda]$ if a, b are real numbers with b nonzero. If the polynomial (3) is not a prime in $\mathbf{R}[\lambda]$, its factorization into a product of monic primes in $\mathbf{R}[\lambda]$ must take the form

$$(\lambda - a)^2 + b^2 = (\lambda - \beta_1)(\lambda - \beta_2) \tag{4}$$

where β_1 and β_2 are real numbers, possibly equal. However, the polynomial (3) has this factorization in $\mathbf{C}[\lambda]$:

$$(\lambda - a)^2 + b^2 = (\lambda - (a + bi))(\lambda - (a - bi)). \tag{5}$$

The polynomials $\lambda - \beta_1$, $\lambda - \beta_2$, $\lambda - (a + bi)$, $\lambda - (a - bi)$, being first degree, are all *primes* in $\mathbf{C}[\lambda]$. By the *unique* factorization theorem, applied to $\mathbf{C}[\lambda]$, the factorizations in (4) and (5) must be the same, apart from the order of the factors. This implies that the real number β_1 equals one of the nonreal numbers $a + bi$, $a - bi$. A contradiction has arisen; therefore the polynomial (3) is indeed a prime in $\mathbf{R}[\lambda]$. □

If we let $z = a + bi$, then its complex conjugate (§2.4) is $\bar{z} = a - bi$, and (5) becomes

$$(\lambda - a)^2 + b^2 = (\lambda - z)(\lambda - \bar{z}). \tag{6}$$

Since b is nonzero, z is not real and $z \neq \bar{z}$.

THEOREM 5.4.5

The factorization of a nonzero polynomial $f(\lambda)$ with real coefficients into products of primes in $\mathbf{R}[\lambda]$ takes this form:

$$f(\lambda) = u(\lambda - \alpha_1)^{e_1} \cdots (\lambda - \alpha_t)^{e_t} p_{t+1}(\lambda)^{e_{t+1}} \cdots p_m(\lambda)^{e_m}, \tag{7}$$

where

$$p_{t+1}(\lambda) = (\lambda - z_{t+1})(\lambda - \bar{z}_{t+1}), \ldots, p_m(\lambda) = (\lambda - z_m)(\lambda - \bar{z}_m). \tag{8}$$

Here $\alpha_1, \ldots, \alpha_t$ are distinct real numbers,

$$z_{t+1}, \bar{z}_{t+1}, \ldots, z_m, \bar{z}_m \tag{9}$$

are distinct nonreal numbers, and u is a real nonzero constant.

PROOF The factorization of f into a product of distinct monic primes in $\mathbf{R}[\lambda]$ takes the form (2) of §5.2. By Lemma 5.4.4 and (6), the primes $p_{t+1}, \ldots, p_m$ have the form indicated in (8), where $z_{t+1}, \bar{z}_{t+1}$ are distinct, $z_{t+2}, \bar{z}_{t+2}$ are distinct, ..., $z_m, \bar{z}_m$ are distinct. To show that no two of the complex numbers (9) are equal, it therefore suffices to show that no two of $p_{t+1}, p_{t+2}, \ldots, p_m$ have a common root. Because p_{t+1}, p_{t+2} are distinct monic primes, their greatest common divisor must be 1. If z were a common root of p_{t+1}, p_{t+2}, both p_{t+1}, p_{t+2} would be divisible by $\lambda - z$, so that their greatest common divisor would also be divisible by $\lambda - z$ and could not be 1. The numbers (9) are therefore distinct. □

COROLLARY 5.4.6

Nonreal roots of a polynomial with real coefficients occur in complex conjugate pairs, the two members of a pair having equal multiplicity.

PROOF From (7) and (8) we see that the factorization of f into distinct monic primes in $\mathbf{C}[\lambda]$ is

$$f = u(\lambda - \alpha_1)^{e_1} \cdots (\lambda - \alpha_t)^{e_t} (\lambda - z_{t+1})^{e_{t+1}} (\lambda - \bar{z}_{t+1})^{e_{t+1}} \cdots \times (\lambda - z_m)^{e_m} (\lambda - \bar{z}_m)^{e_m}. \tag{10}$$

A complex root z of f must be one of the numbers (9), say $z = z_{t+1}$. Then, by (10), e_{t+1} is the multiplicity of both z and $\bar{z}$. □

THEOREM 5.4.7

Let $f(\lambda)$ be a polynomial with real coefficients. If r_1, r_2 are real numbers such that $f(r_1), f(r_2)$ have opposite signs, i.e., *one positive, one negative, then f has a real root between r_1 and r_2.*

PROOF Since $f(\lambda)$ gives results of opposite sign when evaluated at r_1, r_2, at least one of its prime factors in $\mathbf{R}[\lambda]$ must have sign at r_1 opposite to that at r_2. This cannot be a second degree prime since a monic second degree prime yields a positive result when evaluated at any real number. Therefore at least one prime factor of first degree, say $\lambda - \alpha_1$, yields results of opposite sign when evaluated at r_1, r_2. This can happen only when α_1 is between r_1 and r_2. □

THEOREM 5.4.8

A polynomial with real coefficients and odd degree has at least one real root.

PROOF If every prime factor of f within $\mathbf{R}[\lambda]$ had degree 2, the degree of f would be even. Therefore $t \geq 1$ in the factorization (7) when $\deg f$ is odd, implying that α_1 is a real root. □

Let r be a positive real number and n a positive integer. A consequence of Theorem 5.4.7 is that r has a positive real nth root, i.e., there is an $s > 0$ such that $s^n = r$. PROOF: Let $f(\lambda) = \lambda^n - r$. Take $r_1 = 0$, r_2 to be a large positive number. Then $f(r_1) < 0$, $f(r_2) > 0$, so that $f(\lambda)$ has a root s with $r_1 < s < r_2$. □

The results above describe all primes in the rings $\mathbf{R}[\lambda]$ and $\mathbf{C}[\lambda]$. No correspondingly simple descriptions are known for the primes in $\mathbf{Q}[\lambda]$. However, there is an easily proved fact that is quite useful when finding the rational roots of a polynomial with rational coefficients. If we have a polynomial

$$f(\lambda) = a_n\lambda^n + a_{n-1}\lambda^{n-1} + \cdots + a_1\lambda + a_0 \tag{11}$$

with rational numbers $a_n, \ldots, a_0$ as coefficients, no generality is lost when seeking its roots if we take $a_n, \ldots, a_0$ to be integers. This is because we can multiply f by a nonzero integer and clear out all denominators, e.g., multiply by the product of the denominators of $a_n, \ldots, a_0$. When seeking a rational root $\alpha = a/b$ of f, where a, b are integers, we may assume a and b are relatively prime since otherwise we can cancel a common factor from the numerator and denominator of α.

THEOREM 5.4.9

Let nonconstant polynomial $f(\lambda)$, given by (11), have integral coefficients $a_n, \ldots, a_0$. If $\alpha = a/b$ is a rational root of f, where a, b are relatively prime integers, then $b \mid a_n$ and $a \mid a_0$.

PROOF Since $f(\alpha) = 0$, we have

$$a_n\left(\frac{a}{b}\right)^n + a_{n-1}\left(\frac{a}{b}\right)^{n-1} + \cdots + a_1\left(\frac{a}{b}\right) + a_0 = 0.$$

Multiplying by b^n produces

$$a_na^n + a_{n-1}a^{n-1}b + \cdots + a_1ab^{n-1} + a_0b^n = 0, \tag{12}$$

which may be rewritten as

$$a_na^n = (-a_{n-1}a^{n-1} - \cdots - a_1ab^{n-2} - a_0b^{n-1})b.$$

Since b divides the right-hand side, we see that $b \mid a_na^n$. By hypothesis, b and a are relatively prime; hence b and a^n are also relatively prime. (Why?—see Exercise 16 below.) Thus b divides the product of a_n and a^n and is relatively prime to a^n. It follows from Lemma 3.4.10 that $b \mid a_n$. If we now rewrite (12) as

$$a_0b^n = (-a_na^{n-1} - a_{n-1}a^{n-2}b - \cdots - a_1b^{n-1})a,$$

we see that a divides the right-hand side; hence $a \mid a_0b^n$. Because a and b are relatively prime, a and b^n are also relatively prime, and so by Lemma 3.4.10 it follows that $a \mid a_0$. □

The significance of Theorem 5.4.9 is that when searching for rational numbers $\alpha = a/b$ which are roots of a polynomial (11) with integral coefficients, it is necessary to try only a finite number of different α's: try all rational numbers a/b whose numerator a is a factor of a_0 and whose denominator b is a factor of a_n. If it happens that the integer coefficients $a_n, \ldots, a_0$ of f have a common factor d, the number of tests to be made will be reduced if we first cancel d from f, i.e., multiply f by $1/d$.

Example 5.4.10
Find the rational roots of $f(\lambda) = 6\lambda^4 - \lambda^3 + 5\lambda^2 - \lambda - 1$.

SOLUTION Here $a_4 = 6$ and $a_0 = 1$. The possible factors of a_0 are $+1$, -1, and the possible factors of a_4 are $6, -6, 3, -3, 2, -2, 1, -1$. The only possible rational roots of this polynomial are therefore

$$\pm\frac{1}{6}, \pm\frac{1}{3}, \pm\frac{1}{2}, \pm 1.$$

By evaluating $f(\lambda)$ at each of these numbers, we find that $1/2$ and $-1/3$ are the only rational roots of f. In this example it now is easy to obtain the complete factorization of f since we know it is divisible by $\lambda - 1/2$ and $\lambda + 1/3$. We get

$$f(\lambda) = 6(\lambda - 1/2)(\lambda + 1/3)(\lambda^2 + 1).$$

Example 5.4.11
Prove that $\sqrt{2} + \sqrt{3}$ is not a rational number.

SOLUTION Write down the polynomial whose roots are $\pm\sqrt{2} \pm \sqrt{3}$. It is

$$f(\lambda) = (\lambda - (\sqrt{2} + \sqrt{3}))(\lambda - (\sqrt{2} - \sqrt{3})) \times (\lambda - (-\sqrt{2} + \sqrt{3}))(\lambda - (-\sqrt{2} - \sqrt{3})).$$

Multiplying out gives

$$f(\lambda) = \lambda^4 - 10\lambda^2 + 1.$$

This polynomial has integral coefficients. The test for a rational root a/b yields $a \mid 1$ and $b \mid 1$. Thus $a = 1$ or -1, $b = 1$ or -1, and so the only possible rational roots are 1 or -1. Since $f(1) \neq 0$, $f(-1) \neq 0$, this polynomial has no rational roots. Because $\sqrt{2} + \sqrt{3}$ is a root, it cannot be a rational number. (See Example 3.5.6 for another method of proof.) □

There is an immediate corollary of Theorem 5.4.9 which has considerable interest.

COROLLARY 5.4.12
A rational root of a monic polynomial with integral coefficients is an integer.

PROOF If $a_n = 1$, from $b \mid a_n$ we get $b = \pm 1$. Thus $\alpha = \pm a$ is an integer. □

Besides its utilitarian value in locating rational roots, Theorem 5.4.9 may be regarded as a result providing information about the first degree factors of a polynomial in $\mathbf{Q}[\lambda]$. In Chapter 7 we shall obtain a generalization which provides a little information about higher degree factors.

EXERCISES 5.4

In these exercises all polynomials are assumed to have real or complex coefficients.

I

1. Find all primes $p(\lambda)$ in $\mathbf{R}[\lambda]$ for which $p(3 - 2i) = 0$.

2. If $3 + 2i$ and $2 - i$ are roots of multiplicity 2 of a monic eighth degree polynomial $f(\lambda) \in \mathbf{R}[\lambda]$, find f and factor it into primes in $\mathbf{R}[\lambda]$ and in $\mathbf{C}[\lambda]$.

3. Factor $4\lambda^4 + 8\lambda^3 + 7\lambda^2 + 8\lambda + 3$ into primes in $\mathbf{Q}[\lambda]$, $\mathbf{R}[\lambda]$, and in $\mathbf{C}[\lambda]$.

4. Factor $\lambda^3 + 2\lambda^2 - 3$ into primes in $\mathbf{R}[\lambda]$ and in $\mathbf{C}[\lambda]$. Do the same for $\lambda^3 + 2\lambda^2 - 1$.

5. Let p_1 and p_2 be monic primes in $\mathbf{C}[\lambda]$. Is $p_1 + p_2$ again a prime in $\mathbf{C}[\lambda]$? How is your answer changed if $\mathbf{C}[\lambda]$ is replaced with $\mathbf{R}[\lambda]$? Answer these questions also for p_1p_2. Also answer these questions for primes in $\mathbf{Z}$.

6. The complex number i is a root of $f(\lambda) = \lambda^4 + \lambda^3 + 2\lambda^2 + \lambda + 1$. Find the factorization of f into primes in $\mathbf{R}[\lambda]$ and in $\mathbf{C}[\lambda]$.

7. The complex number i is a root of $\lambda^6 + 4\lambda^4 + 5\lambda^2 + 2$ of multiplicity two. Find the factorization of this polynomial into primes in $\mathbf{R}[\lambda]$.

8. Factor the derivative of the polynomial in Exercise 7 into primes in $\mathbf{R}[\lambda]$.

9. Find all rational roots of the following polynomials, then factor them into products of primes in $\mathbf{Q}[\lambda]$, $\mathbf{R}[\lambda]$ and $\mathbf{C}[\lambda]$: $\lambda^4 + 2\lambda^3 + \lambda^2 - 2\lambda - 2$, $6\lambda^3 - 7\lambda^2 + 7\lambda + 6$.

II

10. Let p_1 and p_2 be primes in $\mathbf{R}[\lambda]$ with $\deg p_1 = 2$, $\deg p_2 = 1$, and let $p_1 = qp_2 + r$, where $q \in \mathbf{R}[\lambda]$ and r is a constant. Prove that q is a prime in $\mathbf{R}[\lambda]$ and $r \neq 0$.

11. Deduce Theorem 5.4.8 from Theorem 5.4.7. (Hint: Consider $f(x)$ for large positive x and large negative x.)

12. Let $\alpha_1, \ldots, \alpha_n$ be n distinct elements of **C**. If $g(\lambda)$ is a polynomial of degree $\leq n - 1$, show that

$$\sum_{s=1}^{n} \frac{(\lambda - \alpha_1)\cdots(\lambda - \alpha_{s-1})(\lambda - \alpha_{s+1})\cdots(\lambda - \alpha_n)}{(\alpha_s - \alpha_1)\cdots(\alpha_s - \alpha_{s-1})(\alpha_s - \alpha_{s+1})\cdots(\alpha_s - \alpha_n)} g(\alpha_s) = g(\lambda).$$

[Hint: Each side is a polynomial of degree at most $n - 1$. Show that the two sides are equal for n distinct values of λ. Then use the fact that a nonzero polynomial of degree at most $n - 1$ cannot have n distinct roots (see Exercise 13, §5.2).]

13. Let $f(\lambda) = \lambda^4 + 2\lambda^3 + a\lambda^2 + b\lambda + 4$, where a, b are real numbers. If $f(1 + i) = 0$, find the factorization of f into primes in **R**[λ] and in **C**[λ]. (Hint: Use a formula in Exercise 14, §5.2.)

14. Prove, using Theorem 5.4.9, that $\sqrt{2}$, $\sqrt{3}$, $\sqrt{2} + \sqrt{5}$, $\sqrt{3} + \sqrt{7}$, $\sqrt{2} + \sqrt{3} + \sqrt{5}$ are irrational numbers.

15. Repeat Exercise 14, using the fundamental theorem of arithmetic for **Z** in place of Theorem 5.4.9.

16. Show that if integer b is relatively prime to integer a then b is relatively prime to a^n for any positive integer exponent n. (Hint: If b and a^n had a common factor, there would be a common prime factor p which would necessarily be a common factor of b and a.)

III

17. Use Exercise 12 above to prove that if $f(\lambda)$ is an nth degree polynomial with distinct roots $\alpha_1, \alpha_2, \ldots, \alpha_n$, and $g(\lambda)$ is any monic polynomial with $\deg g = n - 1$, then

$$\sum_{s=1}^{n} \frac{g(\alpha_s)}{f'(\alpha_s)} = 1$$

Here f' is the derivative of f.

18. Using the formulas in Exercise 14, §5.2, find the factorization of $f(\lambda) = \lambda^3 + \lambda^2 - 2\lambda + a$ (a is a real constant) into primes in **R**[λ] and **C**[λ], assuming that one root of f is the negative of another root.

19. Let $f(\lambda)$ be a polynomial of degree n, and let a be a root of multiplicity e. Show that $f(a) = f'(a) = f^{(2)}(a) = \cdots = f^{(e-1)}(a) = 0$, $f^{(e)}(a) \neq 0$, the superscripts denoting repeated differentiation. Show that the Taylor series of $f(\lambda)$ about a has the form

$$f(\lambda) = d_e(\lambda - a)^e + d_{e+1}(\lambda - a)^{e+1} + \cdots + d_n(\lambda - a)^n$$

where $d_e, \ldots, d_n$ are constants, and $d_e \neq 0$, $d_n \neq 0$.

5.5 ROOTS OF SPECIAL POLYNOMIALS

It is generally a difficult task to find the roots of a polynomial with real or complex coefficients. In this section we discuss root finding for several special types of polynomial, presenting some methods of practical interest and some others of theoretical interest only.

Extraction of nth roots

Our first theorem provides a method for finding the nth roots of a complex number. Recall (§2.4) that $e^{i\varphi} = \cos\varphi + i\sin\varphi$ when φ is a real number. In particular, $e^{2\pi ji} = 1$ when j is an integer.

THEOREM 5.5.1
Let a be a nonzero complex number, represented in polar form as

$$a = re^{i\theta}$$

where $r = |a|$, $\theta =$ angle a. Let n be a positive integer. Then the roots of the polynomial $\lambda^n - a$ are the numbers $z_0, z_1, \ldots, z_{n-1}$ given by

$$z_j = r^{1/n}e^{i(\theta+2\pi j)/n}, \qquad \text{for } j = 0, 1, \ldots, n-1.$$

Here $r^{1/n}$ denotes the positive real nth root of r.

PROOF We know, by Corollary 5.4.3, that the polynomial $\lambda^n - a$ has precisely n roots, counting multiplicities. If we can show that the numbers $z_0, \ldots, z_{n-1}$ are distinct and are roots of $\lambda^n - a$, it will follow that $z_0, \ldots, z_{n-1}$ are *all* the roots of $\lambda^n - a$. (It will also follow that each root of $\lambda^n - a$ is a simple root.)

Using the laws of exponents for complex exponentials (§2.4), we get $z_j^n = re^{i(\theta+2\pi j)} = re^{i\theta}e^{2\pi ji} = re^{i\theta} = a$. Thus $z_0, z_1, \ldots, z_{n-1}$ are roots of $\lambda^n - a$. These roots are distinct since, for example, if $z_s = z_t$ (where $0 \le s < t \le n-1$), then angle z_s and angle z_t would have to agree to within a multiple of 2π, implying

$$\frac{\theta + 2\pi t}{n} = \frac{\theta + 2\pi s}{n} + 2\pi u$$

where u is an integer. This equation yields $t - s = nu$, a contradiction since $0 < t - s < n$. □

It is instructive to look at Theorem 5.5.1 geometrically. We have $|z_0| = |z_1| = \cdots = |z_{n-1}| = r^{1/n}$, so that $z_0, \ldots, z_{n-1}$ lie on a circle C of radius $r^{1/n}$ centered at the origin. The angle of z_0 is obtained as the nth part of the angle of a, and then the angles of $z_1, \ldots, z_{n-1}$ are obtained by repeatedly adding $(1/n)$th of a full revolution to angle z_0. In other words:

$z_0, z_1, \ldots, z_{n-1}$ form the vertices of a regular polygon inscribed in the circle C. (A *regular* polygon is a polygon for which all sides have the same length and all pairs of adjacent sides meet at the same angle.) Figure 5.5.1 illustrates this when $n = 4$, when the regular polygon is a square.

A particularly interesting and useful special case arises when $a = 1$. In this case $\theta = 0$. We thus have:

***COROLLARY* 5.5.2**
The roots of the polynomial $\lambda^n - 1$ *are the numbers* $\omega_0, \ldots, \omega_{n-1}$ *given by*

$$\omega_j = e^{2\pi ij/n}, \qquad j = 0, 1, \ldots, n-1.$$

These numbers are the vertices of the regular polygon inscribed in the unit circle having $\omega_0 = 1$ *as a vertex.*

(The *unit circle* is, by definition, the circle of radius 1 centered at the origin.) Figure 5.5.2 illustrates this polygon when $n = 5$. It's then a pentagon.

The numbers ω_j in Corollary 5.5.2 are called *the* nth *roots of unity.* Simplifying notation by setting $\omega = \omega_1$, we have

$$\omega = e^{2\pi i/n}$$

and

$$\omega^j = e^{2\pi ij/n} = \omega_j, \qquad j = 0, 1, \ldots, n-1.$$

Thus all nth roots of 1 are powers of ω.

Returning to the more general polynomial $\lambda^n - a$, it is easy to see that its roots z_j are given by the formula

$$z_j = z_0\omega^j, \qquad j = 0, 1, \ldots, n-1.$$

That is: *we obtain all roots of* $\lambda^n - a$ *by multiplying one of the roots,* z_0, *by the* n *different* nth *roots of unity.*

The nth roots of unity also have this interesting property:

the n nth *roots of unity, under multiplication, form a group.*

PROOF The set of nth roots of unity is a subset of the multiplicative group of nonzero complex numbers. We need therefore (see Theorem 1.6.10) simply verify that products and inverses of nth roots of unity are again nth roots of unity. This is quickly done: If $w_1^n = 1$ and $w_2^n = 1$, then $(w_1w_2)^n = w_1^nw_2^n = 1$ and $(w_1^{-1})^n = (w_1^n)^{-1} = 1$. □

This group is, of course, a commutative group.

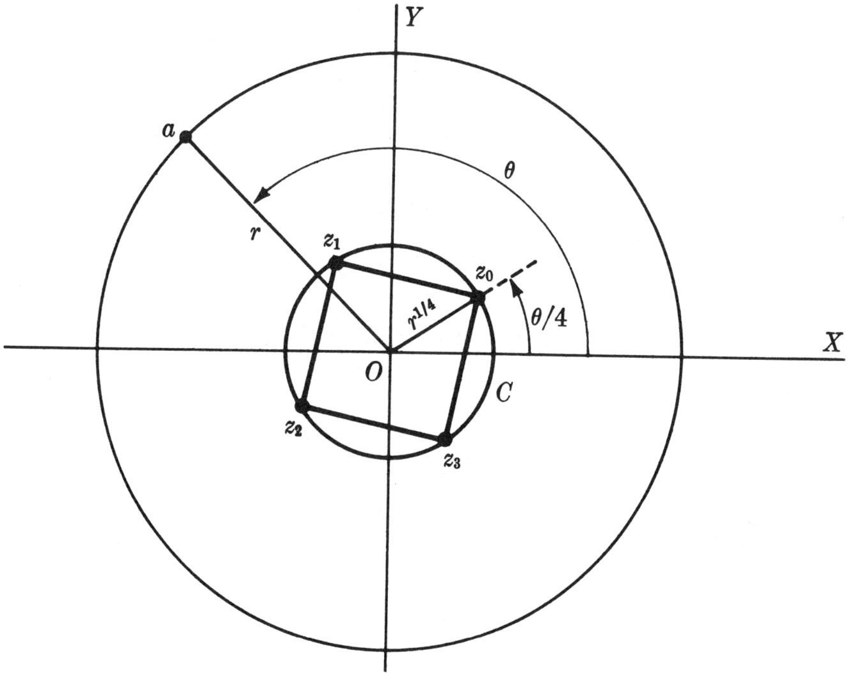

Figure 5.5.1

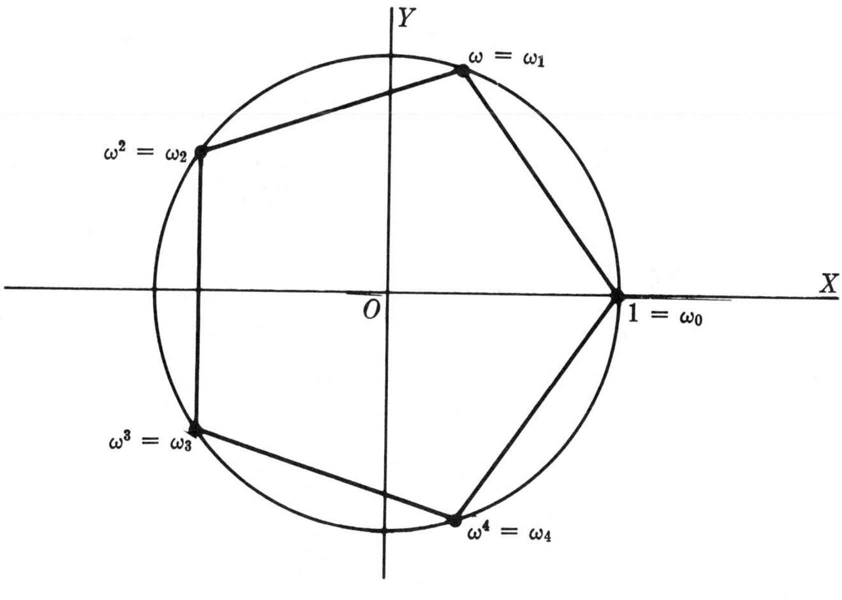

Figure 5.5.2

REMARK A group in which all elements are powers of one fixed element is said to be *cyclic*. The group of the nth roots of unity is, therefore, an example of a cyclic group having precisely n elements. A closely related example of a cyclic group with precisely n elements is the group C_n of rotations encountered in Exercise 12 of §2.5. In fact the rotation $r_{2\pi/n}$ of the complex plane through angle $2\pi/n$ is accomplished by multiplication by $\omega = e^{2\pi i/n}$, that is, $r_{2\pi/n}(z) = \omega z$ for each complex number z; and $r_{2\pi j/n}(z) = \omega^j z$ for any integer j.

Roots of polynomials of degrees 1, 2, 3, 4 (Optional)*

In this subsection we shall give formulas for the roots of a general polynomial $f(\lambda)$ of degree n, for some values of n. Without affecting the roots, we may make $f(\lambda)$ monic by multiplying it by a nonzero constant. So let

$$f(\lambda) = \lambda^n + a_{n-1}\lambda^{n-1} + \cdots + a_1\lambda + a_0 , \tag{1}$$

where $a_{n-1}, \ldots, a_0$ are complex numbers. The roots are complex numbers $\alpha_1, \alpha_2, \ldots, \alpha_n$ for which $f(\lambda)$ factors as

$$f(\lambda) = (\lambda - \alpha_1)(\lambda - \alpha_2)\cdots(\lambda - \alpha_n), \tag{2}$$

and we wish to obtain formulas which express $\alpha_1, \alpha_2, \ldots, \alpha_n$ in terms of $a_0, a_1, \ldots, a_{n-1}$ using finitely many additions, subtractions, multiplications, divisions, and kth root extractions $\sqrt[k]{\ }$ for perhaps several values of k. This type of formula for the roots of f is called a *solution by radicals*. We are able to give a solution by radicals for the general nth degree polynomial only when n is 1, 2, 3, or 4. See the historical remarks beginning on page 217. We now discuss these four values of n individually.

The case $n = 1$. Here $f(\lambda) = \lambda + a_0$ and the root is $\alpha_1 = -a_0$.

The case $n = 2$. Here $f(\lambda) = \lambda^2 + a_1\lambda + a_0$ is a quadratic polynomial. Define a quantity d [called the *discriminant* of $f(\lambda)$] by

$$d = a_1^2 - 4a_0 . \tag{3}$$

Letting $\sqrt{d}$ be one of the square roots of d, set

$$\alpha_1 = (-a_1 + \sqrt{d})/2, \qquad \alpha_2 = (-a_1 - \sqrt{d})/2. \tag{4}$$

We claim that

$$\lambda^2 + a_1\lambda + a_0 = (\lambda - \alpha_1)(\lambda - \alpha_2).$$

This polynomial equation is valid if the right-hand side, when multiplied out, has the same coefficients as the left-hand side. Multiplying out shows this to be the case if

$$a_1 = -(\alpha_1 + \alpha_2), \qquad a_2 = \alpha_1\alpha_2 . \tag{5}$$

* The remainder of this section may be omitted.

A short calculation, using (3) and (4), verifies (5). Thus α_1 and α_2 are indeed the roots of $f(\lambda)$. Subtracting the second of equations (4) from the first, we get $\sqrt{d} = \alpha_1 - \alpha_2$, so that

$$d = (\alpha_1 - \alpha_2)^2. \tag{6}$$

From (3), (4), (5), (6) we deduce these additional facts:

(i) *A quadratic polynomial has simple roots if and only if its discriminant is nonzero;* and

(ii) *When the coefficients are real numbers, the roots are real numbers if and only if the discriminant is nonnegative.*

Preliminary reduction of the cases $n = 3, 4$. The following argument is valid for any n, but will be used only when n is 3 or 4.

Let $f(\lambda)$ be the polynomial (1). Set

$$g(\lambda) = f\left(\lambda - \frac{a_{n-1}}{n}\right).$$

From the binomial theorem (§4.2) we get

$$(\lambda - a_{n-1}/n)^n = \lambda^n - a_{n-1}\lambda^{n-1} + \text{terms in } \lambda^{n-2}, \lambda^{n-3}, \ldots,$$

$$a_{n-1}(\lambda - a_{n-1}/n)^{n-1} = a_{n-1}\lambda^{n-1} + \text{terms in } \lambda^{n-2}, \lambda^{n-3}, \ldots.$$

Furthermore, $(\lambda - a_{n-1}/n)^{n-2}$, $(\lambda - a_{n-1}/n)^{n-3}$, $\ldots$, when multiplied out, yield no power of λ higher than λ^{n-2}. Thus the λ^{n-1} term in $g(\lambda)$ cancels,

$$g(\lambda) = \lambda^n + A_{n-2}\lambda^{n-2} + \cdots + A_1\lambda + A_0, \tag{7}$$

where $A_{n-2}, \ldots, A_1, A_0$ are constants depending in rather complicated ways on $a_{n-1}, a_{n-2}, \ldots, a_1, a_0$. Evaluating (2) at $\lambda - a_{n-1}/n$ yields

$$g(\lambda) = (\lambda - (\alpha_1 + a_{n-1}/n))\cdots(\lambda - (\alpha_n + a_{n-1}/n)),$$

and therefore the roots of $g(\lambda)$ are the numbers $\beta_1, \ldots, \beta_n$ given by

$$\beta_1 = \alpha_1 + a_{n-1}/n, \ldots, \qquad \beta_n = \alpha_n + a_{n-1}/n.$$

Thus: *if we calculate the roots* $\beta_1, \ldots, \beta_n$ *of the polynomial* $g(\lambda)$ *in* (7), *we get the roots* $\alpha_1, \ldots, \alpha_n$ *of* $f(\lambda)$ *from*

$$\alpha_1 = \beta_1 - a_{n-1}/n, \ldots, \qquad \alpha_n = \beta_n - a_{n-1}/n.$$

It will therefore be sufficient to give formulas for the roots $\beta_1, \ldots, \beta_n$ of $g(\lambda)$.

The case $n = 3$. Simplifying notation, let

$$g(\lambda) = \lambda^3 + p\lambda + q,$$

where p, q are complex numbers. If $p = 0$, the roots of $g(\lambda)$ are the cube roots of $-q$ (which may be calculated using Theorem 5.5.1) and the discussion is ended. So let $p \neq 0$. The formulas for the roots $\beta_1, \beta_2, \beta_3$ of $g(\lambda)$ are now the following:

Set $\omega = (-1 + \sqrt{3}i)/2$, *let* w *be any solution of the quadratic equation*

$$w^2 + qw - p^3/27 = 0, \tag{8}$$

and take Δ_1 *to be any cube root of* w. *Let*

$$\Delta_2 = -\frac{p}{3\Delta_1}. \tag{9}$$

For nonzero p, *the roots of* $\lambda^3 + p\lambda + q$ *are the numbers* $\beta_1, \beta_2, \beta_3$ *given by*

$$\begin{aligned} \beta_1 &= \Delta_1 + \Delta_2, \\ \beta_2 &= \Delta_1\omega + \Delta_2\omega^2, \\ \beta_3 &= \Delta_1\omega^2 + \Delta_2\omega. \end{aligned} \tag{10}$$

(Formulas (10) are called *Cardan's formulas*. See remarks on page 217.)

PROOF First observe that $w \neq 0$ since $w = 0$ and (8) imply the contradiction $p = 0$. Therefore $\Delta_1 \neq 0$, so that (9) does yield a number Δ_2. Since $\Delta_1^3 = w$, from (8) we get

$$\Delta_1^6 + q\Delta_1^3 - p^3/27 = 0. \tag{11}$$

Also observe that $\omega^2 + \omega = -1$, $\omega^3 = 1$. In fact, $\omega = e^{2\pi i/3}$.

We must prove that $\lambda^3 + p\lambda + q = (\lambda - \beta_1)(\lambda - \beta_2)(\lambda - \beta_3)$; multiplying out the right-hand side, we see we must verify that

$$\begin{aligned} 0 &= \beta_1 + \beta_2 + \beta_3, \\ p &= \beta_1\beta_2 + \beta_1\beta_3 + \beta_2\beta_3, \\ q &= -\beta_1\beta_2\beta_3. \end{aligned} \tag{12}$$

The verification of the first of equations (12) follows by adding equations (10) and using $\omega^2 + \omega = -1$. Adding and multiplying the last two of equations (10) yields $\beta_2 + \beta_3 = -\Delta_1 - \Delta_2$ and $\beta_2\beta_3 = \Delta_1^2 + \Delta_2^2 - \Delta_1\Delta_2$. Then [using (9)]

$$\begin{aligned} \beta_1\beta_2 + \beta_1\beta_3 + \beta_2\beta_3 &= \beta_1(\beta_2 + \beta_3) + \beta_2\beta_3 \\ &= (\Delta_1 + \Delta_2)(-\Delta_1 - \Delta_2) + (\Delta_1^2 + \Delta_2^2 - \Delta_1\Delta_2) \\ &= -3\Delta_1\Delta_2 \\ &= p, \end{aligned}$$

and [using (9) and (11)]

$$\begin{aligned} \beta_1\beta_2\beta_3 &= (\Delta_1 + \Delta_2)(\Delta_1^2 + \Delta_2^2 - \Delta_1\Delta_2) \\ &= \Delta_1^3 + \Delta_2^3 \\ &= \Delta_1^3 - p^3/27\Delta_1^3 \\ &= -q. \end{aligned}$$

All equations (12) are now verified and the proof is complete. □

Example: Take $g(\lambda) = \lambda^3 - 3\lambda - 1$. Observing that $g(-2)$ is negative, $g(-1)$ is positive, $g(0)$ is negative, and $g(2)$ is positive, we see that $g(\lambda)$ has a root between -2 and -1, another between -1 and 0, and a third between 0 and 2. Therefore $g(\lambda)$ has distinct real roots. However, for this g, the roots of (8) are $w = \frac{1}{2} \pm \frac{1}{2}\sqrt{3}i$. The cube root Δ_1 of w is therefore a nonreal number, so that Cardan's formulas present the *real* roots of $g(\lambda)$ in terms of *nonreal* numbers.

To discuss this and other aspects of formulas (10), we introduce the *discriminant* D of $\lambda^3 + p\lambda + q$. Let $d = q^2 + 4p^3/27$ be the discriminant of the quadratic equation (8). Then we set

$$D = -27d = -27q^2 - 4p^3. \tag{13}$$

We may express D directly in terms of $\beta_1, \beta_2, \beta_3$ in the following way. Combining the first two of equations (12), we get $p = -\beta_2^2 - \beta_3^2 - \beta_2\beta_3$; and combining the first and third equations, we get $q = \beta_2^2\beta_3 + \beta_2\beta_3^2$. Then, by direct calculation,

$$-27q^2 - 4p^3 = 4\beta_2^6 + 12\beta_2^5\beta_3 - 3\beta_2^4\beta_3^2 - 26\beta_2^3\beta_3^3 - 3\beta_2^2\beta_3^4 + 12\beta_2\beta_3^5 + 4\beta_3^6.$$

Furthermore, using $\beta_1 = -\beta_2 - \beta_3$, we get (after some calculation),

$$\begin{aligned}[(\beta_1 - \beta_2)(\beta_1 - \beta_3)(\beta_2 - \beta_3)]^2 &= [(-2\beta_2 - \beta_3)(-2\beta_3 - \beta_2)(\beta_2 - \beta_3)]^2 \\ &= 4\beta_2^6 + 12\beta_2^5\beta_3 - 3\beta_2^4\beta_3^2 - 26\beta_2^3\beta_3^3 \\ &\quad - 3\beta_2^2\beta_3^4 + 12\beta_2\beta_3^5 + 4\beta_3^6.\end{aligned}$$

Comparing these expressions, we see that

$$D = [(\beta_1 - \beta_2)(\beta_1 - \beta_3)(\beta_2 - \beta_3)]^2. \tag{14}$$

An immediate consequence of (14) is that the *roots of* $g(\lambda)$ *are simple if and only if its discriminant* $D = -27q^2 - 4p^3$ *is nonzero.*

Let $g(\lambda)$ have real coefficients. Then its nonreal roots occur in complex conjugate pairs (Corollary 5.4.6), and therefore the roots are either: (i) all real, not necessarily distinct; or (ii) one is real and the other two are a pair of conjugate nonreal numbers. In the second case the roots must be distinct.

When the roots $\beta_1, \beta_2, \beta_3$ of $g(\lambda)$ are real and distinct, (14) shows that $D > 0$. Thus [from (13)] the discriminant d of (8) is negative, and hence w is a nonreal number. The cube root Δ_1 of w is then also nonreal. Therefore: *when* $g(\lambda)$ *has distinct real roots, Cardan's formulas* (10) *present these roots in terms of nonreal numbers* Δ_1 *and* Δ_2. Because of this, the case when $g(\lambda)$ has distinct real roots is called the *irreducible case.*

Continuing to suppose that $g(\lambda)$ has real coefficients, we now assume that one root is real and the other two are a pair of conjugate nonreal numbers. If we denote the real root by β and the conjugate nonreal roots by $x + yi$, $x - yi$, with real x, real nonzero y, (14) becomes

$$D = -4[(\beta - x)^2 + y^2]y^2 < 0$$

(independently of whether β_1, β_2, or β_3 is β). Thus $D < 0$, hence $d > 0$, and so the solution w of (8) is a real number. If we take Δ_1 to be the real cube root of w, then both Δ_1 and Δ_2 are real, and then Cardan's formulas (10) give the roots without unexpected difficulty, the real root being β_1.

Cardan's formulas also give the roots without difficulty when $g(\lambda)$ has nondistinct real roots. In this case $D = 0$, yielding $d = 0$, and hence $w = -q/2$. Take Δ_1 to be the real value of $(-q/2)^{1/3}$. Since $D = 0$, $27q^2 + 4p^3 = 0$; using this, we compute that $\Delta_1^2 = -p/3$. Since $\Delta_1\Delta_2 = -p/3$, we see that $\Delta_1 = \Delta_2$, and then formulas (10) simplify to: *When* $\lambda^3 + p\lambda + q$ *has nondistinct real roots, the roots are*

$$\beta_1 = -2(q/2)^{1/3}, \qquad \beta_2 = \beta_3 = (q/2)^{1/3}.$$

In the irreducible case, Cardan's formulas present the roots in a form involving complex numbers which may be computationally difficult to use. We therefore now give trigonometric formulas for the roots that avoid the necessity of computing with complex numbers. Since $D > 0$, we get $p < 0$ [from (13)]. We take the solution w of (8) to be $w = -q/2 + \sqrt{D/108}i$, and when written in polar form, this becomes $w = re^{i\theta}$ where $r = \sqrt{-p^3/27}$ and

$$\cos\theta = \frac{q\sqrt{27}}{2p\sqrt{-p}}, \qquad \sin\theta = -\frac{\sqrt{D}}{2p\sqrt{-p}}. \tag{15}$$

(The $\sqrt{\ }$ symbols denote positive square roots here.) For the cube root Δ_1 of w we take

$$\Delta_1 = \sqrt{-p/3}\,e^{i\theta/3}.$$

Then $\Delta_1\overline{\Delta_1} = -p/3$. Since $\Delta_1\Delta_2 = -p/3$, we obtain $\Delta_2 = \overline{\Delta_1}$. Because $\omega = e^{2\pi i/3}$, $\omega^2 = \bar{\omega} = e^{4\pi i/3}$ and we now get

$$\Delta_1\omega = \sqrt{-p/3}\,e^{(\theta+2\pi)i/3}, \qquad \Delta_1\omega^2 = \sqrt{-p/3}\,e^{(\theta+4\pi)i/3},$$
$$\Delta_2\omega^2 = \overline{\Delta_1\omega}, \qquad \Delta_2\omega = \overline{\Delta_1\omega^2}.$$

Inserting these expressions into (10), we find: *In the irreducible case, the roots are*

$$\begin{aligned} \beta_1 &= 2\sqrt{-p/3}\cos\theta/3, \\ \beta_2 &= 2\sqrt{-p/3}\cos(\theta + 2\pi)/3, \\ \beta_3 &= 2\sqrt{-p/3}\cos(\theta + 4\pi)/3, \end{aligned} \tag{16}$$

where the angle θ *is defined by* (15).

The case $n = 4$. Let

$$g(\lambda) = \lambda^4 + p\lambda^2 + q\lambda + r,$$

where p, q, r are real or complex constants. We attempt to factor g as

$$g(\lambda) = (\lambda^2 + z\lambda + w)(\lambda^2 - z\lambda + v) \tag{17}$$

where z, w, v are unknown coefficients. If we can find values of z, w, v for which this is a correct factorization, the roots of $g(\lambda)$ will be those of $\lambda^2 + z\lambda + v$ together with those of $\lambda^2 - z\lambda + v$. Multiplying out the right-hand side of (17), then comparing the coefficients with those of g, we get these equations for z, w, v:

$$p = w - z^2 + v, \qquad q = -z(w - v), \qquad r = wv.$$

If $q = 0$, a solution is $z = 0$ and w, v given by $\lambda^2 - p\lambda + r = (\lambda - w)(\lambda - v)$, i.e., roots of a quadratic polynomial. If $q \neq 0$, we get $w - v = -q/z$, $w + v = p + z^2$, yielding

$$w = \frac{1}{2}\left(p + z^2 - \frac{q}{z}\right), \qquad v = \frac{1}{2}\left(p + z^2 + \frac{q}{z}\right), \tag{18}$$

and now $r = wv$ holds if and only if z satisfies

$$z^2(p + z^2)^2 - q^2 = 4rz^2.$$

Thus z^2 must be a root of the *cubic* polynomial

$$\lambda(p + \lambda)^2 - q^2 - 4r\lambda. \tag{19}$$

Using the methods outlined above, we may find a root (only one is needed) of this cubic polynomial; this gives a value for z^2 from which we compute a value for z, and then we get w and v from (18). For this z, w, v, the factorization (17) holds, and the roots of $g(\lambda)$ can then be calculated.

This method for finding the roots of a quartic polynomial will usually involve a great deal of labor in numerical examples.

This ends the discussion of general polynomials of various degrees n.

REMARK Formulas (5) and (12) are special cases of general formulas expressing the coefficients of a monic polynomial in terms of its roots. See Exercise 14 of §5.2.

Historical remarks

Solutions by radicals of first and second degree polynomials were known to the ancients. In the early 1500's solutions by radicals of cubic and quartic polynomials were discovered in Italy, and were published in a highly influential book by the Italian mathematician Cardan (1501–1576). This success, the first time the accomplishments of the ancients had been surpassed, had an enormous effect on the subsequent development of mathematics. For the next 200 or more years, many important mathematicians attempted to extend these results to polynomials of degrees 5, 6, 7, etc. None succeeded.

The first systematic approach to the problem of solution by radicals was made by the French mathematician Lagrange (1736–1813), who examined all known results for polynomials of degrees 1, 2, 3, 4 and showed, in each case, that the solution was made possible by an accidental fact that did not hold in general. The solutions of the cubic and quartic polynomials presented above depend on the solution of an auxiliary equation [the quadratic (8) or the cubic (19)], and Lagrange's methods also lead to an auxiliary equation that must be solved. However, when applied to the fifth degree polynomial, his methods lead to an auxiliary equation of sixth degree, more difficult to solve than the original quintic. This was the first hint that solutions by radicals of higher degree polynomials might not exist.

In 1824, Abel (Norwegian; 1802–1829) proved that the general quintic polynomial does not have a solution by radicals. This is the reason our discussion above of the general nth degree polynomial stopped at $n = 4$. Some higher degree polynomials do possess solutions by radicals, e.g., $\lambda^5 - 2$, and therefore it is natural to ask when a polynomial has a solution by radicals. The French mathematician Galois (1811–1832) provided a full answer for this question. The theory that Galois created, now called Galois theory, is sometimes taught at the advanced undergraduate level and usually taught in the first years of graduate school. It involves an intimate interweaving of group theory with the theory of fields. The tools of Galois theory are sufficient to prove that, in the irreducible case of a cubic polynomial, a solution by radicals cannot present the real roots in terms of real numbers only. Galois theory is still an active branch of mathematics, the most recent major advance being made by the Russian mathematician Safarevic in 1954.

The reader interested in mathematical history will find these facts outlined in greater detail in the following extremely readable books:

Cardano, The Gambling Scholar, by O. Ore (Dover, 1965)

A Concise History of Mathematics, by D. J. Struik (Dover, 1967)

Men of Mathematics, by E. T. Bell (Simon and Schuster, 1965)

The Development of Mathematics, by E. T. Bell (McGraw-Hill, 1945)

Mathematics, Its Content, Methods, and Meaning, Vol. 1, edited by A. D. Aleksandrov, A. N. Kolmogorov, M. A. Lavrent'ev (M. I. T. Press, 1963)

Numerical methods

We now turn to numerical methods for calculating roots of polynomials. Let $f(\lambda)$ be a polynomial with real coefficients, and consider the graph of the function $t = f(x)$. (See Figure 5.5.3.) We wish to calculate the number α for which $f(\alpha) = 0$. Letting x_0 be an initial estimate of α, we produce a new estimate in the following manner. Set $t_0 = f(x_0)$. The straight line through (x_0, t_0) tangent to the graph of $t = f(x)$ has equation $t - t_0 = (x - x_0)f'(x_0)$. We take the number x_1 at which this line crosses the X axis to be the new estimate of α. Setting $t = 0$ and $x = x_1$ in the equation of this line, we get

$$x_1 = x_0 - \frac{f(x_0)}{f'(x_0)}. \tag{20}$$

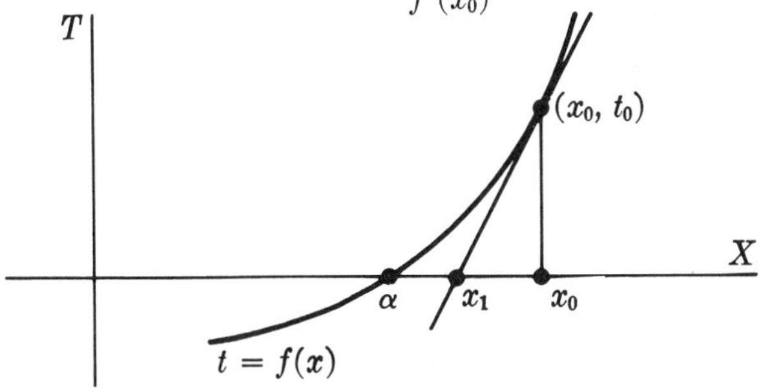

Figure 5.5.3

If the graph of $t = f(x)$ is increasing and concave upwards, and $x_0 > \alpha$, as in Figure 5.5.3, then x_1 lies between α and x_0 and therefore is a better estimate of α than x_0 was. Applying this procedure to x_1, we get x_2, a yet better approximation to α. Continuing this way, we produce a sequence $x_0, x_1, x_2, x_3, \ldots$ of real numbers converging towards α, and in numerical examples we may compute these approximations until we have obtained α to desired accuracy (i.e., correct to a desired number of decimal places). This procedure, called *Newton's method,* frequently works extremely well, but whether it works at all depends on the precision of the initial approximation x_0 of α. Figure 5.5.4 illustrates a situation in which $x_0 = x_2 = x_4 = \cdots$ and $x_1 = x_3 = x_5 = \cdots$; here Newton's method oscillates and does not converge.

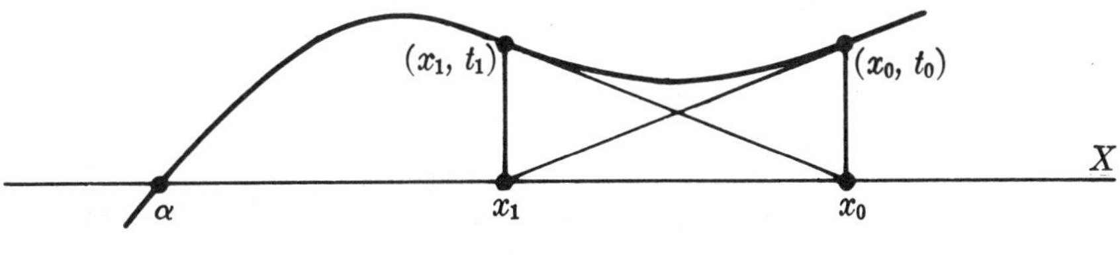

Figure 5.5.4

Newton's method is very effective for computing square roots. (See Figure 5.5.5.) If $f(x) = x^2 - a$ (with $a > 0$) then (20) simplifies to

$$x_1 = \frac{1}{2}\left(x_0 + \frac{a}{x_0}\right). \tag{21}$$

The process clearly converges to $\sqrt{a}$, and does so with great rapidity; in fact, if $a > 1$ and x_0 agrees with $\sqrt{a}$ to t decimal places, then x_1 will agree with $\sqrt{a}$ to $2t$ decimal places. For (21) and simple algebra yield

$$x_1 - \sqrt{a} = \frac{(x_0 - \sqrt{a})^2}{2x_0}. \tag{22}$$

If $x_0 > \sqrt{a}$ and $x_0 - \sqrt{a} < 10^{-t}$, from (22) we get $0 < x_1 - \sqrt{a} < 10^{-2t}/2x_0 < 10^{-2t}$. This proves our claim.

Here is a numerical illustration of the rapidity of convergence. If $a = 2$, and we begin with $x_0 = 2$, then $x_1 = 1.5$, $x_2 = 1.416666666$, $x_3 = 1.414215686$, $x_4 = 1.414213562$, and $x_4{}^2 = 1.99999999$. In fact, $\sqrt{2}$ lies between 1.414213562 and 1.414213563 since 2 lies between squares of these two numbers.

Newton's method may be used to calculate both real and nonreal roots of a polynomial $f(\lambda) \in \mathbf{C}[\lambda]$. In Exercise 15 we sketch a proof that the method will always converge to a real or complex root α of $f(\lambda)$ *provided* it is begun with a sufficiently accurate approximation of α. This convergence can be clearly seen if the method is interpreted geometrically using the surface S described in §5.3. (See Figure 5.5.6.) We are trying to find the point α in the XY plane at which S reaches down to this plane. Let z_0 be an approximation to α. Go vertically from z_0 to point P on S. At P draw the plane π *tangent* to S, and let z_1 be the point in both π and XY plane nearest P. (It turns out that z_1 is also the point in both π and the XY plane nearest z_0.) If z_0 is near enough to α so that P belongs to the "trough" which has α at the bottom, then it is geometrically reasonable that z_1 will be nearer α than z_0. When the expression for z_1 is calculated, it turns out to be

$$z_1 = z_0 - \frac{f(z_0)}{f'(z_0)}, \tag{23}$$

i.e., equation (20), but for complex numbers.

The proof that this geometric description of z_1 does lead to formula (23) for z_1 in terms of z_0 involves a small amount of partial differentiation—normally encountered in a second calculus course—and therefore is not given here. It may be found in Exercise 16.

We can use the tangent plane interpretation of Newton's method to modify the procedure for use when good initial estimates of the roots are not at hand. We are, of course, trying to find the lowest points on S, so that if we have a point P on S above z_0, we wish to find another point $\tilde{P}$

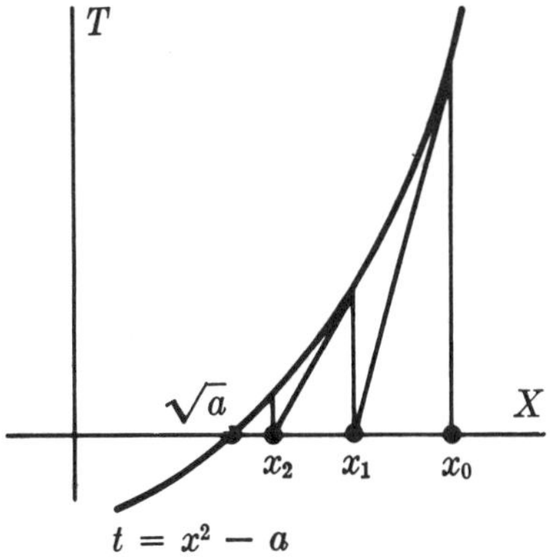

Figure 5.5.5

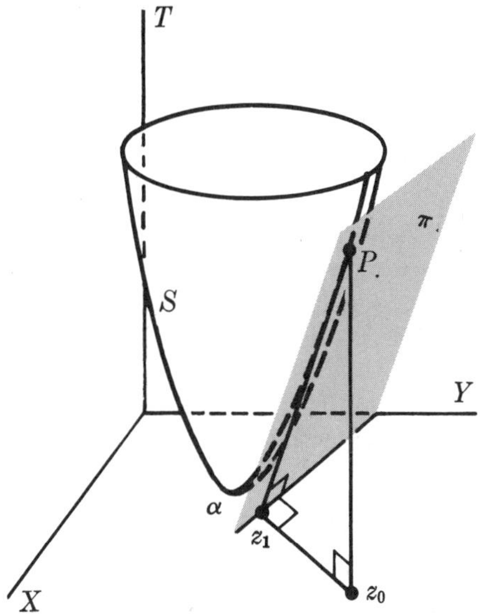

Angles marked ⌉ are right angles

Figure 5.5.6

on S lower than P. By Lemma 5.3.2, there always is a point on S near P but lower than P. Drawing the tangent plane to S at P and following it down to the XY plane, we obtain z_1, and therefore a direction to go from z_0 (namely, towards z_1) in order to find a complex number at which S is lower than at z_0. If z_0 is not a close approximation to a root of $f(\lambda)$, then P will be high above the XY plane, leading therefore to a long journey down the tangent plane to z_1, and consequently to z_1 being far from z_0. Unfortunately, if z_1 is far from z_0, there is no assurance that S is lower at z_1 than at z_0. If S is as high at z_1 as at z_0, we should select as the next approximation to the root not z_1 but a point part way from z_0 towards z_1. Let L be the straight line segment joining z_0 and z_1. If S is lower at the midpoint $z_0 + (z_1 - z_0)/2$ of L than at z_0, we may use this point (rather than z_1) as the next approximation to a root of $f(\lambda)$. If S is still as high at this midpoint as at z_0, we may try the one-quarter point $z_0 + (z_1 - z_0)/4$ on L. If S is as high here as at z_0, then we may try the one-eighth point $z_0 + (z_1 - z_0)/8$ on L. Repeatedly halving L in this manner eventually leads to a complex number near z_0 at which S is lower than at z_0. We then use this number as the next estimate in (modified) Newton's method.

If we have an estimate z_0 for which $f'(z_0) = 0$, then the tangent plane is parallel to the XY plane and we cannot use Newton's formula. In this case we may use the method given in the proof of Lemma 5.3.2 to find a complex number near z_0 at which S is lower than at z_0, and this number may be used as the next approximation to a root of $f(\lambda)$.

In this way, we can compute a succession of complex numbers at which S becomes progressively lower. This method can be used in conjunction with electronic computing equipment, and with due care in its implementation, gives a procedure whereby roots of polynomials with real or complex coefficients may be calculated.

A particularly clear and simple discussion of several methods for calculating roots of polynomials may be found in the Scheid book (pages 313–328) listed on page 455. In §5.7 we shall describe a very interesting method for isolating the real roots of a polynomial with real coefficients.

EXERCISES 5.5

I

1. Find the roots of the following polynomials, using the methods outlined in the text.
   (a) $2\lambda - 3$; (b) $\lambda^2 - \lambda - 6$; (c) $\lambda^2 - 2\lambda + 2$;
   (d) $\lambda^3 - 3\lambda + 2$; (e) $\lambda^3 - 3\lambda^2 - 3$; (f) $\lambda^3 - 2\lambda - 4$;
   (g) $\lambda^3 + \lambda$; (h) $\lambda^3 + 8$; (i) $\lambda^3 - 162^{1/3}\lambda + 5$;
   (j) $\lambda^3 - 3\cdot 10^{1/3}\lambda - 7$; (k) $\lambda^3 - 9\lambda - 12$;
   (l) $\lambda^3 - 54\lambda + 108$; (m) $\lambda^3 - 3\lambda + 2^{1/2}$; (n) $\lambda^8 - 1$;
   (o) $\lambda^4 + 2$; (p) $\lambda^4 + 2\lambda^2 + \lambda + 2$; (q) $\lambda^4 - 4\lambda^2 - \lambda + 2$.
   [(e), (f), (i), (j), (k), (l), (m), (p), (q) are optional exercises.]

2. The polynomial $\lambda^3 + 5\lambda^2 + 8\lambda + 4$ has a double root. Find all roots.

3. Find all roots of the polynomials $\lambda^3 + i$, $\lambda^6 - 64$, $\lambda^4 + 16$.

4. Find the roots of $\lambda^3 + \lambda^2 + \lambda + 1$ by
   (a) Cardan's formulas; (b) Theorem 5.5.1.

5. (*Optional*) Borrow a desk calculating machine, and use Newton's method to calculate $\sqrt{3}$ correct to nine decimal places (or as many places as your machine will permit).

II

6. Show that the cubic polynomial $\lambda^3 + p\lambda + q$ has a multiple root if and only if $27q^2 + 4p^3 = 0$. Do this three ways: (a) by Cardan's formulas; (b) in the case of real p, q, by showing that the X axis is a tangent line to $y = x^3 + px + q$ if and only if $27q^2 + 4p^3 = 0$; (c) by computing the g.c.d. of the polynomial and its derivative. Show also that the auxiliary equation (8) for the cubic has a double root if and only if the cubic itself has a multiple root.

7. (*Optional*) Let m be a positive integer, and $a > 1$ a real number. Show that the iteration

$$x_{k+1} = \frac{m-1}{m} x_k + \frac{a}{m x_k^{m-1}}, \qquad k = 0, 1, 2, \ldots,$$

converges to $a^{1/m}$. (Hint: This is Newton's method for a certain polynomial.) Give a geometric argument to prove convergence.

8. Let the quartic polynomial $\lambda^4 + p\lambda^2 + q\lambda + r$ in the text have roots $\beta_1, \beta_2, \beta_3, \beta_4$. Show that $-(\beta_1 + \beta_2)$, $-(\beta_1 + \beta_3)$, $-(\beta_1 + \beta_4)$, $-(\beta_2 + \beta_3)$, $-(\beta_2 + \beta_4)$, $-(\beta_3 + \beta_4)$ are the possible values of z. (Hint: The factor $\lambda^2 + z\lambda + w$ must be $(\lambda - \beta_i)(\lambda - \beta_j)$ for some $i < j$.) Hence, if the quartic has all roots real, prove that the roots of the auxiliary cubic will all be real.

9. Let $\omega = e^{2\pi i/n}$. Find the sum $1 + \omega + \omega^2 + \cdots + \omega^{n-1}$ of the nth roots of unity. Do this two ways: (a) using Lemma 2.2.3; (b) using Exercise 14, §5.2. Find the value of the product of the nth roots of unity.

10. A polynomial of odd degree with real coefficients always has a real root. What is the real root of $\lambda^n + 1$ when n is odd?

III

11. Let $f(\lambda) = \lambda^n + a_{n-1}\lambda^{n-1} + \cdots + a_1\lambda + a_0$ have roots $\alpha_1, \ldots, \alpha_n$. Show that $g(\lambda) = \lambda^n - a_{n-1}\lambda^{n-1} + a_{n-2}\lambda^{n-2} - a_{n-3}\lambda^{n-3} + \cdots + (-1)^n a_0$ has $-\alpha_1, \ldots, -\alpha_n$ as its roots. Hence show that $f(\lambda)g(\lambda)$ is a polynomial involving only even powers of λ. In $f(\lambda)g(\lambda)$ replace λ^2 throughout by λ, and let $F(\lambda)$ be the resulting polynomial. Show that the roots of $F(\lambda)$ are $\alpha_1^2, \ldots, \alpha_n^2$.

12. Find the factorization of $\lambda^n - 1$ into primes in $\mathbf{R}[\lambda]$.

13. Attempt a solution of the quintic polynomial $\lambda^5 + p\lambda^3 + q\lambda^2 + r\lambda + s$ as follows: Write $\lambda^5 + p\lambda^3 + q\lambda^2 + r\lambda + s = (\lambda^3 + e\lambda^2 + f\lambda + g) \times (\lambda^2 + h\lambda + l)$, where e, f, g, h, l are unknowns. By comparing coefficients on the two sides, set up a system of simultaneous equations in e, f, g, h, l. Attempt to solve these equations by eliminating unknowns one by one. Show that the final polynomial equation in a single unknown (one of e, f, g, h, l) is at least as complicated as the original quintic. Explain why this must be so. Can you predict, without doing the eliminations, what the degree of the final equation will be? (It will not be 6.)

14. Consider the irreducible case of the cubic polynomial $\lambda^3 + p\lambda + q$ when the roots are all real. Show that if θ is any angle for which $\cos\theta$ is given by the first equation in (15), then the formulas (16) will give the roots of the cubic. In other words: the second formula

in (15) [the one for $\sin\theta$] may be ignored. (Hint: If the cosine formula in (15) is satisfied, so is the sine formula for one of the two choices of $\sqrt{D}$.)

15. Let nonzero polynomial $f(\lambda) \in \mathbf{C}[\lambda]$, and let α be a simple root of $f(\lambda)$, possibly complex. Write $f(\lambda)$ in base $\lambda - \alpha$:

$$f(\lambda) = d_1(\lambda - \alpha) + d_2(\lambda - \alpha)^2 + \cdots + d_n(\lambda - \alpha)^n,$$

where $d_1, \ldots, d_n$ are constants. Set

$$g(\lambda) = d_2 + 2d_3(\lambda - \alpha) + 3d_4(\lambda - \alpha)^2 + \cdots + (n-1)d_n(\lambda - \alpha)^{n-2}.$$

Suppose z_0 is an initial (possibly complex) approximation of α. Let

$$z_1 = z_0 - f(z_0)/f'(z_0)$$

be the new approximation of α given by Newton's method.

(i) By differentiation of $f(\lambda)$ and evaluation at a complex number z, obtain $d_1 = f'(z) - 2d_2(z - \alpha) - \cdots - nd_n(z - \alpha)^{n-1}$. Apply the triangle inequality and transpose terms to get $|f'(z)| \geq |d_1| - 2|d_2||z - \alpha| - \cdots - n|d_n||z - \alpha|^{n-1}$. Use this inequality and $d_1 = f'(\alpha) \neq 0$ to argue that there is a positive number M and a small circle C centered at α with a positive radius ϵ such that $|g(z)/f'(z)| \leq M$ whenever $z \in C$. Reduce, if necessary, the radius of C so that $\epsilon M < 1$.

(ii) Evaluate the base $\lambda - \alpha$ representations of $f(\lambda)$, $f'(\lambda)$ at z_0 and substitute in the expression for z_1 to get

$$z_1 - \alpha = (z_0 - \alpha)^2 g(z_0)/f'(z_0).$$

(iii) If $z_0 \in C$, use the formula in (ii) to show that $|z_1 - \alpha| \leq |z_0 - \alpha|^2 M$.
(iv) Use $|z_0 - \alpha| \leq \epsilon$, $\epsilon M < 1$, and (iii), to show that $z_1 \in C$.

(v) If $z_0 \in C$, apply Newton's method to obtain a succession of approximations $z_0, z_1, z_2, z_3, \ldots$, all in C. Then use (iii) to prove that $|z_k - \alpha| \leq (|z_0 - \alpha| M)^{2^k}/M$, $\quad k = 1, 2, 3, \ldots$.

(vi) If $z_0 \in C$, use $\epsilon M < 1$ and (v) to conclude that Newton's method, when applied to z_0, produces a sequence of complex numbers converging to α.
(vii) If α is a root of multiplicity $e > 1$, write $f(\lambda) = d_e(\lambda - \alpha)^e + \cdots + d_n(\lambda - \alpha)^n$, $d_e \neq 0$, and by modifying the discussion in (i)–(vi), show that the approximations $z_0, z_1, \ldots$ given by Newton's method still converge to α, provided z_0 is sufficiently near α. (Note that when α is a multiple root, this method may not be satisfactory in practice since the calculation of $f(z)/f'(z)$ for z near α will involve division of two nearly zero numbers.)

16. Let $f(\lambda)$ be a nonzero polynomial with complex coefficients, and let $\bar{f}(\lambda)$ denote the polynomial obtained by replacing each coefficient of f by its complex conjugate. If $z = x + yi = (x, y)$, the surface S in §5.3 has equation $|f(x + yi)| - t = 0$, which may be rewritten as

$$f(x + yi)\bar{f}(x - yi) - t^2 = 0.$$

For simplicity, denote the left-hand side of this equation by $F(x, y, t)$.

(i) Let $P = (x_0, y_0, t_0)$ be a point on S. The equation of the tangent plane π to S at P is

$$(x - x_0)A + (y - y_0)B + (t - t_0)C = 0,$$

where A, B, C are, respectively, the partial derivatives

$$\frac{\partial F(x, y, t)}{\partial x}, \quad \frac{\partial F(x, y, t)}{\partial y}, \quad \frac{\partial F(x, y, t)}{\partial t}$$

evaluated at x_0, y_0, t_0.

(ii) Write the equation of π as $Ax + By + Ct = D$, where $D = Ax_0 + By_0 + Ct_0$.

(iii) Let $(x, y, 0) \in \pi \cap (XY \text{ plane})$. Use derivatives to minimize the distance $[(x - x_0)^2 + (y - y_0)^2 + t_0^2]^{1/2}$ between $(x, y, 0)$ and P. Show that the coordinates of the minimizing point $(x_1, y_1, 0)$ satisfy

$$Bx_1 - Ay_1 = Bx_0 - Ay_0, \qquad Ax_1 + By_1 = D.$$

(Hint: Use $Ax + By = D$ to treat one of x, y as a function of the other.) Show also that minimizing the distance between $(x, y, 0)$ and $(x_0, y_0, 0)$ leads to the same pair of equations for the minimizing point.

(iv) Solve the equations obtained in (iii) for x_1, y_1 to get

$$(x_1, y_1) = (x_0, y_0) + \frac{1}{A^2 + B^2}(DA - BAy_0 - A^2x_0, DB - ABx_0 - B^2y_0).$$

Then use $D = Ax_0 + By_0 + Ct_0$ to get

$$(x_1, y_1) = (x_0, y_0) + \frac{Ct_0}{A^2 + B^2}(A, B).$$

(v) If $z_1 = x_1 + y_1 i$, $z_0 = x_0 + y_0 i$, rewrite the last equation in part (iv) as

$$z_1 = z_0 + \frac{Ct_0}{A - Bi}.$$

(vi) If $z = x + yi$, use the product and chain laws for differentiation to calculate that

$$\frac{\partial F(x, y, t)}{\partial x} = \frac{\partial f(x + yi)\bar{f}(x - yi)}{\partial x} = f(z)\bar{f}'(\bar{z}) + f'(z)\bar{f}(\bar{z}),$$

$$\frac{\partial F(x, y, t)}{\partial y} = \frac{\partial f(x + yi)\bar{f}(x - yi)}{\partial y} = -if(z)\bar{f}'(\bar{z}) + if'(z)\bar{f}(\bar{z}),$$

$$\frac{\partial F(x, y, t)}{\partial t} = \frac{\partial(-t^2)}{\partial t} = -2t.$$

Evaluate at (x_0, y_0, t_0) to obtain

$$A = f(z_0)\bar{f}'(\bar{z}_0) + f'(z_0)\bar{f}(\bar{z}_0),$$

$$B = -if(z_0)\bar{f}'(\bar{z}_0) + if'(z_0)\bar{f}(\bar{z}_0),$$

$$Ct_0 = -2t_0^2 = -2f(z_0)\bar{f}(\bar{z}_0).$$

(vii) Substitute these values for A, B, C into the last formula in part (v) to obtain

$$z_1 = z_0 - f(z_0)/f'(z_0), \qquad \text{that is, equation (23).}$$

17. If $f(\lambda)$ is a nonzero polynomial with real coefficients, Newton's method (20) when begun with a real number can never converge to a nonreal root of f. Give an explanation for this in terms of the shape of the surface S.

18. Let a be a nonzero real or complex number, and consider $f(\lambda) = \lambda^2 - a$. The surface S for this polynomial is shown in the sketch below. Prove this, then discuss the convergence of Newton's method to the roots of $f(\lambda)$. Give both analytic and geometric arguments. Repeat for $f(\lambda) = \lambda^n - a$.

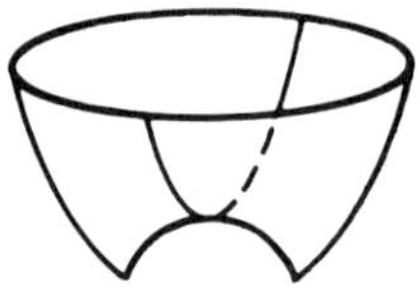

5.6 THE ROOTS OF THE DERIVATIVE (OPTIONAL)

If the roots of a polynomial $f(\lambda) \in \mathbf{C}[\lambda]$ are known, and are plotted as points in the complex plane, in what part of the complex plane will the derivative $f'(\lambda)$ have its roots? We already know one fact (Theorem 5.2.6): a root of f of multiplicity e is a root of f' of multiplicity $e - 1$.

Besides this, many beautiful facts of a geometric nature are known, and we now offer a selection of these having relatively easy proofs. Throughout this section the symbol $f'(\lambda)$ will denote the derivative of $f(\lambda)$.

We begin by introducing the concept of *convexity*.

DEFINITION 5.6.1.
A nonempty subset S of the complex plane is convex *if, whenever P and Q are points in S, the entire straight line segment joining P and Q is also in S.*

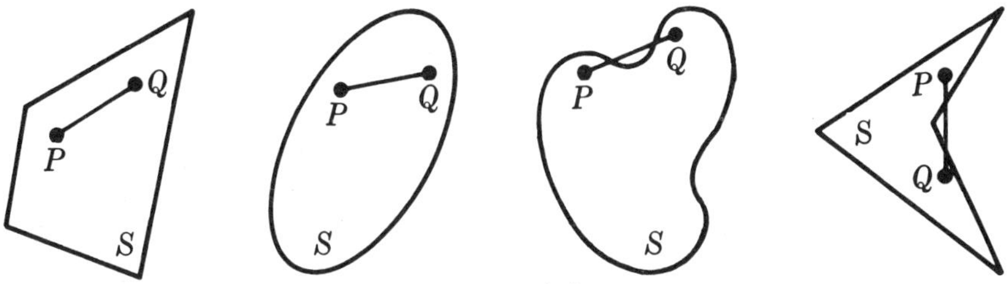

Figure 5.6.1

The first two sets in Figure 5.6.1 are convex, the last two are not. It is possible to cast this geometric definition of convexity into analytic terms. See Exercises 17 and 18 below.

DEFINITION 5.6.2.
The smallest convex set containing complex numbers $z_1, \ldots, z_n$ is called the convex hull of $z_1, \ldots, z_n$.

Figure 5.6.2 illustrates the convex hull S of points $z_1, \ldots, z_8$. In general, the convex hull of $z_1, \ldots, z_n$ consists of the interior and boundary of a polygon for which the vertices are some of the z_j, the remaining z_j being either inside the polygon or on a side. There is an analytic method that describes all complex numbers z in the convex hull of $z_1, \ldots, z_n$. Since we will not need it, we give it in Exercise 19 below.

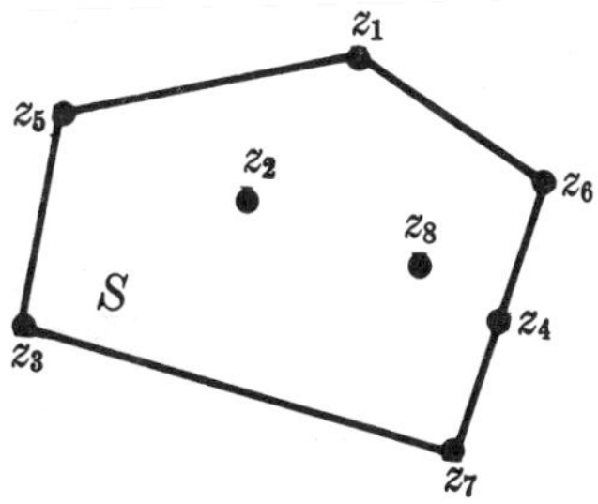

Figure 5.6.2

A basic fact about the roots of f' is given in Theorem 5.6.3.

THEOREM 5.6.3
If f is a nonconstant polynomial with complex coefficients, the convex hull of its roots contains the roots of its derivative f'.

PROOF Let $z_1, \ldots, z_n$ be the roots of f, not necessarily distinct, so that

$$f(\lambda) = c(\lambda - z_1)(\lambda - z_2)\cdots(\lambda - z_n) \tag{1}$$

where c is a nonzero constant. We shall use the following differentiation rule:

$$(g_1g_2\cdots g_n)' = \sum_{j=1}^{n} g_1\cdots g_{j-1}g_j'g_{j+1}\cdots g_n .$$

That is, to differentiate a product, differentiate in turn each of its factors. This rule is a generalization of the product rule $(g_1g_2)' = g_1'g_2 + g_1g_2'$, and its proof (by induction on n) is left as an exercise for the reader (Exercise 16, §3.2). Applying this differentiation rule to (1), we obtain

$$f'(\lambda) = \sum_{j=1}^{n} c(\lambda - z_1)\cdots(\lambda - z_{j-1})(\lambda - z_{j+1})\cdots(\lambda - z_n). \tag{2}$$

Let C be the convex hull of $z_1, \ldots, z_n$ (see Figure 5.6.3), and suppose z lies outside of C. Then z is not equal to any of $z_1, \ldots, z_n$; hence $f(z) \neq 0$. To prove the theorem we must show that z cannot be a root of f'. Substituting z for λ in (1) and (2), then dividing, we get

$$\frac{f'(z)}{f(z)} = \sum_{j=1}^{n} \frac{1}{z - z_j}. \tag{3}$$

We shall show that z is not a root of f' by proving that

$$\sum_{j=1}^{n} \frac{1}{z - z_j} \neq 0. \tag{4}$$

Since z is outside of C, there is a cone K with apex at z and apex angle $\alpha < \pi$ such that C lies entirely within K. (See Figure 5.6.3.) For the moment let ζ be any point in K, and set $Z = \zeta - z$. (See Figure 5.6.4.)

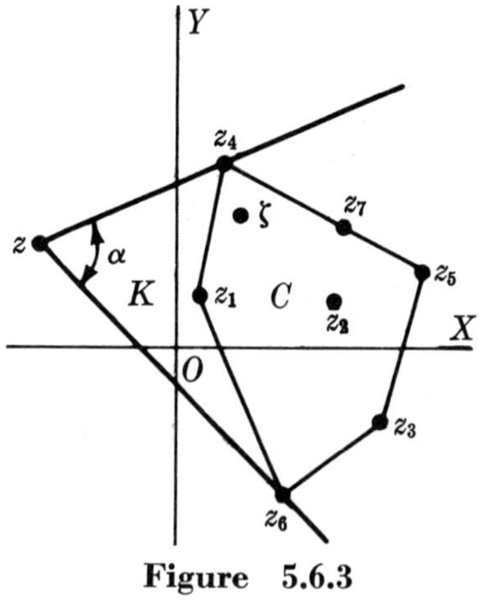

Figure 5.6.3

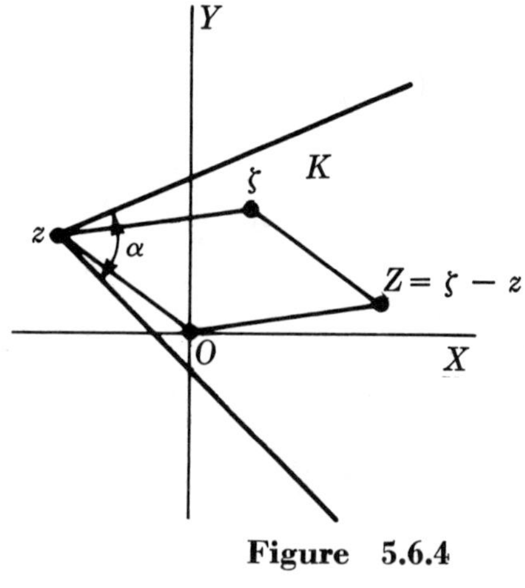

Figure 5.6.4

Because ζ is in K and z is the apex of K, the angle of inclination of the straight line connecting z and ζ is between the angles of inclinations of the sides of K. The angle of inclination of the parallel line connecting O and Z is therefore also between the angles of inclinations of these sides. Therefore Z lies in a cone K_1 with apex at the origin and sides parallel to the sides of K. (See Figure 5.6.5.) This holds for any ζ in K and therefore holds when ζ is $z_1, z_2, \ldots, z_n$. Thus the numbers $Z_1, \ldots, Z_n$ defined by $Z_1 = z_1 - z$, $Z_2 = z_2 - z$, $\ldots$, $Z_n = z_n - z$ all lie in K_1. (The reader familiar with §2.5 may view this argument this way: The translation T_{-z} of the complex plane moves z to the origin, and moves points in K to points in K_1.)

To prove (4), it suffices to prove that

$$\sum_{j=1}^{n} \frac{1}{Z_j} \neq 0. \tag{5}$$

Let $w = e^{i\theta}$, where θ is a real number. Multiplying $Z_1, \ldots, Z_n$ by w adds θ to their angles. Because the apex angle α of K_1 satisfies $\alpha < \pi$, we may choose θ such that $wZ_1, \ldots, wZ_n$ all lie in the upper half-plane, i.e., above the X axis. (Expressed in concepts from §2.5, a rotation r_θ moves all of $Z_1, \ldots, Z_n$ into the upper half-plane.) Letting $\rho_1 = wZ_1, \ldots, \rho_n = wZ_n$, to prove (5) it will suffice to prove that

$$\sum_{j=1}^{n} \frac{1}{\rho_j} \neq 0. \tag{6}$$

Since angle $1/\rho_j = -(\text{angle } \rho_j)$, and since $\rho_1, \ldots, \rho_n$ are all in the upper half-plane, $1/\rho_1, \ldots, 1/\rho_n$ are all in the lower half-plane; that is, each has the form $x + yi$ with real x and negative real y. (See Figure 5.6.6.) Numbers of this form, however, cannot add to zero since their imaginary terms yi cannot add to zero. Therefore (6) holds, ending the proof. □

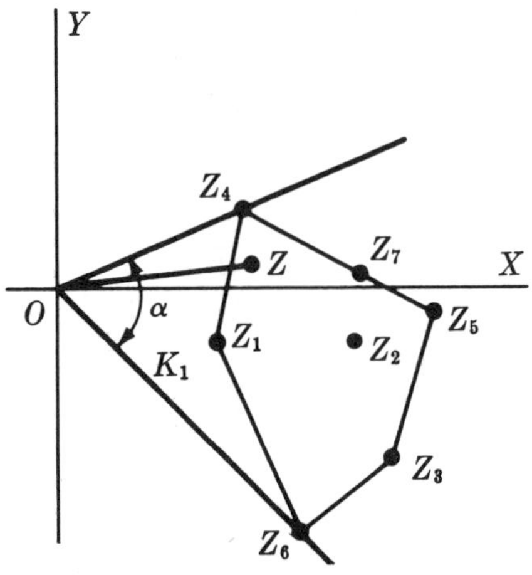

Figure 5.6.5

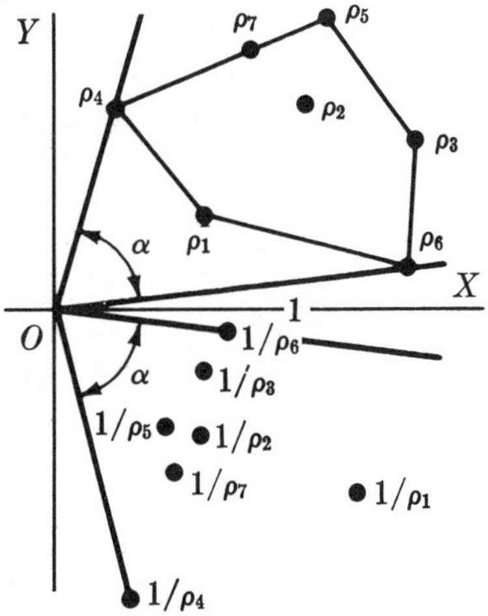

Figure 5.6.6

The above theorem is valid for polynomials with real coefficients and for polynomials with complex coefficients. We next study the roots of f' in the special case when f is a polynomial with real coefficients.

THEOREM 5.6.4 Rolle's Theorem
Let $f(\lambda)$ be a nonconstant polynomial with real coefficients. If a, b are real roots of $f(\lambda)$, with $a < b$, there is at least one root c of $f'(\lambda)$ with $a < c < b$.

Rolle's theorem is usually studied in calculus, both for polynomials and other types of functions. We now give a geometric discussion, for polynomials only. Let $[a, b]$ denote the interval consisting of all real numbers x with $a \leq x \leq b$. Draw the graph of $y = f(x)$ for $x \in [a, b]$. Since $f(a) = 0$, $f(b) = 0$, this graph meets the X axis when x is a or b. Because $f(x)$ is a sum of a finite number of terms of the type (constant times x^t, for integral $t \geq 0$), it depends continuously on x, so that the graph must connect the points $(a, 0)$, $(b, 0)$ with an unbroken curve. If the graph goes above the X axis, it will have a highest point above, and if it goes below the X axis, it will have a lowest point below. In either case, there will be a number c, with $a < c < b$, such that $f(x)$ attains an extreme value when x is c. (See Figure 5.6.7.) Because $f'(x)$ is also a polynomial, it depends continuously on x as well, implying that the graph of $y = f(x)$ has a smoothly turning tangent line. A smoothly turning tangent line must, at an extreme value $f(c)$, with $a < c < b$, be horizontal. That is, $f'(c) = 0$, completing the discussion.

If $f(x)$ did not depend continuously on x, for example, if its graph had a jump as in the first part of Figure 5.6.8, Rolle's theorem would not be valid. It would also not be valid if $f'(x)$ failed to exist for some x in $[a, b]$, for example, if the graph of f had a corner as in the second part of Figure 5.6.8. Jumps and corners can never appear on the graph of a polynomial $f(x)$ since both $f(x)$ and $f'(x)$ depend continuously on x.

There may be more than one real root of f' between two real roots of f, since f may have more than one horizontal tangent line. (See again Figure 5.6.7.)

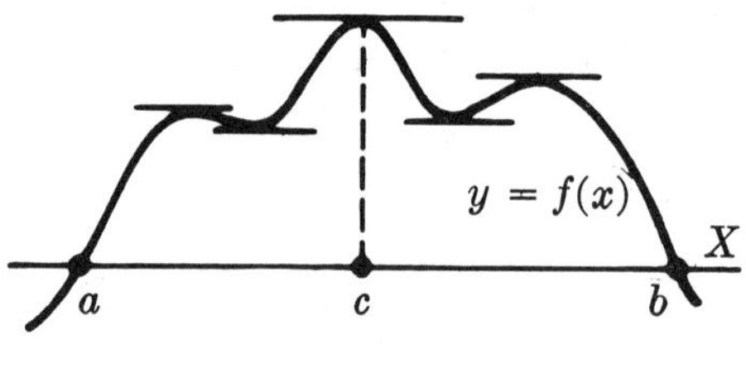

Figure 5.6.7

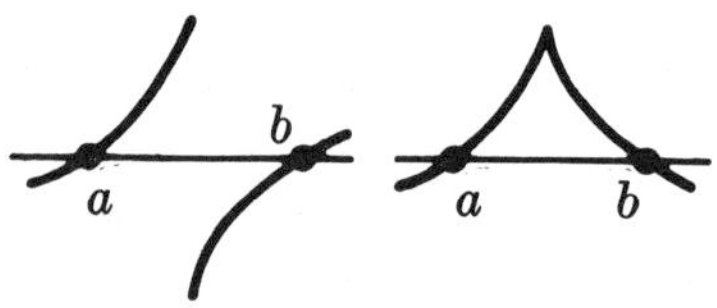

Figure 5.6.8

A nongraphical proof of Rolle's theorem may be given this way. If $f(\lambda)$ has real roots strictly between a and b, replace a with the largest of these. We may then assume that $f(\lambda)$ has no real roots strictly between a and b. Suppose a, b have multiplicities e_1, e_2 respectively. Then

$$f(\lambda) = (\lambda - a)^{e_1}(\lambda - b)^{e_2}g(\lambda),$$

where $g(\lambda)$ has real coefficients and no root between a and b. Then $g(a)$, $g(b)$ are nonzero numbers with the same sign since, if the signs were opposite, Theorem 5.4.7 would imply that $g(\lambda)$ has a root between a and b. Differentiation gives $f'(\lambda) = (\lambda - a)^{e_1-1}(\lambda - b)^{e_2-1}h(\lambda)$, where

$$h(\lambda) = (e_1(\lambda - b) + e_2(\lambda - a))g(\lambda) + (\lambda - a)(\lambda - b)g'(\lambda).$$

Then

$$h(a) = e_1(a - b)g(a), \qquad h(b) = e_2(b - a)g(b).$$

Since $g(a)$, $g(b)$ have the same signs, these formulas show that $h(a)$, $h(b)$ have opposite (nonzero) signs. Theorem 5.4.7, applied to $h(\lambda)$, implies that $h(\lambda)$ has a root c with $a < c < b$. But then c is a root of $f'(\lambda)$. □

***COROLLARY* 5.6.5**

If f is a nonconstant polynomial with real coefficients, the number of nonreal roots of f' (counting multiplicities) is never more than the number of nonreal roots of f (also counting multiplicities).

PROOF Let the distinct real roots of f be: α_1 with multiplicity $e_1, \ldots, \alpha_t$ with multiplicity e_t, and with $\alpha_1 < \cdots < \alpha_t$. The following are among the real roots of f':

$$\begin{gathered}\alpha_1, \text{ with multiplicity } e_1 - 1,\\ \alpha_2, \text{ with multiplicity } e_2 - 1,\\ \cdots\cdots\cdots\cdots\\ \alpha_t, \text{ with multiplicity } e_t - 1,\\ \text{at least one root } c_1 \text{ with } \alpha_1 < c_1 < \alpha_2,\\ \text{at least one root } c_2 \text{ with } \alpha_2 < c_2 < \alpha_3,\\ \cdots\cdots\cdots\cdots\cdots\cdots\\ \text{at least one root } c_{t-1} \text{ with } \alpha_{t-1} < c_{t-1} < \alpha_t.\end{gathered}$$

This list shows that f' has at least $(e_1 - 1) + \cdots + (e_t - 1) + (t - 1) = e_1 + \cdots + e_t - 1$ real roots. If $n = \deg f$, the number N of nonreal roots of f is $n - (e_1 + \cdots + e_t)$. The number of nonreal roots of f' is

$$\begin{aligned} n - 1 - (&\text{the number of real roots of } f')\\ &\le n - 1 - (e_1 + \cdots + e_t - 1)\\ &= n - (e_1 + \cdots + e_t) = N. \qquad \square \end{aligned}$$

Although Corollary 5.6.5 restricts the number of nonreal roots that f' may have, it is still of interest to ask where (in the complex plane) the nonreal roots can be. We need additional terminology to give the answer (Theorem 5.6.7) to this question.

DEFINITION 5.6.6.
Let z be a nonreal complex number. The Jensen circle based on z is the *set consisting of the interior and boundary of the circle for which the straight line joining z and $\bar{z}$ is a diameter.* (See Figure 5.6.9.)

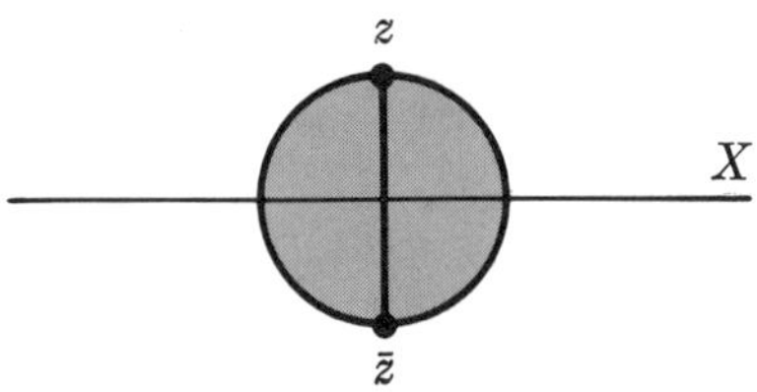

Figure 5.6.9

THEOREM 5.6.7
Let f be a nonconstant polynomial with real coefficients. Then a nonreal root of its derivative f' must belong to at least one Jensen circle based on a nonreal root of f.

Figure 5.6.10 illustrates a set of Jensen circles based on z_1, z_2, z_3, z_4, z_5, and shows also a point z outside the union of these Jensen circles.

PROOF Let $\deg f = n$, and let $z_1, \ldots, z_n$ be the roots of f, not necessarily distinct, and including both real and nonreal roots. Let z be a nonreal complex number outside every Jensen circle. We wish to prove that $f'(z) \neq 0$. For this it suffices to prove that $f'(z)/f(z) \neq 0$, and as in the proof of Theorem 5.6.3 it suffices to prove that

$$\sum_{j=1}^{n} \frac{1}{z - z_j} \neq 0. \qquad (7)$$

The left-hand side of (7) is a sum of two types of terms:

(i) terms of the type $\dfrac{1}{z - z_1}$ in which $z_1 = x_1$ is a real number;

(ii) terms of the type $\dfrac{1}{z - z_1} + \dfrac{1}{z - \bar{z}_1}$ in which $z_1 = x_1 + y_1 i$ is a nonreal number (x_1, y_1 real; $y_1 \neq 0$).

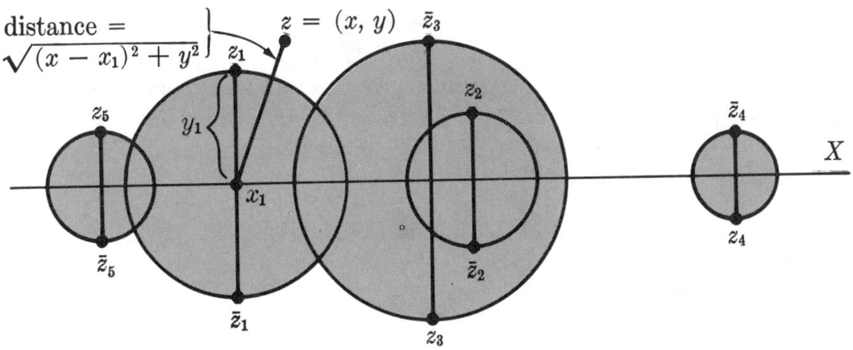

Figure 5.6.10

First consider type (i) terms. Let $z = x + yi$, with x, y real, and $y \neq 0$. Then

$$\frac{1}{z - z_1} = \frac{1}{x - x_1 + yi} = \frac{x - x_1 - yi}{(x - x_1)^2 + y^2} = A + Bi$$

where A and B are real numbers; in fact

$$B = -\frac{y}{(x - x_1)^2 + y^2}$$

and therefore has the sign of $-y$. That is: every type (i) term has the form $A + Bi$, where A, B are real, with B nonzero and carrying the sign of $-y$.

Consider type (ii) terms. If $z = x + yi$ is as above, then

$$\frac{1}{z - z_1} + \frac{1}{z - \bar{z}_1} = \frac{1}{(x - x_1) + (y - y_1)i} + \frac{1}{(x - x_1) + (y + y_1)i}$$

$$= \frac{x - x_1 - (y - y_1)i}{(x - x_1)^2 + (y - y_1)^2} + \frac{x - x_1 - (y + y_1)i}{(x - x_1)^2 + (y + y_1)^2}$$

$$= \frac{[x - x_1 - (y - y_1)i][(x - x_1)^2 + (y + y_1)^2] + [(x - x_1)^2 + (y - y_1)^2][x - x_1 - (y + y_1)i]}{[(x - x_1)^2 + (y - y_1)^2][(x - x_1)^2 + (y + y_1)^2]}$$

$$= A + Bi,$$

for real A, B, with

$$B = \frac{-2y[(x - x_1)^2 + y^2 - y_1^2]}{[(x - x_1)^2 + (y - y_1)^2][(x - x_1)^2 + (y + y_1)^2]}. \tag{8}$$

The quantity $(x - x_1)^2 + y^2 - y_1^2$ in the numerator of the right-hand side of (8) is positive because $z = (x, y)$ lies outside the Jensen circle based on z_1 (see Figure 5.6.10). Thus in case (ii), B is again a nonzero real number carrying the sign of $-y$.

It follows that the left-hand side of (7) is a sum of terms of the form $A + Bi$ with real A, B, and with every B nonzero and carrying the sign of $-y$. Such terms cannot add to zero since the imaginary Bi terms cannot add to zero. The sum in (7) is therefore nonzero, and the proof is complete. □

EXERCISES 5.6

I

1. Sketch the Jensen circles for each of the following polynomials. Sketch also the convex hulls of the roots.
   (i) $\lambda^2 + 1$; (ii) $(\lambda^2 + 1)(\lambda^2 - 10\lambda + 26)$;
   (iii) $(\lambda - 5)(\lambda^2 + 1)(\lambda + 6)$.
   In each case estimate where the roots of the derivative are located.

2. Construct an example of a fourth degree real polynomial f with exactly two real roots (call them a, b with $a < b$) such that f' has a real root between a and b, and another real root not between a and b.

3. Which of the following sets S are convex?
   (i) $S = \{(x, y) \mid x^2 + y^2 \le 1\}$
   (ii) $S = \{(x, y) \mid 2x + 3y - 1 = 0\}$
   (iii) $S = \{(x, y) \mid |x| + |y| = 1\}$
   (iv) $S = \{(x, y) \mid -1 \le x \le 2\}$
   (v) $S = \{(x, y) \mid x = 0, \text{ or } y = 0, \text{ or both}\}$
   (vi) $S = \{(x, y) \mid x^2 + y^2 = 1\}$.

4. Let a polynomial f have real coefficients. Must the real roots of f' belong to a Jensen circle for f? Supply either a proof or a counter-example.

5. Let $f(\lambda) = \lambda^n - 1$. Compute the roots of f' and show that they belong to the convex hull of the roots of f.

6. Estimate the location of the roots of the derivative of

$$\lambda^2(\lambda - 1)(\lambda^2 + 1)(\lambda^2 + \lambda + 4)(\lambda^2 + 10\lambda - 10).$$

II

7. Let f be a polynomial with real coefficients, and suppose all nonreal roots z of f satisfy $|z| = 1$. Let $w = x + yi$ (x, y real) be a nonreal root of f'. Prove: $|x| \le \sqrt{2}$. (Hint: Use Jensen circles.)

8. Let $z_1, \ldots, z_n$ be complex numbers, and let $m_1, \ldots, m_n$ be positive real numbers. By imitating the proof of Theorem 5.6.3, show that

a complex number z satisfying

$$\sum_{j=1}^{n} \frac{m_j}{z - z_j} = 0$$

must belong to the convex hull of $z_1, \ldots, z_n$.

9. Let $z_1, \ldots, z_n, m_1, \ldots, m_n$ be as in Exercise 8. Suppose that the nonreal numbers among $z_1, \ldots, z_n$ occur in complex conjugate pairs, and that $m_s = m_t$ when $z_s = \bar{z}_t$. By imitating the proof of Theorem 5.6.7, show that a nonreal number z satisfying

$$\sum_{j=1}^{n} \frac{m_j}{z - z_j} = 0$$

must belong to a Jensen circle based on $z_1, \ldots, z_n$.

10. Let f be a quadratic polynomial, not necessarily with real coefficients, and suppose a, b are roots of f. Is it always true that the root of f' is the midpoint of the straight line segment joining a and b?

11. Prove: The intersection of convex sets is again convex. What about the union?

12. Let a nonconstant degree-n polynomial f have real roots only, and let them be $x_1 \leq x_2 \leq \cdots \leq x_n$. Show that f' also has real roots only. Let them be $x_1' \leq x_2' \leq \cdots \leq x_{n-1}'$. Show that $x_1 \leq x_1' \leq x_2 \leq x_2' \leq x_3 \leq \cdots \leq x_{n-1} \leq x_{n-1}' \leq x_n$.

13. Suppose that f is a nonconstant polynomial with real coefficients and that its derivative f' has no real roots. Prove that f has exactly one real root, and this real root is simple.

III

14. Let S be the convex hull of numbers $z_1, \ldots, z_n$. If a and b are constants, not necessarily real, show that as z varies over S, the maximum and minimum of the real part of $az + b$ occurs when z is one of $z_1, z_2, \ldots, z_n$.

15. Describe all polynomials f for which the convex hull of the roots of f is a regular n-gon (i.e., a polygon with all sides of equal length and adjacent sides meeting in a common angle). Assume $\deg f = n$.

16. Under what conditions can a point on the boundary of the portion of the complex plane covered by the Jensen circles can be a nonreal root of the derivative?

17. Let z_1 and z_2 be complex numbers. Show that z lies on the straight line segment connecting z_1 and z_2 if and only if $z = \alpha z_1 + (1 - \alpha)z_2$ for a real number α with $0 \leq \alpha \leq 1$. Show also that as α varies from 0 to 1, the number z given by this formula moves from one end of the line segment to the other end.

(Hint: If $z_1 = z_2$, the result is trivial. Let z_1 and z_2 be distinct. As α varies over all real numbers, the equation $z = z_2 + \alpha(z_1 - z_2)$ describes all points z on the straight line L passing through z_2 and z_1; and z lies between z_2 and z_1 if and only if $0 \le \alpha \le 1$.)

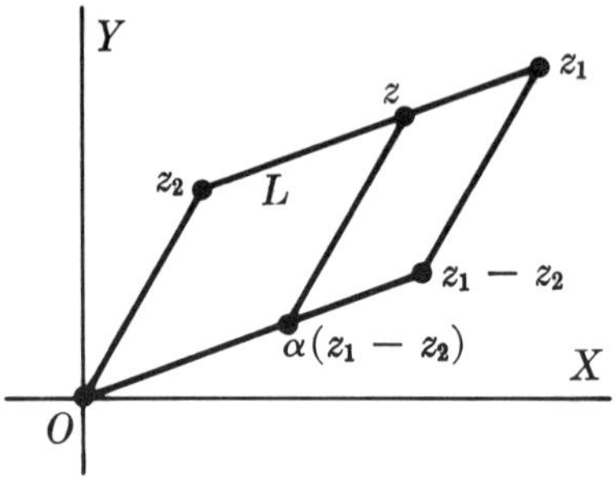

18. Show that a nonempty set S of complex numbers is convex if and and only if: whenever z_1 and z_2 are in S, so is $\alpha z_1 + (1 - \alpha)z_2$ for all real α with $0 \le \alpha \le 1$. (Use Exercise 17.)

19. Show that z belong to the convex hull of $z_1, \ldots, z_n$ if and only if

$$z = \alpha_1 z_1 + \alpha_2 z_2 + \cdots + \alpha_n z_n$$

for nonnegative real numbers $\alpha_1, \alpha_2, \ldots, \alpha_n$ with $\alpha_1 + \alpha_2 + \cdots + \alpha_n = 1$.

Outline of proof: If $n = 1$ this is clear; if $n = 2$ this is simply Exercise 17. Let $n \ge 3$.

Suppose that z belongs to the convex hull of $z_1, \ldots, z_n$. Connect z to z_1 by a straight line L, and extend L until it meets a line segment connecting a couple of other z's, say, meets the line segment l joining z_s and z_t at point P. Use Exercise 17 to write

$$z = \alpha_1 z_1 + (1 - \alpha_1)P$$

where $0 \le \alpha_1 \le 1$, and to write $P = \beta z_s + (1 - \beta)z_t$, where $0 \le \beta \le 1$. Then

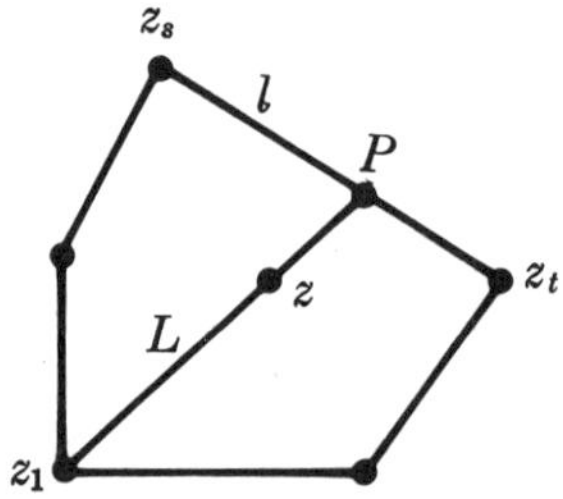

$$z = \alpha_1 z_1 + \alpha_s z_s + \alpha_t z_t$$

where $\alpha_s = (1 - \alpha_1)\beta$, $\alpha_t = (1 - \alpha_1)(1 - \beta)$.

To go in the converse direction, use induction on n, and assume $z' = \beta_2 z_2 + \cdots + \beta_n z_n$ is in the convex hull of $z_2, \ldots, z_n$ if $\beta_2, \ldots, \beta_n$ are nonnegative numbers adding to 1. Write

$$z = \alpha_1 z_1 + (1 - \alpha_1)z'$$

by taking

$$\beta_2 = \frac{\alpha_2}{1 - \alpha_1}, \quad \beta_3 = \frac{\alpha_3}{1 - \alpha_1}, \quad \ldots, \quad \beta_n = \frac{\alpha_n}{1 - \alpha_1}.$$

Combine the induction hypothesis and Exercise 17.

20. The abstract definition of a cone $\mathcal{K}$ in the complex plane with apex at z is this: $\mathcal{K}$ is a nonempty set such that whenever z' and z'' are in $\mathcal{K}$, so is $z + \alpha(z' - z) + \beta(z'' - z)$ for all nonnegative real numbers α and β. Show that the cones K, K_1 in the text meet this definition, the latter with $z = 0$.

21. Let $\mathcal{K}$ be a cone with apex angle α. What can you say about α if $\mathcal{K}$ is convex?

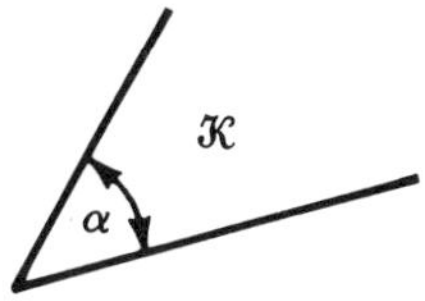

5.7 STURM'S THEOREM (OPTIONAL)

Let $f(\lambda)$ be a nonconstant polynomial with real coefficients. In §5.5 we gave theoretical methods that may be used to compute its roots when $\deg f \leq 4$, and a numerical method (Newton's method) applicable to polynomials of any degree. We now give a rather different numerical method for locating the *real* roots, which also is applicable to polynomials of any degree. This method, discovered by C. Sturm (1803–1855) and named after him, involves a simple modification of Euclid's algorithm described in §3.4. Sturm's method is employed today because it is suited for use by electronic computing machines.

Suppose a and b are real numbers, neither a root of $f(\lambda)$, with $a < b$. Let $[a, b]$ denote the interval consisting of all real x with $a \leq x \leq b$. Sturm's result enables the number of distinct roots of $f(\lambda)$ in $[a, b]$ to be calculated *without knowing the roots*. We now describe the method, and state Sturm's theorem. Then we give a numerical example, and after that the proof.

Let $f_0(\lambda) = f(\lambda)$, $f_1(\lambda) = f'(\lambda)$, the prime denoting differentiation, and suppose $f_2(\lambda), f_3(\lambda), \ldots, f_m(\lambda), q_1(\lambda), q_2(\lambda), \ldots, q_m(\lambda)$ are further polynomials with real coefficients, satisfying these equations:

$$
\begin{aligned}
f_0(\lambda) &= q_1(\lambda)f_1(\lambda) - f_2(\lambda),\\
f_1(\lambda) &= q_2(\lambda)f_2(\lambda) - f_3(\lambda),\\
&\cdots\cdots\cdots\cdots\cdots\\
f_j(\lambda) &= q_{j+1}(\lambda)f_{j+1}(\lambda) - f_{j+2}(\lambda),\\
&\cdots\cdots\cdots\cdots\cdots\\
f_{m-2}(\lambda) &= q_{m-1}(\lambda)f_{m-1}(\lambda) - f_m(\lambda),\\
f_{m-1}(\lambda) &= q_m(\lambda)f_m(\lambda).
\end{aligned}
\tag{1}
$$

[Compare equations (7) of §3.4.] Polynomials $f_2, \ldots, f_m, q_1, \ldots, q_m$ may be found satisfying these equations if we modify Euclid's algorithm for computing the greatest common divisor (f_0, f_1) in this way: When f_j and f_{j+1} have been computed, take f_{j+2} to be the negative of the remainder obtained upon dividing f_j by f_{j+1}. Thus, starting with f_0 and f_1, by long division and changing the signs of the remainders, we get in succession f_2, f_3, f_4, etc., the process continuing until we obtain a zero remainder.

For any real number x, let $V(x)$ denote the number of changes of sign in this sequence of real numbers:

$$f_0(x), f_1(x), f_2(x), \ldots, f_m(x). \tag{2}$$

(For example, if $m = 15$ and the signs of the terms in the sequence (2) are

$$+ + + - - + - + - - - + - + + +,$$

then $V(x) = 8$.) If a term in (2) happens to be zero, give it either sign, $+$ or $-$, as you please. We call $V(x)$ the number of *variations of sign* in (2).

We are now ready to state the main result of this section.

THEOREM 5.7.1

Let $f(\lambda)$ be a nonconstant polynomial with real coefficients, and let real polynomials $f_0(\lambda), f_1(\lambda), f_2(\lambda), \ldots, f_m(\lambda), q_1(\lambda), q_2(\lambda), \ldots, q_m(\lambda)$ satisfy equations (1), with $f_0(\lambda) = f(\lambda)$ and $f_1(\lambda) = f'(\lambda)$. If a and b are real numbers, neither a root of $f(\lambda)$, with $a < b$, then the number of distinct real roots of $f(\lambda)$ in the interval $[a, b]$ is $V(a) - V(b)$.

In other words: The number of distinct roots of $f(\lambda)$ in $[a, b]$ may be found simply by evaluating the polynomials $f_0, f_1, f_2, \ldots, f_m$ at the endpoints of this interval, and counting variations in sign. Notice that Sturm's theorem gives the number of distinct real roots of f in this interval, but does not give their multiplicities.

Example Locate the real roots of $f(\lambda) = \lambda^3 - 3\lambda^2 - 4\lambda + 13$.

SOLUTION Setting $f_0(\lambda) = f(\lambda)$ and $f_1(\lambda) = f'(\lambda) = 3\lambda^2 - 6\lambda - 4$, by long division we obtain:

$$\begin{aligned} f_0(\lambda) &= (\tfrac{1}{3}\lambda - \tfrac{1}{3})f_1(\lambda) - f_2(\lambda), \\ f_1(\lambda) &= (\tfrac{9}{14}\lambda + \tfrac{9}{28})f_2(\lambda) - f_3(\lambda), \\ f_2(\lambda) &= \tfrac{28}{3}(2\lambda - 5)f_3(\lambda), \end{aligned} \tag{3}$$

with $\deg f_0(\lambda) > \deg f_1(\lambda) > \deg f_2(\lambda) > \deg f_3(\lambda)$. [In fact, $f_2(\lambda) = \frac{7}{3}(2\lambda - 5)$, $f_3(\lambda) = \frac{1}{4}$.] We are going to evaluate

$$f_0(x), \quad f_1(x), \quad f_2(x), \quad f_3(x)$$

at various real numbers x. This can be efficiently done using equations (3), in this way: Knowing $f_3(x) = 1/4$, compute $f_2(x)$ using the last equation in (3), then use the middle equation to compute $f_1(x)$, and finally the first equation to compute $f_0(x)$.

Evaluating first at a large positive x, and afterwards at a large negative x, from (3) we get the entries in the columns headed $+\infty$ and $-\infty$ in Table 5.7.1. (For this, remember that the sign of a polynomial in x is determined by the term involving the highest power of x, if $|x|$ is sufficiently large.) Thus $V(-\infty) - V(+\infty) = 3$, implying that $f(\lambda)$ has three distinct real roots. Next, evaluate at $x = 0$, obtaining thereby the signs in the column headed 0 in Table 5.7.1. Thus $V(-\infty) - V(0) = 1$ and $V(0) - V(+\infty) = 2$, so that $f(\lambda)$ has one negative and two positive roots. To locate these roots more accurately, we may give x the integral values 1, 2, 3, etc., continuing until we have isolated the positive roots; then negative integral values -1, -2, -3, etc., continuing until we have isolated the negative root. In this example (see Table 5.7.1) we find the positive roots are between 2 and 3, and the negative root between -2 and -3. To separate the positive roots, evaluate at $x = 5/2$, thereby finding one root between 2 and 5/2, and one between 5/2, and 3. Thus f has one root in each of the intervals $[-3, -2]$, $[2, 5/2]$, $[5/2, 3]$. If still closer estimates are required, we may evaluate f at the midpoints of these intervals, then at the midpoints of resulting intervals, etc. In this manner we obtain small intervals, each containing a single root of f.

Having obtained good estimates of the roots from Sturm's theorem, we may then apply Newton's method and rapidly calculate the roots to any desired number of decimal places. In fact, beginning at -3, 2, and 3, respectively, in six iterations Newton's method gives each root correctly to fourteen decimal places: -2.048917339522305, 2.356895867892210, 2.692021471630096. (In general, of course, Newton's method should not be applied carelessly, since it may not converge to a desired root if an inappropriate initial estimate is used.)

$x =$	$-\infty$	-3	-2	-1	0	1	2	5/2	3	$+\infty$
$f_0(x)$	$-$	$-$	$+$	$+$	$+$	$+$	$+$	$-$	$+$	$+$
$f_1(x)$	$+$	$+$	$+$	$+$	$-$	$-$	$-$	$-$	$+$	$+$
$f_2(x)$	$-$	$-$	$-$	$-$	$-$	$-$	$-$	0	$+$	$+$
$f_3(x)$	$+$	$+$	$+$	$+$	$+$	$+$	$+$	$+$	$+$	$+$
Variations	3	3	2	2	2	2	2	1	0	0

Table 5.7.1

Proof of Sturm's theorem: By imitating the proof of Lemma 3.4.8, we deduce from equations (1) that f_m is a common divisor of f_0, f_1, and is divisible by every common divisor. Thus f_m is an associate of (f_0, f_1).

The proof of Lemma 3.4.8, applied to equations (1) above, also shows that each of f_m, f_{m-1}, ..., f_1, f_0 is divisible by f_m. Let $d(\lambda) = (f_0(\lambda), f_1(\lambda)) = (f(\lambda), f'(\lambda))$. Then

$$f_j(\lambda) = d(\lambda)F_j(\lambda), \qquad j = 0, 1, 2, \ldots, m,$$

where $F_j(\lambda)$ is a polynomial with real coefficients, and $F_m(\lambda)$ is a nonzero constant. Substituting into (1), then cancelling $d(\lambda)$, we get:

$$\begin{aligned}
F_0(\lambda) &= q_1(\lambda)F_1(\lambda) - F_2(\lambda),\\
F_1(\lambda) &= q_2(\lambda)F_2(\lambda) - F_3(\lambda),\\
&\cdots\cdots\cdots\cdots\cdots\\
F_j(\lambda) &= q_{j+1}(\lambda)F_{j+1}(\lambda) - F_{j+2}(\lambda),\\
&\cdots\cdots\cdots\cdots\cdots\\
F_{m-2}(\lambda) &= q_{m-1}(\lambda)F_{m-1}(\lambda) - F_m(\lambda),\\
F_{m-1}(\lambda) &= q_m(\lambda)F_m(\lambda).
\end{aligned} \tag{4}$$

Because $F_0 = f/d$, where $d = (f, f')$, from Corollary 5.2.7 we know that $F_0(\lambda)$ has simple roots (only), coinciding with the distinct roots of $f(\lambda)$.

We now give the proof of Sturm's theorem in a succession of lemmas.

LEMMA 5.7.2

If x is a real number, for no j, $0 \leq j \leq m - 1$, do we have $F_j(x) = F_{j+1}(x) = 0$.

PROOF If $F_j(x) = F_{j+1}(x) = 0$, substituting in (4) gives $F_{j+2}(x) = 0$. Thus $F_{j+1}(x) = F_{j+2}(x) = 0$; hence we may increase j. Eventually we get $F_{m-1}(x) = F_m(x) = 0$. But this is impossible because $F_m(\lambda)$ is a nonzero constant. □

LEMMA 5.7.3

If $F_{j+1}(x) = 0$ for a real number x, and a fixed j with $0 \leq j \leq m - 2$, then $F_j(x)$ and $F_{j+2}(x)$ are nonzero numbers of opposite signs.

PROOF From Lemma 5.7.2 we know that both $F_j(x)$ and $F_{j+2}(x)$ are nonzero. From (4) and $F_{j+1}(x) = 0$, we get $F_j(x) = -F_{j+2}(x)$, so that $F_j(x)$ and $F_{j+2}(x)$ must have opposite signs. □

If x is a real number for which $F_0(x)$ is nonzero, let $V_F(x)$ be the number of variations of sign in the sequence

$$F_0(x), F_1(x), F_2(x), \ldots, F_m(x), \tag{5}$$

selecting arbitrarily the signs (+ or −) for any zero terms in (5).

LEMMA 5.7.4

If $F_0(x) \neq 0$, $V_F(x)$ does not depend on the signs given to the zero terms (if any) in (5).

PROOF If a term in (5) is zero, it cannot be the first term $F_0(x)$ (by hypothesis) nor the last term $F_m(x)$ [since $F_m(\lambda)$ is a nonzero constant], and so must be $F_{j+1}(x) = 0$ for some j with $0 \leq j \leq m - 2$. Then $F_j(x)$ and $F_{j+2}(x)$ are nonzero and of opposite signs, by Lemma 5.7.3. Thus the terms

$$F_j(x), \quad F_{j+1}(x), \quad F_{j+2}(x)$$

in (5) carry either the signs $+$, 0, $-$ or the signs $-$, 0, $+$, and so display exactly one variation whether 0 is given the sign $+$ or the sign $-$. □

Letting x now be a variable real number, we let it increase from a to b. As x increases, $V_F(x)$ may change, but can do so only when one or more of $F_0(x), F_1(x), \ldots, F_m(x)$ changes sign.

LEMMA 5.7.5
As x increases, $V_F(x)$ does not change when $F_{j+1}(x)$ changes sign, for any j with $0 \leq j \leq m - 1$.

PROOF Suppose $F_{j+1}(x)$ changes sign when x increases past α. Thus $F_{j+1}(\alpha) = 0$, by Theorem 5.4.7. We have $j < m - 1$ since $F_m(x)$ is a nonzero constant. By Lemma 5.7.3, $F_j(\alpha)$ and $F_{j+2}(\alpha)$ are nonzero numbers of opposite signs. When x is sufficiently near α, no root of either F_j or F_{j+2} lies between x and α, so that $F_j(x)$, $F_{j+2}(x)$ are nonzero numbers carrying the signs of $F_j(\alpha)$, $F_{j+2}(\alpha)$ respectively. The graphs of $F_j(x)$, $F_{j+1}(x)$, $F_{j+2}(x)$ for x near α then have the forms sketched in Figure 5.7.1, the various cases arising according as $F_j(\alpha)$ is positive or negative, and as $F_{j+1}(x)$ changes sign from $+$ to $-$, or from $-$ to $+$. Table 5.7.2 shows the signs of $F_j(x)$, $F_{j+1}(x)$, $F_{j+2}(x)$ as x increases past α in each of these cases. Inspecting this table, we see that the portion $F_j(x)$, $F_{j+1}(x)$, $F_{j+2}(x)$ of the sequence (5) always presents one variation (as x increases past α) even though $F_{j+1}(x)$ changes its sign. This change of sign therefore does not cause $V_F(x)$ to change. □

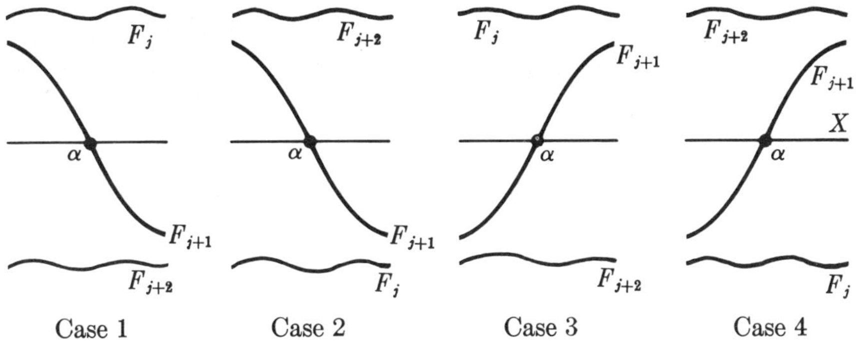

Figure 5.7.1

	Case 1			Case 2			Case 3			Case 4		
x:	$<\alpha,$	$=\alpha,$	$>\alpha;$	$<\alpha,$	$=\alpha,$	$>\alpha;$	$<\alpha,$	$=\alpha,$	$>\alpha;$	$<\alpha,$	$=\alpha,$	$>\alpha;$
$F_j(x)$:	+	+	+	−	−	−	+	+	+	−	−	−
$F_{j+1}(x)$:	+	0	−	+	0	−	−	0	+	−	0	+
$F_{j+2}(x)$:	−	−	−	+	+	+	−	−	−	+	+	+
Variations	1	1	1	1	1	1	1	1	1	1	1	1

Table 5.7.2

A consequence of Lemma 5.7.5 is that $V_F(x)$ can change only when $F_0(x)$ changes its sign.

LEMMA 5.7.6

As x increases, $F_0(x)$ changes its sign precisely when x passes a real root of $f(\lambda)$.

PROOF Recall that $F_0(\lambda)$ has simple roots (only), coinciding with the distinct roots of $f(\lambda)$.

If $F_0(x)$ changes its sign as x increases past α, then $F_0(\alpha) = 0$ and therefore α is a real root of $f(\lambda)$. On the other hand, if α is a real root of $f(\lambda)$, then it is a simple root of $F_0(\lambda)$, so that

$$F_0(\lambda) = (\lambda - \alpha)G(\lambda),$$

where $G(\lambda)$ is a polynomial with real coefficients and $G(\alpha) \neq 0$. If x is sufficiently near α, no root of $G(\lambda)$ lies between x and α, so that (by Theorem 5.4.7) $G(x)$, $G(\alpha)$ have the same sign. As x increases past α, $F_0(x)$ then successively has the signs of $-G(\alpha)$, 0, $G(\alpha)$, and therefore does change its sign. □

LEMMA 5.7.7

If α is a root of $f(\lambda)$ with multiplicity e, then $F_1(\alpha) = eF_0'(\alpha)$.

PROOF Since α is a root of $f(\lambda)$ of multiplicity e, it is a root of $f'(\lambda)$ of multiplicity $e - 1$ and therefore a root of $d(\lambda)$ of multiplicity $e - 1$. Thus

$$d(\lambda) = (\lambda - \alpha)^{e-1}h(\lambda), \qquad f_0(\lambda) = (\lambda - \alpha)^e g(\lambda),$$

where $h(\lambda)$, $g(\lambda)$ are polynomials with $h(\alpha) \neq 0$, $g(\alpha) \neq 0$. From $f_1 = F_1 d$, $f_1 = f_0'$, and $f_0 = (\lambda - \alpha)^e g$, we get

$$F_1(\lambda)(\lambda - \alpha)^{e-1}h(\lambda) = e(\lambda - \alpha)^{e-1}g(\lambda) + (\lambda - \alpha)^e g'(\lambda),$$

so that

$$F_1(\lambda)h(\lambda) = eg(\lambda) + (\lambda - \alpha)g'(\lambda).$$

Evaluating at α gives

$$F_1(\alpha)h(\alpha) = eg(\alpha). \tag{6}$$

However, from $f_0 = (\lambda - \alpha)^e g$, $f_0 = F_0 d$, and $d = (\lambda - \alpha)^{e-1}h$, we get

$$(\lambda - \alpha)^e g(\lambda) = F_0(\lambda)(\lambda - \alpha)^{e-1}h(\lambda);$$

hence

$$(\lambda - \alpha)g(\lambda) = F_0(\lambda)h(\lambda).$$

Differentiating this equation, then evaluating at α, we obtain $g(\alpha) = F_0'(\alpha)h(\alpha)$, using $F_0(\alpha) = 0$. Substituting this in (6) gives

$$F_1(\alpha)h(\alpha) = eF_0'(\alpha)h(\alpha).$$

Since $h(\alpha) \neq 0$, we deduce that $F_1(\alpha) = eF_0'(\alpha)$, as claimed. □

LEMMA 5.7.8

As x increases, $V_F(x)$ changes precisely when x crosses a real root of $f(\lambda)$, and there it decreases by 1.

PROOF We already know that $V_F(x)$ can change only when x increases past a root of $f(\lambda)$. We now show that $V_F(x)$ does change when x increases past a root α of $f(\lambda)$, and that the change is a decrease of 1. From Lemma 5.7.6, we know that $F_0(x)$ changes sign at α. If the sign change is from $+$ to $-$, $F_0(x)$ is then decreasing at α, hence $F_0'(\alpha)$ is negative (by §4.2), therefore $F_1(\alpha) = eF_0'(\alpha)$ also is negative, and hence $F_1(x)$ is negative for x near α. Similarly, if the sign of $F_0(x)$ goes from $-$ to $+$, $F_0'(\alpha)$ is positive, therefore $F_1(\alpha) = eF_0'(\alpha)$ is positive, and hence $F_1(x)$ is positive for x near α. The signs of the initial terms $F_0(x)$, $F_1(x)$ in (5) in these two cases are then as shown in Table 5.7.3.

	Case 1			Case 2		
x:	$<\alpha$,	$=\alpha$,	$>\alpha$;	$<\alpha$,	$=\alpha$,	$>\alpha$;
$F_0(x)$:	$+$	0	$-$	$-$	0	$+$
$F_1(x)$:	$-$	$-$	$-$	$+$	$+$	$+$
Variations:	1		0	1		0

Table 5.7.3

In both cases, the initial portion $F_0(x)$, $F_1(x)$ of the sequence (5) shows one variation for $x < \alpha$ and no variation for $x > \alpha$. This means: $V_F(x)$ decreases by 1 as x increases past α. □

LEMMA 5.7.9

The number of distinct real roots of $f(\lambda)$ in $[a, b]$ is $V_F(a) - V_F(b)$.

PROOF As x increases from a to b, $V_F(x)$ changes precisely when x crosses a root of $f(\lambda)$, and there it decreases by 1. Thus

$$V_F(b) = V_F(a) - (\text{the number of distinct roots of } f(\lambda) \text{ in } [a, b]). \quad \square$$

LEMMA 5.7.10
$V_F(a) = V(a)$ and $V_F(b) = V(b)$.

PROOF We have $f_j(x) = F_j(x)d(x)$, for any x. Taking x equal to a and $j = 0$ gives $f(a) = F_0(a)d(a)$. Consequently, $d(a)$ is nonzero since $f(a)$ is nonzero. Each term in the sequence

$$f_0(a), f_1(a), \ldots, f_m(a) \tag{7}$$

is therefore the nonzero number $d(a)$ times the corresponding term in the sequence

$$F_0(a), F_1(a), \ldots, F_m(a). \tag{8}$$

Thus, term by term, the signs in (7) coincide with the signs in (8) [when $d(a)$ is positive] or are opposite to the signs in (8) [when $d(a)$ is negative]. In either case, (7) and (8) show the same number of variations of sign. Since $V(a)$ is the number of variations of sign in (7), and $V_F(a)$ is the number of variations of sign in (8), we get $V(a) = V_F(a)$. Similarly, $V(b) = V_F(b)$. □

Combining Lemmas 5.7.9 and 5.7.10, we see that the number of distinct real roots of $f(\lambda)$ in $[a, b]$ is $V(a) - V(b)$, completing the proof of Sturm's theorem. □

EXERCISES 5.7

All polynomials in these exercises are to have real coefficients.

I

1. Use Sturm's theorem to obtain as much information as you can about the roots of the following polynomials:
   (a) $\lambda^3 - 3\lambda^2 + 6\lambda - 2$ (b) $\lambda^3 + 6\lambda^2 + 10\lambda - 1$
   (c) $\lambda^3 - 6\lambda^2 - 6\lambda + 52$ (d) $\lambda^4 + 2\lambda^2 - 4\lambda + 8$
   (e) $\lambda^4 - 4\lambda^3 + 2\lambda^2 + 4\lambda + 2$ (f) $\lambda^4 - 4\lambda^3 + 2\lambda^2 + 4\lambda$

2. Use Sturm's theorem to determine when the roots of a quadratic polynomial $\lambda^2 + p\lambda + q$ are real.

3. Locate the roots of $\lambda^4 - 2\lambda^3 + 2\lambda^2 - 2\lambda + 1$. Find the factorization of this polynomial into primes in $\mathbf{R}[\lambda]$.

4. Determine, using Sturm's theorem, how many real roots the polynomials $\lambda^n - 1$ and $\lambda^n + 1$ have, then prove (again by Sturm's theorem) that all real roots belong to the interval $[-1, 1]$.

II

5. Use Sturm's theorem to determine when the roots of a cubic polynomial $\lambda^3 + p\lambda + q$ are all real, then determine (again by Sturm's theorem) when the roots are all positive.

6. Prove, using Sturm's theorem, that a cubic polynomial with real coefficients must have a real root.

7. Prove, using Sturm's theorem, that a polynomial with real coefficients and odd degree must have a real root.

8. Adjust equations (1) in this way: multiply the polynomials $f_2, f_3, \ldots$ as they arise by positive constants so that $f_0 = f$, $f_1 = f'$; and then

$$\begin{aligned} f_0 &= q_1 f_1 - c_2 f_2, \\ f_1 &= q_2 f_2 - c_3 f_3, \\ &\cdots\cdots\cdots\cdots \\ f_{m-2} &= q_{m-1} f_{m-1} - c_m f_m, \\ f_{m-1} &= q_m f_m + 0. \end{aligned}$$

Here $c_2, \ldots, c_m$ are positive constants. Show that Sturm's theorem applies to $f_0, f_1, f_2, \ldots, f_m$. (By this device one may sometimes simplify the arithmetic involved in calculating $f_2, f_3, \ldots, f_m$.)

9. When calculating the polynomials f_0, f_1, f_2, etc., suppose we come to a polynomial f_k for which $f_k(x) \neq 0$ for all x in $[a, b]$. Show that Sturm's theorem may be applied to the truncated series $f_0, f_1, \ldots, f_k$ to obtain the number of real roots of f in $[a, b]$.

III

10. Apply Sturm's theorem to the quartic polynomial $\lambda^4 + p\lambda^2 + q\lambda + r$, and determine necessary and sufficient conditions in order that all its roots be real.

11. (a) Let $F(\lambda) = a_k\lambda^k +$ lower powers of λ, and $G(\lambda) = b_{k-1}\lambda^{k-1} +$ lower powers of λ. Show that the two-term sequence $F(x)$, $G(x)$ presents either exactly one variation for $x = -\infty$ and no variation for $x = +\infty$, or it presents no variations for $x = -\infty$ and one variation for $x = +\infty$.

(b) Suppose that $\deg f = n$ and that in equations (1) the degree drops exactly one at each division so that $\deg f_1 = n - 1$, $\deg f_2 = n - 2, \ldots, \deg f_{n+1} = 0$. (This is what happens in most numerical examples.) Prove: $V(-\infty) + V(+\infty) = n$. Hence prove: The number of real roots of f is $n - 2V(+\infty)$.

5.8 REVIEW

In this chapter a careful study has been made of roots of polynomials, when the coefficients are rational, real, or complex numbers. A basic result is that a nonconstant polynomial with coefficients in one of these fields always has a complex root. It then follows that the number of roots in **C** (counting multiplicities) is the degree of the polynomial. If the coefficients are real, the nonreal roots occur in complex conjugate pairs. We found trigonometric formulas for roots of polynomials of the form $\lambda^n - a$, and gave algebraic methods for finding the roots of general polynomials of degree at most 4. For degrees 3 or 4 these algebraic methods are not very useful. Sturm's theorem gives a method for isolating the real roots of a polynomial with real coefficients, and then Newton's method may be applied to calculate these real roots to high precision. With due care, Newton's method may also be used to calculate nonreal roots. There are useful facts about the roots of the derivative f' when the roots of the nonconstant polynomial f are known: the roots of f' are in the convex hull of the roots of f, and when f has real coefficients, the number of nonreal roots of f' does not exceed the number of nonreal roots of f. We gave also other quantitative facts (Rolle's theorem, Jensen circles) about the roots of f'.

REVIEW EXERCISES 5.8

1. The monic sixth degree polynomial $f(\lambda)$ has real coefficients. It satisfies $f(2 + i) = 0$, $f(3 - i) = 0$, $f'(2 + i) = 0$. Find f, and find all of its roots. Give as much information as you can about the roots of f'. What is the factorization of f into primes in $\mathbf{R}[\lambda]$?

2. The polynomials $\lambda^4 + 2\lambda^3 + 4\lambda^2 + 3\lambda + 2$ and $\lambda^3 + 2\lambda^2 + 3\lambda + 2$ are not relatively prime. Use this fact to find all roots of each polynomial.

3. Let a be a positive real number. Suppose one tries to use Newton's method to compute $a^{1/m}$, beginning with an initial estimate x_0 which is a negative number. See Exercise 7, §5.5. Will Newton's process converge, and if so, to what? (Consider separately the cases of even or odd m. Use graphical arguments.)

4. Let $w = z^2 - 2iz + 3$. If z travels once about the origin on the circle $|z| = 2$, how many times does w travel about the origin?

5. The roots of the derivative f' of f lie in the convex hull of the roots of f. Under what conditions is a vertex of this convex hull a root of f'?

6. Let f be a polynomial with real coefficients. Suppose a and b are real numbers, with $a < b$. If $f(a) < 0$ and $f(b) > 0$, must f have a real root between a and b? What about f'?

6

THE FIELD OF FRACTIONS. RATIONAL NUMBERS AND FUNCTIONS

6.1 INTRODUCTION

We described the digital representations of an integer or polynomial with respect to a base in Chapter 4, and there we showed how these representations lead to familiar arithmetic procedures and important theorems (Taylor's theorem, the binomial theorem). We now study the digital representations with respect to a base of a quotient of integers (i.e., a rational number) and a quotient of polynomials. This study will give, for numbers, a deepened understanding of the real numbers, and for quotients of polynomials, a corresponding deepened understanding of types of series encountered in calculus. Many of the concepts introduced in this chapter will be encountered again in Chapter 8, where they will lead to some new types of number systems and a further study of series.

We begin the Chapter (§§6.2, 6.3) with an abstract version of the discussion of quotients (i.e., fractions) already given in Example 1.5.3. The reader may think that our abstract treatment is unnecessary. We justify it by remarking that it will provide a mathematical framework in §8.6 within which differentiation becomes a fraction.

6.2 EQUIVALENCE RELATIONS

Equivalence relations occur in every branch of mathematics, and have (without explicit statement) occurred in this book, when we introduced rational numbers (Example 1.5.3) and congruences (§3.6). They will occur again, in §6.3 and Chapter 7. The following discussion proceeds in stages, beginning with *Cartesian products*, going on to *relations*, and then to *equivalence relations*.

Cartesian Products

The Cartesian product of sets S and T, denoted by $S \times T$, is the set

$$S \times T = \{(s, t) \mid s \in S, t \in T\}$$

of all ordered pairs with entries $s \in S$, $t \in T$, under this definition of equality for ordered pairs:

$$(s, t) = (s', t') \quad \text{if and only if} \quad s = s' \text{ and } t = t'.$$

The special case $\mathbf{R} \times \mathbf{R}$ therefore consists of all ordered pairs (x, y) with real number entries, and $(x, y) = (x', y')$ if and only $x = x'$ and $y = y'$. Consequently, $\mathbf{R} \times \mathbf{R}$ *is just the Cartesian coordinate plane* as described in §2.3. The Cartesian product of arbitrary sets S,T is therefore an abstract generalization of the familiar XY coordinate plane. If S and T are subsets of $\mathbf{R}$, we may think of S as a subset of the X axis, T as a subset of the Y axis, and then $S \times T$ is a subset of the XY plane. When S, T are intervals, $S \times T$ is a rectangle; Figure 6.2.1 shows this when S,T are intervals of positive numbers. When S is all integral points on the X axis, T all integral points on the Y axis, $S \times T$ is then a grid of points (called a *lattice*) consisting of all (x, y) with integral x, y.

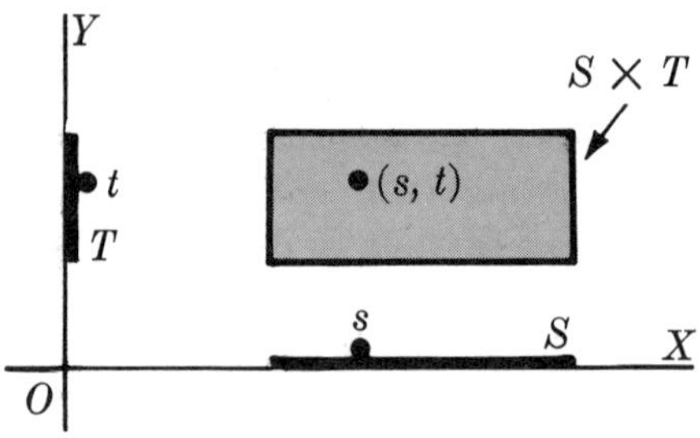

Figure 6.2.1

The Cartesian product $S \times T$ has meaning even when S, T are not sets of numbers. For example, we may take S and T to be finite sets having m and n elements, respectively. Then $S \times T$ is also finite and has mn elements. (The proof is left to the reader.)

Relations

Often in mathematics, we have nonempty sets S, T and a condition connecting elements s of S and t of T which, for specific s, t, either is satisfied or is not satisfied. Those ordered pairs (s, t) for which s, t satisfy the condition then form a subset of $S \times T$, which we shall denote by R. We now give several examples.

Example 6.2.1
Let S, T both be $\mathbf{R}$, and let the condition be "$s > t$." Here $S \times T$ is the whole Cartesian plane $\mathbf{R} \times \mathbf{R}$, and R consists of all points strictly to the right and below the line bisecting the first and third quadrants (the shaded region in Figure 6.2.2).

Example 6.2.2
Let S, T again both be $\mathbf{R}$, but take the condition to be "s, t have like signs (i.e., both positive, both zero, or both negative)." Again $S \times T$ is the whole Cartesian plane, but now R is the first and third quadrants, excluding the coordinate axes, but including the origin (Figure 6.2.3).

Example 6.2.3
Let S be all real numbers, T all nonnegative real numbers, and let the condition be "$t = |s|$." Here $S \times T$ is the "half" coordinate plane of all points on or above the X axis, and R is all points on the lines bisecting the first and second quadrants (Figure 6.2.4).

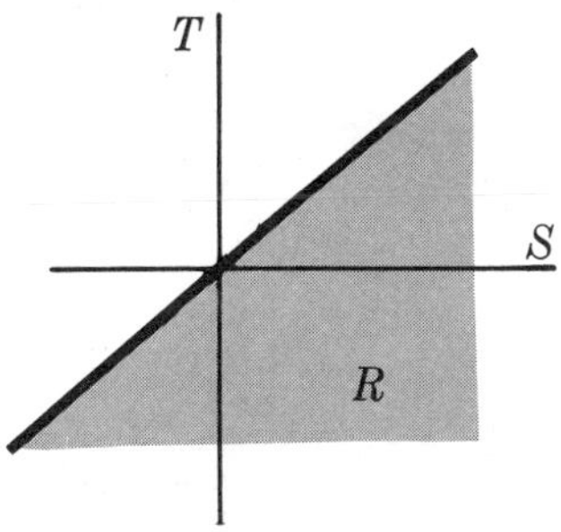

Figure 6.2.2

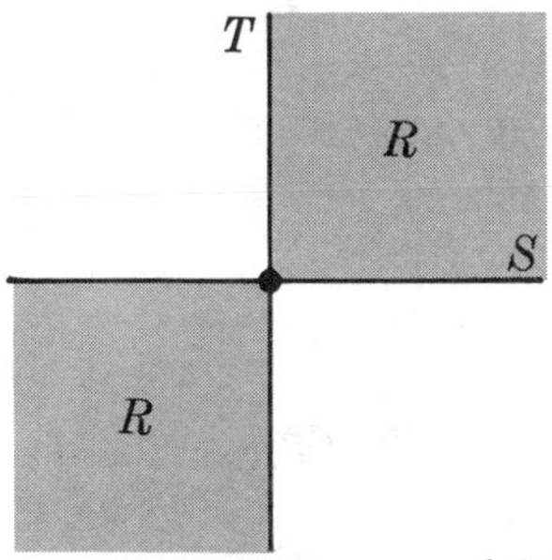

Figure 6.2.3

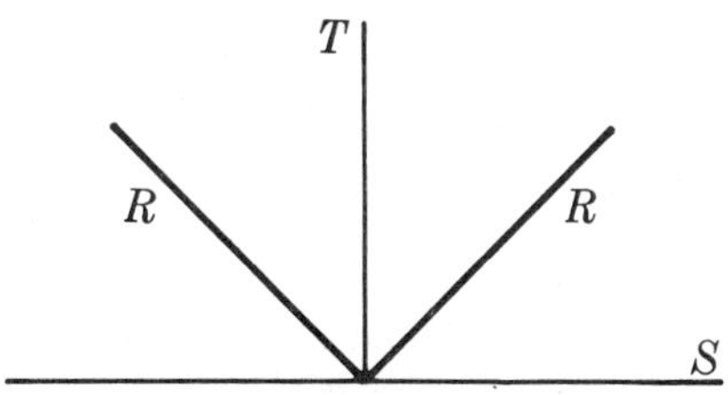

Figure 6.2.4

Example 6.2.4

Let S and T be $\mathbf{Z}$, let m be a fixed integer, and let the condition be "$s \equiv t \pmod{m}$." This time $S \times T$ is the lattice of points having integral coordinates, and R consists of all points (s, t) in this lattice whose coordinates s, t differ by an integral multiple of m. Figure 6.2.5 illustrates the case $m = 3$.

Example 6.2.5

Again let S, T be $\mathbf{Z}$, but now let the condition be "$s \mid t$." Draw all straight lines through the origin with integral slope. Then R consists of the lattice points on these lines (Figure 6.2.6).

Example 6.2.6

Once more let S and T be $\mathbf{Z}$, and this time let the condition be "s and t are relatively prime." Then R consists of all lattice points "visible" from the origin, i.e., such that the line segment connecting the lattice point and the origin does not pass through a nonzero lattice point closer to the origin. See Figure 6.2.7 and Exercise 9.

Example 6.2.7

Let S be the polynomial ring $\mathbf{R}[\lambda]$, let T be $\mathbf{C}$, and take the condition to be "t is a root of s." Here $S \times T$ is not contained in the Cartesian plane, and a geometric description of it and R cannot be given. Nevertheless, $S \times T$ exists as an abstract set and R is a subset of it.

Various conditions connecting elements $s \in S$, $t \in T$ yield therefore a variety of corresponding subsets R of $S \times T$. Any subset R of $S \times T$ arises in this manner since, if R is given, the condition "s and t are to be such that the ordered pair (s, t) is an element of R" has R as its corresponding subset. The technical term used to describe an abstract condition connecting elements $s \in S$, $t \in T$ is *relation.* To specify a condition or relation amounts to specifying those ordered pairs (s, t) whose components satisfy the condition, i.e., to specify a subset R of $S \times T$. We are therefore led to the following definition. (See Figure 6.2.8.)

DEFINITION 6.2.8

If S,T are sets, a relation *linking S to T is a subset R of $S \times T$. If $(s, t) \in R$, we write $s \sim t$ and say that s, t satisfy relation R. If $(s, t) \notin R$, we write $s \nsim t$ and say that s,t do not satisfy relation R.*

When $T = S$, we shorten "relation linking S to T" to "relation on S."

NOTE The symbol $\sim$ is often encountered in mathematics, but has no generally accepted name. It may be called "wave," "curl," "swiggle," or any name of your own invention.

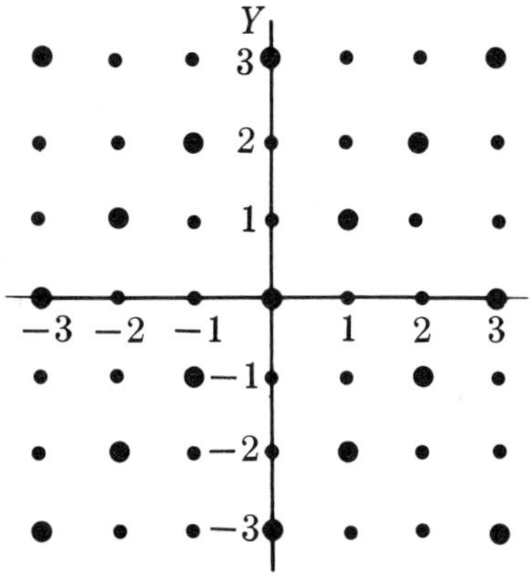

Lattice points in R are shown as heavier dots

Figure 6.2.5

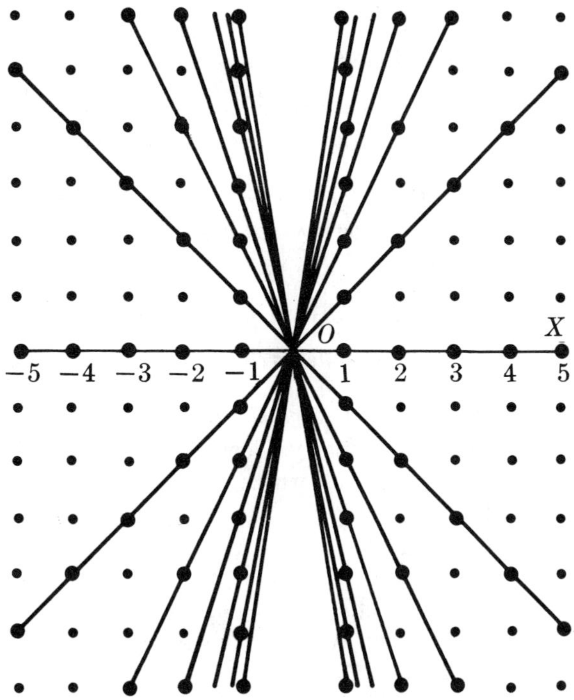

Lattice points in R are shown as heavier dots

Figure 6.2.6

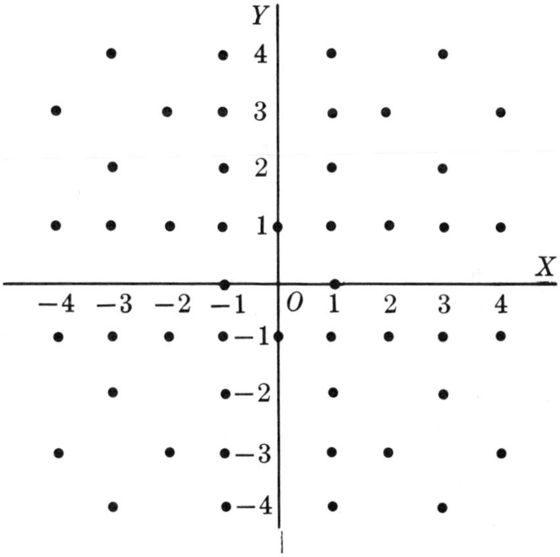

The only lattice points shown are those in R

Figure 6.2.7

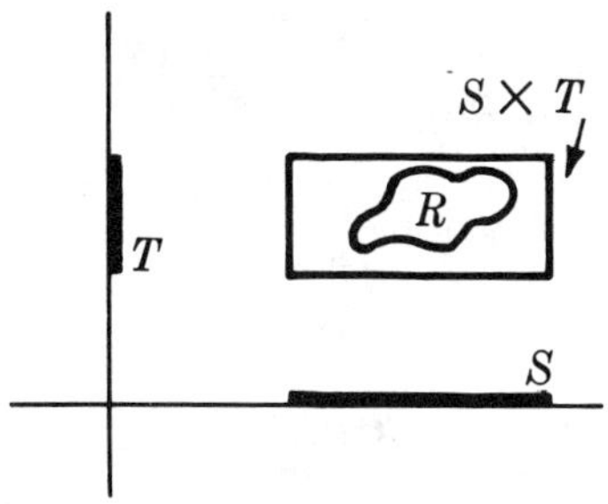

Figure 6.2.8

Equivalence relations

The relation concept is too general to be of great value. Taking $T = S$, we now introduce the much more useful *equivalence relation*.

DEFINITION 6.2.9
Let S be a nonempty set. A relation on S is an equivalence relation if:

(i) $s \sim s$ *for all* s *in* S;

(ii) *When* a, b *are elements of* S *with* $a \sim b$, *then* $b \sim a$;

(iii) *When* a, b, c *are elements of* S *with* $a \sim b$, $b \sim c$, *then* $a \sim c$.

We call (i) the *reflexive* property, (ii) the *symmetric* property, and (iii) the *transitive* property. Geometric interpretations of these properties are shown in Figure 6.2.9, where, to make easily drawn diagrams, we have taken S to be an interval of positive real numbers. (Of course, not every set S is of this type.)

When the relation R is an equivalence relation, we shall read "$a \sim b$" as "a is equivalent to b." The relations in Examples 6.2.2 and 6.2.4 are equivalence relations; those in Examples 6.2.1, 6.2.3, 6.2.5, 6.2.6, 6.2.7 are not.

Recall (§1.2) that a collection of subsets $T, U, V, \ldots$ of a set S *partitions* S if any element of S belongs to precisely one of $T, U, V, \ldots$. This amounts to any two of $T, U, V, \ldots$ being disjoint and the union of $T, U, V, \ldots$ being S. We call $T, U, V, \ldots$ the *components* of the partition.

Example 6.2.10
A partition of $\mathbf{Z}$ having two components is obtained by taking T to be all even integers and U to be all odd integers.

Example 6.2.11
A partition of $\mathbf{Z}$ having three components is obtained by taking T to be all positive integers, U to consist solely of zero, V to be all negative integers.

Example 6.2.12
A partition of $\mathbf{C}$ having infinitely many components is obtained by letting a component be the points on the circle C_r of radius r centered at the origin. We then have a component C_r for each nonnegative real number r.

We now give the basic theorem about equivalence relations.

THEOREM 6.2.13
An equivalence relation on a set S induces a partition of S such that elements of S are equivalent if and only if they belong to the same component of the partition.

Reflexive property

Property (i) says that R must contain the entire "diagonal" of $S \times S$ consisting of all pairs (a, a) for a in S.

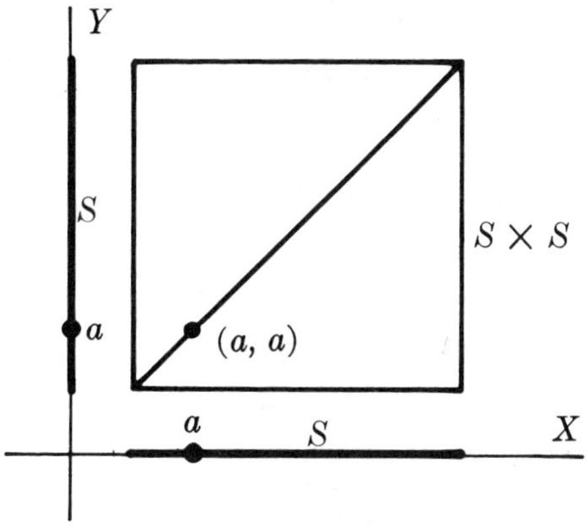

Symmetric property

If $(a, b) \in R$, property (ii) says that the mirror reflection (b, a) of (a, b) with respect to the diagonal of $S \times S$ must also be in R. Thus R is symmetric with respect to the diagonal of $S \times S$; i.e., this diagonal "bisects" R. (See Exercise 11.)

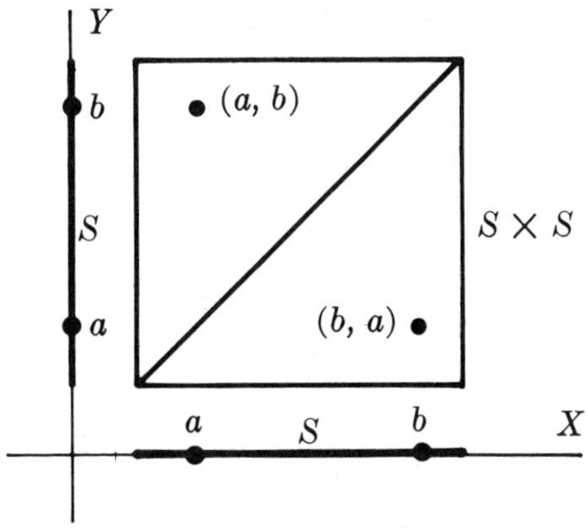

Transitive property

Property (iii) may be interpreted as: Let $P(a, b)$ and $Q(b, c)$ be points in R. Drop lines horizontally and vertically from P and Q as shown, meeting in T and U. The point T is on the diagonal. The statement that property (iii) holds is simply the statement that U is also in R. (As we saw, property (i) implies that T is in R.) So R must be closed under the operation of finding the fourth vertex U of a rectangle $PTQU$, when vertices P, Q are in R, and T is on the diagonal.

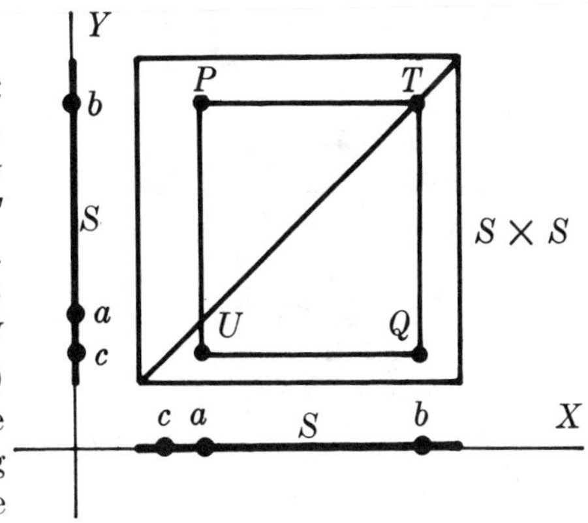

Figure 6.2.9

PROOF If a is an element of S, define a subset S_a of S by

$$S_a = \{s \mid a \sim s\}.$$

Note that $a \in S_a$ since $a \sim a$ [by (i)].

If α is any element of S_a, this definition when applied to α yields the same set S_a; that is, if $\alpha \in S_a$, then $S_\alpha = S_a$. To prove this, first note that $\alpha \in S_a$ implies $a \sim \alpha$, and therefore [by (ii)] $\alpha \sim a$. To show that $S_a \subseteq S_\alpha$, let $x \in S_a$. Then $a \sim x$. From $\alpha \sim a$, $a \sim x$, we have $\alpha \sim x$ [by (iii)], implying that $x \in S_\alpha$. Therefore $S_a \subseteq S_\alpha$. Since $\alpha \sim a$, we have $a \in S_\alpha$ and therefore we also have $S_\alpha \subseteq S_a$. Hence $S_a = S_\alpha$.

The subset S_a is therefore defined by any of its elements α. Since the definition no longer depends solely on a, denote this subset by T, and now any element of T may be used to obtain T. Doing this for every element of S, we obtain sets T, U, V, etc.

Let b be an element of S not in T. The above procedure, when applied to b, yields a subset U of S containing b. We claim that $U \cap T = \varnothing$. If not, let $s \in U \cap T$. Then $s \in T$, hence our definition above when applied to s yields T. Since $s \in U$, the definition when applied to s yields U. Therefore $U = T$. Since $b \in U$, we get $b \in T$. But this contradicts the fact that b is not in T. Hence $U \cap T = \varnothing$.

Thus T, U, V, ... are subsets of S, no two having a common element, and such that any element of S belongs to one of these subsets. The collection T, U, V, ... of subsets of S is therefore a partition of S.

It remains to show that elements a, c of S are in the same component of the partition if and only if they are equivalent. If a and c are both in the component T, then $a \sim c$ since $c \in T = S_a$. Conversely, if $a \sim c$, then $a \in S_a$, $c \in S_a$ and therefore a, c are in the same component. □

TERMINOLOGY The components of this partition of S are called *the equivalence classes under the equivalence relation R.*

Example 6.2.14

The relation "$a \sim b$ if and only if a, b have the same parity (i.e., both even or both odd)" is an equivalence relation on $\mathbf{Z}$. The equivalence classes are the sets T, U in Example 6.2.10.

Example 6.2.15

The relation "$a \sim b$ if and only if a, b have like signs" is an equivalence relation on $\mathbf{Z}$. The equivalence classes are the sets T, U, V in Example 6.2.11.

Example 6.2.16

The relation "$z_1 \sim z_2$ if and only if $|z_1| = |z_2|$" is an equivalence relation on $\mathbf{C}$. The equivalence classes are the sets C_r in Example 6.2.12.

Representatives. Systems of distinct representatives

Suppose subsets T, U, V, ... of a set S are equivalence classes under an equivalence relation on S. A *representative* of an equivalence class T is simply an element $t \in T$. Any element of an equivalence class is therefore a representative of the class, and if $t \in T$, we say T is *represented* by t. A *system of distinct representatives* is a collection of elements of S consisting of precisely one representative from each equivalence class. A Venn diagram illustration is given in Figure 6.2.10. The figure shows S, the equivalence classes T, U, V, ..., and elements t, u, v, ..., one element in each equivalence class.

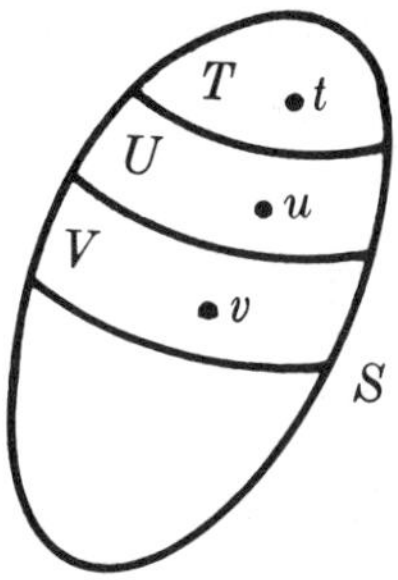

Figure 6.2.10

Example 6.2.17
The integers 0, 1 are a system of distinct representatives for the equivalence relation on $\mathbf{Z}$ in Example 6.2.14.

Example 6.2.18
The integers 1, 0, -1 are a system of distinct representatives for the equivalence relation on $\mathbf{Z}$ in Example 6.2.15.

Example 6.2.19
The nonnegative real numbers are a system of distinct representatives for the equivalence relation on $\mathbf{C}$ in Example 6.2.16.

Example 6.2.20

Let $\mathcal{R}$ be the integer or polynomial rings $\mathbf{Z}$, $\mathcal{F}[\lambda]$, where $\mathcal{F}$ is a field. Take S to be the set of all primes in $\mathcal{R}$, and define a relation $\sim$ on S by $p_1 \sim p_2$ if and only if p_1, p_2 are associates; i.e., $p_1 = up_2$ for some unit u in $\mathcal{R}$. (See Definition 3.4.3 and §3.5.) Then $\sim$ is an equivalence relation, and the positive or monic primes constitute a system of distinct representatives.

EXERCISES 6.2

I

1. On $\mathbf{Z}$, which of the following relations are equivalence relations?
   (a) $a \sim b$ if and only if $3 \mid (a - b)$.
   (b) $a \sim b$ if and only if $a^2 = b^2$.
   (c) $a \sim b$ if and only if exactly one of a or b is zero.
   (d) $a \sim b$ if and only if $a - b$ is 1, 0, or -1.
   (e) $a \sim b$ if and only if both a and b are nonzero.
   (f) $a \sim b$ if and only if $a \geq b$.
   Whenever the relation is an equivalence relation, determine how many equivalence classes there are, describe the elements in each equivalence class, and find a system of distinct representatives.

2. Suppose a relation on S satisfies:
   (ii) $a \sim b$ implies $b \sim a$;
   (iii) $a \sim b$ and $b \sim c$ imply $a \sim c$.
   Show that it need not satisfy
   (i) $a \sim a$ for every a in S,
   by furnishing an example of a relation for which (ii) and (iii) are true but (i) is false. (Hint: Try one of the relations in Exercise 1.)

3. Extend Exercise 2 to show that the conditions (i), (ii), (iii) of Definition 6.2.9 are independent. Do this by finding examples of relations on $\mathbf{Z}$ satisfying any two of (i), (ii), (iii) but not the third. (Study the examples in Exercise 1.)

4. Find examples of relations on $\mathbf{Z}$ satisfying any one (but not the other two) of conditions (i), (ii), (iii) of Definition 6.2.9. Find also an example of a relation satisfying none of (i), (ii), (iii).

5. Define relation $\sim$ on the set of positive integers by $a \sim b$ if and only if a and b have the same lowest digit in their base ten representations. Is this an equivalence relation? Describe the equivalence classes. Find a system of distinct representatives.

II

6. By Exercise 2 above we know that (ii) and (iii) together (of Definition 6.2.9) do not imply (i). What is wrong with the following "proof"? By (ii), we have $a \sim b$ and $b \sim a$, hence by (iii) [with $c = a$ in (iii)] we get $a \sim a$.

7. Prove: $S \times T$ is the empty set if and only if at least one of S, T is empty.

8. Define relation $\sim$ on $\mathbf{C}$ by $z_1 \sim z_2$ if and only if $z_1 = rz_2$ for some positive real number r. Prove that this is an equivalence relation. Describe the equivalence classes and find a system of distinct representatives. Note: r is not fixed.

9. Let a, b be integers, not both zero, and let d be their greatest common divisor. If $d > 1$, show that the lattice point (a, b) is hidden from a viewer at the origin by the lattice point $(a/d, b/d)$. Conversely, if (a, b) is hidden by a lattice point (α, β), show that it is hidden by a lattice point (α, β) with α, β relatively prime. Then show $a = t\alpha$, $b = t\beta$ for an integer $t > 1$. (Look ahead to, and use, Theorem 6.4.1.) Hence deduce that lattice point (a, b) is visible precisely when a, b are relatively prime.

10. Let $\mathcal{G}$ be a group with operation $\circ$, and let $\mathcal{H}$ be a subgroup of $\mathcal{G}$. Define a relation $\sim$ on $\mathcal{G}$ by $g_1 \sim g_2$ if and only if $g_1 \circ g_2^{-1} \in \mathcal{H}$. Show that this relation is an equivalence relation. Describe the equivalence classes under this relation and find a system of distinct representatives when $\mathcal{G}$ is the multiplicative group of nonzero complex numbers and $\mathcal{H}$ is the multiplicative group of positive real numbers.

11. Let $P = (a, b)$ and $Q = (b, a)$ be points in the Cartesian plane. Show that Q is the mirror reflection of P in the line $y = x$. That is, prove that the line $y = x$ is the perpendicular bisector of the straight line segment joining P and Q.

III

12. Let $\mathcal{G}$ be the multiplicative group of nonzero complex numbers and $\mathcal{H}$ the multiplicative group of positive real numbers. Let $\sim$ be the equivalence relation of Exercise 10. Define multiplication of equivalence classes by this rule: If C_1 and C_2 are equivalence classes, select any g_1 in C_1 and any g_2 in C_2. Let $C_1 \circ C_2$ be the equivalence class containing g_1g_2.
    (a) Prove that this multiplication depends only on C_1 and C_2 and not on the elements g_1, g_2 selected from these classes. (In other words, if $g_1' \in C_1$ and $g_2' \in C_2$, show that g_1g_2 and $g_1'g_2'$ lie in the same equivalence class.)
    (b) Prove that the equivalence classes under this multiplication form a group.
    (c) Show that this group is identical, except for the notation used, to the multiplicative group of all complex numbers z on the unit circle (i.e., satisfying $|z| = 1$). Do this by utilizing a system of distinct representatives for the equivalence classes.

13. Let $\mathcal{G}$ be a finite group, $\mathcal{H}$ a subgroup of $\mathcal{G}$. Suppose that $\mathcal{G}$ contains precisely m elements and $\mathcal{H}$ precisely n elements. Let $\sim$ be the equivalence relation of Exercise 10.
    (a) Show that each equivalence class contains precisely n elements.
    (b) Deduce that $n \mid m$.
    (That $n \mid m$ is *Lagrange's theorem*. The number of elements in a finite group is called the *order* of the group. Lagrange's theorem therefore asserts that the order of a subgroup of $\mathcal{G}$ divides the order of $\mathcal{G}$.)

6.3 THE FIELD OF FRACTIONS

Throughout this section $\mathcal{R}$ will be a commutative ring with at least two elements and without zero divisors. Thus $\mathcal{R}$ is almost an integral domain; it is an integral domain if it has a multiplicative identity. Examples are: (i) the ring of even integers, (ii) the ring of polynomials with integral coefficients and even constant term, (iii) any integral domain, (iv) any field. We want to study quotients of elements of $\mathcal{R}$, and to do so we shall imitate the discussion of quotients of integers in Example 1.5.3, giving greater detail; because $\mathcal{R}$ is an abstract ring, the generality will be greater also. The results obtained will be applied to polynomials in §§6.4 and 8.6 to study *rational functions* and to introduce the *operational calculus.*

Let $\mathcal{S}$ be the set of all ordered pairs (m, n) in which $m, n \in \mathcal{R}$ and $n \neq 0$. Thus $\mathcal{S}$ is a subset of $\mathcal{R} \times \mathcal{R}$. *Particularly notice that the second component n is nonzero.* We shall introduce a relation $\sim$ on $\mathcal{S}$, prove it is an equivalence relation, then introduce addition and multiplication of the *equivalence classes,* finally obtaining a field. The reader should, step by step, compare with Example 1.5.3 and observe how the discussion there becomes the present one when the fraction symbol m/n is replaced with the ordered pair symbol (m, n), and fraction equality $m/n = m'/n'$ is replaced with equivalence $(m, n) \sim (m', n')$ of ordered pairs.

Define $\sim$ on $\mathcal{S}$ by

$$(m, n) \sim (m', n') \quad \text{if and only if} \quad mn' = nm'. \tag{1}$$

The motivation behind this definition is that our objective is to obtain fractions but, unfortunately, we do not have any natural meaning to attribute to a fraction symbol m/n. Ordered pairs, however, have a perfectly clear meaning, so we study them, and introduce $\sim$ with the expectation that it will be an equivalence relation and bring into a single equivalence class all ordered pairs (m, n) which should correspond to "equal fractions m/n." The reader should therefore regard the roles played by the first and second components m, n in (m, n) to be the roles of numerator and denominator, respectively.

We now prove that $\sim$ is indeed an equivalence relation.

Proof of reflexivity: Clearly $(m, n) \sim (m, n)$ since $mn = nm$, $\mathcal{R}$ being commutative. □

Proof of symmetry: If $(m, n) \sim (m', n')$, then $mn' = nm'$; hence $m'n = n'm$, and therefore $(m', n') \sim (m, n)$. □

Proof of transitivity: If $(m, n) \sim (m', n')$ and $(m', n') \sim (m'', n'')$, then $mn' = nm'$ and $m'n'' = n'm''$. Multiplying the first of these equations by n'' and the second by n gives $mn'n'' = nm'n'' = nn'm''$. Hence $n'(mn'' - nm'') = 0$. Because $n' \neq 0$ and $\mathcal{R}$ does not have zero divisors, $mn'' - nm'' = 0$, so that $mn'' = nm''$ and therefore $(m, n) \sim (m'', n'')$. □

Since $\sim$ is an equivalence relation, $\mathcal{S}$ partitions into subsets, the equivalence classes under $\sim$. Because pairs are equivalent precisely when they belong to the same equivalence class,

$$(m, n) \text{ and } (m', n') \text{ are in the same equivalence class}$$
$$\text{if and only if } mn' = nm'.$$

Thus an equivalence class does contain all pairs which would be equal if pairs were interpreted as fractions.

We wish now to introduce addition and multiplication of equivalence classes corresponding to the natural operations with fractions. If a and b are *equivalence classes*, let (m, n), (m', n') be representatives of a, b; i.e.,

$$(m, n) \in a, \qquad (m', n') \in b. \tag{2}$$

Then $n \neq 0$, $n' \neq 0$, hence $nn' \neq 0$ (since $\mathcal{R}$ has no zero divisors), and therefore $(mn' + nm', nn')$ and (mm', nn') are ordered pairs in $\mathcal{S}$. Define $a + b$ and ab by:

$$\begin{aligned} a + b &\text{ is the equivalence class containing } (mn' + nm', nn'), \\ ab &\text{ is the equivalence class containing } (mm', nn'). \end{aligned} \tag{3}$$

As in Example 1.5.3, there is a logical difficulty here which must be cleared up before proceeding further. If we used different representatives, say

$$(m_1, n_1) \in a, \qquad (m_1', n_1') \in b, \tag{4}$$

the definition (3) of $a + b$ and ab, when applied to these pairs, becomes

$$\begin{aligned} a + b &\text{ is the equivalence class containing } (m_1n_1' + n_1m_1', n_1n_1') \\ ab &\text{ is the equivalence class containing } (m_1m_1', n_1n_1'). \end{aligned} \tag{5}$$

It must be shown that the calculations of $a + b$ and ab in (3) yield equivalence classes equal to those obtained in (5). This will be the case only if the pairs in (3) are equivalent to the corresponding pairs in (5); that is, if

$$(mn' + nm')n_1n_1' = nn'(m_1n_1' + n_1m_1'), \tag{6}$$

$$(mm')(n_1n_1') = (nn')(m_1m_1'). \tag{7}$$

Because $(m, n) \sim (m_1, n_1)$ and $(m', n') \sim (m_1', n_1')$, we have

$$mn_1 = nm_1 \quad \text{and} \quad m'n_1' = n'm_1'. \tag{8}$$

Precisely as in Example 1.5.3, equations (6) and (7) are consequences of (8), as the reader may show for himself. The sum and product of equivalence classes a, b depends therefore on a, b only and not on the representatives used to perform the calculations.

Under the addition and multiplication just introduced, we now verify that the equivalence classes form a field. Throughout this verification, a, b, c are equivalence classes of ordered pairs, and (m, n), (m', n'), (m'', n'') are ordered pairs with

$$(m, n) \in a, \qquad (m', n') \in b, \qquad (m'', n'') \in c.$$

First we verify axioms R1–R5 of a commutative ring.

R1(i)–*Closure under addition:* No further verification is needed since the sum of equivalence classes is an equivalence class. □

R1(ii)–*Associativity of addition:* The class $a + b$ contains the ordered pair $(mn' + nm', nn')$; hence $(a + b) + c$ is the class containing the pair

$$((mn' + nm')n'' + nm'', (nn')n''). \tag{9}$$

Similarly $a + (b + c)$ is the class containing

$$(m(n'n'') + n(m'n'' + n'm''), n(n'n'')). \tag{10}$$

Since the pairs (9), (10) are equivalent, $(a + b) + c = a + (b + c)$. □

R1(iii)–*Presence of zero:* If ν_1, ν_2 are any nonzero elements of $\mathcal{R}$, observe that $(0, \nu_1) \sim (0, \nu_2)$. All pairs in $\mathcal{S}$ of the form $(0, \nu)$ therefore lie in a single equivalence class. Denote this class by $\mathcal{O}$. From $(0, \nu) \in \mathcal{O}$, $(m, n) \in a$, we get $(\nu m, \nu n) \in \mathcal{O} + a$. Since $(\nu m, \nu n) \sim (m, n)$, and $(m, n) \in a$, we have $\mathcal{O} + a = a$. Similarly, $a + \mathcal{O} = a$. The equivalence class $\mathcal{O}$ thus plays the role of zero. (This is not surprising if $(0, \nu)$ is regarded as a fraction with zero as numerator and ν as denominator.) □

R1(iv)–*Presence of additive inverses:* Let α be the equivalence class containing the pair $(-m, n)$. Then $a + \alpha$ contains the pair $(mn + n(-m), n^2) = (0, n^2) \in \mathcal{O}$. Therefore $a + \alpha = \mathcal{O}$ and similarly $\alpha + a = \mathcal{O}$. The equivalence class α is therefore the additive inverse of a. (Thinking in terms of numerators and denominators, this again is not surprising.) □

R1(v)–*Commutativity of addition:* Both $a + b$ and $b + a$ contain $(mn' + nm', nn')$; hence $a + b = b + a$. □

We now know that the equivalence classes form an additive commutative group. Thus Axiom R1 is verified.

R2–*Closure under multiplication:* No further verification is needed since the product of equivalence classes is an equivalence class. □

R3–*Associativity of multiplication:* Equivalence class ab contains (mm', nn'), hence $(ab)c$ contains $((mm')m'', (nn')n'')$. Similarly $a(bc)$ contains $(m(m'm''), n(n'n''))$. These are equivalent pairs; hence $(ab)c = a(bc)$. □

R4–*Distributive laws:* A short calculation shows that the equivalence class $ab + ac$ contains the pair

$$(mm'nn'' + nn'mm'', nn'nn''),$$

whereas $a(b + c)$ contains

$$(m(m'n'' + n'm''), nn'n'').$$

Since these are equivalent pairs, $ab + ac = a(b + c)$. Similarly one verifies that $ba + ca = (b + c)a$. □

R5–*Commutativity of multiplication:* Both ab and ba contain (mm', nn'); hence $ab = ba$. □

The equivalence classes therefore form a commutative ring. We now prove that this ring is a field. For this we verify F1 and F2 of Theorem 1.6.7.

F1–*Presence of a nonzero multiplicative identity:* If ν_1, ν_2 are nonzero elements of $\mathcal{R}$, observe that $(\nu_1, \nu_1) \sim (\nu_2, \nu_2)$. All pairs of the form (ν, ν) [with ν nonzero] therefore belong to a single equivalence class. Denote this class by e. Since $(\nu, \nu) \in e$, $(0, \nu) \in \mathcal{O}$, and $(\nu, \nu) \not\sim (0, \nu)$, clearly $e \neq \mathcal{O}$. Furthermore, ea is the equivalence class containing the pair $(\nu m, \nu n) \sim (m, n)$. Hence $ea = a$. Thus e is the required multiplicative identity. (Thinking of (ν, ν) as a fraction with equal numerator and denominator, it is not surprising that such pairs act as a multiplicative identity.) □

F2–*Multiplicative inverses of nonzero equivalence classes:* If $a \neq \mathcal{O}$, then $(m, n) \not\sim (0, \nu)$, implying that $m\nu \neq 0$, and hence $m \neq 0$. Thus (n, m) is an ordered pair with second component nonzero. It therefore lies in some equivalence class; say, in α. Then $a\alpha$ contains (mn, nm) which, having equal components, is in e. Thus $a\alpha = e$. The class α is the multiplicative inverse of a. (Again, this is expected if one thinks in terms of numerators and denominators.) □

We have now proved that the equivalence classes, under their addition and multiplication operations, form a field. This field is called the *field of fractions* of $\mathcal{R}$; let us denote it by $\mathcal{K}$. The term "fraction" is used to indicate an element of $\mathcal{K}$, that is, an equivalence class of ordered pairs. Often, however, (as above) "fraction" is used for a representative of an element of $\mathcal{K}$, i.e., for an ordered pair. Both uses will occur in this book.

The reader knows that **Q**, the field of fractions of **Z**, contains **Z** as a subring; to regard an integer m as a fraction, write it as $m\nu/\nu$ for any nonzero integer ν. We therefore expect that the field $\mathcal{K}$ will contain a subring which, apart from notation, is just $\mathcal{R}$. This we now demonstrate. The reader may find it helpful to keep the example $\mathcal{R} = \mathbf{Z}$, $\mathcal{K} = \mathbf{Q}$ in mind when reading this demonstration.

Observe first, if $m \in \mathcal{R}$ and ν_1, ν_2 are any nonzero elements of $\mathcal{R}$, that $(m\nu_1, \nu_1) \sim (m\nu_2, \nu_2)$. Thus all pairs $(m\nu, \nu)$, with ν any nonzero element of $\mathcal{R}$, lie in a single equivalence class. Denote this equivalence class by $\varphi(m)$, so that

$$(m\nu, \nu) \in \varphi(m) \quad \text{for all nonzero } \nu \in \mathcal{R}.$$

We think of φ as a function or mapping which acts on an element m of $\mathcal{R}$ and produces an element $\varphi(m)$ of $\mathcal{K}$ as functional value. This function φ has the following properties: If m, m' are elements of $\mathcal{R}$, then

$$\varphi(m) = \varphi(m') \quad \text{if and only if} \quad m = m'; \tag{11}$$

$$\varphi(m + m') = \varphi(m) + \varphi(m'); \tag{12}$$

$$\varphi(mm') = \varphi(m)\varphi(m'). \tag{13}$$

These facts, if verified, enable an identification compatible with addition and multiplication to be made between elements m of $\mathcal{R}$ and those elements of $\mathcal{K}$ of the form $\varphi(m)$. The elements $\varphi(m)$, for $m \in \mathcal{R}$, then form a subset of $\mathcal{K}$ which may be regarded as $\mathcal{R}$ in different notation. In fact, this subset of $\mathcal{K}$ is a subring of $\mathcal{K}$. (Exercise 7.)

Proof of (11): If ν is a nonzero element of $\mathcal{R}$, $\varphi(m)$ and $\varphi(m')$ are, respectively, equivalence classes containing $(m\nu, \nu)$ and $(m'\nu, \nu)$. These equivalence classes are the same if and only if $(m\nu, \nu) \sim (m'\nu, \nu)$, and this holds if and only if $m\nu^2 = m'\nu^2$, i.e., if and only if $m = m'$, since $\mathcal{R}$ has no zero divisors. □

Since $\varphi(0)$ contains the pair $(0\nu, \nu) = (0, \nu) \in \mathcal{O}$, we have $\varphi(0) = \mathcal{O}$. From (11) we get $\varphi(m) = \varphi(0)$ if and only if $m = 0$. Therefore we also have:

$$\varphi(m) = \mathcal{O} \quad \text{if and only if} \quad m = 0.$$

Proof of (12) *and* (13): We have $(m\nu, \nu) \in \varphi(m)$, $(m'\nu, \nu) \in \varphi(m')$; therefore $(m\nu^2 + m'\nu^2, \nu^2) \in \varphi(m) + \varphi(m')$ and $(mm'\nu^2, \nu^2) \in \varphi(m)\varphi(m')$. Since $((m + m')\nu^2, \nu^2) \in \varphi(m + m')$ and $(mm'\nu^2, \nu^2) \in \varphi(mm')$, we deduce that $\varphi(m + m') = \varphi(m) + \varphi(m')$ and $\varphi(mm') = \varphi(m)\varphi(m')$. □

Notice that an element a of $\mathcal{K}$ may now be written as

$$a = \varphi(m)\varphi(n)^{-1} \tag{14}$$

where (m, n) is a pair in a. (Here $\varphi(n)^{-1}$ denotes the multiplicative inverse, in $\mathcal{K}$, of its element $\varphi(n)$.) *Proof:* $(m\nu, \nu) \in \varphi(m)$, $(n\nu, \nu) \in \varphi(n)$; since $n\nu \neq 0$, the verification of F2 shows that $(\nu, n\nu) \in \varphi(n)^{-1}$; hence $(m\nu^2, n\nu^2) \in \varphi(m)\varphi(n)^{-1}$. Because $(m\nu^2, n\nu^2) \sim (m, n)$ and $(m, n) \in a$, we obtain (14).

Equation (14) displays an element a of $\mathcal{K}$ as a *true* quotient of elements of $\mathcal{K}$ which are identified with elements of $\mathcal{R}$. Furthermore, if $(m', n') \in b$, then

$$b = \varphi(m')\varphi(n')^{-1},$$

and it is easy to see that the following natural rules for fractions hold:

$$\varphi(m)\varphi(n)^{-1} = \varphi(m')\varphi(n')^{-1} \quad \text{if and only if} \quad \varphi(m)\varphi(n') = \varphi(n)\varphi(m'),$$

$$\varphi(m)\varphi(n)^{-1} + \varphi(m')\varphi(n')^{-1} = (\varphi(m)\varphi(n') + \varphi(n)\varphi(m'))\varphi(n)^{-1}\varphi(n')^{-1},$$

$$\varphi(m)\varphi(n)^{-1}\cdot\varphi(m')\varphi(n')^{-1} = \varphi(m)\varphi(m')\cdot\varphi(n)^{-1}\varphi(n')^{-1}.$$

Since these are equations involving elements $\varphi(m)$, $\varphi(m')$, $\varphi(n)$, $\varphi(n')$ of a field $\mathcal{K}$, simple algebra (at most) is needed to verify them.

We have completed the proof of the following basic theorem.

THEOREM 6.3.1

If $\mathcal{R}$ is a commutative ring with at least two elements and without zero divisors, there is a field $\mathcal{K}$, containing a subring identified with $\mathcal{R}$, and consisting of quotients of those elements identified with $\mathcal{R}$ ("fractions"), such that the natural laws of fraction equality, addition, and multiplication hold.

Very often we simplify notation and say that the subring of $\mathcal{K}$ identified with $\mathcal{R}$ is $\mathcal{R}$. This amounts to agreeing to denote an equivalence class of pairs $(m\nu, \nu)$ by m; i.e., write m instead of $\varphi(m)$. We then say that $\mathcal{R}$ is *embedded* in $\mathcal{K}$, and write $\varphi(m)\varphi(n)^{-1}$ as mn^{-1} or as m/n. This is analogous to the agreements made (when introducing complex numbers) to write a point $(x, 0)$ concisely as x and (when introducing polynomials) to write an infinite sequence $(a, 0, 0, 0, \ldots)$ concisely as a.

The function φ above is called an *isomorphism*. Isormorphisms will be studied in some detail in Chapter 7.

Since elements of $\mathcal{K}$ are equivalence classes (of ordered pairs), it is natural to ask if a system of distinct representatives for these equivalence classes can be found. When $\mathcal{R}$ is $\mathbf{Z}$ or $\mathcal{F}[\lambda]$, a system of distinct representatives will be given in Theorem 6.4.1.

EXERCISES 6.3

I

1. Let $\mathcal{R}$ be $\mathbf{Z}$. Find all pairs in the equivalence class containing the pair $(3, 5)$. Repeat, for the pair $(6, 8)$.

2. Let $\mathcal{R}$ be $\mathbf{R}[\lambda]$. Find all pairs in the equivalence class containing the pair $(\lambda + 1, \lambda^2 - 1)$. Repeat, for the pair $(\lambda + 1, \lambda^2 + 1)$.

3. What fields are the fields of fractions of: (i) $\mathbf{Z}$, (ii) the ring of even integers, (iii) the ring of polynomials with integral coefficients and even constant term, (iv) $\mathbf{Q}$, (v) $\mathbf{R}$, (vi) $\mathbf{C}$?

4. Suppose $\sim$ is defined, as in the text, but for all pairs in $\mathcal{R} \times \mathcal{R}$, i.e., including pairs with zero second component. By taking $\mathcal{R}$ to be $\mathbf{Z}$, demonstrate that $\sim$ is now *not* an equivalence relation.

5. Let $\mathcal{R}$ be $\mathbf{C}[\lambda]$. Is there a fraction belonging to the field of fractions of $\mathbf{R}[\lambda]$ equal to $(\lambda + i)/(\lambda^2 - 1)$?

6. The hypothesis that $\mathcal{R}$ does not have zero divisors was used several times in the discussion leading to Theorem 6.3.1. Not all of these were indicated explicitly. Identify each place in the discussion where this hypothesis is used.

II

7. In the notation of the text, prove that $\varphi(-m) = -\varphi(m)$. Then prove that the set of all $\varphi(m)$, for $m \in \mathcal{R}$, is a subring of $\mathcal{K}$.

8. If $\mathcal{R}$ is a commutative ring with at least two elements and without zero divisors, and if $m, n, \mu, \nu \in \mathcal{R}$, with n, μ, ν nonzero, prove that

$$\frac{m\nu}{n\nu} = \frac{m}{n}, \qquad \frac{m}{n} \Big/ \frac{\mu}{\nu} = \frac{m\nu}{n\mu}.$$

9. Prove: Every pair in $\mathcal{O}$ has the form $(0, \nu)$. [The text merely proved that $(0, \nu) \in \mathcal{O}$.]

10. Prove: Every pair in e has the form (ν, ν). [The text merely proved that $(\nu, \nu) \in e$.]

11. If $m \in \mathcal{R}$, prove that every pair in $\varphi(m)$ has the form $(m\nu, \nu)$. [The text merely proved that $(m\nu, \nu) \in \varphi(m)$.]

12. If $\mathcal{R}$ has a multiplicative identity, denoted by 1, prove that $\varphi(1) = e$.

13. If $\mathcal{R}$ has a multiplicative identity 1, and m is a unit in $\mathcal{R}$, so that $1 = mm_1$ for an element $m_1 \in \mathcal{R}$, prove that $\varphi(m_1) = \varphi(m)^{-1}$, i.e., $\varphi(m^{-1}) = \varphi(m)^{-1}$.

14. If a is an equivalence class and $-a$ is its additive inverse, prove that pairs $(m, n) \in a$ and $(p, q) \in -a$ are related by $mq + np = 0$.

15. If a is a nonzero equivalence class and a^{-1} its multiplicative inverse, prove that pairs $(m, n) \in a$ and $(p, q) \in a^{-1}$ are related by $mp = nq$.

III

16. Suppose $\mathcal{R}$ has zero divisors, but it has a nonempty subset S such that $0 \notin S$ and $s_1 s_2 \in S$ when $s_1 \in S$ and $s_2 \in S$. (We call S a multiplicative system.) Suppose also that no element of S is a zero divisor. By considering pairs of the form (m, s) for $m \in \mathcal{R}$ and $s \in S$, construct a ring of fractions with denominators from S only, and such that each element of S has a multiplicative inverse in this ring of fractions. Adapt the discussion in the text.

17. (Construction of negative integers) Let P be the set of natural numbers, i.e., positive integers. The set P is closed under addition and multiplication, but not subtraction. To get a set closed under

subtraction, a larger number system, **Z**, needs to be invented. Assuming knowledge of P *only*, and not of **Z**, this exercise outlines the procedure for obtaining **Z**. Let p, n, p', n', etc., denote elements of P.
(i) On $P \times P$, define a relation $\sim$ by

$$(p, n) \sim (p', n') \quad \text{if and only if} \quad p + n' = n + p'.$$

(Motivation: We want to invent the difference $p - n$; once invented, we expect $p - n$ to equal $p' - n'$ if and only if $p + n' = n + p'$. Therefore we introduce $\sim$ with the expectation it is an equivalence relation, and segregates in a single equivalence class all pairs (p, n) with components having a common difference.) Prove: $\sim$ is an equivalence relation.
(ii) If a and b are equivalence classes, with $(p, n) \in a$ and $(p', n') \in b$, define addition and multiplication of equivalence classes by:

$$a + b \text{ is the equivalence class containing } (p + p', n + n'),$$
$$ab \text{ is the equivalence class containing } (pp' + nn', pn' + np').$$

If also $(p_1, n_1) \in a$, $(p_1', n_1') \in b$, prove that this definition, when applied to (p_1, n_1) and (p_1', n_1'), yields the same sum $a + b$ and product ab. Conclude that $a + b$ and ab depend only on a, b and not on the pairs selected from a, b to do the calculations.
(iii) Let **Z** be the set of equivalence classes. Prove that **Z** is an integral domain.
(iv) If $p \in P$, define $\varphi(p)$ to be the equivalence class containing all pairs $(p + \nu, \nu)$, for ν any element of P. Prove that:

$$\varphi(p) = \varphi(p') \quad \text{if and only if} \quad p = p',$$
$$\varphi(p + p') = \varphi(p) + \varphi(p'), \qquad \varphi(pp') = \varphi(p)\varphi(p').$$

Conclude that an identification may be made between P and those equivalence classes of the form $\varphi(p)$.
(v) Show that the order axioms (§1.3) hold for **Z**, the positive elements of **Z** being those equivalence classes identified with elements of P in (iv).

6.4 RATIONAL NUMBERS AND RATIONAL FUNCTIONS

If we take the ring $\mathfrak{R}$ in §6.3 to be **Z**, the field of fractions obtained is **Q**, the field of rational numbers. If we take $\mathfrak{R}$ to be the polynomial ring $\mathfrak{F}[\lambda]$, where $\mathfrak{F}$ is a field, the field of fractions then obtained is called the *field of rational functions in the indeterminate* λ, *with coefficients from* $\mathfrak{F}$, and is denoted by $\mathfrak{F}(\lambda)$. (Note: round brackets, i.e., parentheses.) Let $m(\lambda), m_1(\lambda), n(\lambda), n_1(\lambda) \in \mathfrak{F}[\lambda]$ be polynomials with $n(\lambda)$, $n_1(\lambda)$ nonzero.

The field $\mathfrak{F}(\lambda)$ consists of all quotients

$$\frac{m(\lambda)}{n(\lambda)}$$

of polynomials, with these rules for equality, addition, and multiplication:

$$\frac{m(\lambda)}{n(\lambda)} = \frac{m_1(\lambda)}{n_1(\lambda)} \quad \text{if and only if} \quad m(\lambda)n_1(\lambda) = n(\lambda)m_1(\lambda),$$

$$\frac{m(\lambda)}{n(\lambda)} + \frac{m_1(\lambda)}{n_1(\lambda)} = \frac{m(\lambda)n_1(\lambda) + n(\lambda)m_1(\lambda)}{n(\lambda)n_1(\lambda)},$$

$$\frac{m(\lambda)}{n(\lambda)} \cdot \frac{m_1(\lambda)}{n_1(\lambda)} = \frac{m(\lambda)m_1(\lambda)}{n(\lambda)n_1(\lambda)}.$$

An element of $\mathfrak{F}(\lambda)$ is called a *rational function* (whereas an element of $\mathfrak{F}[\lambda]$ is, of course, called a polynomial). Note that we no longer use the ordered pair notation (m, n) for fractions.

Throughout the remainder of this chapter, the symbol $\mathfrak{R}$ is to denote either $\mathbf{Z}$ or $\mathfrak{F}[\lambda]$, and then $\mathfrak{K}$ is to denote $\mathbf{Q}$ or $\mathfrak{F}(\lambda)$, respectively. The term *fraction* will denote an element of $\mathfrak{K}$. In this and the next section we discuss several basic properties of $\mathfrak{K}$ and its elements. Throughout this discussion, m, m_1, m_2, n, n_1, n_2 denote elements of $\mathfrak{R}$, with n, n_1, n_2 nonzero. When $\mathfrak{R}$ is $\mathfrak{F}[\lambda]$, we shall occasionally write $m(\lambda)$, $m_1(\lambda)$, $n(\lambda)$, etc.

Fractions in lowest terms

The numerator m and denominator n of a fraction m/n need not always be relatively prime. Let $d = (m, n)$ be the greatest common divisor of m and n. Then $m = sd$, $n = td$, where $s = m/d$, $t = n/d$ are relatively prime elements of $\mathfrak{R}$, and

$$m/n = s/t.$$

Multiplying numerator and denominator of s/t by ± 1 (when $\mathfrak{R}$ is $\mathbf{Z}$) or by a nonzero constant (when $\mathfrak{R}$ is $\mathfrak{F}[\lambda]$), we may make t positive or monic. We say a fraction is in *lowest terms* if its numerator and denominator are relatively prime and the denominator is positive (for rational numbers) or monic (for rational functions). We have therefore shown that any fraction m/n in $\mathfrak{K}$ equals a fraction s/t in lowest terms. Conversely, suppose s/t is in lowest terms, and that m_1/n_1 is any fraction satisfying $m_1/n_1 = s/t$. Then $m_1 t = n_1 s$, so that $t \mid n_1 s$; since $(s, t) = 1$, Lemma 3.4.10 shows that $t \mid n_1$. Hence $n_1 = tw$, where w is a nonzero element of $\mathfrak{R}$; and then $m_1 t = tws$, yielding $m_1 = sw$. Thus the numerator and denominator of m_1/n_1 coincide, apart from the common factor w, with the numerator and denominator of s/t. In particular, if m_1/n_1 should be in lowest terms, then

w must be 1. Thus fractions in lowest terms are equal only when they have the same numerator and the same denominator. Fractions in lowest terms therefore constitute a system of distinct representatives for the elements of $\mathcal{K}$.

THEOREM 6.4.1
A system of distinct representatives for the elements of $\mathcal{K}$ are the fractions s/t in lowest terms, i.e., *with s, t relatively prime and t positive (for rational numbers) or monic (for rational functions). Any fraction equal to a fraction s/t in lowest terms is given by*

$$sw/tw, \qquad w \in \mathcal{R};\ w \neq 0,$$

so is obtained by inserting a common nonzero factor w into numerator and denominator.

Unique factorization

The unique factorization theorem (3.5.3) states that a nonzero element α of $\mathcal{R}$ is factorable, uniquely apart from the order of the factors, as

$$\alpha = up_1^{e1}p_2^{e2}\cdots p_k^{ek}, \tag{1}$$

where u is a unit in $\mathcal{R}$, $p_1, \ldots, p_k$ are distinct positive or monic primes in $\mathcal{R}$, and $e_1, \ldots, e_k$ are positive integers. There is a similar unique factorization theorem if α is an element of $\mathcal{K}$, except that in it the integer exponents $e_1, \ldots, e_k$ may be positive or negative.

THEOREM 6.4.2
A nonzero element α of $\mathcal{K}$ is factorable, uniquely apart from the order of the factors, as a product (1), where u is a unit in $\mathcal{R}$, $p_1, \ldots, p_k$ are distinct positive or monic primes in $\mathcal{R}$, and $e_1, \ldots, e_k$ are nonzero integers.

PROOF Write $\alpha = s/t$, in lowest terms, and apply Theorem 3.5.3 to s and t, to get

$$s = u_1p_1^{e1}\cdots p_r^{er}, \qquad t = u_2p_{r+1}^{er+1}\cdots p_k^{ek}.$$

Here u_1, u_2 are units in $\mathcal{R}$, $p_1, \ldots, p_r$ are distinct positive or monic primes in $\mathcal{R}$, $p_{r+1}, \ldots, p_k$ are also distinct positive or monic primes in $\mathcal{R}$, and $e_1, \ldots, e_k$ are positive integers. Since s, t are relatively prime, $p_1, \ldots, p_r$ are distinct from $p_{r+1}, \ldots, p_k$. Then

$$\alpha = up_1^{e1}\cdots p_r^{er}p_{r+1}^{-er+1}\cdots p_k^{-ek}, \tag{2}$$

where $u = u_1/u_2$ is a unit in $\mathcal{R}$, $p_1, \ldots, p_k$ are distinct positive or monic primes in $\mathcal{R}$, and $e_1, \ldots, e_r, -e_{r+1}, \ldots, -e_k$ are nonzero. Apart from a trivial notational change, this is (1). The existence part of the proof is finished, and now we prove uniqueness. Suppose also

$$\alpha = vq_1^{f_1}\cdots q_s^{f_s}q_{s+1}^{-f_{s+1}}\cdots q_l^{-f_l}, \tag{3}$$

where v is a unit in $\mathcal{R}$, $q_1, \ldots, q_l$ are distinct positive or monic primes in $\mathcal{R}$, and $f_1, \ldots, f_l$ are positive integers. From (2) and (3) we get

$$up_1^{e_1}\cdots p_r^{e_r}q_{s+1}^{f_{s+1}}\cdots q_l^{f_l} = vq_1^{f_1}\cdots q_s^{f_s}p_{r+1}^{e_{r+1}}\cdots p_k^{e_k}. \tag{4}$$

Because $p_1, \ldots, p_k$ are distinct, and $q_1, \ldots, q_l$ are distinct, unique factorization for elements of $\mathcal{R}$, when applied to the two sides of (4), yields $u = v$, $r = s$, $k = l$, and (possibly after renumbering $q_1, \ldots, q_r$ and $q_{r+1}, \ldots, q_k$):

$$p_1 = q_1, \ldots, \quad p_r = q_r, \quad p_{r+1} = q_{r+1}, \ldots, \quad p_k = q_k,$$
$$e_1 = f_1, \ldots, \quad e_r = f_r, \quad e_{r+1} = f_{r+1}, \ldots, \quad e_k = f_k.$$

Uniqueness of the factorization (2), apart from order, is now clear. □

Absolute values for rational numbers and functions

In previous chapters (especially Chapters 3 and 4) we have seen a strong correspondence between properties of integers and polynomials. Does this extend to a correspondence between properties of quotients of integers and quotients of polynomials? It does, in a variety of ways, and one that the reader perhaps does not expect will now be described.

The field **Q** is endowed with an absolute value function satisfying these properties: If $q, q_1, q_2 \in \mathbf{Q}$, then

V1 $|q|$ *is a nonnegative real number;*

V2 $|q| = 0$ *if and only if* $q = 0$;

V3 $|q_1q_2| = |q_1|\,|q_2|$;

V4 $|q_1 + q_2| \leq |q_1| + |q_2|$.

Of course, V4 is the triangle inequality. In the next definition, we introduce the corresponding *absolute value for rational functions.*

DEFINITION 6.4.3
Let ρ be a real number, with $0 < \rho < 1$. The absolute value *of the rational function $q(\lambda) = m(\lambda)/n(\lambda)$ is defined to be*

$$|q| = 0 \qquad \text{if } q = 0, \tag{5}$$
$$|q| = \rho^{\deg n - \deg m}, \quad \text{if } q \neq 0. \tag{6}$$

REMARK 1 The number ρ is held fixed throughout the discussion. At the outset, however, it may be chosen at will, provided $0 < \rho < 1$. A particularly convenient value is $\rho = \frac{1}{2}$.

REMARK 2 Let q be nonzero. Because q may be represented as a quotient of polynomials in many ways, it is necessary to show that $|q|$ is independent of the particular representation used. If $q(\lambda) = m(\lambda)/n(\lambda) = m_1(\lambda)/n_1(\lambda)$ are two representations of q as a quotient of polynomials, then $m(\lambda)n_1(\lambda) = n(\lambda)m_1(\lambda)$. Computing the degree of each side gives $\deg m + \deg n_1 = \deg n + \deg m_1$. Hence $\deg n - \deg m = \deg n_1 - \deg m_1$, so that

$$\rho^{\deg n - \deg m} = \rho^{\deg n_1 - \deg m_1}.$$

All ways of writing q as a quotient of polynomials therefore yield equal $|q|$.

REMARK 3 Recalling (from §3.2) the conventions that the degree of the zero polynomial is $-\infty$ and $\rho^\infty = 0$, applying (6) to $q = 0/n$ gives $|q| = \rho^{\deg n - \deg 0} = \rho^\infty = 0$. Thus (5) may be regarded as a special case of (6).

REMARK 4 Since $0 < \rho < 1$, $\rho^{\deg n - \deg m}$ is a small real number when $\deg n$ is much larger than $\deg m$, and is a large real number when $\deg m$ is much larger than $\deg n$. This absolute value therefore regards q as "small" when its denominator has degree much higher than the numerator, and as "large" when the reverse is true.

REMARK 5 If c is a nonzero constant, this absolute value, when applied to c, gives $|c| = 1$. When applied to any nonzero polynomial $m(\lambda)$, it gives $|m(\lambda)| \geq 1$.

To justify giving the name "absolute value" to $|q(\lambda)|$, it is necessary to show that it has suitable properties. We take V1, V2, V3, V4 above to be the properties expected of an absolute value, and we now show that these properties hold when $q(\lambda), q_1(\lambda), q_2(\lambda) \in \mathfrak{F}(\lambda)$.

Proof of V1: Clear from the definition of $|q(\lambda)|$. □

Proof of V2: Also clear from the definition of $|q(\lambda)|$. □

Proof of V3: If $q_1 = 0$ or $q_2 = 0$, each side of the formula in V3 is zero, so that the formula is valid. If q_1, q_2 are nonzero, let $q_1(\lambda) = m_1(\lambda)/n_1(\lambda)$, $q_2(\lambda) = m_2(\lambda)/n_2(\lambda)$. Then $q_1q_2 = m_1m_2/n_1n_2$, and

$$|q_1| = \rho^{\deg n_1 - \deg m_1}, \qquad |q_2| = \rho^{\deg n_2 - \deg m_2},$$

$$|q_1q_2| = \rho^{\deg n_1n_2 - \deg m_1m_2}.$$

Since $\deg n_1n_2 = \deg n_1 + \deg n_2$, $\deg m_1m_2 = \deg m_1 + \deg m_2$, it is clear that $|q_1q_2| = |q_1|\,|q_2|$. □

Before establishing V4, we give first a stronger property V4′.

V4′ $|q_1 + q_2| \leq \max(|q_1|, |q_2|)$, *if* $q_1(\lambda), q_2(\lambda)$ *are rational functions.*

[$\max(|q_1|, |q_2|)$ denotes the larger of the real numbers $|q_1|, |q_2|$.]

Proof of V4′: We first show that V4′ holds for polynomials by proving that

$$|m_1(\lambda) + m_2(\lambda)| \le \max(|m_1(\lambda)|, |m_2(\lambda)|).$$

From §3.2 we have $\deg(m_1 + m_2) \le \max(\deg m_1, \deg m_2)$. Using this and $0 < \rho < 1$, we get:

$$|m_1 + m_2| = \rho^{-\deg(m_1+m_2)} \le \rho^{-\max(\deg m_1,\ \deg m_2)}$$
$$= \max(\rho^{-\deg m_1}, \rho^{-\deg m_2}) = \max(|m_1|, |m_2|).$$

(This calculation is valid even when some or all of m_1, m_2, $m_1 + m_2$ are zero.) Therefore, V4′ holds for polynomials. Applying it to the polynomials m_1n_2 and m_2n_1, we obtain

$$|m_1n_2 + m_2n_1| \le \max(|m_1n_2|, |m_2n_1|).$$

Multiplying both sides of this inequality by $|1/n_1n_2|$, and using V3, we then find that

$$\left|\frac{m_1}{n_1} + \frac{m_2}{n_2}\right| \le \max\left(\left|\frac{m_1}{n_1}\right|, \left|\frac{m_2}{n_2}\right|\right);$$

thus *V*4′ holds for $q_1 = m_1/n_1$, $q_2 = m_2/n_2$. □

We now use V4′ to establish V4, in this way:

$$|q_1 + q_2| \le \max(|q_1|, |q_2|) = \textit{either } |q_1| \textit{ or } |q_2| \le |q_1| + |q_2|. \qquad \square$$

Property V4′ is usually called the *strong triangle law*, whereas V4 is called the *weak triangle law* or (usually) simply the *triangle law*. The absolute value for rational functions therefore satisfies both triangle laws. However, the absolute value for rational, real, or complex numbers described in Chapter 2 satisfies only the weak triangle law (Exercise 3).

The inequality $|m(\lambda)| < |n(\lambda)|$ *between polynomials* $m(\lambda)$, $n(\lambda)$ *simply means that* $\deg m < \deg n$. *This important fact will often be used.* Furthermore, if $|m(\lambda)| < |n(\lambda)|$, then $|m + n| = |n|$ since $\deg(m + n) = \deg n$ when $m(\lambda)$ has smaller degree than $n(\lambda)$. This fact may be extended to a basic property of the absolute value of rational functions, called the *principle of domination*. We state this principle in the next theorem, which also summarizes the facts proved above.

THEOREM 6.4.4
The absolute value for rational functions obeys rules V1, V2, V3, V4′, V4 *above. It also satisfies this principle of domination: if* $q_1(\lambda)$, $q_2(\lambda)$ *are rational functions with* $|q_1| < |q_2|$, *then* $|q_1 + q_2| = |q_2|$.

PROOF Only the principle of domination remains to be proved. Since $|-1| = 1$, $|-q| = |q|$, for any rational function q. From V4′ we get

$$\begin{aligned} |q_2| &= |(q_1 + q_2) - q_1| \leq \max\,(|q_1 + q_2|, |-q_1|) \\ &= \text{either } |q_1 + q_2| \quad \text{or} \quad |q_1|. \end{aligned}$$

The maximum here cannot be $|q_1|$, for this would give $|q_2| \leq |q_1|$, contrary to hypothesis. Thus the maximum is $|q_1 + q_2|$, implying that

$$|q_2| \leq |q_1 + q_2|.$$

However, by V4′, $|q_1 + q_2| \leq \max\,(|q_1|, |q_2|) = |q_2|$. Combining these estimates, we get

$$|q_2| \leq |q_1 + q_2| \leq |q_2|,$$

implying that $|q_1 + q_2| = |q_2|$. □

CAUTIONARY REMARK If the field $\mathfrak{F}$ is **Q**, **R**, or **C**, and if q is an element of $\mathfrak{F}$, the symbol $|q|$ may be interpreted in two ways: (i) It may mean the mean the absolute value described in §2.2 and 2.4, i.e., if $q = x + yi$ with real x, y, then $|q| = (x^2 + y^2)^{1/2}$. (ii) Or it may mean that q is to be regarded as a constant element of $\mathbf{Q}(\lambda)$, $\mathbf{R}(\lambda)$, or $\mathbf{C}(\lambda)$, and then $|q| = 1$ if q is nonzero, $= 0$ if q is zero. Therefore, when speaking of the absolute value of a constant, it must be specified which absolute value is intended. If we mean (i), we will say that *the usual absolute value for rational, real, or complex numbers is intended.*

TERMINOLOGY Absolute values satisfying properties V1, V2, V3, V4 above are frequently called *valuations*. We now have two examples of valuations: that just introduced for rational functions, satisfying the strong triangle law; and the usual one for rational, real, or complex numbers, not satisfying the strong triangle law. Infinitely many other valuations, all satisfying the strong triangle law, will be seen in §8.2.

Order relations (Optional)

A field is *ordered* if it contains a subset $\mathcal{P}$ such that:

P1 *An element α satisfies exactly one of these statements:*

(i) $\alpha \in \mathcal{P}$; (ii) $\alpha = 0$; (iii) $-\alpha \in \mathcal{P}$.

P2 *If α and β are in $\mathcal{P}$, so are $\alpha + \beta$ and $\alpha\beta$.*

These are just the *order axioms* already seen in §1.3. We say that α is positive if $\alpha \in \mathcal{P}$, and negative if $-\alpha \in \mathcal{P}$. Using the order axioms, inequalities comparing elements may be introduced and manipulated, as in §1.3.

We know (§§1.3 and 2.2) that both **Q** and **R** are ordered fields. Both are also endowed with an absolute value function, and it is possible to rewrite a statement involving absolute values as a set of equivalent statements solely involving inequalities. For example, if q and q_1 are real numbers, with q_1 positive, the condition

$$|q| < q_1 \tag{7}$$

is equivalent to the *two* conditions

$$uq < q_1, \qquad \text{for } u = 1 \text{ and } u = -1. \tag{8}$$

Consider now the field $\mathbf{R}(\lambda)$ of rational functions with real number coefficients. The reader will perhaps be surprised to learn that this field is also an ordered field. To explain this, let $q(\lambda) = m(\lambda)/n(\lambda)$ be a nonzero element of $\mathbf{R}(\lambda)$, and let M, N be the leading coefficients of the polynomials $m(\lambda)$, $n(\lambda)$ respectively. *For the set $\mathcal{P}$ of positive rational functions, we take all $q(\lambda)$ for which M/N is a positive real number.* For example, all of

$$\frac{2\lambda^2 - 1}{\lambda + 3}, \qquad \frac{-\lambda}{-\lambda^2 + 1}, \qquad \frac{3\lambda^2 - 1}{4\lambda^3 - \lambda^2}$$

are positive rational functions.

(Note that if $q(\lambda)$ has constant nonzero numerator and denominator, it is positive as a rational function if and only if it is positive as a real number.)

Conceivably this definition of positivity depends on the manner in which $q(\lambda)$ is written as a quotient of polynomials. Suppose also $q(\lambda) = m_1(\lambda)/n_1(\lambda)$, where the leading coefficients of $m_1(\lambda)$, $n_1(\lambda)$ are M_1, N_1 respectively. Then $mn_1 = nm_1$. The leading coefficient of mn_1 is MN_1, that of nm_1 is NM_1, hence $MN_1 = NM_1$, and therefore $M/N = M_1/N_1$. Whether $q(\lambda)$ is positive depends therefore solely on q and not on its representation as a quotient of polynomials.

We now show that this definition of positivity satisfies the order axioms P1, P2. Clearly, if $q(\lambda) = m(\lambda)/n(\lambda)$ is a nonzero rational function, either it or $-q(\lambda)$ is positive, but not both, since precisely one of M/N, $-M/N$ is a positive real number. To verify P2, let $q = m/n$, $q_1 = m_1/n_1 \in \mathcal{P}$, with M, N, M_1, N_1 the leading coefficients of m, n, m_1, n_1, as before. Then both M/N and M_1/N_1 are positive real numbers. From

$$q + q_1 = \frac{mn_1 + nm_1}{nn_1}, \qquad qq_1 = \frac{mm_1}{nn_1},$$

we deduce that $q + q_1 \in \mathcal{P}$ and $qq_1 \in \mathcal{P}$, in this way: If $\deg mn_1 = \deg nm_1$, the sum $MN_1 + NM_1$ of the leading coefficients of mn_1 and nm_1 is nonzero (if not, then $M/N = -M_1/N_1$, a contradiction) and therefore $MN_1 + NM_1$ is the leading coefficient of $mn_1 + nm_1$. If $\deg mn_1 \neq \deg nm_1$, the leading coefficient of $mn_1 + nm_1$ is either MN_1 or NM_1. The leading coefficient of mm_1 is MM_1, and that of nn_1 is NN_1. Then $q + q_1$ and qq_1 are in $\mathcal{P}$ because all of

$$\frac{MN_1 + NM_1}{NN_1} = \frac{M}{N} + \frac{M_1}{N_1}, \qquad \frac{MN_1}{NN_1}, \qquad \frac{NM_1}{NN_1}, \qquad \frac{MM_1}{NN_1}$$

are positive real numbers.

Therefore $\mathbf{R}(\lambda)$ is an ordered field. Consequently, exactly as in §1.3, inequalities may be introduced and manipulated. For example, if q_1 and q_2 are rational functions, the symbol $>$ is defined by $q_1 > q_2$ if and only $q_1 - q_2$ is a positive rational function. Thus

$$\frac{\lambda^2}{\lambda - 1} > \frac{(\lambda - 1)^3}{\lambda^2}, \qquad \lambda > 1{,}000.$$

Just as with numbers, the function field $\mathbf{R}(\lambda)$ is endowed with an absolute value. Is it possible to rewrite a statement involving absolute values of rational functions as a set of equivalent statements solely involving inequalities? It is, but infinitely many inequalities may be needed. For example, parallel to (7) and (8), we have this equivalence: *Let $q, q_1 \in \mathbf{R}(\lambda)$ be rational functions, with q_1 positive. Then*

$$|q| < |q_1| \tag{9}$$

if and only if

$$uq < q_1 \qquad \textit{for each nonzero constant } u. \tag{10}$$

PROOF There is nothing to prove if $q = 0$. Let $q \neq 0$, and set $q = m/n$, $q_1 = m_1/n_1$, where polynomials m, n, m_1, n_1 have M, N, M_1, N_1 as their respective leading coefficients. Suppose (9) holds. Then

$$\rho^{\deg n - \deg m} < \rho^{\deg n_1 - \deg m_1}, \tag{11}$$

hence $\deg n - \deg m > \deg n_1 - \deg m_1$, and therefore $\deg m_1 n > \deg n_1 m$. For any constant u,

$$q_1 - uq = \frac{m_1 n - u n_1 m}{n n_1}. \tag{12}$$

The leading coefficient in the numerator on the right-hand side is $M_1 N$, (since $\deg m_1 n > \deg n_1 m$), that of the denominator is NN_1, and so $q_1 - uq$ is positive because $M_1N/NN_1 = M_1/N_1$ is a positive real number. Thus $q_1 > uq$. Conversely, suppose $q_1 > uq$ for any nonzero constant u. If $\deg n_1 m \geq \deg m_1 n$, the number u could be chosen such that the leading coefficient in the numerator of the right-hand side of (12) is $-NN_1$. (Exercise 11.) Since $-NN_1/NN_1 = -1$, it would then follow that $q_1 - uq < 0$. This is contrary to hypothesis. Therefore $\deg m_1 n > \deg n_1 m$. Thus $\deg n - \deg m > \deg n_1 - \deg m_1$, and so (11) and therefore (9) holds. □

Property of Archimedes (Optional)

This property comes in two versions, one for absolute values (valuations), one for order relations. Let 1 denote the multiplicative identity of the field $\mathcal{K}$.

> *An absolute value* $|\ |$ *on a field is* archimedean *if, for any element q, there is a natural number k (dependent on q) such that*
>
> $$|\underbrace{1 + 1 + \cdots + 1}_{k \text{ terms}}| > |q|. \tag{13}$$

(The symbol $>$ here means that $|1 + 1 + \cdots + 1| - |q|$ is a positive real number.)

> *An order relation* $>$ *on a field is* archimedean *if, for any element q, there is a natural number k (dependent on q) such that*
>
> $$\underbrace{1 + 1 + \cdots + 1}_{k \text{ terms}} > q. \tag{14}$$

(Here $>$ now means that the element $1 + 1 + \cdots + 1 - q$ of $\mathcal{K}$ belongs to the set $\mathcal{P}$ of positive elements of $\mathcal{K}$.)

An absolute value or order relation which is not archimedean is said to be *nonarchimedean*. To prove that an absolute value or order is nonarchimedean, it suffices to find one element q for which (13) or (14) cannot be satisfied for any natural number k.

Evidently the usual absolute value on the fields **Q**, **R**, **C** *is archimedean, and the order on* **Q**, **R**, *is also archimedean.* More surprising are the following:

> *The absolute value* $|\ |$ *on the rational function field* $\mathfrak{F}(\lambda)$ *is nonarchimedean.*

PROOF Because $1 + 1 + \cdots + 1$ (k terms) is a constant in $\mathfrak{F}(\lambda)$, its absolute value $|1 + 1 + \cdots + 1|$ is either the (real number) 0 or 1. Since $|\lambda| = \rho^{-1} > 1$, we never have

$$|1 + 1 + \cdots + 1| > |\lambda|. \qquad \square$$

> *The order relation described above on the rational function field* **R**(λ) *is nonarchimedean.*

PROOF For every natural number k, $\lambda > k$ since $\lambda - k$ belongs to $\mathcal{P}$. $\square$

The usual absolute value on **Q** (or **R**, or **C**) is archimedean but does not satisfy the strong triangle law. The absolute value on $\mathfrak{F}(\lambda)$, however, is nonarchimedean and does satisfy the strong triangle law. These facts are not unrelated: it can be proved that an absolute value on a field satisfies the strong triangle law if and only if it is nonarchimedean. One of the main themes of this book is the strong correspondence between properties of rational numbers and those of rational functions. That this correspondence is not fully exact is shown by the archimedean nature of the absolute value on **Q** and the nonarchimedean nature of that on $\mathfrak{F}(\lambda)$. In fact, advanced work shows this to be a crucial distinction between **Q** and $\mathfrak{F}(\lambda)$.

Substitution into a rational function

In §3.2 we discussed the evaluation of a polynomial $a(\lambda)$ at another polynomial $\alpha(\lambda) \in \mathfrak{F}[\lambda]$, to obtain $a(\alpha(\lambda))$, and we showed that when $a(\lambda)$, $b(\lambda)$ are polynomials,

$$\begin{aligned}(a+b)(\alpha(\lambda)) &= a(\alpha(\lambda)) + b(\alpha(\lambda)),\\ (ab)(\alpha(\lambda)) &= a(\alpha(\lambda))b(\alpha(\lambda)).\end{aligned}$$

We point out that these substitution formulas continue to hold when a, b are polynomials and $\alpha(\lambda) \in \mathfrak{F}(\lambda)$ is a rational function. The proofs in §3.2 may be applied without change to demonstrate this. From this we wish to deduce that these substitution rules apply to rational functions, as well as to polynomials, provided that the substitution of $\alpha(\lambda)$ for λ does not produce a zero denominator.

Specifically, if $q(\lambda) = m(\lambda)/n(\lambda)$ and $\alpha(\lambda) \in \mathfrak{F}(\lambda)$, the symbol $q(\alpha(\lambda))$ denotes the rational function $m(\alpha(\lambda))n(\alpha(\lambda))^{-1} = m(\alpha(\lambda))/n(\alpha(\lambda))$.

(i) Suppose $m(\lambda)/n(\lambda) = m_1(\lambda)/n_1(\lambda)$. Then $m(\lambda)n_1(\lambda) = n(\lambda)m_1(\lambda)$. This is an equation involving polynomials, so in it we may substitute $\alpha(\lambda)$ for λ, obtaining $m(\alpha)n_1(\alpha) = n(\alpha)m_1(\alpha)$. If $n(\alpha)$, $n_1(\alpha)$ are nonzero, multiplying by $n(\alpha)^{-1}n_1(\alpha)^{-1}$ gives $m(\alpha)/n(\alpha) = m_1(\alpha)/n_1(\alpha)$. *Substitution into a rational function is therefore compatible with equality of rational functions, provided no denominator becomes zero.*

(ii) If

$$\frac{m(\lambda)}{n(\lambda)} + \frac{m_1(\lambda)}{n_1(\lambda)} = \frac{m_2(\lambda)}{n_2(\lambda)} \quad \text{or} \quad \frac{m(\lambda)}{n(\lambda)}\cdot\frac{m_1(\lambda)}{n_1(\lambda)} = \frac{m_2(\lambda)}{n_2(\lambda)},$$

then

$$\begin{aligned}m(\lambda)n_1(\lambda)n_2(\lambda) + m_1(\lambda)n(\lambda)n_2(\lambda) &= m_2(\lambda)n(\lambda)n_1(\lambda),\\ m(\lambda)m_1(\lambda)n_2(\lambda) &= m_2(\lambda)n(\lambda)n_1(\lambda).\end{aligned}$$

These are equations between polynomials; in them we may therefore substitute $\alpha(\lambda)$ for λ. If $n(\alpha)$, $n_1(\alpha)$, $n_2(\alpha)$ are nonzero, in the resulting equations we may multiply by $n(\alpha)^{-1}n_1(\alpha)^{-1}n_2(\alpha)^{-1}$, obtaining

$$\frac{m(\alpha)}{n(\alpha)} + \frac{m_1(\alpha)}{n_1(\alpha)} = \frac{m_2(\alpha)}{n_2(\alpha)} \quad \text{and} \quad \frac{m(\alpha)}{n(\alpha)}\cdot\frac{m_1(\alpha)}{n_1(\alpha)} = \frac{m_2(\alpha)}{n_2(\alpha)}.$$

Therefore: *Substitution into a rational function is compatible with addition and multiplication of rational functions, provided no denominator becomes zero.*

EXERCISES 6.4

I

1. Let $q(\lambda)$ be an element of $\mathbf{C}(\lambda)$. Are we permitted to evaluate q at α for every $\alpha \in \mathbf{C}$? Explain.

2. Let $q(\lambda) \in \mathbf{Q}(\lambda)$ and $\alpha \in \mathbf{Q}$. Is it true that $|q(\lambda)| = |q(\alpha)|$ always holds (assuming that the denominator of $q(\alpha)$ is nonzero)? Furnish either a proof or a counterexample. (Interpret the symbol $|q(\alpha)|$ in either of two ways and answer the question for each interpretation.)

3. Let q_1 and $q_2 \in \mathbf{Q}$. Is it always true that the strong triangle law $|q_1 + q_2| \leq \max(|q_1|, |q_2|)$ is valid? Furnish either a proof or a counterexample. Here the usual absolute value of rational numbers is intended. Show by an example that the principle of domination does not hold.

4. Elements q_1 and q_2 of $\mathcal{K}$ are said to be *close* if $|q_1 - q_2|$ is a small real number. Let $\mathcal{K} = \mathbf{Q}(\lambda)$ and take $q_1 = 2\lambda/(\lambda + 1)$. Which, in your opinion, of the following rational functions q_2 are close to q_1?

$$q_2 = \frac{2\lambda - \lambda^8 - 1}{\lambda + 1}; \qquad q_2 = \frac{\lambda - 1}{\lambda + 1};$$

$$q_2 = \frac{2\lambda - 1}{\lambda + 1}; \qquad q_2 = \frac{2\lambda^3 + 4\lambda^2 + 2\lambda - 1}{(\lambda + 1)^3}.$$

5. We say that the sequence $q_1(\lambda), q_2(\lambda), q_3(\lambda), \ldots$ of elements of $\mathbf{Q}(\lambda)$ *approaches zero* if, whenever ϵ is a positive rational number, there is a natural number K (dependent on ϵ) such that $|q_j| < \epsilon$ for all $j \geq K$. (This simply says that $|q_j|$ is less than a tolerance ϵ if we take j sufficiently large, i.e., $j \geq K$.) Which of the following sequences approach zero?
   (i) $1, \lambda, \lambda^2, \lambda^3, \ldots, \lambda^j, \ldots$;
   (ii) $1, \lambda^{-1}, \lambda^{-2}, \lambda^{-3}, \ldots, \lambda^{-j}, \ldots$.

6. Factor $-15/28$ into primes in $\mathbf{Q}$. Factor $(\lambda^2 + 4)/(\lambda^2 + 5\lambda + 6)$ into primes in $\mathbf{R}(\lambda)$ and in $\mathbf{C}(\lambda)$.

II

7. Let a and b be elements of $\mathfrak{F}[\lambda]$, with $b \neq 0$. Show that the division algorithm for polynomials (Theorem 3.3.3) may be expressed this way: polynomials q and r in $\mathfrak{F}[\lambda]$ exist such that $a = qb + r$, with $|r| < |b|$. In fact show that $|r| \leq \rho\,|b|$. The corresponding fact for integers is Theorem 3.3.1 or Theorem 3.3.2. In Theorem 3.3.2 it is shown that we can achieve $|r| \leq \frac{1}{2}|b|$. Is it possible to do this well with polynomials, namely to have $|r| \leq \frac{1}{2}|b|$?

8. Show, by an example, that if $|q_1(\lambda)| = |q_2(\lambda)|$, then $|q_1(\lambda) + q_2(\lambda)|$ need not equal $|q_2(\lambda)|$.

9. Let $\mathcal{S}$ be the subset of $\mathbf{Q}(\lambda)$ consisting of all rational functions $q(\lambda)$ with $|q| \leq 1$. Prove that $\mathcal{S}$, under addition and multiplication of rational functions, is a ring. Give a verbal description of the elements of $\mathcal{S}$. Is the corresponding fact true for rational numbers, namely, is the set of rational numbers q with $|q| \leq 1$ (usual absolute value) a ring under addition and multiplication of rational numbers?

10. Let $\mathcal{K}$ be a field. Define a new absolute value on $\mathcal{K}$ by this rule: $|\alpha|_{\text{new}} = 0$ if α is the zero element of $\mathcal{K}$ and $|\alpha|_{\text{new}} = 1$ if α is a nonzero element of $\mathcal{K}$. Here 0 and 1 denote the real numbers *zero* and *one*. Show that $|\alpha|_{\text{new}}$ satisfies Axioms V1, V2, V3, V4. Does it satisfy V4′?

11. (*Optional*) Referring to equation (12) and the discussion of order relations, suppose $\deg n_1m \geq \deg m_1n$. If $\deg n_1m = \deg m_1n$, let $u = (NN_1 + M_1N)/N_1M$. If $\deg n_1m > \deg m_1n$, let $u = N/M$. Show that $u \neq 0$ and that the leading term in the numerator of the right-hand side of (12) has coefficient $-NN_1$.

12. This exercise outlines an alternative procedure for defining the absolute value on rational functions, by defining it first on polynomials. If $m(\lambda)$ is a polynomial, define $|m(\lambda)|$ by

$$|m(\lambda)| = 0 \qquad \text{if } m(\lambda) \text{ is the zero polynomial,}$$

$$|m(\lambda)| = \rho^{-\deg m} \qquad \text{if } m(\lambda) \text{ is nonzero.}$$

Prove that V1, V2, V3, V4′ are valid when applied to polynomials. If $q(\lambda) = m(\lambda)/n(\lambda)$, now define

$$|q(\lambda)| = |m(\lambda)|/|n(\lambda)|.$$

If also $q(\lambda) = m_1(\lambda)/n_1(\lambda)$, then $mn_1 = nm_1$; from this show that $|m|/|n| = |m_1|/|n_1|$. Deduce, under this definition, that $|q(\lambda)|$ does not depend on the representation of q as a quotient of polynomials. Now verify V1, V2, V3, V4′ for rational functions, and that $|q| = \rho^{\deg n - \deg m}$.

13. Let $\mathcal{G}$ be a multiplicative group. A set of elements $g_1, g_2, g_3, \ldots$ in $\mathcal{G}$ is said to *generate* $\mathcal{G}$ if each element of $\mathcal{G}$ can be written as a product of powers of finitely many of $g_1, g_2, g_3, \ldots$, i.e., as a product $g_1^{e_1}g_2^{e_2}g_3^{e_3}\cdots$ in which the exponents $e_1, e_2, e_3, \ldots$ are integers (positive, negative, or zero) with almost all zero. The nonzero elements of $\mathbf{Q}$ form a group under multiplication. Find generators for this group. Do the same for the multiplicative group of nonzero elements in $\mathbf{C}(\lambda)$.

III

14. Let $q_1, q_2, \ldots, q_n, \ldots$ be in $\mathfrak{F}(\lambda)$. We say that the sequence $q_1, \ldots, q_n, \ldots$ converges to q if, whenever ϵ is a positive real number, there is a natural number K (dependent on ϵ) such that $|q_k - q| < \epsilon$ for all $k > K$. This simply says that we can make $|q_k - q|$ smaller than a tolerance ϵ if we take k sufficiently large, i.e., $k \geq K$. Let x be an element of $\mathfrak{F}(\lambda)$, and let $S_k = 1 + x + \cdots + x^{k-1}$. Under what circumstances does the sequence $S_1, S_2, \ldots, S_k, \ldots$ converge to $1/(1 - x)$? (See Lemma 2.2.3.) Show that $1 + \lambda^{-1} + \lambda^{-2} + \lambda^{-3} + \cdots$ converges to $\lambda/(\lambda - 1)$.

15. (*Differentiation of rational functions*) The derivative $D(a)$ of a polynomial $a(\lambda)$ was described in §3.2. In this exercise we establish the familiar principles governing differentiation of rational functions, i.e., elements of $\mathfrak{F}(\lambda)$. The field $\mathfrak{F}$ is arbitrary.
(i) Let $m(\lambda)$, $n(\lambda)$, $m_1(\lambda)$, $n_1(\lambda)$ be polynomials with $mn_1 = nm_1$. Assume n, n_1 are nonzero. Show that

$$n_1D(m) + mD(n_1) = nD(m_1) + m_1D(n).$$

(ii) Multiply each side of this formula by nn_1, then use $mnn_1 = m_1n^2$, $m_1nn_1 = mn_1^2$ to show that

$$n_1^2nD(m) + n^2m_1D(n_1) = n^2n_1D(m_1) + n_1^2mD(n).$$

(iii) Using the formula established in (ii), show that

$$\frac{nD(m) - mD(n)}{n^2} = \frac{n_1D(m_1) - m_1D(n_1)}{n_1^2}.$$

Define the derivative of a rational function m/n by

$$D\left(\frac{m}{n}\right) = \frac{nD(m) - mD(n)}{n^2}.$$

If $m/n = m_1/n_1$, show that $D(m/n) = D(m_1/n_1)$. Deduce that the derivative of a rational function is independent of the particular way it is represented as a quotient of polynomials. Using this definition of derivative, prove the following differentiation rules for rational functions.

(iv) $D(q) = 0$ if $q(\lambda)$ is a constant.
(v) $D(q + q_1) = D(q) + D(q_1)$.
(vi) $D(qq_1) = qD(q_1) + q_1D(q)$.

(v) For any integral exponent k, positive or negative, show that $D(q^k) = q^{k-1}D(q)$ if $q \neq 0$.
(vi) Establish the chain law for differentiation: $D(q(\alpha(\lambda))) = (Dq)(\alpha(\lambda))D\alpha(\lambda)$.

[Here $(Dq)(\alpha(\lambda))$ means Dq evaluated at $\alpha(\lambda)$.] Do this in stages: first when $q(\lambda) = \lambda^k$ is a polynomial with a single term [use (v)]; next when $q(\lambda)$ is any polynomial; finally when q is any rational function.

16. If we attempted to define the absolute value of a rational function as in Definition 6.4.3, using however a real number ρ with $\rho > 1$, show that neither V4′ nor V4 would hold. If we take $\rho = 1$, show that we obtain the absolute value described in Exercise 10.

17. If the validity of the inequality (10) in the text is assumed for a fixed finite set of nonzero constants u, show that it is generally not possible to deduce (9) as a consequence. Do this by finding q, q_1 with q_1 positive such that (10) holds for these values of u but (9) does not hold.

18. Let $\mathfrak{F}$ be an ordered field. A nonempty subset S of $\mathfrak{F}$ is said to have an *upper bound* if $\mathfrak{F}$ contains an element b such that $s \leq b$ for all $s \in S$. An upper bound B for S is said to be a *least upper bound* if $B \leq b$ for every upper bound b of S. A basic property of the real numbers is that any nonempty subset of $\mathbf{R}$ having an upper bound has a least upper bound. Show that this is not true of $\mathbf{R}(\lambda)$ by showing that the set of natural numbers has an upper bound in $\mathbf{R}(\lambda)$ but no least upper bound.

19. If every nonempty subset of an ordered field $\mathfrak{F}$ having an upper bound has a least upper bound (see Exercise 18), prove that the ordering is archimedean. (Hint: If B is a least upper bound for the set of all natural numbers n, consider $B - 1$.)

20. Verify that the order relation on $\mathbf{Q}$ and $\mathbf{R}$ is unique, i.e., that each of these fields can be ordered in only one way.

21. Verify that the set of all real numbers of the form $x + y\sqrt{2}$ (where $x, y \in \mathbf{Q}$) is a field which can be ordered in two distinct ways.

6.5 REPRESENTATIONS RELATIVE TO A BASE

Let b be an integer, with $b > 1$. We saw in §4.2 that any nonnegative integer is representable as a sum $d_l b^l + d_{l-1}b^{l-1} + \cdots + d_1 b + d_0$ of powers of b, the integral coefficients $d_l, \ldots, d_0$ being $0, 1, \ldots,$ or $b - 1$. We now extend this representation to rational numbers, then to real numbers, and afterwards we give the corresponding discussion for rational functions.

Base b representations of rational numbers

If m, n are integers, with n nonzero, we call m/n a *proper fraction* if $0 \leq m < n$.

Let q_0 be a proper fraction. Beginning with q_0, we give a recursive procedure for computing integers $d_{-1}, d_{-2}, d_{-3}, \ldots$ and proper fractions $q_{-1}, q_{-2}, q_{-3}, \ldots$, such that

$$0 \leq d_t < b \tag{1}$$

for all t, and

$$q_0 = d_{-1}b^{-1} + d_{-2}b^{-2} + d_{-3}b^{-3} + \cdots + d_{-k}b^{-k} + q_{-k}b^{-k} \tag{2.k}$$

for every integer $k \geq 0$. (We use negative integers as subscripts for a reason to become clear below.) Setting $q_0 = m_0/n$, where m_0, $n \in \mathbf{Z}$, $0 \leq m_0 < n$, we shall find that

$$q_t = m_t/n, \qquad m_t \in \mathbf{Z}; 0 \leq m_t < n, \tag{3}$$

for all t; that is, the proper fractions $q_0, q_{-1}, q_{-2}, \ldots$ have n as a common denominator.

If, for an integral $k > 0$, we have obtained equation (2.k), where integers $d_{-1}, \ldots, d_{-k}$ satisfy (1) and $q_{-k} = m_{-k}/n$ is a proper fraction with integral numerator m_{-k} and n as denominator, we obtain $d_{-(k+1)}$ and $q_{-(k+1)}$ in this way: Divide $m_{-k}b$ by n, obtaining a quotient, denoted by $d_{-(k+1)}$, and least nonnegative remainder, denoted by $m_{-(k+1)}$, such that

$$m_{-k}b = d_{-(k+1)}n + m_{-(k+1)}, \qquad 0 \leq m_{-(k+1)} < n.$$

Multiplying by $1/n$ and setting $q_{-(k+1)} = m_{-(k+1)}/n$, we find that

$$q_{-k}b = d_{-(k+1)} + q_{-(k+1)}, \qquad 0 \leq q_{-(k+1)} < 1. \tag{4}$$

Then $d_{-(k+1)} = q_{-k}b - q_{-(k+1)} \leq q_{-k}b < b$ (since $q_{-k} < 1$) and also $d_{-(k+1)} \geq -q_{-(k+1)} > -1$. Therefore $-1 < d_{-(k+1)} < b$; because $d_{-(k+1)}$ is an integer, we have $0 \leq d_{-(k+1)} < b$. Thus $d_{-(k+1)}$ satisfies (1); clearly $q_{-(k+1)}$ satisfies (3). From (4) we get

$$q_{-k} = d_{-(k+1)}b^{-1} + q_{-(k+1)}b^{-1}, \tag{5}$$

and substituting this into (2.k) gives

$$q_0 = d_{-1}b^{-1} + \cdots + d_{-k}b^{-k} + d_{-(k+1)}b^{-(k+1)} + q_{-(k+1)}b^{-(k+1)}. \tag{2.(k + 1)}$$

We may therefore increase k; consequently (2.k) holds for any nonnegative integer k since it clearly holds for $k = 0$.

If q is any nonnegative rational number, say $q = m/n$, where m, $n \in \mathbf{Z}$, $m \geq 0$, $n > 0$, divide m by n to obtain a quotient, denoted by M, and a least nonnegative remainder, denoted by m_0, such that

$$m = Mn + m_0, \qquad 0 \leq m_0 < n.$$

Then

$$q = M + q_0\,, \qquad 0 \le q_0 < 1,$$

where integer M is nonnegative and $q_0 = m_0/n$ is a proper fraction. Writing M in base b, as in §4.2, we get

$$M = d_l b^l + d_{l-1}b^{l-1} + \cdots + d_1 b + d_0\,, \tag{6}$$

for some integer $l \ge 0$ and digits $d_l\,, \ldots, d_0$ satisfying (1). Combining this representation of M with the representation (2.k) of q_0, we get the following theorem.

THEOREM 6.5.1
If q is a nonnegative number, there is a subscript $l \ge 0$, integers d_l, $d_{l-1}\,, \ldots, d_0\,, d_{-1}\,, \ldots$ satisfying $0 \le d_t < b$ for each t, and numbers $q_0\,, q_{-1}\,, q_{-2}\,, \ldots$ with $0 \le q_t < 1$ for each t, such that

$$q = d_l b^l + d_{l-1}b^{l-1} + \cdots + d_1 b + d_0 + d_{-1}b^{-1} + \cdots + d_{-k}b^{-k} + q_{-k}b^{-k} \tag{7}$$

for any integer $k \ge 0$. If $q = m/n$ where $m,\ n \in \mathbf{Z},\ n > 0$, then $q_{-k} = m_{-k}/n$ where $m_{-k} \in \mathbf{Z}$ and $0 \le m_{-k} < n$.

We may also derive (7) by applying (2.(k + l + 1)) to q/b^{l+1}, where l is chosen so large that q/b^{l+1} is a proper fraction. (See Exercise 9.)

The integer k is at our disposal. Since $0 \le q_{-k} < 1$, the last term in (7) satisfies $0 \le q_{-k}b^{-k} < b^{-k}$. If we let k grow large, b^{-k} approaches zero, therefore so does $q_{-k}b^{-k}$, and then, omitting $q_{-k}b^{-k}$, the portion

$$d_l b^l + \cdots + d_0 + d_{-1}b^{-1} + \cdots d_{-k}b^{-k} \tag{8}$$

of the right-hand side of (7) will be a close approximation to q. Specifically, if we are given a tolerance ϵ (a positive real number), there will be a natural number K (dependent on ϵ) such that $b^{-K} < \epsilon$. Then $b^{-k} < \epsilon$ will hold for every $k \ge K$, and hence also $0 \le q_{-k}b^{-k} < \epsilon$. From (7) we then have

$$|q - (d_l b^l + \cdots + d_{-k}b^{-k})| < \epsilon \qquad \text{for every } k \ge K. \tag{9}$$

Thus (8) approximates q within this tolerance ϵ, for all $k \ge K$. See Figure 6.5.1. (The bars $|\ |$ denote the usual absolute for rational or real numbers.)

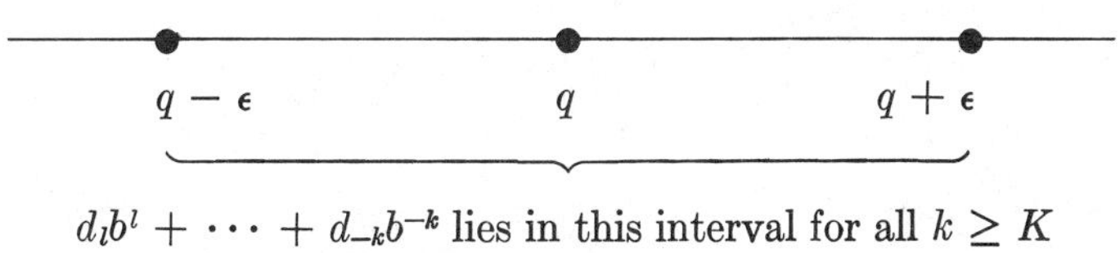

Figure 6.5.1

Because the term $q_{-k}b^{-k}$ in (7) approaches zero as k grows large, we write

$$q = d_l b^l + \cdots + d_1 b + d_0 + d_{-1}b^{-1} + d_{-2}b^{-2} + \cdots, \tag{10}$$

without a last term on the right-hand side. The sum on the right-hand side is an *infinite series*, and the meaning conveyed by equation (10) is precisely the approximation property given above: whenever a tolerance ϵ is given, there is a natural number K (dependent on ϵ) such that, for any $k \geq K$, the *truncated series* (8) obtained from (10) by stopping at $d_{-k}b^{-k}$, satisfies (9). In other words: sufficiently many terms in (10) give an approximation of q within tolerance ϵ such that this precision of approximation is retained when more terms are used. We say that the right-hand side of (10) is a *convergent* infinite series and that it *converges* to q.

It is customary to rewrite (10) in abbreviated form as

$$q = d_l d_{l-1} \ldots d_1 d_0 . d_{-1} d_{-2} d_{-3} \ldots, \tag{11}$$

to call either (10) or (11) the *representation of* q *in base* b, and to call $d_l, d_{l-1}, \ldots, d_1, d_0, d_{-1}, \ldots$ the *digits of* q *in base* b. These digits, of course, are integers satisfying (1). When b is ten, the heavy dot between d_0 and d_{-1} in (11) is called the *decimal point*; in base b we may call it the *b-point*. It separates the terms involving nonnegative powers of b from the terms involving negative powers. Let us emphasize *that* (11) *is simply abbreviated notation for* (10).

When it is necessary to indicate the base in (11), we attach it as a subscript, in this way:

$$d_l d_{l-1} \ldots d_0 . d_{-1} d_{-2} \ldots {}_b .$$

For convenience we will always give the subscript specifying the base in familiar base ten notation.

When b is ten, the fact that sufficiently many terms in the infinite series (10) give an approximation of q within a prescribed tolerance, such that the precision of approximation is retained when more terms are used, is just this property known to everybody: A number may be given in decimal notation to desired accuracy by using sufficiently many digits to the right of the decimal point; and using more digits does not impair the accuracy.

We wish to study the digits $d_{-1}, d_{-2}, \ldots$ in (10) and (11) more closely. Since the integers $m_0, m_{-1}, m_{-2}, \ldots$ satisfy $0 \leq m_t < n$ for every t, these integers cannot all be different. In fact, because $0, 1, 2, \ldots, n-1$ comprise only n distinct integers, at least two of $m_0, m_{-1}, \ldots, m_{-n}$ must be equal. Thus subscripts r, s exist with $0 \leq r < s \leq n$ such that $m_{-r} = m_{-s}$. Then $q_{-r} = q_{-s}$. Using (5) for $k = r, r+1, \ldots, s-1$, we obtain

$$\begin{aligned} q_{-r} &= d_{-(r+1)}b^{-1} + q_{-(r+1)}b^{-1} \\ &= d_{-(r+1)}b^{-1} + d_{-(r+2)}b^{-2} + q_{-(r+2)}b^{-2} \\ &= \cdots\cdots\cdots\cdots\cdots\cdots \\ &= d_{-(r+1)}b^{-1} + d_{-(r+2)}b^{-2} + \cdots + d_{-s}b^{-(s-r)} + q_{-s}b^{-(s-r)}. \end{aligned}$$

Setting $P = s - r$ and using $q_{-r} = q_{-s}$, we then get

$$q_{-r} = d_{-(r+1)}b^{-1} + \cdots + d_{-(r+P)}b^{-P} + q_{-r}b^{-P}. \tag{12}$$

Beginning with (7), for $k = r$, then using (12) repeatedly, we now obtain:

$$\begin{aligned} q - \{d_l b^l + \cdots + d_{-r}b^{-r}\} &= q_{-r}b^{-r} \\ &= \{d_{-(r+1)}b^{-(r+1)} + \cdots + d_{-(r+P)}b^{-(r+P)}\} + q_{-r}b^{-(r+P)} \\ &= \{d_{-(r+1)}b^{-(r+1)} + \cdots + d_{-(r+P)}b^{-(r+P)}\} \\ &\quad + \{d_{-(r+1)}b^{-(r+P+1)} + \cdots + d_{-(r+P)}b^{-(r+2P)}\} + q_{-r}b^{-(r+2P)} \\ &= \cdots\cdots\cdots \\ &= \{d_{-(r+1)}b^{-(r+1)} + \cdots + d_{-(r+P)}b^{-(r+P)}\} + \cdots \\ &\quad + \{d_{-(r+1)}b^{-(r+(t-1)P+1)} + \cdots + d_{-(r+P)}b^{-(r+tP)}\} + q_{-r}b^{-(r+tP)}, \end{aligned}$$

for any integer $t \geq 1$. Letting t grow arbitrarily large, we find that

$$\begin{aligned} q = \{d_l b^l + \cdots + d_{-r}b^{-r}\} &+ \{d_{-(r+1)}b^{-(r+1)} + \cdots + d_{-(r+P)}b^{-(r+P)}\} \\ &+ \{d_{-(r+1)}b^{-(r+P+1)} + \cdots + d_{-(r+P)}b^{-(r+2P)}\} + \cdots. \end{aligned}$$

This is the infinite series (10), but the digits in it now show a repetitive pattern. Rewriting this series in its abbreviated form (11), it becomes

$$q = d_l \ldots d_0 . d_{-1} \ldots d_{-r} \underbrace{d_{-(r+1)} \ldots d_{-(r+P)}}_{P \text{ digits}} \underbrace{d_{-(r+1)} \ldots d_{-(r+P)}}_{P \text{ digits}} \ldots \tag{13}$$

In other words: *the base b representation of the nonnegative rational number q is repetitive from some digit onwards.* We say the representation is *ultimately periodic.* The *smallest* integer P for which q has an ultimately periodic representation, as in (13), is called the *period* of q, and the digits $d_{-(r+1)}$, $\ldots$, $d_{-(r+P)}$ are then called the *periodic digits.* Furthermore $P \leq n$, i.e., *the period is at most the denominator n of $q = m/n$.* For conciseness, we write (13) as

$$q = d_l \ldots d_0 . d_{-1} \ldots d_{-r} \overline{d_{-(r+1)} \ldots d_{-(r+P)}}, \tag{14}$$

a bar being placed over the periodic digits. For example, in base ten,

$$1/3 = .\overline{3}.$$

(We could also write $1/3 = .\overline{33}$, but then we would not be using the smallest value of P.) If

$$q = .\overline{d_{-1} \ldots d_{-P}},$$

we say q is *purely periodic.*

Base *b* representations of real numbers

Suppose now that q is a nonnegative *real* number, not necessarily rational. Without any greater difficulty we prove that q is representable in base b, as in (10) or (11), but not necessarily in an ultimately periodic way. The proof is the same. Let

$$q = M + q_0$$

where M is an integer and $0 \leq q_0 < 1$. Then M is the largest integer not exceeding q. We call M the *integral part* of q, or the *greatest integer in* q, and we call q_0 the *fractional part* of q (even though q_0 may not be a rational number). Write M in base b, as in (6). Beginning with q_0, we give a recursive procedure for obtaining integers $d_{-1}, d_{-2}, \ldots$ satisfying (1) and real numbers $q_{-1}, q_{-2}, \ldots$ satisfying

$$0 \leq q_t < 1$$

for all t, such that (2.k) holds for every integer $k \geq 0$. If we have found integers $d_{-1}, \ldots, d_{-k}$ and real q_{-k}, $0 \leq q_{-k} < 1$, such that (2.k) holds, let $d_{-(k+1)}$ be the integral part and $q_{-(k+1)}$ the fractional part of $q_{-k}b$. Then (4) holds, $0 \leq q_{-(k+1)} < 1$, and exactly as in the discussion below (4), we have $0 \leq d_{-(k+1)} < b$. From (4) we get (5); and substituting (5) into (2.k), we get (2.(k + 1)). Thus we may raise k indefinitely and obtain (2.k) for every integer $k \geq 0$.

From $q = M + q_0$, (6), and (2.k), we obtain (7). [Or we may obtain (7) by applying (2.(k + l + 1)) to q/b^{l+1} for sufficiently large l; see Exercise 9.] Therefore: *Theorem 6.5.1 applies to any nonnegative real number* q.

As before, we let k grow arbitrarily large, and then express q as the infinite series (10). The meaning of (10) is the same as before: if a tolerance ϵ is given, sufficiently many terms in (10) provide, within this tolerance, an approximation of q which continues to hold if more terms are used; i.e., (9) holds for all $k \geq$ some natural number K (dependent on ϵ).

We know, for rational numbers, that the base b representation is ultimately periodic. We now prove the converse, namely, if q is a real number having an ultimately periodic base b representation (14), then q is rational. From (14), we get

$$\begin{aligned} q &= \{d_l b^l + \cdots + d_{-r}b^{-r}\} + \{d_{-(r+1)}b^{-(r+1)} + \cdots + d_{-(r+P)}b^{-(r+P)}\} \\ &\qquad + \{d_{-(r+1)}b^{-(r+P+1)} + \cdots + d_{-(r+P)}b^{-(r+2P)}\} + \cdots \\ &= \{d_l b^l + \cdots + d_{-r}b^{-r}\} \\ &\qquad + \{d_{-(r+1)}b^{-(r+1)} + \cdots + d_{-(r+P)}b^{-(r+P)}\}\{1 + b^{-P} + b^{-2P} + \cdots\}. \end{aligned}$$

Using Corollary 2.2.4 to sum the geometric series $1 + b^{-P} + b^{-2P} + \cdots$, we get

$$\begin{aligned} q &= \{d_l b^l + \cdots + d_{-r}b^{-r}\} \\ &\qquad + \{d_{-(r+1)}b^{-(r+1)} + \cdots + d_{-(r+P)}b^{-(r+P)}\}\{1 - b^{-P}\}^{-1}. \end{aligned}$$

Since the right-hand side here consists of sums, differences, products, and quotients of rational numbers, it is a rational number. Therefore q is rational if its base b representation is ultimately periodic.

THEOREM 6.5.2
Any nonnegative real number q may be expressed in base b as

$$q = d_l b^l + \cdots + d_0 + d_{-1}b^{-1} + \cdots = d_l \ldots d_0 . d_{-1} \ldots,$$

where each $d_t = 0, 1, \ldots, b-1$. *This representation is ultimately periodic if and only if q is a rational number.*

Example 6.5.3
If $q = 1507/200$ (numerator and denominator are given in base ten), express q in base five, as in Theorem 6.5.2.

SOLUTION We do all calculations in base ten. Write $q = M + q_0$, where $0 \leq q_0 < 1$. Then $M = 7$, and expressed in base five, $M = 12_5$. According to equation (4), d_{-1} is the integral part and q_{-1} is the fractional part of $5q_0$. Then d_{-2} is the integral part and q_{-2} is the fractional part of $5q_{-1}$. Then d_{-3} is the integral part and q_{-3} is the fractional part of $5q_{-2}$. And so on. We continue until periodicity emerges. Here $q_0 = 107/200$. Then:

$$\begin{aligned} 5q_0 &= 535/200 = d_{-1} + q_{-1}, & d_{-1} &= 2, & q_{-1} &= 135/200; \\ 5q_{-1} &= 675/200 = d_{-2} + q_{-2}, & d_{-2} &= 3, & q_{-2} &= 75/200; \\ 5q_{-2} &= 375/200 = d_{-3} + q_{-3}, & d_{-3} &= 1, & q_{-3} &= 175/200; \\ 5q_{-4} &= 875/200 = d_{-4} + q_{-4}, & d_{-4} &= 4, & q_{-4} &= 75/200. \end{aligned}$$

Thus $q_{-4} = q_{-2}$. Since

$$q_{-2} = d_{-3}5^{-1} + q_{-3}5^{-1} = d_{-3}5^{-1} + d_{-4}5^{-2} + q_{-4}5^{-2},$$

we have

$$q_{-2} = 1 \cdot 5^{-1} + 4 \cdot 5^{-2} + q_{-2}5^{-2}. \tag{15}$$

Beginning with q_0 and using (15) repeatedly, we get

$$\begin{aligned} q_0 &= 2 \cdot 5^{-1} + q_{-1}5^{-1} = 2 \cdot 5^{-1} + 3 \cdot 5^{-2} + q_{-2}5^{-2} \\ &= 2 \cdot 5^{-1} + 3 \cdot 5^{-2} + \{1 \cdot 5^{-3} + 4 \cdot 5^{-4}\} + q_{-2}5^{-4} \\ &= 2 \cdot 5^{-1} + 3 \cdot 5^{-2} + \{1 \cdot 5^{-3} + 4 \cdot 5^{-4}\} + \{1 \cdot 5^{-5} + 4 \cdot 5^{-6}\} + q_{-2}5^{-6} \\ &= \cdots \end{aligned}$$

Thus $q_0 = .23\overline{14}_5$, and hence $q = 12.23\overline{14}_5$.

Example 6.5.4 Express $\sqrt{2}$ in base five.

SOLUTION Again we calculate in base ten. From §5.5 we know that $\sqrt{2} = 1.414213562 + \epsilon$, where the error ϵ satisfies $0 < \epsilon < 10^{-9}$. Thus $q_0 = \sqrt{2} - 1 = .414213562 + \epsilon$. Then $5q_0 = 2.071067810 + 5\epsilon$, so that $d_{-1} = 2$, $q_{-1} = .071067810 + 5\epsilon$. Then $5q_{-1} = .355339050 + 5^2\epsilon$, so that $d_{-2} = 0$, $q_{-2} = .355339050 + 5^2\epsilon$. Then $5q_{-2} = 1.776695250 + 5^3\epsilon$, so that $d_{-3} = 1$, $q_{-3} = .776695250 + 5^3\epsilon$. We may continue in this way as long as the error terms ϵ, 5ϵ, $5^2\epsilon$, $5^3\epsilon$, ... do not become too large. We get $d_{-4} = 3$, $d_{-5} = 4$, $d_{-6} = 2$, $d_{-7} = 0$, $d_{-8} = 2$, $d_{-9} = 0$, and $d_{-10} = 4$, with $q_{-10} = .31640625 + 5^{10}\epsilon$. Then $5q_{-10} = 1.58203125 + 5^{11}\epsilon$. Since $5^{11}\epsilon < 5^{11}/10^9 = 5^2(5/10)^9 = 25/512 < 1/20$, we are certain that $d_{-11} = 1$ and $q_{-11} = .58203125 + 5^{11}\epsilon$. Then $5q_{-11} = 2.91015625 + 5^{12}\epsilon$. We now only know that $5^{12}\epsilon < 5^3(5/10)^9 = 125/512 < 1/4$, so that we only deduce that $5q_{-11}$ lies between 2.9 and 3.2. Therefore d_{-12} is either 2 or 3. This uncertainty prevents calculation of q_{-12} and of further digits. However, we have proved that

$$1.201342020412_5 < \sqrt{2} < 1.201342020414_5\,.$$

Uniqueness

We have not established that the base b representation of a nonnegative rational or real number q is unique. In fact, it isn't. For example, in base ten,

$$\begin{aligned} .999\ldots &= \frac{9}{10} + \frac{9}{10^2} + \frac{9}{10^3} + \cdots = \frac{9}{10}\left(1 + \frac{1}{10} + \frac{1}{10^2} + \frac{1}{10^3} + \cdots\right) \\ &= \frac{9}{10}\frac{1}{1 - 1/10} = 1, \end{aligned}$$

using Corollary 2.2.4 to sum the geometric series $1 + 10^{-1} + 10^{-2} + 10^{-3} + \cdots$. Thus $1 = 1.0000\ldots = .999999\ldots$. More generally, suppose that in base b,

$$q = d_l \ldots d_0 . d_{-1} \ldots d_{-k}(b-1)(b-1)(b-1)\ldots,$$

that is, every digit from some point onwards equals $b - 1$. Here d_{-k} is the last digit not equal to $b - 1$. Possibly k is negative; this will happen when every digit to the right of the b point is $b - 1$. There always is a digit not equal to $b - 1$, since we may take $d_l = 0$ by adding a zero digit at the left. Then we have

$$\begin{aligned} q &= d_l b^l + \cdots + d_{-k}b^{-k} + (b-1)b^{-(k+1)} + (b-1)b^{-(k+2)} + \cdots \\ &= d_l b^l + \cdots + d_{-k}b^{-k} + (b-1)b^{-(k+1)}(1 + b^{-1} + b^{-2} + \cdots) \\ &= d_l b^l + \cdots + d_{-k}b^{-k} + (b-1)b^{-(k+1)}(1 - b^{-1})^{-1} \\ &= d_l b^l + \cdots + d_{-(k-1)}b^{-(k-1)} + d_{-k}b^{-k} + b^{-k} \\ &= d_l b^l + \cdots + d_{-(k-1)}b^{-(k-1)} + (d_{-k} + 1)b^{-k} \\ &= d_l \ldots d_0 . d_{-1} \ldots d_{-(k-1)}(d_{-k} + 1)000000\ldots. \end{aligned}$$

(Again we used Corollary 2.2.4, to sum the geometric series $1 + b^{-1} + b^{-2} + \cdots$.) Therefore, if every base b digit in q from some point onwards equal $b - 1$, then q has a second, *terminating*, representation in base b in which all digits from this point onwards are 0.

Now we can give a uniqueness theorem.

THEOREM 6.5.5

Let the nonnegative real number q be represented in base b as in Theorem 6.5.2, *subject to the condition that from no point onward are all digits equal to $b - 1$. Then these digits are uniquely determined by q.*

PROOF As we saw above, in base b

$$.(b-1)(b-1)(b-1)\ldots = 1.0000\ldots = 1.$$

Therefore, if at least one digit in $.d_{-1}d_{-2}d_{-3}\ldots$ is less than $b - 1$, then

$$.d_{-1}d_{-2}d_{-3}\ldots < .(b-1)(b-1)(b-1)\ldots = 1.$$

Suppose that q has two base b representations of the type described in Theorem 6.5.2. Inserting zero digits at the left, we may assume l is the same for both representations. Thus

$$q = d_l b^l + d_{l-1} b^{l-1} + \cdots = d_l' b^l + d'_{l-1} b^{l-1} + \cdots,$$

where $0 \le d_t < b$, $0 \le d_t' < b$ for all t, infinitely many d_t are different from $b - 1$, and infinitely many d_t' are different from $b - 1$. Then

$$d_l + d_{l-1}b^{-1} + d_{l-2}b^{-2} + \cdots = d_l' + d'_{l-1}b^{-1} + d'_{l-2}b^{-2} + \cdots,$$

$$d_l.d_{l-1}d_{l-2}\ldots = d_l'.d'_{l-1}d'_{l-2}\ldots. \tag{16}$$

By the remark above, $.d_{l-1}d_{l-2}\ldots < 1$ and $.d'_{l-1}d'_{l-2}\ldots < 1$. Thus d_l is the integral part of the left-hand side of (16) and d_l' is the integral part of the right-hand side. Therefore $d_l = d_l'$. Subtracting d_l from each side of (16), then multiplying by b, we get

$$d_{l-1}.d_{l-2}d_{l-3}\ldots = d'_{l-1}.d'_{l-2}d'_{l-3}\ldots.$$

As with (16), this implies $d_{l-1} = d'_{l-1}$. Subtracting d_{l-1} from each side, then multiplying by b gives

$$d_{l-2}.d_{l-3}d_{l-4}\ldots = d'_{l-2}.d'_{l-3}d'_{l-4}\ldots,$$

implying $d_{l-2} = d'_{l-2}$. Continuing in this fashion, we prove $d_t = d_t'$ for every t. □

Base *b* representations of rational functions.
Laurent series (Optional)

Let $b(\lambda) \in \mathfrak{F}[\lambda]$ be a nonconstant polynomial. We saw in §4.2 that any polynomial $\in \mathfrak{F}[\lambda]$ is representable in the form $d_l b^l + d_{l-1} b^{l-1} + \cdots + d_0$, where polynomials $d_l(\lambda), d_{l-1}(\lambda), \ldots, d_0(\lambda) \in \mathfrak{F}[\lambda]$ and satisfy

$$|d_t| < |b| \tag{1'}$$

for every t. (The bars now denote the absolute value of rational functions described in §6.4.) Parallel to the discussion of rational numbers, we now extend this representation of polynomials to rational functions.

If $m(\lambda)$, $n(\lambda)$ are polynomials, with $n(\lambda)$ nonzero, we call m/n a *proper fraction* if $|m/n| < 1$; this simply means $\deg m < \deg n$.

Let $q_0(\lambda) = m_0(\lambda)/n(\lambda)$ be a proper fraction, where $m_0(\lambda), n(\lambda) \in \mathfrak{F}[\lambda]$. We now give a recursive procedure for computing polynomials $d_{-1}(\lambda)$, $d_{-2}(\lambda)$, $d_{-3}(\lambda)$, $\ldots \in \mathfrak{F}[\lambda]$, all satisfying (1′), and proper fractions $q_{-1}(\lambda)$, $q_{-2}(\lambda)$, $q_{-3}(\lambda)$, …, with

$$q_t(\lambda) = m_t(\lambda)/n(\lambda), \qquad m_t(\lambda) \in \mathfrak{F}[\lambda], \tag{3′}$$

for each t, such that (2.k) above holds for any nonnegative integer k.

If we have found polynomials $d_{-1}, \ldots, d_{-k}$ and proper fraction $q_{-k} = m_{-k}/n$ such that (2.k) holds, divide $m_{-k}b$ by n to get polynomials $d_{-(k+1)}$, $m_{-(k+1)}$ with

$$m_{-k}b = d_{-(k+1)}n + m_{-(k+1)}, \qquad \deg m_{-(k+1)} < \deg n.$$

Setting $q_{-(k+1)} = m_{-(k+1)}/n$, it is clear that $q_{-(k+1)}$ is a proper fraction, and also that

$$q_{-k}b = d_{-(k+1)} + q_{-(k+1)}, \qquad |q_{-(k+1)}| < 1, \tag{4′}$$

$$q_{-k} = d_{-(k+1)}b^{-1} + q_{-(k+1)}b^{-1}. \tag{5}$$

Since $|b(\lambda)| > 1$, we have $|q_{-(k+1)}| < |b|$, so that $|q_{-(k+1)}/b| < 1$. If we had $|d_{-(k+1)}| \geq |b|$, then $|d_{-(k+1)}/b| \geq 1$ and the principle of domination would give

$$|q_{-k}| = |d_{-(k+1)}/b + q_{-(k+1)}/b| = |d_{-(k+1)}/b| \geq 1,$$

contradicting $|q_{-k}| < 1$. Thus polynomial $d_{-(k+1)}$ satisfies (1′). Substituting (5) into (2.k) then gives $(2.(k+1))$. Thus we may increase k; hence (2.k) holds for every $k \geq 0$.

If $q(\lambda) = m(\lambda)/n(\lambda)$ is any rational function, we divide to get $m(\lambda) = M(\lambda)n(\lambda) + m_0(\lambda)$, where $M(\lambda)$, $m_0(\lambda)$ are polynomials with $\deg m_0 < \deg n$. Setting $q_0(\lambda) = m_0(\lambda)/n(\lambda)$, we have $q(\lambda) = M(\lambda) + q_0(\lambda)$. Writing $M(\lambda)$ in base b, as in equation (6) above, and combining this with (2.k), we obtain the following rational function version of Theorem 6.5.1. [Or else we may apply $(2.(k + l + 1))$ to q/b^{l+1} for sufficiently large l, as in Exercise 9.]

THEOREM 6.5.1 **(*Rational function version*)**
If $q(\lambda) \in \mathfrak{F}(\lambda)$ is a rational function, there is a subscript $l \geq 0$, polynomials $d_l(\lambda)$, $d_{l-1}(\lambda)$, …, $d_0(\lambda)$, $d_{-1}(\lambda)$, $\ldots \in \mathfrak{F}[\lambda]$ satisfying $|d_t| < |b|$ for each t, and rational functions $q_0(\lambda), q_{-1}(\lambda), q_{-2}(\lambda), \ldots \in \mathfrak{F}(\lambda)$ satisfying $|q_t(\lambda)| < 1$ for each t, such that equation (7) above holds for any integer $k \geq 0$. If $q(\lambda) = m(\lambda)/n(\lambda)$, where $m(\lambda)$, $n(\lambda) \in \mathfrak{F}[\lambda]$, $n(\lambda) \neq 0$, then $q_{-k}(\lambda) = m_{-k}(\lambda)/n(\lambda)$, where $m_{-k}(\lambda) \in \mathfrak{F}[\lambda]$, for each $k \geq 0$.

As with rational numbers, we may let k grow indefinitely large in (7), and so obtain an infinite series (10) for $q(\lambda)$. What is the meaning of this infinite series? The meaning is as before: If a tolerance ϵ (*a* positive real number) is given, there is an integer K, dependent on ϵ, such that the truncated series

$$d_l(\lambda)b(\lambda)^l + \cdots + d_{-k}(\lambda)b(\lambda)^{-k} \tag{8'}$$

provides an approximation to $q(\lambda)$ within tolerance ϵ for all $k \geq K$; i.e.,

$$|q(\lambda) - (d_l(\lambda)b(\lambda)^l + \cdots + d_{-k}(\lambda)b(\lambda)^{-k})| < \epsilon \quad \text{for all } k \geq K. \tag{9'}$$

The proof of (9′) is the same as the proof of (9) and uses $|q_{-k}(\lambda)| < 1$, $|b(\lambda)| > 1$. We now say that the right-hand side of (10) is a *convergent series of rational functions,* and that it *converges in the field* $\mathfrak{F}(\lambda)$ *to* $q(\lambda)$. Parallel to the rational number case, we may call $d_l(\lambda)$, $d_{l-1}(\lambda)$, ... the *digits* of $q(\lambda)$ in base $b(\lambda)$ [although, in fact, the term "digit" is not often used when working with rational functions].

The discussion for rational functions is, so far, parallel to the discussion for rational numbers. There are some differences, however, and we now describe two.

The first difference is this: The base b representation of $q(\lambda)$,

$$q(\lambda) = d_l(\lambda)b(\lambda)^l + \cdots + d_{-k}(\lambda)b(\lambda)^{-k} + \cdots, \quad |d_t| < |b| \text{ for all } t, \tag{10}$$

is unique, and not just "almost" unique, as was the case for numbers; i.e., if also

$$q(l) = d_l(\lambda)'b(\lambda)^l + \cdots + d_{-k}(\lambda)'b(\lambda)^{-k} + \cdots, \quad |d_t'| < |b| \text{ for all } t,$$

then the polynomials $d_t(\lambda)$, $d_t(\lambda)'$ are equal, for each t. The proof is outlined in Exercise 19 and uses the principle of domination.

A second difference is this: Although $q_0(\lambda) = m_0(\lambda)/n(\lambda)$, $q_{-1}(\lambda) = m_{-1}(\lambda)/n(\lambda)$, ... are proper fractions with the same denominator, it does not follow that $q_{-r}(\lambda) = q_{-s}(\lambda)$ for a pair of distinct subscripts r, s. This is because the polynomials $m_0(\lambda)$, $m_{-1}(\lambda)$, ... may be distinct [even though all have degree $< \deg n(\lambda)$], since in general there are infinitely many polynomials in $\mathfrak{F}[\lambda]$ of fixed degree. For most fields $\mathfrak{F}$, therefore, the representation of $q(\lambda)$ in base $b(\lambda)$ need not be ultimately periodic. For example, when $\mathfrak{F}$ is $\mathbf{C}$, the representation of $\lambda/(\lambda - 1)^2$ in base λ is

$$\lambda(\lambda - 1)^{-2} = \lambda^{-1} + 2\lambda^{-2} + 3\lambda^{-3} + 4\lambda^{-4} + \cdots,$$

with digit $d_{-k} = k$ for every natural number k. However, if $\mathfrak{F}$ is a *finite field,* i.e., a field containing only finitely many elements, then $m_{-r}(\lambda) = m_{-s}(\lambda)$ must hold for distinct subscripts r, s. This is because there are only a finite number of polynomials of fixed degree with coefficients in a finite field, since only finitely many values are possible for each coefficient. Thus the base b representation of a rational function $q(\lambda) \in \mathfrak{F}(\lambda)$ must be ultimately periodic when $\mathfrak{F}$ is a finite field, and in this case the analogy between $\mathbf{Q}$ and $\mathfrak{F}(\lambda)$ is strongest.

In other ways, too, the analogy between **Q** and $\mathfrak{F}(\lambda)$ is strongest when $\mathfrak{F}$ is a finite field; in recent years this analogy has been exploited by mathematicians at the research level in branches of algebra and geometry known as *algebraic function theory* and *algebraic geometry.*

Many examples of finite fields will be seen in Chapter 7. One example has already been seen, in Exercise 3 of §1.5.

Allowing $\mathfrak{F}$ to be an infinite field again, let us now take $b(\lambda) = \lambda - \alpha$, where α is a *constant*, i.e., $\alpha \in \mathfrak{F}$. Since a digit d_t then satisfies $|d_t| < 1$, each polynomial $d_t(\lambda)$ must be a *constant*. Formulas (7) and (10) then become

$$q(\lambda) = d_l(\lambda - \alpha)^l + \cdots + d_{-k}(\lambda - \alpha)^{-k} + q_{-k}(\lambda)(\lambda - \alpha)^{-k}, \quad (17)$$

$$q(\lambda) = d_l(\lambda - \alpha)^l + \cdots + d_{-k}(\lambda - \alpha)^{-k} + \cdots, \quad (18)$$

where each $d_t \in \mathfrak{F}$, and where $q_{-k}(\lambda) = m_{-k}(\lambda)/n(\lambda)$ is a proper fraction with the same denominator $n(\lambda)$ as $q(\lambda)$. The series (18) is frequently encountered when $\mathfrak{F}$ is the complex number field **C**, and then it is a special case of a type of series called *Laurent series.* To be specific, we call (18) a *Laurent expansion of* $q(\lambda)$ *about* α. Notice that Taylor series for polynomials, as explained in §4.2, are special cases of (17) and (18) in which $d_{-k} = 0$ and $q_{-k}(\lambda) = 0$ for every natural number k.

Laurent series in another form will be encountered in §8.4.

The reader, if he takes a course in the *theory of a complex variable,* will encounter Laurent series, not only for rational functions, but also for many other types of functions. Applications of these series occur in mathematical analysis, engineering, and advanced branches of algebra and number theory. One application, often seen in introductory complex variable theory, is to develop procedures for evaluating numerous definite integrals of real-valued functions.

Substitution in Laurent series (Optional)

Let the field $\mathfrak{F}$ be **C**, and let z be a complex number. If we substitute z for λ in the Laurent series (18), do we obtain

$$q(z) = d_l(z - \alpha)^l + \cdots + d_{-k}(z - \alpha)^{-k} + \cdots? \quad (19)$$

Assume that $n(z) \neq 0$ and $z \neq \alpha$, i.e., the substitution does not produce a zero denominator. Before answering this, we must specify the meaning to be given to (19). We adopt the definition of convergence used before except that, because the terms in (19) are complex numbers, we take as the appropriate absolute value the usual absolute value of complex numbers. Letting the bars $|\ |$ now denote this absolute value, we interpret (19) in this sense: If a positive number ϵ is given, there is an integer K (dependent on ϵ) such that

$$|q(z) - (d_l(z - \alpha)^l + \cdots + d_{-k}(z - \alpha)^{-k})| < \epsilon \quad \text{for all } k \geq K.$$

In other words: sufficiently many terms from the right-hand side of (19) provide, within tolerance ϵ, an approximation of $q(z)$ which continues to hold with this precision when more terms are used. (See Figure 6.5.2.)

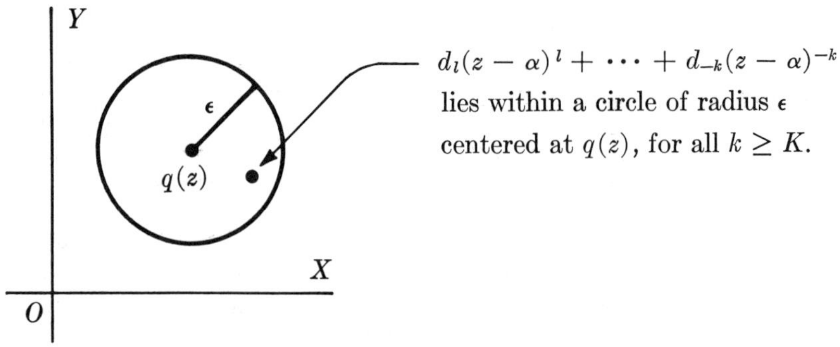

Figure 6.5.2

Does the series $d_l(z - \alpha)^l + \cdots$ converge to $q(z)$ in this sense? Sometimes it does, and sometimes it does not. Let C be the smallest circle in the complex plane centered at α and enclosing all roots of the denominator $n(\lambda)$ of $q(\lambda)$. The result is this: The series on the right-hand side of (19) converges to $q(z)$ if z belongs to the exterior of C, and does not converge if z belongs to the interior of C. If z is on the circumference of C, convergence may or may not occur. (Note: "Exterior" means strictly outside; "interior," strictly inside.) The proof that the right-hand side of (19) does not converge when z is inside C involves ideas somewhat too advanced to be included in this book. In some special cases easy proofs may be given using Lemma (2.2.3) giving the sum of a finite geometric series; see Exercise 15. The proof of convergence when z is outside C involves technical difficulties best circumvented by using *partial fractions*, to be explained in §8.3. However, a special case has a straightforward proof, which we give here. If $q(\lambda) = m(\lambda)/n(\lambda)$ is in lowest terms, we call a root of $n(\lambda)$ a *pole* of $q(\lambda)$, and a simple root of $n(\lambda)$ a *simple pole* of $q(\lambda)$. We prove here the following Theorem 6.5.6 in the special case when all poles of $q(\lambda)$ are simple poles; the proof in the general case will be given in §8.3.

THEOREM 6.5.6

Let α be a complex number, $q(\lambda)$ a rational function with complex coefficients, and C the smallest circle centered at α enclosing all poles of $q(\lambda)$. If z is a complex number belonging to the exterior of C, the Laurent series (18) of $q(\lambda)$ about α, when evaluated at z, is a series of complex numbers converging to $q(z)$.

PROOF (when $q(\lambda)$ has simple poles only) Let $q(\lambda) = m(\lambda)/n(\lambda)$ be in lowest terms. If $n(\lambda)$ has degree T and roots $z_1, \ldots, z_T$, then

$$n(\lambda) = (\lambda - z_1)\cdots(\lambda - z_T), \tag{20}$$

and $z_1, \ldots, z_T$ are (by hypothesis) distinct.

Our proof makes use of an identity called the *Lagrange interpolation formula*. For the moment let $A(\lambda)$ be a polynomial with $\deg A(\lambda) < T$. The Lagrange interpolation formula is this polynomial equality:

$$A(\lambda) = \sum_{s=1}^{T} \frac{(\lambda - z_1)\cdots(\lambda - z_{s-1})(\lambda - z_{s+1})\cdots(\lambda - z_T)}{(z_s - z_1)\cdots(z_s - z_{s-1})(z_s - z_{s+1})\cdots(z_s - z_T)} A(z_s).$$

To prove this, let $B(\lambda)$ be the polynomial on the right-hand side of this equation. Its degree is plainly at most $T - 1$. If $B(\lambda)$ is evaluated at z_j, all terms in the sum become zero except the term with $s = j$, and it becomes $A(z_j)$. Thus $A(z_j) = B(z_j)$ for $j = 1, \ldots, T$. The polynomial $A - B$ therefore has degree at most $T - 1$, and it has roots $z_1, \ldots, z_T$. Since the number of roots exceeds the degree, it must be the zero polynomial. Thus $A(\lambda) = B(\lambda)$, establishing the Lagrange interpolation formula.

For any natural number k, we have (7), i.e.,

$$q(\lambda) = \frac{m(\lambda)}{n(\lambda)} = d_l(\lambda - \alpha)^l + \cdots + \frac{d_{-k}}{(\lambda - \alpha)^k} + \frac{m_{-k}(\lambda)}{(\lambda - \alpha)^k n(\lambda)},$$

with $\deg m_{-k}(\lambda) < \deg n(\lambda)$. Multiplying through by $(\lambda - \alpha)^k n(\lambda)$ leads to

$$m(\lambda)(\lambda - \alpha)^k = P(\lambda)n(\lambda) + m_{-k}(\lambda), \qquad \deg m_{-k}(\lambda) < \deg n(\lambda),$$

where $P(\lambda) = d_l(\lambda - \alpha)^{l+k} + \cdots + d_{-k}$ is a polynomial. Evaluating each side of this polynomial equation at z_s and using $n(z_s) = 0$, we get

$$m(z_s)(z_s - \alpha)^k = m_{-k}(z_s), \qquad s = 1, \ldots, T. \tag{21}$$

Setting $A(\lambda) = m_{-k}(\lambda)$, the Lagrange interpolation formula and equation (21) give

$$m_{-k}(\lambda) = \sum_{s=1}^{T} \frac{(\lambda - z_1)\cdots(\lambda - z_{s-1})(\lambda - z_{s+1})\cdots(\lambda - z_T)}{(z_s - z_1)\cdots(z_s - z_{s-1})(z_s - z_{s+1})\cdots(z_s - z_T)} m(z_s)(z_s - \alpha)^k.$$

Using (20) and this expression for $m_{-k}(\lambda)$, we get

$$\frac{m_{-k}(\lambda)}{(\lambda - \alpha)^k n(\lambda)}$$

$$= \sum_{s=1}^{T} \frac{m(z_s)}{(z_s - z_1)\cdots(z_s - z_{s-1})(z_s - z_{s+1})\cdots(z_s - z_T)} \left(\frac{z_s - \alpha}{\lambda - \alpha}\right)^k \frac{1}{\lambda - z_s}. \tag{22}$$

Evaluating each side of (22) at z yields

$$\frac{q_{-k}(z)}{(z-\alpha)^k} = \sum_{s=1}^{T} C_s \left(\frac{z_s - \alpha}{z - \alpha}\right)^k,$$

where C_s is a complex number dependent on $z, z_1, \ldots, z_T$, but *independent of* k, for $s = 1, \ldots, T$. Hence, by the triangle inequality of §2.4,

$$\left|\frac{q_{-k}(z)}{(z-\alpha)^k}\right| \le \sum_{s=1}^{T} |C_s| \left|\frac{z_s - \alpha}{z - \alpha}\right|^k.$$

Because z lies outside the circle C, we have $|z_s - \alpha| < |z - \alpha|$, and hence $|z_s - \alpha|/|z - \alpha| < 1$. Therefore the nonnegative real number $(|z_s - \alpha|/|z - \alpha|)^k$ approaches zero as k increases, and this holds for each $s = 1, \ldots, T$. If we are given a tolerance ϵ, there will therefore be an integer K (dependent on ϵ) such that

$$\left|\frac{q_{-k}(z)}{(z-\alpha)^k}\right| < \epsilon \qquad \text{for all } k \ge K.$$

For any $k \ge K$ we then have

$$|q(z) - (d_l(z-\alpha)^l + \cdots + d_{-k}(z-\alpha)^{-k}| = |q_{-k}(z)(z-\alpha)^{-k}| < \epsilon. \quad \square$$

EXERCISES 6.5

I

1. Find base b representations of 1/3, for $b = 2, 3, 4, 5, 6$. Do the same for 1/2, 3/4, 2/5, 1/6, 13/3. Find, in general, the base b representation of $(b - 1)^{-1}$.

2. What rational number is q if, in base five, $q = 2.12\overline{234}$? Repeat, if, in base seven, $q = 55.\overline{13}$?

3. If a nonnegative real number is written in base b, at least two of its digits must be equal. Why? Does this prove that the representation will be ultimately periodic? Why is the base b representation of $\sqrt{2}$ not ultimately periodic? (*Optional:* When a rational function is written in base b, must at least two of its digits be equal?)

4. Find two representations in base ten for 3/4. Do the same in base twelve. Can this be done in base nine?

5. Use the method outlined in Example 6.5.4 to express $\sqrt{2}$ in bases two, six, eight, nine, sixteen. Are any of these representations ultimately periodic?

6. (*Optional*) Find the Laurent series for $1/\lambda$ in the base $\lambda + 1$. In this Laurent series substitute the complex number z for λ. For which z does the resulting series of complex numbers converge to $1/z$? Repeat, for $\lambda/(\lambda - 1)^2$, in base λ.

7. (*Optional*) Express $3/4$ and $\sqrt{2}$ in base λ in the form described in Theorem 6.5.1 (rational function version). Are these representations ultimately periodic? The base field is $\mathbf{R}$.

II

8. Let the field $\mathcal{K}$ be either rational numbers or rational functions. If the sum of two proper fractions in $\mathcal{K}$ is always a proper fraction, which field is $\mathcal{K}$?

9. Let q be a nonnegative rational (or real) number.
(i) Show, for a sufficiently large natural number l, that $0 \leq q/b^{l+1} < 1$.
(ii) Let $Q_0 = q/b^{l+1}$. Apply (2.k) to Q_0, to get integers $\delta_{-1}, \delta_{-2}, \ldots$ and rational (or real) numbers $Q_{-1}, Q_{-2}, \ldots$ satisfying $0 \leq \delta_t < b$ and $0 \leq Q_t < 1$, for each t, such that

$$Q_0 = \delta_{-1}b^{-1} + \delta_{-2}b^{-2} + \cdots + \delta_{-k}b^{-k} + Q_{-k}b^{-k}$$

holds for each integer $k \geq 0$.
(iii) Set $d_l = \delta_{-1}$, $d_{l-1} = \delta_{-2}$, etc., $q_l = Q_{-1}$, $q_{l-1} = Q_{-2}$, etc. From the displayed formula in (ii) deduce that

$$q = d_l b^l + d_{l-1}b^{l-1} + \cdots + d_{l-k+1}b^{l-k+1} + q_{l-k+1}b^{l-k+1},$$

for each integer $k \geq 0$.
(iv) Let $q = m/n$. If $k \leq l$, rewrite the formula in (iii) as

$$m - (d_l b^l + \cdots + d_{l-k+1}b^{l-k+1})n = q_{l-k+1}b^{l-k+1}n.$$

Deduce that the proper fraction q_{l-k+1} has denominator $b^{l-k+1}n$.
(v) Replace k in the formula in (iii) with $k + l + 1$, then deduce equation (7) in the text. If $q = m/n$, multiply equation (7) by b^k, and deduce that q_{-k} has denominator n.
(vi) Repeat (i)–(v) for rational functions.

10. Let $q = m/n$. Deduce (7) for rational numbers (and rational functions) in this way: Use the division algorithm to write $mb^k = Pn + m_{-k}$, where $P \in \mathbf{Z}$, $m_{-k} \in \mathbf{Z}$, $0 \leq m_{-k} < n$. Write P in base b by the methods of §4.2, say as $P = d_l b^{l+k} + d_{l-1}b^{l+k-1} + \cdots + d_{-k}$. Then deduce (7).

11. Using the method outlined in Exercise 9, recalculate the base five representation of $q = 1507/200$, i.e., apply the method in Example 6.5.3 to $q/5^2$.

12. (i) Suppose q is a nonnegative rational or real number, written in

two ways as in equation (7):

$$q = d_l b^l + \cdots + d_{-k} b^{-k} + q_{-k} b^{-k} = d_l' b^l + \cdots + d_{-k}' b^{-k} + q_{-k}' b^{-k},$$

where d_t, d_t' are integers with $0 \leq d_t, d_t' < b$ for each t, and $0 \leq q_{-k}$, $q_{-k}' < 1$. Deduce that $d_t = d_t'$ for each t. Imitate the uniqueness proof in the text.
(ii) If $q = d_l \ldots d_0 . d_{-1} \ldots d_{-k}(b-1)(b-1)(b-1)\ldots$, with all digits to the right of d_{-k} equal to $b - 1$, show that $q = d_l b^l + \cdots + d_{-k} b^{-k} + q_{-k} b^{-k}$ with $q_{-k} = 1$.
(iii) If nonnegative real number q is representable in some base b in more than one way, prove that it is rational. Conversely, if q is rational, prove that it is representable in some base b in two ways.

13. Show that it is theoretically impossible to calculate $\sqrt{2}$ in base five to fourteen digits, knowing only the ten digit base ten approximation $\sqrt{2} = 1.414213562$. (Hint: If we could calculate fourteen digits of $\sqrt{2}$ in base five, we would have a rational approximation of $\sqrt{2}$ within a tolerance of 5^{-13}. Converting this to base ten would give a decimal approximation of $\sqrt{2}$ within tolerance 5^{-13}. Since $5^{-13} < 10^{-9}$, we would have a decimal approximation of $\sqrt{2}$ more accurate than 1.414213562.)

14. (*Optional*) If $q \in \mathbf{C}$, prove that all digits $d_{-1}, d_{-2}, d_{-3}, \ldots$ in Theorem 6.5.1 (rational function version over $\mathbf{C}$, in base λ) are zero.

15. (*Optional*) Let α, β be distinct complex numbers. Write out the Laurent series (18) for $(\lambda - \alpha)^{-1}$ in the base $\lambda - \beta$. If complex number z is substituted for λ in this series, show that the resulting series of complex numbers converges to $(z - \alpha)^{-1}$ if $|\alpha - \beta| < |z - \beta|$, and does not converge if $|\alpha - \beta| > |z - \beta|$. Use Lemma 2.2.3.

16. Prove: If a nonnegative real number has an ultimately periodic representation in one base, it has an ultimately periodic representation in every base.

17. If rational number $q = .d_{-1}d_{-2}\ldots d_{-k}\overline{d_{-(k+1)}\ldots d_{-(k+P)}}$, in base b, prove that

$$q = \frac{A(b^P - 1) + B}{b^k(b^P - 1)},$$

where integers A, B, in base b, are:

$$A = d_{-1}d_{-2}\ldots d_{-k}, \qquad B = d_{-(k+1)}d_{-(k+2)}\ldots d_{-(k+P)}.$$

Conclude that the denominator n of q satisfies $n \mid b^k(b^P - 1)$. In particular, if q is purely periodic, deduce that the period P and denominator n of q are related by $b^P \equiv 1 \pmod{n}$.

III

18. (For readers who have reached infinite series in calculus) If $b > 1$, $0 \leq d_t < b$ for each t, prove that $d_{-1}b^{-1} + d_{-2}b^{-2} + \cdots$ is always a convergent series in the real numbers. Use a comparison test, comparing with $b^{-1} + b^{-2} + \cdots$.

19. Let q be a rational function, b a nonconstant polynomial, both q, b in $\mathfrak{F}(\lambda)$. If

$$q = d_l b^l + d_{l-1}b^{l-1} + \cdots = d_l'b^l + d'_{l-1}b^{l-1} + \cdots$$

are two representations of q in base b, prove that $d_t = d_t'$ for every t. (Hint: Rewrite as $(d_l - d_l') = (d'_{l-1} - d_{l-1})b^{-1} + \cdots$. Use the strong triangle law, $|d_t| < |b|$, $|d_t'| < |b|$, for each t, to deduce that the absolute value of the right-hand side is <1. If the left-hand side is nonzero, show that its absolute value is ≥ 1. From this deduce $d_l = d_l'$. Cancel the term involving d_l, then deduce $d_{l-1} = d'_{l-1}$. Continue in this manner.)

20. If $\mathfrak{F}$ is a finite field having precisely N elements, estimate the length of the period when rational function $q(\lambda) \in \mathfrak{F}(\lambda)$ is written in base $b(\lambda) \in \mathfrak{F}[\lambda]$. The estimate should involve N and the degree of the denominator of $q(\lambda)$.

21. Suppose $q(\lambda) \in \mathbf{R}(\lambda)$ and that every digit in the Laurent expansion (18) for $q(\lambda)$ is positive in the sense of the ordering of $\mathbf{R}(\lambda)$ in §6.4. Is $q(\lambda)$ then positive in this ordering?

6.6 BASIC ARITHMETIC

Addition and subtraction of rational numbers with terminating base b representations are the same as for integers. We add or subtract digits, position by position, then reduce and carry, beginning at the extreme right-hand position. The discussion of addition and subtraction in §4.3 demonstrates this without difficulty, if we modify it slightly to apply to numbers of the form

$$d_l b^l + \cdots + d_0 + d_{-1}b^{-1} + \cdots + d_{-k}b^{-k} = d_l \ldots d_0 . d_{-1} \ldots d_{-k}$$

having a fixed number of digits to the right of the b point.

Suppose α and γ are nonnegative numbers with terminating base b representations, having k, k' digits respectively to the right of the b point. Note that if $\alpha = d_l \ldots d_0 . d_{-1} \ldots d_{-k}$, then $a = b^k \alpha = d_l \ldots d_0 d_{-1} \ldots d_{-k}$ is an integer. Thus $\alpha = ab^{-k}$ and similarly $\gamma = cb^{-k'}$, where a, c are integers. Then

$$\alpha\gamma = acb^{-(k+k')}. \tag{1}$$

Calculate the base b representation of the integer product ac, as in §4.3. Then (1) shows that $\alpha\gamma$ is got from ac by shifting the b point $k + k'$ places to the left. This proves the following rule: *if α, γ have respectively k, k' digits to the right of the b point, then $\alpha\gamma$ has $k + k'$ digits to the right of the b point.* This is the rule taught in grade school for multiplying decimals: multiply, treating the numbers as integers, then place the b point in the product by counting the number of digits to the right of the b point in the factors.

If α, β are positive rational numbers with terminating base b representations, how is a base b representation for α/β calculated? We wish to find base b numbers q, r with $\alpha = q\beta + r$, and r is to be small. Let $B = \beta b^s$, $A = \alpha b^s$, where the exponent s is chosen so that B is an integer, i.e., s shifts to the right of the b point in both α, β place it to the right of all digits in β. Suppose $A = \alpha b^s$ still has t digits to the right of the b point. Then Ab^t is an integer. Apply long division to the integers Ab^t, B to obtain integers Q, R with $Ab^t = QB + R$, $0 \leq R < B$. That is, apply long division to A, B treating A as an integer by ignoring its b point. Setting $q = Qb^{-t}$, $r = Rb^{-s-t}$, we have $\alpha = q\beta + r$ and $0 \leq r < \beta b^{-t}$. Note that q is obtained by moving the b point to the left in Q until it and A have the same number of digits after the b point, and r is obtained by moving the b point to the left in R until it has as many digits after the b point as A and β together have.

Thus we have the following rule, usually taught in grade school:

To compute q, r in $\alpha = q\beta + r$, shift the b point to the right of the digits in β, to get integer B, and shift it the same number of places to the right in α, to get A. Apply long division for integers to A, B, treating A as an integer by ignoring its b point, to get integral quotient Q and remainder R. Then shift the b point to the left in Q, R, obtaining q by giving Q as many digits after the b point as A has, and obtaining r by giving R as many digits after the b point as A and β together have.

If we increase the number of digits in α by adding zero digits at the right, then t increases. Thus βb^{-t} decreases, and since $0 \leq r < \beta b^{-t}$, r is forced to grow small. *We can therefore make the division as nearly exact as we please by adding sufficiently many zero digits to the right of α.*

Example Find q, r upon dividing $\alpha = 43.311$ by $\beta = 4.22$. (The base is ten.)

First step: In

$$4.22\overline{)43.311}$$

shift the decimal two places to the right to convert β into an integer. Thus $B = 422$, $A = 4331.1$.

Second step: Calculate quotient Q and remainder R, treating A as an integer. The details of the long division are:

$$\begin{array}{r|l} 422\overline{)4331.1} & \\ 4220\ 0 & 100 \\ \hline 111\ 1 & \\ 84\ 4 & 2 \\ \hline 26\ 7 & \end{array}$$

$$Q = 100 + 2 = 102, \quad R = 267.$$

Third step: Give Q the number of decimal places in A, getting $q = 10.2$, and give R the number of decimal places in A and β together, getting $r = .267$.

Thus $\alpha = 10.2\beta + .267$, or $\alpha/\beta = 10.2$ plus an error of $.267/\beta$. If we wish to calculate the quotient more exactly, write α as 43.31100. Still using $\beta = 4.22$, we get $A = 4331.100$, $B = 422$. Long division produces $Q = 10263$, $R = 114$. Giving Q the number of decimal places in A yields $q = 10.263$, and giving R the number of decimal places in A and β combined yields $r = .00114$. Thus now $\alpha/\beta = 10.263$ plus a much smaller error of $.00114/\beta$.

An alternative procedure for calculating $1/\beta$ can be based on Newton's method (§5.5). See Exercise 9.

How is arithmetic done with nonterminating base b representations? No "farthest to the right" positions then exist in which to begin adding, subtracting, multiplying digits, reducing and carrying. The procedure is to "round off" by deleting all digits from some point onwards, and perhaps raising the last retained digit by 1; e.g., rounding to two decimal places, .2341... becomes .23, and .2352... becomes .24. Sometimes one merely truncates instead of rounding, e.g. both .2341... and .2352... become .23. The numbers we work with are then approximately equal to those given, and hopefully performing the desired arithmetic on the rounded numbers gives results approximately equal to the true results. Usually, but not always, this will be the case. The following examples show how the triangle inequality may be used to give an *error analysis* estimating the maximum possible size of the error. The examples are in base ten.

(1) Calculate $\sqrt{2} + \sqrt{3}$ and $\sqrt{2}\cdot\sqrt{3}$. Rounded to the nearest second decimal place, $\sqrt{2} = 1.41$ and $\sqrt{3} = 1.73$. This means $\sqrt{2} = 1.41 + \epsilon_1$, $\sqrt{3} = 1.73 + \epsilon_2$, where $|\epsilon_1| < 5\cdot 10^{-3}$ and $|\epsilon_2| < 5\cdot 10^{-3}$. Adding and multiplying, and rounding the product 1.41×1.73 to two decimal places by writing it as $1.41 \times 1.73 = 2.44 + \epsilon_3$, $|\epsilon_3| < 5\cdot 10^{-3}$, we get

$$\sqrt{2} + \sqrt{3} = 3.14 + E_1, \qquad \sqrt{2}\cdot\sqrt{3} = 2.44 + E_2,$$

where $E_1 = \epsilon_1 + \epsilon_2$ and $E_2 = 1.41\epsilon_2 + 1.73\epsilon_1 + \epsilon_3$. The triangle inequality gives

$$|E_1| \leq |\epsilon_1| + |\epsilon_2| < 10^{-2},$$

$$|E_2| \leq 1.41\,|\epsilon_2| + 1.73\,|\epsilon_1| + |\epsilon_3| < (1.41 + 1.73 + 1)\cdot 5\cdot 10^{-3} < 3\cdot 10^{-2}.$$

The sum and product therefore are 3.14, 2.44, with errors, respectively, of at most 1, 3 units in the second decimal place.

(2) Calculating $27\sqrt{2} - 22\sqrt{3}$ in the same way, we get

$$27\sqrt{2} - 22\sqrt{3} = .01 + E_3\,, \quad \text{where } E_3 = 27\epsilon_1 - 22\epsilon_2\,.$$

Thus $|E_3| \leq 27\,|\epsilon_1| + 22\,|\epsilon_2| < 49\cdot 5\cdot 10^{-3} < .25$. In this case, the approximate answer, .01, is much smaller than the estimated maximum size of the error, suggesting that the approximate answer may not be very precise. In fact, correct to five decimal places, $27\sqrt{2} - 22\sqrt{3} = .07865$, so that the approximate answer, .01, differs from the true result by more than 780%. This example illustrates a general property of subtraction: if nearly equal numbers are subtracted, most digits cancel, giving a result very sensitive to round-off error.

Conceivably, when numerous additions, multiplications, etc., are preformed, each subject to a round-off error, these errors may accumulate and lead to inaccurate results. In fact this sometimes happens.

Because digital computing machines employ only a finite number of digits, they must use rounded or truncated numbers, and furthermore, the numbers they employ have a maximum permissible size. Large electronic digital machines are very powerful tools, both for business and science, because of their capability of performing many thousands of arithmetic operations in very short time periods. However, the programmers of these machines have to be alert to the possibility of loss of accuracy from accumulation of round-off or truncation error, subtraction of nearly equal numbers (underflow), or the growth of numbers beyond the maximum size the computer can handle (overflow).

It is worth noting that algebra of rounded or truncated numbers does not satisfy some of the normal algebraic laws. For example, if $[\alpha]_k$ denotes the number obtained from α by rounding at the kth decimal place, then the product of k-decimal place numbers α, β, γ may satisfy

$$[[\alpha\beta]_k\gamma]_k \neq [\alpha[\beta\gamma]_k]_k\,.$$

That is, rounded number multiplication is *not* associative. The reader may furnish an example for himself.

EXERCISES 6.6

I

1. Let $a = 23.412_b$, $c = 14.11_b$. Find $a + c$, $a - c$, ac and a/c, all correct to two b-places, when the base b is:
(i) five; (ii) six; (iii) nine; (iv) ten; (v) twelve.

2. Let $\alpha = 21.23$ and $\beta = 34.15$, both being base six numbers. Find the base six representation of α/β, correct to within one unit in the fourth place to the right of the six-point. Find the remainder also.

3. Use Newton's method (see §5.5) to calculate $\sqrt{2}$ in base two, accurate to one unit in the fifth place to the right of the two-point.

4. Suppose we calculate $15\sqrt{2} - 7\sqrt{3}$ using the (base ten) approximations 1.41 and 1.73 for $\sqrt{2}$ and $\sqrt{3}$, then rounded the result to two decimal places. Estimate the accuracy of the result.

II

5. Suppose α and β are positive real numbers, and that $\tilde{\alpha}$, $\tilde{\beta}$ are t-decimal place approximations to α and β respectively, where $t \geq 1$. Thus $\alpha = \tilde{\alpha} + \epsilon_1$, $\beta = \tilde{\beta} + \epsilon_2$, with $|\epsilon_1| \leq 5 \cdot 10^{-(t+1)}$, $|\epsilon_2| \leq 5 \cdot 10^{-(t+1)}$. Prove that the discrepancy between $\tilde{\alpha} + \tilde{\beta}$ and $\alpha + \beta$ is at most 10^{-t}. Show that the discrepancy between $\alpha\beta$ and $\tilde{\alpha}\tilde{\beta}$ is at most $(\tilde{\alpha} + \tilde{\beta} + 1) \cdot 5 \cdot 10^{-(t+1)}$.

6. Suppose we calculate $\sqrt{3}$ by dividing 2.45 by 1.41. (These are the 2-decimal place approximations to $\sqrt{6}$ and $\sqrt{2}$ respectively.) If we calculate 2.45/1.41 by long division, then round the quotient obtained to two decimal places, estimate how close the result would be to $\sqrt{3}$. Then do the arithmetic, and compare the actual error with your estimated error. To make the error estimate, use either $1/(1 + x) = 1 - x + x^2 - x^3 + \cdots$ or $|1/(a + b) - 1/a| \leq |b|/|a|^2$ for positive a, nonnegative b.

7. The 5-decimal place approximations to $\sqrt{2}$ and $\sqrt{3}$ are 1.41421 and 1.73205. Show that there are positive integers x and y such that $x(1.41421) - y(1.73205) = .00001$. For this x and y, if $x\sqrt{2} - y\sqrt{3}$ is calculated using these 5-decimal place approximations, estimate the error.

III

8. Let a be a positive real number, and suppose x is an approximation to $\sqrt{a}$, in fact, that $x^2 = a + \epsilon$ where ϵ is positive and small. (This is the situation encountered when using Newton's method for calculating square roots.) Prove that $\sqrt{a}$ lies between $x - \epsilon/x$ and x. Using $x = 1.42$, prove that $\sqrt{2}$ lies between 1.408 and 1.42.

9. Let $0 < \beta < 1$. Define an iterative scheme by $x_{n+1} = x_n(2 - \beta x_n)$, for $n = 0, 1, 2, \ldots$. Show that if $0 < x_0 < 1$, then x_n converges to $1/\beta$ as $n \to \infty$. (This is Newton's method applied to function $f(x) = x^{-1} - \beta$.) Use a graphical argument. Show that if x_n agrees with $1/\beta$ to an accuracy to t decimal places, then x_{n+1} agrees with $1/\beta$ to an accuracy of at least $2t$ decimal places. Further show that if the initial estimate $x_0 > 1$, the iteration may not converge.

6.7 REVIEW

The field of fractions of a commutative ring $\mathcal{R}$ without zero divisors was formally constructed in the opening sections of the chapter, using an equivalence relation on the set of ordered pairs (m, n) with entries from $\mathcal{R}$ and nonzero second entry. The equivalence classes under this equivalence relation consist of ordered pairs regarded as equal when an ordered pair (m, n) is interpreted as a fraction m/n. Operations of addition and multiplication of these equivalence classes were then defined, leading to the natural operations with fractions. In this way a "field of fractions" of $\mathcal{R}$ is obtained, each element of this field having the form mn^{-1}, for $m, n \in \mathcal{R}$. The fields of fractions of the integer and polynomial rings, namely the rational number and rational function fields, were then closely studied. The rational function field has many properties parallel to properties of the rational number field, and in particular is endowed with an absolute value function. The absolute value for rational functions satisfies the strong triangle law and the principle of domination, whereas that for rational numbers satisfies only the weak triangle law. Base b representations of rational or real numbers, and of rational functions, were described; these generalize the base b representations of integers and polynomials described in Chapter 4. A real number has an ultimately periodic base b representation if and only if it is rational.

REVIEW EXERCISES 6.7

1. Let $\mathcal{S}$ consist of all real numbers α for which $|\alpha|$ has a terminating representation in base ten. Endow $\mathcal{S}$ with the usual operations of addition and multiplication of numbers. Is $\mathcal{S}$ a ring? a field?

2. Define relation $\sim$ on $\mathbf{R}$ by $\alpha \sim \beta$ if and only if $\alpha - \beta \in \mathcal{S}$, where $\mathcal{S}$ is the set described in Exercise 1. Is $\sim$ an equivalence relation on $\mathbf{R}$?

3. If m is a fixed positive integer, let $\mathcal{R}$ consist of all integers r for which $m \mid r$. Show that $\mathcal{R}$ is a commutative ring without zero divisors. What field is the field of fractions of $\mathcal{R}$?

4. Does the absolute value on $\mathbf{Q}$ have the property of Archimedes? What about the absolute value on $\mathbf{Q}(\lambda)$? What about the absolute value on $\mathfrak{F}(\lambda)$, where $\mathfrak{F}$ is any field? What about the absolute value on $\mathbf{C}$?

5. If $p > 0$ and $q > 0$ are distinct primes in $\mathbf{Z}$, then $\sqrt{pq}$ does not have a terminating representation in base ten. Prove this. Does it have an ultimately periodic representation in base ten?

6. Let $\alpha = 1.21$, $\beta = 2.11$, $\gamma = 1.11$, all in base b. Find $\dfrac{(\alpha - 1)\beta}{\gamma}$ in base b, when b is three and five.

7

MAPPINGS OF RINGS

7.1 INTRODUCTION

We have seen in previous chapters that rings and fields are useful algebraic structures. The rings of most interest to us have been **Z**, the integers, and $\mathfrak{F}[\lambda]$, the polynomials with coefficients in $\mathfrak{F}$. The fields of most interest to us have been **Q**, the rational numbers, **R**, the real numbers, **C**, the complex numbers, and $\mathfrak{F}(\lambda)$, the rational functions with coefficients in $\mathfrak{F}$. Here $\mathfrak{F}$ itself has been a field, generally one of **Q**, **R**, or **C**. We begin this chapter by constructing many additional examples of rings and fields, including *finite* rings and fields, making use of the fact that congruence of integers or polynomials (§3.6) is an equivalence relation. We then go on to a discussion of functions, in particular, functions which act on elements of a ring and produce functional values in the same or some other ring. We shall be particularly interested in special types of functions called homomorphisms; polynomial evaluation at a fixed polynomial $\alpha(\lambda)$ is an example of a homomorphism. Some results about prime elements in the polynomial ring **Q**$[\lambda]$ will then be obtained by applying the discussion of finite fields and homomorphisms. In the last (optional) section, theorems classifying homomorphisms will be given. For example, it will be proved that evaluation at a fixed polynomial $\alpha(\lambda)$ is basically the only type of homomorphism of the polynomial ring **R**$[\lambda]$ to itself.

The theory of the polynomial ring $\mathfrak{F}[\lambda]$ given in Chapter 3 will be applied in Chapter 7 when $\mathfrak{F}$ is a field other than **Q**, **R**, or **C**. In particular, it will be applied when $\mathfrak{F}$ is one of the finite fields mentioned above.

7.2 THE RING OF CONGRUENCE CLASSES

In this section the symbol $\mathcal{R}$ denotes either the integer ring **Z** or the polynomial ring $\mathfrak{F}[\lambda]$, where $\mathfrak{F}$ is a field.

Let $m \in \mathcal{R}$. Recall from §3.6 that elements a, $b \in \mathcal{R}$ are congruent modulo m if and only if $m \mid (a - b)$. Congruence modulo m is written as

$$a \equiv b \pmod{m}, \tag{1}$$

and is an equivalence relation on $\mathcal{R}$; see (i), (ii), (iii) in §3.6. It therefore induces a partition of $\mathcal{R}$. (See §§1.2, 6.2.) Although the components of a partition of $\mathcal{R}$ induced by an abstract equivalence relation are usually called equivalence classes, in the present case, when congruence is the name of the equivalence relation, the components are called *congruence classes* or *congruence classes modulo m* (mod m). Elements a, b of $\mathcal{R}$ therefore belong to (i.e., represent) the same congruence class (mod m) if and only if the congruence (1) is satisfied. Figure 7.2.1 gives a Venn diagram illustrating the partitioning of $\mathcal{R}$ into congruence classes, and shows a few elements in several classes.

We now introduce addition and multiplication operations for congruence classes. We use a procedure similar to that used to introduce addition and multiplication of equivalence classes of ordered pairs in §6.3, namely, we define these operations by using representatives of the congruence classes; then we show that the resulting operations are independent of the particular representatives used. Let A and B be congruence classes and select representatives $a \in A$, $b \in B$. Let S be the congruence class represented by $a + b$, and P the congruence class represented by ab. We then define the sum $A + B$ and product AB by

$$A + B = S, \qquad AB = P. \tag{2}$$

If $a' \in A$, $b' \in B$ are other representatives of A, B, then $a \equiv a' \pmod{m}$ and $b \equiv b' \pmod{m}$. By (iv) of §3.6, these congruences may be added and multiplied, giving $a + b \equiv a' + b' \pmod{m}$ and $ab \equiv a'b' \pmod{m}$. Thus $a + b$, $a' + b'$ represent the same congruence class, and ab, $a'b'$ represent the same congruence class. Congruence class addition and multiplication therefore depends solely on the congruence classes themselves and not on the representatives used to calculate the sum and product. We say that these congruence class operations are *induced* by the corresponding operations in $\mathcal{R}$.

We now prove that the set of congruence classes, under addition and multiplication as just described, forms a ring. Throughout this proof, A, B, C are congruence classes and a, b, c are elements of $\mathcal{R}$ representing, respectively, A, B, C. One by one, we give the verifications of the ring axioms.

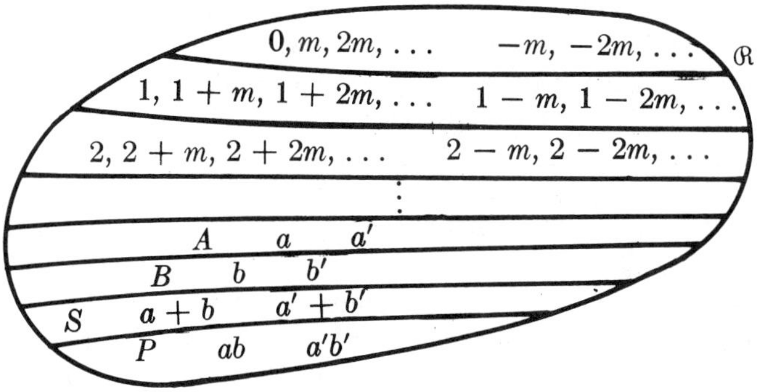

Figure 7.2.1

R1(i)–*Closure under addition:* This is evident from the description above of congruence class addition. □

R1(ii)–*Associativity of addition:* Since $a + b$ represents $A + B$, $(a + b) + c$ represents $(A + B) + C$. Similarly, $a + (b + c)$ represents $A + (B + C)$. Since associativity of addition holds for elements of $\mathcal{R}$, both $(A + B) + C$ and $A + (B + C)$ are represented by $a + b + c$, implying that $(A + B) + C = A + (B + C)$. □

R1(iii)–*Presence of zero:* Let $\mathcal{O}$ be the congruence class represented by the zero element 0 of $\mathcal{R}$. Then $\mathcal{O} + A$ is represented by $0 + a = a \in A$. Hence $\mathcal{O} + A = A$; similarly, $A + \mathcal{O} = A$. The congruence class $\mathcal{O}$ is therefore the additive identity. □

R1(iv)–*Additive inverses:* Let D be the congruence class represented by $-a$. Then $A + D$ is the congruence class represented by $a + (-a) = 0 \in \mathcal{O}$. Thus $A + D = \mathcal{O}$ and similarly $D + A = \mathcal{O}$. Congruence class D is therefore the additive inverse of A. That is: $-A$ is the congruence class containing $-a$, for $a \in A$. □

R1(v)–*Commutativity of addition:* Both $A + B$ and $B + A$ are represented by $a + b = b + a$, implying that $A + B = B + A$. □

At this point we have completed the proof that congruence classes under addition form a commutative group. Ring axiom R1 is thus verified.

R2–*Multiplicative closure:* This is evident from the discussion of congruence class multiplication above. □

R3–*Associativity of multiplication:* Element $(ab)c$ represents $(AB)C$, and $a(bc)$ represents $A(BC)$. Since multiplicative associativity holds for elements of $\mathcal{R}$, both $(AB)C$ and $A(BC)$ are represented by abc, so that $(AB)C = A(BC)$. □

R4–*Distributive laws:* Since the distributive laws hold for elements of $\mathcal{R}$, $a(b + c) = ab + ac$ represents both $A(B + C)$ and $AB + AC$. Therefore $A(B + C) = AB + AC$. Similarly, $A(B + C) = AB + AC$. □

Since the ring axioms are now verified, we know that the congruence classes modulo m constitute a ring when endowed with the addition and multiplication operations introduced above. Since $\mathfrak{R}$ is commutative, $ab = ba$ represents both AB and BA, so that $AB = BA$, implying that the congruence class ring is also commutative. Let E be the congruence class containing the multiplicative identity 1 of $\mathfrak{R}$. Then EA contains $1a = a \in A$; therefore $EA = A$ and similarly $AE = A$. Consequently E is the multiplicative identity of the congruence class ring.

Conceivably the congruence class ring consists solely of the zero class $\mathcal{O}$. When is this the case? Precisely when m is a unit in $\mathfrak{R}$. To prove this, assume first that $\mathcal{O}$ is the only congruence class. Then $E = \mathcal{O}$, hence $1 \equiv 0 \pmod{m}$, so that $m \mid 1$, implying that m is a unit. Conversely, suppose m is a unit in $\mathfrak{R}$. Then $m \mid 1$, hence $1 \equiv 0 \pmod{m}$, so that $E = \mathcal{O}$. For any congruence class A we then have $A = EA = \mathcal{O}A = \mathcal{O}$. Thus the congruence class ring has at least two congruence classes precisely when m is not a unit in $\mathfrak{R}$, and when this is so $E \neq \mathcal{O}$.

NOTATION The symbol $\mathfrak{R}_m$ will denote the ring of congruence classes $(\text{mod } m)$. When $\mathfrak{R}$ is $\mathbf{Z}$, we write this as $\mathbf{Z}_m$, and when $\mathfrak{R}$ is $\mathfrak{F}[\lambda]$, as $\mathfrak{F}[\lambda]_m$.

We summarize the results proved so far in the following theorem.

THEOREM 7.2.1
The set $\mathfrak{R}_m$ of congruence classes modulo m is a commutative ring with identity under the addition and multiplication operations induced by $\mathfrak{R}$. The ring $\mathfrak{R}_m$ has least two elements if and only if m is not a unit in $\mathfrak{R}$.

When working with an equivalence relation on a set S, it usually is important that a *system of distinct representatives* be found for the equivalence classes. This is easily done for the relation congruence $(\text{mod } m)$ on $\mathfrak{R}$. First let $m = 0$. Then $a \equiv b \pmod{m}$ holds if and only if $a = b$. In this case a congruence class consists of a single element of $\mathfrak{R}$, and therefore the elements of $\mathfrak{R}$ themselves are a system of distinct representatives. For nonzero m, the discussion is slightly different when $\mathfrak{R}$ is $\mathbf{Z}$ or $\mathfrak{F}[\lambda]$.

THEOREM 7.2.2
Assume $m \neq 0$. The integers

$$0, 1, 2, 3, \ldots, |m| - 1, \tag{3}$$

are a system of distinct representatives for the congruence classes modulo m in $\mathbf{Z}$.

PROOF Let A be a congruence class and suppose $a \in A$. Dividing a by m, we get $a = qm + r$ with $q, r \in \mathbf{Z}$ and $0 \leq r < |m|$. Then $m \mid (a - r)$, so that $a \equiv r \pmod{m}$, and therefore $r \in A$. Thus a congruence class contains at least one integer, namely r, from the list (3). If it contained another integer from (3), say s, then $r \equiv s \pmod{m}$, so that $m \mid (r - s)$. Because $0 \leq r < |m|$, $0 \leq s < |m|$, we have $-|m| < r - s < |m|$. Since the only integral multiple of m strictly between $-|m|$ and $|m|$ is 0, we obtain $r - s = 0$. Thus $r = s$, implying that a congruence class contains a unique integer from (3). □

The system (3) of distinct representatives uses least nonnegative remainders, modulo m. A system using numerically least remainders may also be given. (See Exercise 11.)

THEOREM 7.2.3
Let $m(\lambda) \neq 0$. The polynomials $r(\lambda)$ in $\mathfrak{F}[\lambda]$ for which

$$|r| <\cdot |m| \tag{4}$$

are a system of distinct representatives for the congruence classes modulo m in $\mathfrak{F}[\lambda]$.

Condition (4) may, of course, be restated as $\deg r < \deg m$.

PROOF Let A be a congruence class, and suppose $a \in A$. Dividing a by m, we get $a = qm + r$, with q, $r \in \mathfrak{F}[\lambda]$, and $|r| < |m|$. Then $a \equiv r \pmod{m}$, so that $r \in A$. A congruence class therefore contains at least one polynomial r satisfying (4). If it contains another polynomial of this type, say s, then $r \equiv s \pmod{m}$, $|r| < |m|$, $|s| < |m|$. Thus $m \mid (r - s)$ and $|r - s| \leq \max(|r|, |-s|) < |m|$. Hence $r - s$ is a multiple of m and has lower degree than m. Since this is possible only when $r - s = 0$, we get $r = s$. A congruence class therefore contains precisely one polynomial r satisfying (4). □

Here is an example. Take $\mathcal{R}$ to be $\mathbf{Z}$ and $m = 4$. There are four congruence classes, and the integers 0, 1, 2, 3 constitute a system of distinct representatives. Figure 7.2.2 illustrates, in the form of a Venn diagram, the partition of $\mathbf{Z}$ into its congruence classes modulo 4, and shows some integers within each class.

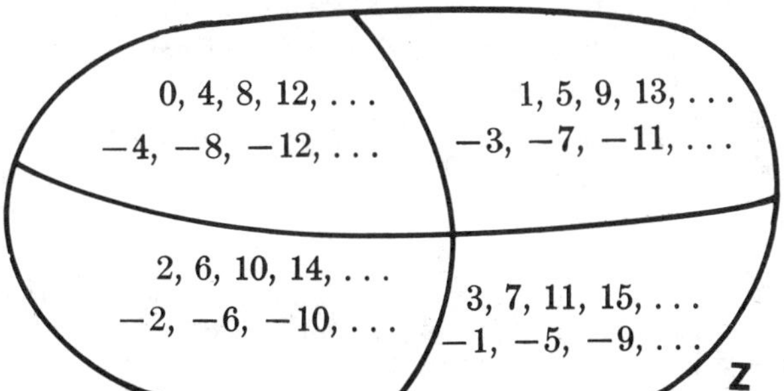

Figure 7.2.2

Let m again be a general nonzero element of $\mathcal{R}$. Since each congruence class contains precisely one element from the lists (3) or (4), and since these elements are all possible remainders after division by m, the term *residue* class is often used in place of congruence class. The word "residue" is meant to suggest the part left over (remainder) after division by m. The congruence class ring is therefore sometimes called the *residue class ring*. Other names given to the ring of congruence classes are *difference ring* or *quotient ring*. (The term quotient ring should not be confused with the field of quotients in Chapter 6.)

Let A, B be congruence classes represented, respectively, by a, b, where a, b belong to the systems of distinct representatives listed in (3) or (4). How do we calculate the representatives for $A + B$, AB in these systems? By long division, find q, r, q_1, $r_1 \in \mathcal{R}$ such that $a + b = qm + r$, $ab = q_1m + r_1$, with $0 \leq r, r_1 < |m|$ when $\mathcal{R}$ is $\mathbf{Z}$, and $|r|, |r_1| < |m|$ when $\mathcal{R}$ is $\mathfrak{F}[\lambda]$. Then $a + b \equiv r \pmod{m}$ and $ab \equiv r_1 \pmod{m}$. Thus $r \in A + B$ and $r_1 \in AB$, and therefore r, r_1 are the required representatives of $A + B$ and AB.

Example: When $\mathcal{R}$ is $\mathbf{Z}$ and $m = 4$, let A be the congruence class represented by 3, and B the congruence class represented by 2. Then $A + B$ is represented by 1 and AB by 2.

NOTATION A frequently used notation is to let $\bar{a}$ be the congruence containing the element $a \in \mathcal{R}$. An advantage of this notation* is that it avoids using too many different letters in discussions involving congruence classes. Thus $\bar{a}$ consists of all elements $\alpha \in \mathcal{R}$ for which $\alpha \equiv a \pmod{m}$. This congruence is equivalent to $m \mid (\alpha - a)$ and hence to $\alpha = a + tm$ for $t \in \mathcal{R}$. Therefore $\bar{a}$ consists of all elements of $\mathcal{R}$ of the form $a + tm$ as t ranges over $\mathcal{R}$. Using this notation, the rules for congruence class addition and multiplication become

$$\bar{a} + \bar{b} = \overline{a + b}, \qquad \bar{a}\,\bar{b} = \overline{ab}. \tag{5}$$

For $a \in \bar{a}$, $b \in \bar{b}$, so that $\bar{a} + \bar{b}$ and $\bar{a}\,\bar{b}$ are the congruence classes containing $a + b$ and ab, respectively, meaning that $\bar{a} + \bar{b} = \overline{a + b}$ and $\bar{a}\,\bar{b} = \overline{ab}$.

In this notation, $\bar{0}$ and $\bar{1}$ are the additive and multiplicative identity elements of $\mathcal{R}_m$.

* In Chapter 2 the symbol $\bar{z}$ was used for the conjugate of a complex number z. That use of the bar over a letter is quite distinct from the use described above. In this chapter the symbol $\bar{a}$ always denotes the congruence class containing a unless an explicit statement to the contrary is made.

EXERCISES 7.2

I In Exercises 1–6 below, express your answers in terms of the systems of distinct representatives described in the text.

1. Find: $\bar{2} + \bar{3}$ and $\bar{2} \cdot \bar{3}$ in $\mathbf{Z}_4$, $\mathbf{Z}_5$, $\mathbf{Z}_6$. What congruence class is $\bar{3}$ in $\mathbf{Z}_2$? in $\mathbf{Z}_3$?

2. Find: $-\bar{2}$ in $\mathbf{Z}_3$, $\mathbf{Z}_4$, $\mathbf{Z}_5$, $\mathbf{Z}_6$. Do this also in $\mathbf{Z}_2$.

3. Find: $\overline{\lambda + 2} + \overline{2\lambda + 3}$ and $\overline{\lambda + 2} \cdot \overline{2\lambda + 3}$ in $\mathbf{R}[\lambda]_{f(\lambda)}$ when $f(\lambda) = \lambda^2 + 1$ and when $f(\lambda) = \lambda^2$.

4. Solve for the congruence class X satisfying $\bar{4}X + \bar{2} = \bar{5}$ in $\mathbf{Z}_6$, in $\mathbf{Z}_7$, in $\mathbf{Z}_8$.

5. Solve for the congruence class X satisfying $\bar{\lambda}X = \bar{1}$ in $\mathbf{R}[\lambda]_{f(\lambda)}$, for $f(\lambda) = \lambda^2 + 1$ and for $f(\lambda) = \lambda^2$.

6. Find all solutions of the quadratic equation $X^2 - \bar{5}X + \bar{6} = \bar{0}$ in $\mathbf{Z}_2$, $\mathbf{Z}_3$, $\mathbf{Z}_4$, $\mathbf{Z}_5$, $\mathbf{Z}_6$.

II

7. Show that the ring $\mathbf{Z}_4$ contains zero divisors. Generalize and show that $\mathcal{R}_m$ contains zero divisors whenever m is composite.

8. Show that there are four roots of the quadratic polynomial $\lambda^2 - \bar{1}$ in the ring $\mathbf{Z}_8$. Find the number of roots of the same polynomial in $\mathbf{Z}_{16}$, and in $\mathbf{Z}_{32}$. Can you determine the number of roots of this polynomial in $\mathbf{Z}_m$ when $m = 2^t$ is a power of 2?

9. Show that the powers 2^t for $t = 1, 2, 3, 4$, are a system of distinct representatives for the nonzero elements of $\mathbf{Z}_5$. Find an integer a such that the powers a^t, $t = 1, 2, \ldots, 6$, are a system of distinct representatives for the nonzero element of $\mathbf{Z}_7$. Can a similar integer a be found for $\mathbf{Z}_8$? or $\mathbf{Z}_{10}$?

10. Find the smallest positive integer k with this property: $k\bar{a} = \bar{0}$ for every $\bar{a}$ in $\mathbf{Z}_m$. (Recall Exercise 12 of §1.4.)

11. Let m be a nonzero integer. Show that the integers r satisfying $-|m|/2 < r \leq |m|/2$ constitute a system of distinct representatives for the congruence classes modulo m in $\mathbf{Z}$.

III

12. An element A in $\mathcal{R}_m$ is a *unit* if there exists an X in $\mathcal{R}_m$ such that $AX = \bar{1}$. Let $a \in A$.
    (i) If A is a unit, prove that $(a, m) = 1$.
    (Hint: Consider $ax \equiv 1 \pmod{m}$, that is, $ax - 1 = ym$.)
    (ii) If $(a, m) = 1$, prove that A is a unit.
    (iii) Show that the units in $\mathcal{R}_m$ under operation multiplication form a commutative group. (*cont.*)

(iv) When $m = p$ is a positive prime integer, show that there are exactly $p - 1$ units in $\mathbf{Z}_p$.

13. Let A be a congruence class in $\mathcal{R}_m$. If both a, a' represent A, prove that $(a, m) = (a', m)$. Is the converse true? That is, if $(a, m) = (a', m)$ must a, a' represent the same congruence class?

14. If $m = 2^t$, find the lowest positive exponent e such that 5^e represents the identity element of $\mathcal{R}_m$.

7.3 THE RING OF CONGRUENCE CLASSES IN SOME SPECIAL CASES

We continue to let the ring $\mathcal{R}$ be either $\mathbf{Z}$, the integers, or $\mathcal{F}[\lambda]$, the polynomials with coefficients in a field $\mathcal{F}$.

The ring $\mathbf{Z}_2$ consists of two congruence classes, $\bar{0}$ and $\bar{1}$. These classes add and multiply according to the following tables.

$+$	$\bar{0}$	$\bar{1}$
$\bar{0}$	$\bar{0}$	$\bar{1}$
$\bar{1}$	$\bar{1}$	$\bar{0}$

$\cdot$	$\bar{0}$	$\bar{1}$
$\bar{0}$	$\bar{0}$	$\bar{0}$
$\bar{1}$	$\bar{0}$	$\bar{1}$

In these tables the entry at the intersection of the row with $\bar{a}$ at the left and the column with $\bar{b}$ at the top is the sum (in the first table) or product (in the second table) of $\bar{a}$ and $\bar{b}$. A convenient simplification of notation is to use the distinct representatives given in Theorems 7.2.2, 7.2.3 for the congruence classes, instead of using the congruence classes themselves. Then we write $\bar{a}$ as a, and a sum $a + b$ or product ab is calculated by finding the remainder after dividing the usual sum or product of a, b by m. Under this simplification, the above tables become the following.

$+$	0	1
0	0	1
1	1	0

$\cdot$	0	1
0	0	0
1	0	1

(1)

The reader may check that this ring is a field. It is the first ring described in Exercise 3 of §1.5.

As another example, the ring $\mathbf{Z}_3$ consists of three congruence classes, denoted by 0, 1, 2, when we use the system of distinct representatives given in Theorem 7.2.2. The following tables give addition and multiplication in this ring, i.e., addition and multiplication for these representatives:

+	0	1	2
0	0	1	2
1	1	2	0
2	2	0	1

·	0	1	2
0	0	0	0
1	0	1	2
2	0	2	1

(2)

The reader may verify directly that $\mathbf{Z}_3$ is a field. For example, $2^{-1} = 2$ in $\mathbf{Z}_3$ because 2^{-1} denotes the element x satisfying $2x = 1$, and the multiplication table shows that $x = 2$. Furthermore, $-2 = 1$ in $\mathbf{Z}_3$ since -2 denotes the element x satisfying $2 + x = 0$ and the addition table shows that $x = 1$.

Addition and multiplication tables for $\mathbf{Z}_4$ are:

+	0	1	2	3
0	0	1	2	3
1	1	2	3	0
2	2	3	0	1
3	3	0	1	2

·	0	1	2	3
0	0	0	0	0
1	0	1	2	3
2	0	2	0	2
3	0	3	2	1

Notice that $2 \cdot 2 = 0$ in $\mathbf{Z}_4$. Thus $\mathbf{Z}_4$ contains zero divisors and is not a field.

Since $\mathbf{Z}_2$ is a field, we may consider the polynomial ring $\mathbf{Z}_2[\lambda]$. This ring consists of polynomials $a_0 + a_1\lambda + a_2\lambda^2 + \cdots$ in which the coefficients are 0 or 1. These coefficients behave under addition and multiplication as indicated in the tables (1). For example, in $\mathbf{Z}_2[\lambda]$ we have $(\lambda^2 + \lambda + 1) + (\lambda^2 + 1) = \lambda$ and $(\lambda + 1)^2 = \lambda^2 + 1$. The quotient $q(\lambda)$ and remainder $r(\lambda)$ obtained when $\lambda^4 + \lambda^3 + \lambda$ is divided by $\lambda^2 + 1$ are $q(\lambda) = \lambda^2 + \lambda + 1$ and $r(\lambda) = 1$. This quotient and remainder may be found using long division in the normal way, but remembering the special rules (1) for manipulating the polynomial coefficients:

$$
\begin{array}{r|l}
\lambda^2 + 1 \overline{\smash{)}\lambda^4 + \lambda^3 \qquad\quad + \lambda} & \\
\lambda^4 \qquad + \lambda^2 \qquad\qquad & \lambda^2 \\
\hline
\lambda^3 + \lambda^2 + \lambda & \\
\lambda^3 \qquad\quad + \lambda & \lambda \\
\hline
\lambda^2 \qquad\qquad & \\
\lambda^2 \qquad + 1 & 1 \\
\hline
1 &
\end{array}
$$

Let $m(\lambda) = \lambda^2 + \lambda + 1$. This polynomial is a prime in $\mathbf{Z}_2[\lambda]$ since, if it were composite, it would factor as a product of two first degree polynomials, and if this were so, it would have a root in $\mathbf{Z}_2$ (see Lemma 5.2.2). But $m(\lambda)$ does not have a root in $\mathbf{Z}_2$ since $m(0) = 1$, $m(1) = 1$. Let us consider the ring $\mathbf{Z}_2[\lambda]_m$. The distinct representatives for the residue classes modulo m are the polynomials $r(\lambda)$ with $|r| < |m|$, that is, $r(\lambda)$ has the form $a + b\lambda$, in which a and b are constants. Since the only constants in $\mathbf{Z}_2$ are 0 and 1, the distinct representatives of the congruence classes in $\mathbf{Z}_2[\lambda]$ modulo $m(\lambda)$ are

$$0, \quad 1, \quad \lambda, \quad \lambda + 1.$$

Addition and multiplication for the congruence classes in $\mathbf{Z}_2[\lambda]$ modulo $m(\lambda)$ are given using these representatives in the following tables.

$+$	0	1	λ	$\lambda+1$
0	0	1	λ	$\lambda+1$
1	1	0	$\lambda+1$	λ
λ	λ	$\lambda+1$	0	1
$\lambda+1$	$\lambda+1$	λ	1	0

$\cdot$	0	1	λ	$\lambda+1$
0	0	0	0	0
1	0	1	λ	$\lambda+1$
λ	0	λ	$\lambda+1$	1
$\lambda+1$	0	$\lambda+1$	1	λ

For example, $\lambda \cdot \lambda$ is $\lambda + 1$ because $\lambda + 1$ is the remainder in $\mathbf{Z}_2[\lambda]$ after dividing λ^2 by $\lambda^2 + \lambda + 1$. These tables may be used to verify that $\mathbf{Z}_2[\lambda]_m$ is a field; e.g., the tables show that $\lambda^{-1} = \lambda + 1$ and $-\lambda = \lambda$.

It is helpful to let α denote the congruence class represented by λ. Then the four elements of this field are

$$0, \quad 1, \quad \alpha, \quad \alpha + 1.$$

Since λ represents α, $\lambda^2 + \lambda + 1$ represents $\alpha^2 + \alpha + 1$. Because $\lambda^2 + \lambda + 1 \equiv 0 \pmod{m(\lambda)}$, $\lambda^2 + \lambda + 1$ represents the zero congruence class, implying that $\alpha^2 + \alpha + 1 = 0$. The field $\mathbf{Z}_2[\lambda]_{m(\lambda)}$ therefore may be described as four distinct elements $0, 1, \alpha, \alpha + 1$, with α satisfying $\alpha^2 + \alpha + 1 = 0$, or equivalently, $\alpha^2 = \alpha + 1$ (because $-1 = 1$ in this field). This point of view simplifies calculations. For example, we may calculate the product $\alpha(\alpha + 1)$ by multiplying out parentheses, then using $\alpha^2 = \alpha + 1$ and $2 = 0$, in this way: $\alpha(\alpha + 1) = \alpha^2 + \alpha = (\alpha + 1) + \alpha = 2\alpha + 1 = 1$.

Observe that the elements 0, 1 in this field add and multiply precisely as in tables (1). Thus $\mathbf{Z}_2[\lambda]_{m(\lambda)}$ contains $\mathbf{Z}_2$ as a subfield.

These examples show that $\mathcal{R}_m$ may be a field. When is this the case? The answer is very simple.

THEOREM 7.3.1
The ring $\mathcal{R}_m$ is a field precisely when m is a prime in $\mathcal{R}$.

PROOF To avoid confusion, in this proof we use the symbol 1 for the multiplicative identity of $\mathcal{R}$ and E for the multiplicative identity of $\mathcal{R}_m$. Thus E is the congruence class represented by 1.

First suppose that m is a prime in $\mathcal{R}$. To show that $\mathcal{R}_m$ is a field, we must show that E is not the zero element $\mathcal{O}$ of $\mathcal{R}_m$ and that each nonzero element of $\mathcal{R}_m$ has a multiplicative inverse in $\mathcal{R}_m$. These are conditions F1, F2 of Theorem 1.6.7.

Since m is not a unit in $\mathcal{R}$, $m \nmid 1$, so that $1 \not\equiv 0 \pmod{m}$, and thus $E \neq \mathcal{O}$. Condition F1 is verified.

Let A be a nonzero congruence class, represented by $a \in A$. Since $A \neq \mathcal{O}$, we get $a \not\equiv 0 \pmod{m}$, implying that $m \nmid a$. Let $d = (a, m)$ be the greatest common divisor of a and m. Since m is a prime and $d \mid m$, d is either a unit or an associate of m. If d were an associate of m, from $d \mid a$ we would get the contradiction $m \mid a$. Therefore d is a unit, and hence is 1. Thus $(a, m) = 1$, so that $ax + my = 1$ for suitable elements $x, y \in \mathcal{R}$. (Corollary 3.4.7.) Thus $ax \equiv 1 \pmod{m}$. Let X be the congruence class represented by x. Then AX is represented by ax. Since $ax \equiv 1 \pmod{m}$ and $1 \in E$, we get $AX = E$. Congruence class X is therefore a multiplicative inverse for A. This verifies F2 and completes the proof that $\mathcal{R}_m$ is a field. [The existence of an element $x \in \mathcal{R}$ satisfying $ax \equiv 1 \pmod{m}$ may also be proved using (viii) of §3.6, since the necessary and sufficient condition $(a, m) \mid 1$ for the solvability of this congruence is satisfied.]

Before continuing the proof, we observe that the last paragraph yields the following procedure for computing the multiplicative inverse of a congruence class A: Select a representative a of A, then find elements x, y in $\mathcal{R}$ such that $ax + my = 1$. The inverse of A is then the congruence class X represented by x. The calculation of x and y (only x is actually needed) may be efficiently performed using the method described at the end of §3.4.

Continuing with the proof, suppose now that $\mathcal{R}_m$ is a field. We know that m is either 0, a unit, composite, or a prime. We shall exclude, one by one, the first three possibilities.

If $m = 0$, each congruence class $(\text{mod } m)$ consists of a single element of $\mathcal{R}$. Since addition and multiplication of congruence classes is performed by adding and multiplying representatives, it is evident that the ring $\mathcal{R}_m$ is simply the ring $\mathcal{R}$. However, $\mathcal{R}$ is not a field. Therefore m cannot be 0.

In the next two parts we shall assume only that $\mathcal{R}_m$ is an integral domain.

If m is a unit, then $\mathcal{R}_m$ consists of a single class, $\mathcal{O}$, by Theorem 7.2.1. However, an integral domain always has a multiplicative identity different from its zero element. Therefore m is not a unit.

Suppose m is composite, say $m = ab$, where $a, b \in \mathcal{R}$ and neither is a unit. Let A, B be congruence classes represented, respectively, by a, b.

Then AB is represented by ab, and $ab \equiv 0 \pmod{m}$. Thus $AB = \mathcal{O}$. Since an integral domain has no zero divisors, either $A = \mathcal{O}$ or $B = \mathcal{O}$; say $A = \mathcal{O}$. From $a \in A$ and $A = \mathcal{O}$, we get $a \equiv 0 \pmod{m}$. Thus $m \mid a$, so that $a = tm$ for some $t \in \mathcal{R}$. From $a = tm$ and $m = ab$, we get $m = mtb$. Since m is a nonzero element of the integral domain $\mathcal{R}$, it may be cancelled to give $1 = tb$. This implies that b is a unit in $\mathcal{R}$, a contradiction. Thus m cannot be composite.

Therefore m must be a prime. □

Example 7.3.2

Take $\mathcal{R}$ to be $\mathbf{Z}$ and m to be a positive integer. The ring $\mathbf{Z}_m$ is a *finite ring* containing precisely m elements, since, by Theorem 7.2.2, the integers

$$0, 1, 2, \ldots, m - 1$$

are a system of distinct representatives for its elements. In particular, when m is a positive prime integer p, the finite ring $\mathbf{Z}_p$, being a field by Theorem 7.3.1, is a *finite field*, a system of distinct representatives for its elements being

$$0, 1, 2, \ldots, p - 1. \tag{3}$$

We shall use this finite field in the next example to obtain more finite rings and fields.

Example 7.3.3

Let p denote a positive prime integer. Since $\mathbf{Z}_p$ is a field, the field $\mathfrak{F}$ in the symbol $\mathfrak{F}[\lambda]$ may be taken to be $\mathbf{Z}_p$. That is, we may study polynomials with coefficients in $\mathbf{Z}_p$. For convenience, we denote the elements of $\mathbf{Z}_p$ by their representatives (3). Suppose $m(\lambda)$ is a nonzero polynomial, of degree k, belonging to the polynomial ring $\mathbf{Z}_p[\lambda]$. Then, we claim, the congruence class ring $\mathbf{Z}_p[\lambda]_{m(\lambda)}$ is another example of a finite ring. To prove this, recall (Theorem 7.2.3) that the polynomials $r(\lambda) \in \mathbf{Z}_p[\lambda]$ with degree less than the degree of $m(\lambda)$ are a system of distinct representatives for the elements of $\mathbf{Z}_p[\lambda]_{m(\lambda)}$. Each polynomial of this type may be written as

$$r(\lambda) = r_0 + r_1\lambda + r_2\lambda^2 + \cdots + r_{k-1}\lambda^{k-1}, \tag{4}$$

where $r_0, r_1, r_2, \ldots, r_{k-1}$ are constants, i.e., elements of $\mathbf{Z}_p$. Since $\mathbf{Z}_p$ consists precisely of the elements (3), each of the k coefficients $r_0, r_1, \ldots, r_{k-1}$ may assume any of these p values (only), so that there are precisely p^k different polynomials (4). Thus, *when $m(\lambda)$ is a nonzero polynomial in $\mathbf{Z}_p[\lambda]$ of degree k, the congruence class ring $\mathbf{Z}_p[\lambda]_{m(\lambda)}$ is a finite ring with p^k elements.* In particular, *when $m(\lambda)$ is a prime polynomial of degree k in $\mathbf{Z}_p[\lambda]$, the congruence class ring $\mathbf{Z}_p[\lambda]_{m(\lambda)}$ is a finite field* (by Theorem 7.3.1) *with p^k elements.*

Continuing Example 7.3.3, consider the *special* congruence classes represented by polynomials (4) in which $r_1 = r_2 = \cdots = r_{k-1} = 0$. Two of these special congruence classes are equal precisely when their representatives (4) have equal elements of $\mathbf{Z}_p$ as the constant term r_0, and furthermore, these special congruence classes are added and multiplied simply by calculating (within $\mathbf{Z}_p$) the sum and product of these constant terms. This means that the special congruence classes constitute a field which is just $\mathbf{Z}_p$ itself. Thus:

The ring $\mathbf{Z}_p[\lambda]_{m(\lambda)}$ contains $\mathbf{Z}_p$ as a subfield.

As in the discussion of $\mathbf{Z}_2[\lambda]_{m(\lambda)}$, it is helpful to let α denote the congruence class represented by λ. Since the polynomials (4) are a system of distinct representatives for the elements of $\mathbf{Z}_p[\lambda]_{m(\lambda)}$, and since λ represents α, an element of this ring is uniquely representable in the form

$$r_0 + r_1\alpha + r_2\alpha^2 + \cdots + r_{k-1}\alpha^{k-1},$$
$$\text{with } r_0, r_1, r_2, \ldots, r_{k-1} \in \mathbf{Z}_p;\ k = \deg m(\lambda). \qquad (5)$$

Suppose that $m(\lambda) = m_0 + m_1\lambda + \cdots + m_k\lambda^k$, where $m_0, m_1, \ldots, m_k \in \mathbf{Z}_p$. Since λ represents α, the polynomial $m_0 + m_1\lambda + \cdots + m_k\lambda^k$ represents the congruence class $m_0 + m_1\alpha + \cdots + m_k\alpha^k$. Because $m_0 + m_1\lambda + \cdots + m_k\lambda^k \equiv 0 \pmod{m(\lambda)}$, this congruence class is the zero congruence class, so that α satisfies

$$m_0 + m_1\alpha + \cdots + m_k\alpha^k = 0.$$

We may write this as $m(\alpha) = 0$.

Taking $p = 7$ as a special case, and noting that $m(\lambda) = \lambda^3 + 2$ is a prime polynomial in $\mathbf{Z}_7[\lambda]$, it follows that $\mathbf{Z}_7[\lambda]_{\lambda^3+2}$ is a field with $7^3 = 343$ elements. (The polynomial $\lambda^3 + 2$ is a prime in $\mathbf{Z}_7[\lambda]$ since if it were not, it would necessarily have a root in $\mathbf{Z}_7$. This is not the case because evaluating $\lambda^3 + 2$ at any of 0, 1, 2, 3, 4, 5, 6 does not give a result equal to 0 in $\mathbf{Z}_7$.) Using α for the congruence class represented by λ, we then know that each element of $\mathbf{Z}_7[\lambda]_{\lambda^3+2}$ is uniquely representable as

$$r_0 + r_1\alpha + r_2\alpha^2, \quad \text{with } r_0, r_1, r_2 \text{ each } 0, 1, 2, 3, 4, 5, \text{ or } 6,$$

and α satisfies

$$\alpha^3 + 2 = 0.$$

Adding 5 to each side yields $\alpha^3 = 5$. We may use this equation to perform calculations in $\mathbf{Z}_7[\lambda]_{\lambda^3+2}$. For example, to calculate $(2 + 4\alpha^2)(3 + 2\alpha + 5\alpha^2)$, use the distributive laws to remove parentheses in the normal way; then use $\alpha^3 = 5$ and $7 = 0$, as follows:

$$\begin{aligned}(2 + 4\alpha^2)(3 + 2\alpha + 5\alpha^2) &= 6 + 4\alpha + 10\alpha^2 + 12\alpha^2 + 8\alpha^3 + 20\alpha^4 \\ &= 6 + 4\alpha + \alpha^2 + \alpha^3 + 6\alpha^4 \\ &= 6 + 4\alpha + \alpha^2 + 5 + 30\alpha \\ &= 4 + 6\alpha + \alpha^2.\end{aligned}$$

This concludes Example 7.3.3.

The finite field $\mathbf{Z}_p$ introduced in Example 7.3.2 and the polynomial ring $\mathbf{Z}_p[\lambda]$ over this finite field will be used in §7.6 to study factorization theory for the polynomial ring $\mathbf{Q}[\lambda]$. We shall use the finite fields $\mathbf{Z}_p[\lambda]_{m(\lambda)}$ later in this section, in the special case $p = 2$, to discuss *algebraic codes.*

In addition to asking when the congruence class ring $\mathcal{R}_m$ is a field, it is equally natural to ask when it is an integral domain. The answer is:

$\mathcal{R}_m$ is an integral domain if and only if m is 0 or a prime in $\mathcal{R}$.

PROOF If $m = 0$, the proof of Theorem 7.3.1 showed that $\mathcal{R}_m$ is $\mathcal{R}$, and therefore an integral domain. If m is a prime in $\mathcal{R}$, then $\mathcal{R}_m$ is a field (Theorem 7.3.1) and therefore an integral domain. Conversely, if $\mathcal{R}_m$ is an integral domain, the last part of the proof of Theorem 7.3.1 showed that m, if it is not 0, must be a prime in $\mathcal{R}$. □

The examples of finite rings and fields above do not include an example of a finite integral domain which is not a field. In fact, no such example exists, as the next theorem shows.

THEOREM 7.3.4
A finite integral domain is a field.

PROOF Let $\mathcal{I}$ be a finite integral domain, and let $a_1, a_2, \ldots, a_n$ be its elements. To prove that $\mathcal{I}$ is a field, we must prove that each nonzero element of $\mathcal{I}$ has a multiplicative inverse. (This is F2 of §1.6; F1 needs no proof since it is the same as I1.) Let a, b be elements of $\mathcal{I}$, with a nonzero. Thus a is one of $a_1, \ldots, a_n$, and so is b. Consider the following elements, all of which are in $\mathcal{I}$: $aa_1, aa_2, aa_3, \ldots, aa_n$. Could two of these be the same, say $aa_1 = aa_2$? Since cancellation of a nonzero element is permissible in an integral domain (Theorem 1.6.5), from $aa_1 = aa_2$ and $a \neq 0$, we get $a_1 = a_2$. This is impossible because $a_1, a_2, \ldots, a_n$ are the (distinct) elements of $\mathcal{I}$. Therefore $aa_1, aa_2, \ldots aa_n$ are n distinct elements in $\mathcal{I}$, and hence must all be elements in $\mathcal{I}$, since $\mathcal{I}$ has only n elements. Since $b \in \mathcal{I}$, there is a subscript t such that $aa_t = b$. In particular, taking b to be the multiplicative identity e of $\mathcal{I}$, we get $aa_t = e$. Thus a has a multiplicative inverse in $\mathcal{I}$, namely a_t. □

In the finite fields above, in each case the number of elements is a power p^k of a positive prime integer p. This is not an accidental circumstance as it can be proved that the number of elements in a finite field must be p^k for a positive prime integer p and positive integer k. Furthermore, for any positive prime integer p and positive integer k, it can also be proved that there is one and (apart from change of notation) only one finite field with p^k elements, this unique field being $\mathbf{Z}_p[\lambda]_{m(\lambda)}$, where $m(\lambda)$ is a prime polynomial in $\mathbf{Z}_p[\lambda]$ of degree k. (The theory shows that $\mathbf{Z}_p[\lambda]$ has at least one prime polynomial of any degree k.). This means, if we let α be the congruence class represented by λ, that an arbitrary finite field consists of a collection of p^k distinct elements (5), with α satisfying $m(\alpha) = 0$. This equation may be used to calculate products of these elements, since it enables powers of α with high exponent to be expressed as combinations of lower powers of α. This procedure was illustrated in the examples worked above.

Because there is only one finite field (apart from notational change) with p^k elements, and because these fields were discovered by the mathematician Galois, the finite field with p^k elements is usually denoted by $GF(p^k)$, the letters GF standing for "Galois Field." Much additional information about these fields may be found in the Peterson-Weldon and Berlekamp books cited on page 455.

The reader may wish to know if finite rings and fields have practical applications. They do. For example, the fields $\mathbf{Z}_p$ are used in a recently developed method for solving systems of many equations in many unknowns using large electronic computers. For some details at an elementary level, see the article by R. T. Gregory cited on page 455. Further comments will be found in Chapter 8. An older application of finite fields is to the design of statistical experiments when, for example, an agricultural scientist needs to determine which of several varieties of a field crop such as wheat are most desirable. For a brief comment on this, see page 375 of the Beckenbach book listed on page 455. A very recent, and increasingly important, application of finite fields is to codes with error-correcting properties. We now discuss this type of code.

Application: Algebraic codes (Optional)

The reader perhaps is familiar with the Morse code, in which a symbol to be transmitted over a telegraph or cable channel is represented as a sequence of "dots" or "dashes," i.e., short and long pulses of sound. If we rewrite a dot as 0, and a dash as 1, then a symbol is represented in the code by a string of a few 0's and 1's. It is customary to give the name *code word* to the string of 0's and 1's representing a single symbol, and *digit* to each 0 or 1 in the code word. To transmit an English word involving several symbols, such as "the," therefore requires three Morse code words, namely, the code words for t, for h, and for e. Morse code unfortunately has the defect that several of its code words can be converted into other code words by changing a single digit. Thus, if a transmission error occurs, and 0 is erroneously transmitted as 1, or 1 as 0, it may not be possible to deduce the transmitted message knowing only the received message. *Error-correcting codes* are codes designed to remedy this defect, and there has been much research into them, especially since 1950. Some of the best error-correcting codes known are derived using finite fields; and furthermore, because of the mathematical nature of their derivation, they have algebraic properties which may be used to facilitate this error correction.

While we cannot give a thorough explanation of these codes here, we do give an example illustrating the role played by finite fields.

Example 7.3.5 A double error-correcting code
The code we shall describe is one of a family of error correcting codes discovered about 1960. The description of the code itself involves only the finite field $\mathbf{Z}_2$, but the procedure for correction of the transmission errors involves a finite field $\mathbf{Z}_2[\lambda]_{m(\lambda)}$ for a certain prime polynomial $m(\lambda)$. We first describe the code, then the finite field used in the error correction, and finally the error correction procedure.

A word in this code is to be a string of 15 symbols $[x_0, x_1, x_2, \ldots, x_{14}]$, with each x_i equal to 0 or 1, i.e., an element of $\mathbf{Z}_2$, such that the polynomial

$$x(\lambda) = x_0 + x_1\lambda + x_2\lambda^2 + \cdots + x_{14}\lambda^{14} \in \mathbf{Z}_2[\lambda]$$

is divisible by

$$g(\lambda) = (\lambda^4 + \lambda + 1)(\lambda^4 + \lambda^3 + \lambda^2 + \lambda + 1) \in \mathbf{Z}_2[\lambda].$$

This amounts to

$$\begin{aligned} x(\lambda) &= (b_0 + b_1\lambda + b_2\lambda^2 + \cdots + b_6\lambda^6)g(\lambda) \\ &= b(\lambda)g(\lambda), \end{aligned}$$

say, where $b(\lambda) \in \mathbf{Z}_2[\lambda]$. How many words are there in this code? We first agree that words are the same precisely when they agree digit by digit. If $b(\lambda)$, $B(\lambda)$ are polynomials in $\mathbf{Z}_2[\lambda]$ with $b(\lambda)g(\lambda) = B(\lambda)g(\lambda)$, then $b(\lambda) = B(\lambda)$, since $\mathbf{Z}_2[\lambda]$ is an integral domain (Theorems 1.6.5, 3.2.1). Distinct elements $b_0, b_1, \ldots, b_6 \in \mathbf{Z}_2$ lead therefore to distinct codewords. Since the seven coefficients $b_0, b_1, \ldots, b_6$ each may be 0 or 1, we deduce that this code has precisely $2^7 = 128$ code words. By selecting values for $b_0, b_1, \ldots, b_6$, the reader may write out some of these code words for himself.

Next we describe the field $\mathbf{Z}_2[\lambda]_{m(\lambda)}$. We take $m(\lambda) = \lambda^4 + \lambda + 1$. To show that this polynomial is a prime element of $\mathbf{Z}_2[\lambda]$, first observe that neither 0 nor 1 is a root. Thus $m(\lambda)$ cannot have a first degree factor, so that, if it is not a prime, it factors as $m(\lambda) = (\lambda^2 + a\lambda + b)(\lambda^2 + c\lambda + d)$, where $a, b, c, d \in \mathbf{Z}_2$. Comparing constant terms on each side yields $bd = 1$, hence $b = d = 1$. Since $\lambda^2 + a\lambda + 1$ has 1 as a root if $a = 0$, it follows that $a = 1$, and similarly $c = 1$. But then $m(\lambda) = (\lambda^2 + \lambda + 1)^2 = \lambda^4 + \lambda^2 + 1 \neq m(\lambda)$. This contradiction shows that $m(\lambda)$ is a prime in $\mathbf{Z}_2[\lambda]$, and therefore that $\mathbf{Z}_2[\lambda]_{m(\lambda)}$ is indeed a finite field. Denote this finite field by $GF(2^4)$. It consists of the 16 distinct expressions

$$r_0 + r_1\alpha + r_2\alpha^2 + r_3\alpha^3, \qquad r_0, r_1, r_2, r_3 \in \mathbf{Z}_2, \tag{6}$$

with α satisfying $m(\alpha) = \alpha^4 + \alpha + 1 = 0$, and therefore satisfying $\alpha^4 = 1 + \alpha$. By straightforward calculation, the reader may verify that 1, α, $\alpha^2, \ldots, \alpha^{14}$ are all the nonzero elements of $GF(2^4)$. In fact,

$$\alpha^0 = 1, \quad \alpha^1 = \alpha, \quad \alpha^2 = \alpha^2, \quad \alpha^3 = \alpha^3, \quad \alpha^4 = 1 + \alpha,$$
$$\alpha^5 = \alpha + \alpha^2, \quad \alpha^6 = \alpha^2 + \alpha^3, \quad \alpha^7 = 1 + \alpha + \alpha^3, \quad \alpha^8 = 1 + \alpha^2,$$
$$\alpha^9 = \alpha + \alpha^3, \quad \alpha^{10} = 1 + \alpha + \alpha^2, \quad \alpha^{11} = \alpha + \alpha^2 + \alpha^3,$$
$$\alpha^{12} = 1 + \alpha + \alpha^2 + \alpha^3, \quad \alpha^{13} = 1 + \alpha^2 + \alpha^3, \quad \alpha^{14} = 1 + \alpha^3,$$

and the right-hand sides of these equations are all the nonzero expressions (6). Thus, each nonzero element of this field is uniquely representable as α^i, with an integral exponent i satisfying $0 \le i \le 14$.

Notice that α satisfies not only $m(\alpha) = \alpha^4 + \alpha + 1 = 0$, but also $\alpha^{12} + \alpha^9 + \alpha^6 + \alpha^3 + 1 = 0$, as a short calculation using the above formulas shows. Thus α is a root of $m(\lambda) = \lambda^4 + \lambda + 1$, and α^3 is a root of $\lambda^4 + \lambda^3 + \lambda^2 + \lambda + 1$, and therefore both α and α^3 are roots of $g(\lambda)$. This is the crucial property of $g(\lambda)$ that makes error correction possible, as we shall next explain.

Suppose that a code word $x = [x_0, x_1, \ldots, x_{14}]$ is transmitted, and two transmission errors are made, in the jth and kth digits, where $0 \le j < k \le 14$, so that the received word is

$$\begin{aligned} y &= [y_0, \ldots, y_{14}] \\ &= [x_0, \ldots, x_{j-1}, x_j + 1, x_{j+1}, \ldots, x_{k-1}, x_k + 1, x_{k+1}, \ldots x_{14}]. \end{aligned}$$

The person receiving y wishes to deduce which digits are incorrect. To test whether y is a code word, he divides

$$y(\lambda) = y_0 + y_1\lambda + \cdots + y_{14}\lambda^{14}$$

by $g(\lambda)$. If the division is exact, i.e., zero remainder, then y would be a code word. Since $y(\lambda) = x(\lambda) + \lambda^j + \lambda^k$, and $\lambda^j + \lambda^k$ is not divisible by $m(\lambda)$, let alone by $g(\lambda)$, this test fails.

[*Proof:* $\lambda^j + \lambda^k$ is not divisible by $m(\lambda)$ since if it were, α would be a root of $\lambda^j + \lambda^k$. However, $\alpha^j + \alpha^k = 0$ leads to $\alpha^j = \alpha^k$, with $0 \le j < k \le 14$, a contradiction since $1, \alpha, \alpha^2, \ldots, \alpha^{14}$ are distinct elements of $GF(2^4)$.]

The person receiving the word y knows only y and $y(\lambda)$; he has to deduce the positions j, k occupied by the incorrect digits. To do this, he evaluates $y(\lambda)$ at α and at α^3. Because $g(\lambda) \mid x(\lambda)$, and $g(\alpha) = 0$, $g(\alpha^3) = 0$, we have $x(\alpha) = 0$, $x(\alpha^3) = 0$, so that

$$\begin{aligned} y(\alpha) &= \alpha^j + \alpha^k, \\ y(\alpha^3) &= \alpha^{3j} + \alpha^{3k}. \end{aligned} \tag{7}$$

To simplify notation, let $s_1 = y(\alpha)$, $s_3 = y(\alpha^3)$. (Code theorists usually give the name *syndrome* to s_1, s_3.) Let $z_1 = \alpha^j$, $z_2 = \alpha^k$. Then we may rewrite (7) as a pair of equations

$$\begin{aligned} s_1 &= z_1 + z_2 \\ s_3 &= z_1{}^3 + z_2{}^3 \end{aligned} \tag{8}$$

in which s_1, s_3 are known quantities and z_1, z_2 are unknowns. Both s_1, s_3 lie in $GF(2^4)$, not necessarily in $\mathbf{Z}_2$. The first equation (8) yields $z_2 = s_1 + z_1$; substituting this into the second equation (8) produces

$$\begin{aligned} s_3 &= z_1^3 + (s_1 + z_1)^3 \\ &= z_1^3 + s_1^3 + 3s_1^2 z_1 + 3s_1 z_1^2 + z_1^3 \\ &= s_1^3 + s_1^2 z_1 + s_1 z_1^2 \end{aligned}$$

(using $-1 = 1$ and $2 = 0$). Thus z_1 satisfies the quadratic equation

$$s_3 + s_1^3 + s_1^2 z + s_1 z^2 = 0, \tag{9}$$

and similarly z_2 also satisfies this equation. By calculating the roots of this quadratic equation in $GF(2^4)$, the person receiving y obtains $z_1 = \alpha^j$, $z_2 = \alpha^k$. Knowing these roots, he then knows the integral exponents j, k and hence knows that digits y_j, y_k in his received word y are incorrect. Correcting these digits, he then obtains the transmitted word x.

If there is only one transmission error, say in the jth digit, then $y(\lambda) = x(\lambda) + \lambda^j$, so that

$$s_1 = y(\alpha) = \alpha^j, \qquad s_3 = y(\alpha^3) = \alpha^{3j}.$$

Thus $s_3 = s_1^3$. In this case, (9) becomes $s_1 z(s_1 + z) = 0$, and its roots are 0 and $s_1 = \alpha^j$. That is, when there is just one transmission error, (9) has two roots: one zero, one nonzero, with the nonzero root giving the location of the single incorrect digit.

Thus, by calculating the roots in $GF(2^4)$ of the quadratic equation (9), the location of the incorrect digits in the received word can be determined, when there are no more than two errors. These errors may then be corrected. More than two errors cannot be corrected by this procedure (for this code.)

This code is a member of the *BCH family* of error correcting codes (the initials are taken from the names of the discoverers). There are BCH codes capable of correcting t errors, for general $t \geq 1$; we use the above example to illustrate the construction of these codes. If we imitate the discussion just given, but using $g(\lambda) = \lambda^4 + \lambda + 1$, we obtain a code with $2^{11} = 2048$ code words. Because α, but not α^3, is now a root of $g(\lambda)$, only single error correction is possible. The error correction is accomplished by evaluating the received word-polynomial $y(\lambda)$ at α, to obtain the syndrome $s_1 = y(\alpha)$. If the single error is in the digit in position j, then $y(\alpha) = \alpha^j$, and thus we obtain j from $s_1 = \alpha^j$. If, instead, we take

$$g(\lambda) = (\lambda^4 + \lambda + 1)(\lambda^4 + \lambda^3 + \lambda^2 + \lambda + 1)(\lambda^2 + \lambda + 1),$$

then α, α^3, α^5 are all roots of $g(\lambda)$. This time the code has $2^5 = 32$ code words, and any combination of three or fewer transmission errors may be corrected. The error correction is accomplished by calculating the syndromes $s_1 = y(\alpha)$, $s_3 = y(\alpha^3)$, $s_5 = y(\alpha^5)$. If the errors are in positions j, k, l, and if $z_1 = \alpha^j$, $z_2 = \alpha^k$, $z_3 = \alpha^l$, then we have

$$s_1 = z_1 + z_2 + z_3\,, \qquad s_3 = z_1{}^3 + z_2{}^3 + z_3{}^3, \qquad s_5 = z_1{}^5 + z_2{}^5 + z_3{}^5.$$

It can be shown that this system of three equations leads to one cubic equation with coefficients involving s_1, s_3, s_5 and roots z_1, z_2, z_3. By calculating the roots of this cubic equation, we obtain z_1, z_2, z_3 and hence the locations j, k, l of the incorrect digits.

By taking codes with more digits, and polynomials $g(\lambda)$ with carefully chosen elements of a finite field as roots, it is possible to construct in this manner codes capable of correcting t transmission errors, for any given t. In general, the error correction will involve calculation of the roots of a tth degree polynomial within this finite field. These calculations may require substantial amounts of computation and therefore require much computation with congruences, mod $m(\lambda)$, within the polynomial ring $\mathbf{Z}_2[\lambda]$. If decoding is to be performed on a computing machine, the machine has to be programmed to perform the requisite polynomial operations of addition, subtraction, multiplication, long division. The calculation of a multiplicative inverse within the finite field involves solving a congruence of the form $a(\lambda)x(\lambda) \equiv 1 \pmod{m(\lambda)}$; for this, the procedure described at the end of §3.4 is very efficient.

The BCH codes are used in modern communication systems, and some practical experience with them is reported in the book of Birkhoff and Bartee listed on page 455. The reader may also be interested in consulting the books of Berlekamp and Peterson-Weldon. These are fascinating books, giving the foundations of algebraic coding theory, with the theory of finite fields and algebraic codes tightly woven into a discussion of the equipment to implement the codes.

EXERCISES 7.3

I

1. Obtain the quotient and remainder in $\mathbf{Z}_2[\lambda]$ when $\lambda^5 + \lambda^2 + 1$ is divided by $\lambda^2 + \lambda$. Do the same in $\mathbf{Z}_3[\lambda]$ and in $\mathbf{Z}_5[\lambda]$.

2. Construct addition and multiplication tables for $\mathbf{Z}_2[\lambda]_m$, where $m(\lambda) = \lambda^2 + \lambda$. Repeat with $m(\lambda) = \lambda^2 + 1$ and with $m(\lambda) = \lambda^2$. Are any of these rings fields? integral domains?

3. Construct addition and multiplication tables for $\mathbf{Z}_3[\lambda]_m$, where $m(\lambda) = \lambda^2 + 1$. Repeat with $m(\lambda) = \lambda^2 + \lambda + 2$. Are the two rings obtained this way fields? Can you see any pattern common to the two pairs of tables?

4. Find the derivative of the polynomial $(\lambda + 1)^2$ in $\mathbf{Z}_2[\lambda]$. Find also the derivative of the polynomial $(\lambda + 1)^3$ in $\mathbf{Z}_3[\lambda]$.

5. If $m(\lambda) = \lambda^2 + \lambda + 1$, show that each element of $\mathbf{Z}_2[\lambda]_m$ is a root of the polynomial $y^4 - y$. Factor this polynomial into primes in $\mathbf{Z}_2[\lambda]_m$.

6. Find all first, second and third degree primes in $\mathbf{Z}_2[\lambda]$.

II

7. For each positive prime integer p, give an example of a field containing exactly p elements. Give an example of a finite ring having m elements, for each positive integer m.

8. Give an example of a field having exactly four elements; exactly eight elements; exactly twenty five elements.

9. Find an example of a ring in which the equation $x^2 = 1$ has four solutions.

10. Give examples of multiplicative groups containing exactly $p - 1$ elements, where p is a positive prime integer.

11. Let $\mathfrak{F}$ be a finite field. Prove, without using Theorem 7.3.4, that $\mathfrak{F}[\lambda]_{m(\lambda)}$ is a finite integral domain if and only if it is a finite field.

12. If $m(\lambda)$ is a composite element of $\mathfrak{F}[\lambda]$, describe all zero divisors in $\mathfrak{F}[\lambda]_m$. Do the same for $\mathbf{Z}_m$ if m is a composite integer.

III

13. Let $m(\lambda)$ be a nonzero polynomial in $\mathfrak{F}[\lambda]$. Using the system of distinct representatives for the congruence classes $(\text{mod}\, m(\lambda))$ given in Theorem 7.2.3, prove that $\mathfrak{F}[\lambda]_{m(\lambda)}$ is a finite ring if and only if $\mathfrak{F}$ is a finite field.

14. Let $\alpha_1, \alpha_2, \ldots, \alpha_{p-1}$ denote the nonzero elements of $\mathbf{Z}_p$. Let α be a fixed one of these elements.
    (a) Show that $\alpha\alpha_1, \alpha\alpha_2, \ldots, \alpha\alpha_{p-1}$ are different elements of $\mathbf{Z}_p$ and are all nonzero.
    (b) Conclude that $\alpha\alpha_1, \alpha\alpha_2, \ldots, \alpha\alpha_{p-1}$ are just $\alpha_1, \ldots, \alpha_{p-1}$ in some order, and hence that $(\alpha\alpha_1)(\alpha\alpha_2)\cdots(\alpha\alpha_{p-1}) = \alpha_1\alpha_2\cdots\alpha_{p-1}$.
    (c) Deduce $\alpha^{p-1} = 1$.
    (d) As an immediate consequence, deduce FERMAT'S LITTLE THEOREM:

    If a is an integer relatively prime to the positive prime integer p, then $a^{p-1} \equiv 1 \pmod p$.

15. By imitating the proof of Theorem 7.3.4, establish the following stronger result: *A finite commutative ring without zero divisors must be a field.*

16. Verify the assertions made in Example 7.3.5 about single and triple error correcting codes.

17. Verify that

$$(x_{14} + x_0\lambda + \cdots + x_{12}\lambda^{13} + x_{13}\lambda^{14}$$
$$= (x_0 + x_1\lambda + \cdots + x_{14}\lambda^{14})\lambda - x_{14}(\lambda^{15} - 1).$$

From this identity, deduce that if $[x_0, x_1, \ldots, x_{14}]$ is a word in any of the codes in Example 7.3.5, so is the word $[x_{14}, x_0, x_1, \ldots, x_{13}]$ obtained by cyclically shifting the digits. [Suggestion: Check that $g(\lambda) \mid (\lambda^{15} - 1)$.]

7.4 FUNCTIONS

The function concept has been encountered in several places in this book (e.g., Example 1.4.8, §2.5.) This important concept will play a major role in the remaining sections of this chapter, and we now give an abstract approach to it.

Let S and T be nonempty sets. Recall, from §6.2, that the *Cartesian product* $S \times T$ consists of all ordered pairs (s, t) in which $s \in S$ and $t \in T$. Ordered pairs (s, t) and (s', t') are equal if and only if $s = s'$ and $t = t'$.

DEFINITION 7.4.1
Let S and T be nonempty sets. A function f, *defined on S and taking values in T, is a subset of $S \times T$ such that*

if $s \in S$, there is precisely one $t \in T$ for which $(s, t) \in f$.

The unique t for which $(s, t) \in f$ is called the functional value *of f at s and is denoted by $f(s)$.*

Of course, the element t for which $(s, t) \in f$ need not be the same for every s.

When introducing new mathematical objects, we must always specify clearly when two of the objects are equal. Since a function is a subset of $S \times T$, it is natural to use equality of sets as the criterion for equality of functions.

DEFINITION 7.4.2
Functions f_1 and f_2, defined on S and taking values in T, are equal *if and only if f_1 and f_2 are the same subset of $S \times T$.*

Although expressed abstractly, Definition 7.4.1 simply states that corresponding to an element s of S, there is an element of T, denoted by $f(s)$, and called the *functional value of f at s*. Furthermore, if f_1, f_2 are functions from S to T, Definition 7.4.2 asserts that

$$f_1 = f_2 \quad \textit{if and only if} \quad f_1(s) = f_2(s) \quad \textit{for every } s \textit{ in } S.$$

Example 7.4.3

Take both S and T to be the real numbers $\mathbf{R}$. Define a function f by this rule: $f(x) = x^2$, for every $x \in \mathbf{R}$. Here $S \times T$ is the Cartesian plane $\mathbf{R} \times \mathbf{R}$, and the subset f of this plane consists of all points (x, y) for which $y = x^2$, that is, all points on the parabola with equation $y = x^2$. (See Figure 7.4.1.) This subset of $\mathbf{R} \times \mathbf{R}$ is thus what one often calls the *graph* of the function. More generally, if f is any function defined on $\mathbf{R}$ and taking values in $\mathbf{R}$, the subset f of the Cartesian plane $\mathbf{R} \times \mathbf{R}$ consists precisely of points (x, y) for which $y = f(x)$ and so is what one commonly regards as the graph of the function. Thus a function is simply a graph and computing a functional value $f(x)$ amounts to locating a point $(x, f(x))$ on the graph. The definition of equality of functions is therefore just this: Functions are equal precisely when their graphs are the same.

It is still worthwhile, when S, T are abstract sets, to think of the subset of $S \times T$ specifying a function f as the graph of f.

When we have sets S, T and wish to speak of a function f defined on S and taking values in T, we often simply say that "f is a function from S to T." If T is S we shorten this still more and say that "f is a function on S."

Example 7.4.4

If $f(\lambda)$ is a polynomial with coefficients in $\mathbf{R}$, we may use f to define a *polynomial function* on $\mathbf{R}$, also denoted by f, by specifying, for any $x \in \mathbf{R}$, that the functional value $f(x)$ is the real number obtained by evaluating the polynomial $f(\lambda)$ at x, in the manner described in §3.2.

If $f_1(\lambda), f_2(\lambda) \in \mathbf{R}[\lambda]$, is it true that f_1 and f_2 are equal *as polynomials* if and only if the associated polynomial functions f_1 and f_2 are equal *as functions*? Equality as polynomials means, of course, that the coefficients of λ^k in $f_1(\lambda)$ and $f_2(\lambda)$ are the same, for each $k = 0, 1, 2, \ldots,$ whereas equality as functions means $f_1(x) = f_2(x)$ for each real number x. The answer to this question is yes.

PROOF If f_1 and f_2 are equal as polynomials, obviously $f_1(x) = f_2(x)$ for each real x. Conversely, suppose $f_1(x) = f_2(x)$ for each real x. Each real number x is then a root of the polynomial $f_1 - f_2$. By corollary 5.2.5, a nonzero polynomial has only finitely many roots. Since $f_1 - f_2$ has infinitely many, it must be the zero polynomial, meaning that each coefficient of $f_1 - f_2$ is zero, so that f_1 and f_2 are equal as polynomials. □

Example 7.4.5

Generalizing Example 7.4.4, let $\mathfrak{F}$ be any field and $f(\lambda)$ an element of the polynomial ring $\mathfrak{F}[\lambda]$. We define a polynomial function on $\mathfrak{F}$, also denoted

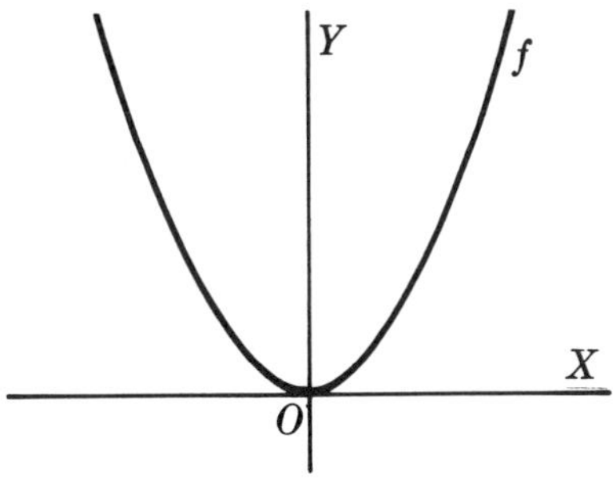

Figure 7.4.1

by f, by specifying the functional value $f(x)$ for each x in $\mathfrak{F}$ to be the value obtained by substituting x for λ in the manner explained in §3.2; i.e., evaluate f at x.

We now repeat the question asked in Example 7.4.4. If $f_1(\lambda), f_2(\lambda) \in \mathfrak{F}[\lambda]$, are f_1 and f_2 equal as polynomials (i.e., same coefficients) if and only if f_1 and f_2 are equal as functions (i.e., $f_1(x) = f_2(x)$ for every x in $\mathfrak{F}$)? The answer depends on the field $\mathfrak{F}$. If $\mathfrak{F}$ has infinitely many elements the answer is indeed yes, but otherwise may be no. Clearly $f_1(x) = f_2(x)$ for every x in $\mathfrak{F}$ when f_1 and f_2 have the same coefficients. Equality as polynomials therefore always implies equality as functions. Conversely, if $f_1(x) = f_2(x)$ for every x in $\mathfrak{F}$, each element of $\mathfrak{F}$ is a root of the polynomial $f_1 - f_2$. When $\mathfrak{F}$ has infinitely many elements, $f_1 - f_2$ will thus have infinitely many roots and (by corollary 5.2.5) must be the zero polynomial. Thus, when $\mathfrak{F}$ has infinitely many elements, equality of f_1 and f_2 as functions on $\mathfrak{F}$ implies equality as elements of $\mathfrak{F}[\lambda]$. That this is not the case when $\mathfrak{F}$ is a finite field is shown by the polynomials

$$f_1(\lambda) = \lambda^2 + \lambda, \qquad f_2(\lambda) = \lambda^3 + \lambda$$

in $\mathbf{Z}_2[\lambda]$. Clearly f_1 and f_2 are different elements of $\mathbf{Z}_2[\lambda]$, yet $f_1(0) = f_2(0)$, $f_1(1) = f_2(1)$, so that $f_1(x) = f_2(x)$ for each x in $\mathbf{Z}_2$.

***THEOREM* 7.4.6**
Let $\mathfrak{F}$ be an infinite field and f_1, f_2 polynomials in $\mathfrak{F}[\lambda]$. Then f_1 and f_2 are equal as polynomials if and only if f_1 and f_2 are equal as functions.

***THEOREM* 7.4.7**
Let $\mathfrak{F}$ be a finite field consisting of n elements $\alpha_1, \alpha_2, \ldots, \alpha_n$. Set

$$m(\lambda) = (\lambda - \alpha_1)(\lambda - \alpha_2)\cdots(\lambda - \alpha_n).$$

Then polynomials $f_1(\lambda)$, $f_2(\lambda)$ with coefficients in $\mathfrak{F}$ yield equal polynomial functions on $\mathfrak{F}$ if and only if

$$f_1(\lambda) \equiv f_2(\lambda) \pmod{m(\lambda)}.$$

PROOF Notice that the roots of $m(\lambda)$ are $\alpha_1, \alpha_2, \ldots, \alpha_n$. Suppose that $f_1(x) = f_2(x)$ for all $x \in \mathfrak{F}$. Every element x of $\mathfrak{F}$ is then a root of $f_1 - f_2$, so that the distinct primes $\lambda - \alpha_1, \lambda - \alpha_2, \ldots, \lambda - \alpha_n$ must all divide $f_1 - f_2$. Consequently, their product $m(\lambda)$ also divides $f_1(\lambda) - f_2(\lambda)$, implying that $f_1(\lambda) \equiv f_2(\lambda) \pmod{m(\lambda)}$. Conversely, if $f_1(\lambda) \equiv f_2(\lambda) \pmod{m(\lambda)}$, then $f_1(\lambda) = f_2(\lambda) + g(\lambda)m(\lambda)$ for some polynomial $g(\lambda)$, so that $f_1(x) = f_2(x)$ for each $x \in \mathfrak{F}$, since $m(x) = 0$ for every element x of $\mathfrak{F}$. □

The polynomial $m(\lambda)$ in this theorem can be explicitly calculated: It is

$$m(\lambda) = \lambda^n - \lambda.$$

The proof is outlined in Exercise 16. By the remarks about finite fields in §7.3, the integer n in Theorem 7.4.7 must have the form $n = p^k$ for some positive prime integer p.

Theorem 7.4.7 shows that the concept of a polynomial as a sequence (§3.2) does not always coincide with the concept of a polynomial function. For this reason the sequence approach to polynomials was used in §3.2, and the function approach usually given in calculus was avoided.

An interesting and easily proved fact is that *any* function on a finite field equals some *polynomial function* on the field. The proof is in Exercise 11.

Frequently functions are added, subtracted, multiplied. To discuss this, let S be an arbitrary nonempty set and $\mathcal{S}$ a ring. We consider functions f_1 and f_2 defined on S and taking values in $\mathcal{S}$.

The sum $f_1 + f_2$ and the difference $f_1 - f_2$ are the functions, defined on S and taking values in $\mathcal{S}$, specified by these rules:

$$\left.\begin{aligned}(f_1 + f_2)(s) &= f_1(s) + f_2(s)\\ (f_1 - f_2)(s) &= f_1(s) - f_2(s)\end{aligned}\right\} \quad \textit{for all } s \textit{ in } S.$$

(NOTE The $+$ and $-$ signs on the left-hand side denote addition and subtraction of functions, whereas the $+$ and $-$ on the right-hand side denote addition or subtraction of elements of the ring $\mathcal{S}$.)

When $S = \mathcal{S} = \mathfrak{F}$ and $f_1(\lambda)$, $f_2(\lambda)$ are polynomials in $\mathfrak{F}[\lambda]$, from §3.2 we get the substitution rules

$$\left.\begin{aligned}(f_1 + f_2)(\alpha) &= f_1(\alpha) + f_2(\alpha)\\ (f_1 - f_2)(\alpha) &= f_1(\alpha) - f_2(\alpha)\end{aligned}\right\} \quad \textit{for all } \alpha \textit{ in } \mathfrak{F}.$$

These substitution rules show that addition or substraction of polynomials does not conflict with addition or subtraction of the corresponding polynomial functions.

Several different kinds of function multiplication occur in mathematics. We now describe two that are particularly important, called the *pointwise product* and *composition*. A third type, called *convolution*, will be seen in §8.6.

The pointwise product

With S a nonempty set, and $\mathcal{S}$ a ring, as above, suppose that f_1 and f_2 are functions from S to $\mathcal{S}$. *The pointwise product of* f_1 *and* f_2 *is the function* f_1f_2 *specified by*

$$(f_1 f_2)(s) = f_1(s)f_2(s) \qquad \text{for each } s \text{ in } S.$$

That is, $f_1 f_2$, when evaluated at s, yields as functional value the product of the elements $f_1(s)$ and $f_2(s)$ of the ring $\mathcal{S}$. Clearly, $f_1 f_2$ is again a function from S to $\mathcal{S}$.

We have already seen a special case of this multiplication. Let S and $\mathcal{S}$ both be $\mathfrak{F}$. If f_1 and f_2 are polynomials in $\mathfrak{F}[\lambda]$, from §3.2 we have

$$(f_1 f_2)(\alpha) = f_1(\alpha)f_2(\alpha) \qquad \text{for all } \alpha \text{ in } \mathfrak{F}.$$

Thus *the multiplication used when working with polynomial functions is the pointwise product.*

THEOREM 7.4.8
Let S be a nonempty set, $\mathcal{S}$ a ring. The collection of all functions from S to $\mathcal{S}$, when endowed as above with addition and pointwise product multiplication, forms a ring. This ring is commutative if $\mathcal{S}$ is commutative.

We ask the reader to supply for himself the easy details of the proof, merely commenting that the zero element is the function f for which

$$f(s) = 0 \qquad \text{for every } s \text{ in } S.$$

We call this function the *zero function,* and denote it by 0. Thus a function f is nonzero if $f(s) \neq 0$ for at least one s.

When $\mathfrak{F}$ is a field, the polynomial functions from $\mathfrak{F}$ to itself also form a ring under function addition and pointwise product multiplication. This ring is a subring of the ring of all functions from $\mathfrak{F}$ to itself (the ring of Theorem 7.4.8 for $S = \mathfrak{F}$). When $\mathfrak{F}$ is a finite field, these rings coincide since any function on $\mathfrak{F}$ is then a polynomial function (Exercise 11).

Composition

Let S, T, U be nonempty sets. Suppose g is a function from S to T, and f a function from T to U. *The composite* fg *is the function specified by*

$$(fg)(s) = f(g(s)) \qquad \text{for each } s \in S.$$

That is, to evaluate fg at s, evaluate g at s to get an element $t = g(s) \in T$; then evaluate f at t. (See Figure 7.4.2.) Clearly fg is a function defined on S and taking values in U. Composition, in the special case $S = T = U = \mathbf{Z}$, was encountered in Example 1.4.8, for a restricted class of functions, and

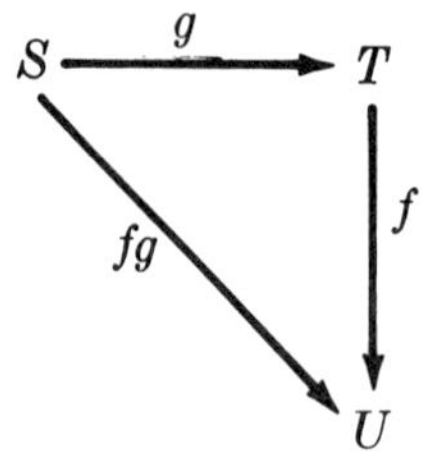

Figure 7.4.2

in the special case $S = T = U = \mathbf{C}$, was also encountered in §2.5. The reader may wish to read Example 1.4.8 and §2.5 now if these sections were omitted when studying Chapters 1 and 2.

When working with composition, it is very important to pay careful attention to the order of factors in a product. The reader will see an illustration of this in Example 1.4.8, where there are functions f, g for which $fg \neq gf$. Other illustrations of noncommutativity may be found in §2.5. It may even happen, when S, T, U are different sets, that fg is defined but gf is not.

Suppose now that R, S, T, U are nonempty sets, and that h is a function from R to S, g a function from S to T, and f a function from T to U. (See Figure 7.4.3.) Then gh is a function from R to T, so that

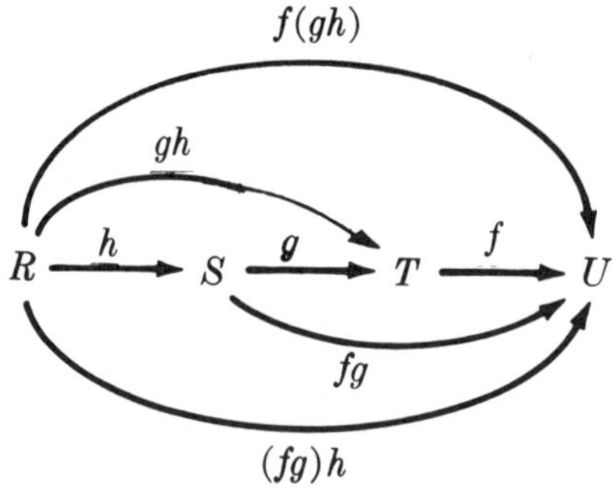

Figure 7.4.3

$f(gh)$ is a function from R to U. Furthermore, fg is a function from S to U, so that $(fg)h$ is also a function from R to U. Are $(fg)h$, $f(gh)$ equal functions? They are, and the very easy proof amounts to showing that both $(fg)h$ and $f(gh)$, when evaluated at an arbitrary $r \in R$, yield $f(g(h(r)))$. [See the proof of associativity in Theorem 2.5.8.] Therefore

composition of functions is associative.

The *identity function* on a set S is the function i_S defined by

$$i_S(s) = s \qquad \text{for all } s \in S.$$

If g is any function from S to T, by evaluating $i_T g$ and $g i_S$ at an arbitrary

$s \in S$, we deduce that

$$i_T g = g \quad \text{and} \quad g i_S = g. \tag{1}$$

These equations show that

composition of an identity function with another function gives the other function.

Take R, S, T all to be a ring $\mathcal{R}$. Does the collection of all functions from $\mathcal{R}$ to itself form a ring under function addition and composition? The answer is no. The reader should check this for himself. (Exercise 10.) He will see that the answer is almost yes in the sense that the only ring axiom not satisfied is one of the distributive laws.

However, if S is an arbitrary nonempty set, the collection of all functions from S to itself forms a monoid under composition. *Proof:* Composition of functions from S to itself gives a function from S to itself, associativity is proved above, and the identity function i_S is the identity of the monoid. □

This monoid is usually noncommutative. For further results, see Exercise 10.

Surjections, injections, bijections

The next definition introduces concepts that will play an important role in subsequent sections of this chapter.

DEFINITION 7.4.9

Let S, T be nonempty sets and f a function defined on S and taking values in T. Then:

(i) *f is* surjective, *or* "onto", *if, whenever $t \in T$, there is at least one $s \in S$ for which $f(s) = t$;*

(ii) *f is* injective, *or* "one-to-one", *if, whenever elements s_1 and s_2 of S satisfy $f(s_1) = f(s_2)$, then $s_1 = s_2$;*

(iii) *f is* bijective *if it is both surjective and injective.*

There is a brief discussion of bijectivity and many examples of bijective functions on $\mathbf{C}$ in §2.5. If f is surjective, we call f a *surjection;* if injective, and *injection;* and if bijective, a *bijection.*

The surjective property (i) requires that each element t of T be a functional value $f(s)$. The injective property (ii) requires that unequal elements s_1, s_2 of S yield functional values $f(s_1)$, $f(s_2)$ which are also unequal. The bijective property (iii) requires this: if $t \in T$, there is precisely one $s \in S$ for which $f(s) = t$.

These ideas may be illustrated graphically in the following way. Imagine set S as an X axis, set T as a Y axis, in a Cartesian coordinate plane. If $s \in S$, the functional value $t = f(s)$ is found moving vertically from s to the graph of f, then from there horizontally to t in T (see Figure 7.4.4).

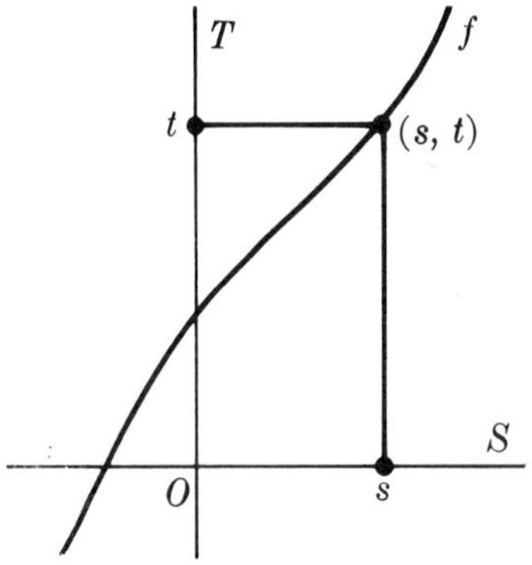

Figure 7.4.4

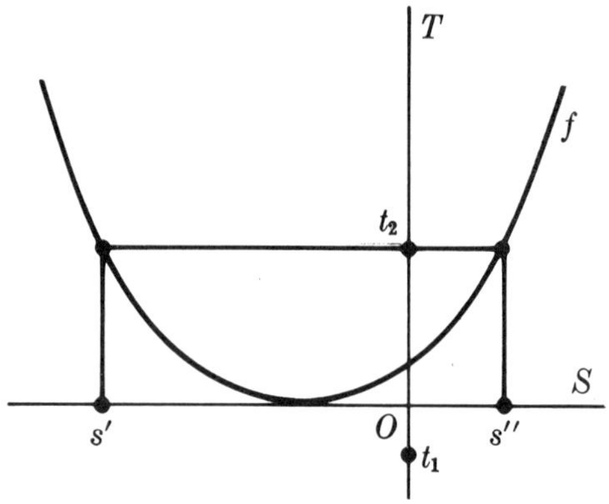

Figure 7.4.5

Surjectivity therefore requires that a horizontal line through an element t on the T axis meet the graph of f at least once. Figure 7.4.5 shows a function f and an element t_1 in T for which there is no s in S such that $f(s) = t_1$. Injectivity requires that for any element t on the T axis, the horizontal line through t meets the graph of f at most once. Figure 7.4.5 shows an element t_2 in T for which $f(s) = t_2$ holds for two elements of S, namely s', s''. Bijectivity then has this graphical interpretation: for each element t on the T axis, the horizontal line through t meets the graph of f exactly once. This is illustrated in Figure 7.4.4.

Composition and inversion of bijections

Let S, T, U be nonempty sets. If g is a bijection from S to T, and f a bijection from T to U, composition produces a function fg from S to U which we claim is also a bijection. PROOF Let $s_1, s_2 \in S$. To prove that fg is injective, suppose $(fg)(s_1) = (fg)(s_2)$. Then $f(g(s_1)) = f(g(s_2))$. Because f is injective, we obtain $g(s_1) = g(s_2)$, and then $s_1 = s_2$ because g is injective. Therefore fg is injective. To prove that it is also surjective, let $u \in U$. Then $u = f(t)$ for some $t \in T$ since f is surjective, and $t = g(s)$ for some $s \in S$, since g is also surjective. But then $(fg)(s) = u$. Thus fg is surjective as well as injective, and therefore is bijective. Consequently,

composition of bijections gives a bijection. □

Evidently the identity function on any nonempty set is a bijection.

Now suppose f is a bijection from T to U. We claim that a unique function F from U to T exists such that

$$Ff = i_T, \qquad fF = i_U. \tag{2}$$

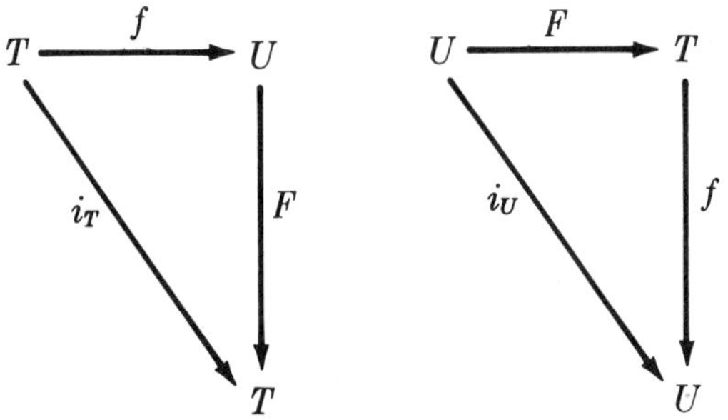

Figure 7.4.6

This function F is called the *inverse* of f, and it is also a bijection, but from U to T. We prove first that there is a function F satisfying (2), afterwards that F is a bijection, and finally the uniqueness of F. Let $u \in U$. Because f is a bijection, there is precisely one element $t \in T$ for which $f(t) = u$. We define F by specifying that $F(u) = t$, *i.e.*,

$$F(u) = t \quad \text{if} \quad f(t) = u; \qquad t \in T; u \in U. \tag{3}$$

(Compare §2.5.) From (3) we get

$$f(F(u)) = u \qquad \text{for any } u \in U.$$

On the other hand, if $t \in T$, by letting $u = f(t)$, from (3) we get

$$F(f(t)) = t \qquad \text{for any } t \in T.$$

Thus $(fF)(u) = i_U(u)$ and $(Ff)(t) = i_T(t)$ for all $t \in T$, $u \in U$. This shows that F satisfies equations (2).

Using (2), we deduce that F is injective and bijective, in this way: Suppose u_1, u_2 are elements of U with $F(u_1) = F(u_2)$. Then

$$u_1 = i_U(u_1) = (fF)(u_1) = f(F(u_1)) = f(F(u_2)) = (fF)(u_2) = i_U(u_2) = u_2.$$

Therefore F is injective. Let $t \in T$. Then

$$t = i_T(t) = (Ff)(t) = F(f(t)) = F(u) \qquad \text{for } u = f(t).$$

Therefore F is also surjective and hence is a bijection.

To prove the uniqueness of F, suppose F_1 is a function from U to T such that

$$F_1 f = i_T, \qquad fF_1 = i_U. \tag{4}$$

Using (1), (2), (4) and the associativity of composition, we get

$$F = i_T F = (F_1 f)F = F_1(fF) = F_1 i_U = F_1.$$

Thus F is unique. □

The unique bijection F satisfying (2) is often denoted by f^{-1}. Note that f^{-1}, when evaluated at $u \in U$, is usually *not* $1/f(u)$.

THEOREM 7.4.10
If f is a bijection from T to U, there is a unique inverse function f^{-1} from U to T, satisfying $ff^{-1} = i_U$ and $f^{-1}f = i_T$. This inverse function is also a bijection.

Example 7.4.11
If $T = U = \mathbf{R}$, the function f given by

$$f(x) = x^3 \qquad \text{for every } x \in \mathbf{R},$$

is a bijection. Its inverse is the function F defined by

$$F(x) = x^{1/3} \qquad \text{for all } x \in \mathbf{R},$$

$x^{1/3}$ denoting the real cube root of x.

For other examples of inverse functions, see Example 1.4.8 and §2.5.

Taking $S = T = U$ produces interesting and usually noncommutative examples of groups.

THEOREM 7.4.12
The collection of all bijections on a nonempty set S is a group under the composition operation. If S has at least three elements, this group is noncommutative.

PROOF Closure holds since composition of bijections gives a bijection; associativity because composition of any functions is associative. The group identity is the identity function, and the group inverse is the inverse function. Suppose S has at least three elements, and let s_1, s_2, s_3 be three elements of S. Define functions f, g on S in this way:

$$f(s_1) = s_2,\ f(s_2) = s_1,\ f(s) = s \text{ for all elements } s \text{ of } S \text{ other than } s_1, s_2;$$

$$g(s_2) = s_3,\ g(s_3) = s_2,\ g(s) = s \text{ for all elements } s \text{ of } S \text{ other than } s_2, s_3.$$

Plainly f, g are bijections on S. Since

$$(gf)(s_1) = s_3, \qquad (fg)(s_1) = s_2,$$

evidently $gf \neq fg$. Noncommutativity is now clear. □

This group is called the *group of bijections on* S, or the *symmetric group* on S. In the special case when S is $\mathbf{C}$, it is the group denoted by $\mathfrak{B}$ in §2.5.

REMARK A function defined on S and taking values in T is often called a *mapping* from S to T. Usually "function" is used in elementary work, e.g., in calculus; whereas "mapping" is often used in more advanced work. In §7.5, we shall study mappings from S to T when both are rings.

EXERCISES 7.4

I

1. Define functions f_1 and f_2 on a ring $\mathcal{R}$ and taking values in $\mathcal{R}$, by specifying that

$$f_1(r) = r^2, \quad f_2(r) = r^3, \qquad \text{for all } r \text{ in } \mathcal{R}.$$

Determine whether the functions f_1, f_2 are equal, when $\mathcal{R}$ is:

(a) $\mathbf{R}$, (b) $\mathbf{C}$, (c) $\mathbf{Z}$, (d) $\mathbf{Z}_2$, (e) $\mathbf{Z}_3$, (f) $\mathbf{Z}_2[\lambda]$,

(g) $\mathbf{Z}_2[\lambda]_m$ with $m(\lambda) = \lambda^2 + \lambda + 1$.

2. For which of the rings $\mathcal{R}$ in Exercise 1 is f_1 surjective? injective? bijective? Repeat, using f_2.

3. Let f_1 and f_2 be the functions given in Exercise 1. If r is an element of $\mathcal{R}$, what are $f_1 + f_2$, $f_1 - f_2$, f_1f_2, f_2f_1 when evaluated at r? Answer this for each of the rings $\mathcal{R}$ listed in Exercise 1 and for the two types of multiplication described in the text.

4. Find nonzero polynomial functions f_1, f_2 on $\mathbf{Z}_2$ for which $f_1 f_2$ is the zero function. Do this for the pointwise product and also for composition.

5. In (a), (b), (c), finite sets S, T of integers are given, together with a function f from S to T. Which of these functions are injections? surjections? bijections?

(a) $S = \{-1, 0, 1\}$, $T = \{-1, 0, 1\}$, $f(s) = s^2$ for all $s \in S$;

(b) $S = \{0, 1\}$, $T = \{-1, 0, 1\}$, $f(s) = s^2$ for all $s \in S$;

(c) $S = \{0, 1\}$, $T = \{0, 1\}$, $f(s) = s^2$ for all $s \in S$.

6. Let $S = \{1, 2, 3, 4\}$ and f the function on S given by $f(1) = 2$, $f(2) = 3$, $f(3) = 4$, $f(4) = 1$. If g is a function on S satisfying $fg = gf$ and $g(1) = 3$, find $g(2)$, $g(3)$, $g(4)$. Is $g = ff$?

7. Let f be a function on $\mathbf{R}$, given by $f(x) = ax + b$, for all $x \in \mathbf{R}$. For which a, b does f have an inverse? When f has an inverse, give a formula for $f^{-1}(x)$.

8. If S is $\mathbf{R}$ and T is all positive real numbers, what is the inverse of the function f from S to T given by $f(x) = e^x$ for all $x \in \mathbf{R}$?

II

9. Let S be a finite set and f a function on S to itself. Prove that f is injective if and only if it is surjective.

10. (i) Supply the proof of Theorem 7.4.8.
 (ii) Show that neither the set of all functions on a field to itself, nor the set of polynomial functions, form a ring under addition and composition. Which ring axiom is not satisfied? *(cont.)*

(iii) Show that the set of all polynomial functions on a field forms a noncommutative monoid under composition but never a group.

11. Let $\mathfrak{F}$ be a finite field consisting of n elements $\alpha_1, \alpha_2, \ldots, \alpha_n$. If g is any function on $\mathfrak{F}$ to itself, show that g equals the polynomial function obtained from the following polynomial:

$$\sum_{s=1}^{n} \frac{(\lambda - \alpha_1)\cdots(\lambda - \alpha_{s-1})(\lambda - \alpha_{s+1})\cdots(\lambda - \alpha_n)}{(\alpha_s - \alpha_1)\cdots(\alpha_s - \alpha_{s-1})(\alpha_s - \alpha_{s+1})\cdots(\alpha_s - \alpha_n)} g(\alpha_s).$$

(Hint: Evaluate this polynomial at α_s, for $s = 1, 2, \ldots, n$.)

12. Let T, U be nonempty sets and f a function from T to U. If there is a function F from U to T satisfying $Ff = i_T$, $fF = i_U$, prove that f is a bijection. Conclude that only bijections have inverse functions.

13. If a set S has precisely one or two elements, show that the group of bijections on S is commutative.

14. Suppose, under composition, that the polynomial function on $\mathbf{R}$ obtained from the polynomial $f(\lambda) \in \mathbf{R}[\lambda]$ has an inverse function which also is a polynomial function. Find $f(\lambda)$.

15. Let $f(\lambda) \in \mathbf{C}[\lambda]$. If $f(\lambda)$ is not a constant, show that the polynomial function on $\mathbf{C}$ obtained from f is surjective. Describe $f(\lambda)$ if this polynomial function is injective.

16. Let $\mathfrak{F}$ be a finite field, with $\alpha_1, \alpha_2, \ldots, \alpha_n$ its elements and $\alpha_n = 0$. Let α be a nonzero element of $\mathfrak{F}$ (one of $\alpha_1, \ldots, \alpha_{n-1}$).
    (i) Show that $\alpha\alpha_1 \neq \alpha\alpha_2$.
    (ii) Conclude that $\alpha\alpha_1, \alpha\alpha_2, \ldots, \alpha\alpha_{n-1}$ are $n - 1$ distinct nonzero elements of $\mathfrak{F}$, and therefore are $\alpha_1, \alpha_2, \ldots, \alpha_{n-1}$ in some order.
    (iii) Deduce that $(\alpha\alpha_1)\cdots(\alpha\alpha_{n-1}) = \alpha_1\cdots\alpha_{n-1}$.
    (iv) Cancel $\alpha_1\cdots\alpha_{n-1}$ and show that $\alpha^{n-1} = 1$ if α is any nonzero element of $\mathfrak{F}$.
    (v) Hence show that any element α of $\mathfrak{F}$, zero or nonzero, satisfies $\alpha^n = \alpha$.
    (vi) Conclude that each of $\alpha_1, \ldots, \alpha_n$ is a root of $\lambda^n - \lambda$, and hence that the polynomial $m(\lambda)$ in Theorem 7.4.7 is $\lambda^n - \lambda$.

III

17. Let $\mathfrak{F}$ be a finite field. Combine exercises 11 and 16 to give explicit formulas for all polynomials $f(\lambda) \in \mathfrak{F}[\lambda]$ for which the associated polynomial function on $\mathfrak{F}$ is a bijection.

18. If S is a finite set with n elements, compute the number of bijections on S.

19. Let $f(\lambda) \in \mathbf{C}[\lambda]$, with $\deg f(\lambda) \geq 2$. Show that the collection of all polynomials $g(\lambda) \in \mathbf{C}[\lambda]$ with $f(g(\lambda)) = g(f(\lambda))$ forms a monoid containing at most a finite number of polynomials of fixed degree. Show that this monoid contains at least one polynomial of degree $(\deg f(\lambda))^k$ for each $k = 0, 1, 2, \ldots$, and hence that this monoid is infinite. Is this monoid a group? (The reader will find further information in a recently published research article by W. M. Boyce, in the *Proceedings of the American Mathematical Society,* volume 33, 1972, pages 229–234. Many university libraries subscribe to this journal.)

20. Are there polynomial functions f_1, f_2 on $\mathbf{Z}_2$ for which $f_1 f_2$ is the zero function for both composition and the pointwise product?

7.5 MAPPINGS OF RINGS

In this section the symbol $\mathcal{R}$ denotes an arbitrary ring. We begin by introducing a type of mapping of rings often encountered in algebra.

DEFINITION 7.5.1
Let $\mathcal{R}$, $\mathcal{S}$ be rings. A mapping φ defined on $\mathcal{R}$ and taking values in $\mathcal{S}$ is a homomorphism *if*

$$\varphi(r_1 + r_2) = \varphi(r_1) + \varphi(r_2), \qquad \varphi(r_1 r_2) = \varphi(r_1)\varphi(r_2) \tag{1}$$

for all elements r_1, r_2 of $\mathcal{R}$.

Notice that the addition and multiplication on the left-hand sides of equations (1) takes place within $\mathcal{R}$, whereas that on the right-hand sides takes place within $\mathcal{S}$.

Example 7.5.2
Let the ring $\mathcal{R}$ be $\mathbf{Z}$, the integers, or $\mathfrak{F}[\lambda]$, the polynomials with coefficients in a field $\mathfrak{F}$. Take m to be an element of $\mathcal{R}$, and let $\mathcal{S}$ be $\mathcal{R}_m$, the congruence class ring modulo m (§7.2). Define a mapping φ from $\mathcal{R}$ to $\mathcal{R}_m$ by specifying its action on an element r of $\mathcal{R}$ to be

$$\varphi(r) = \bar{r},$$

$\bar{r}$ denoting the congruence class (mod m) represented by r. Equations (5) of §7.2 show that this mapping is a homomorphism. Specifically, if r_1, $r_2 \in \mathcal{R}$, then

$$\varphi(r_1 + r_2) = \overline{r_1 + r_2} = \bar{r}_1 + \bar{r}_2 = \varphi(r_1) + \varphi(r_2),$$
$$\varphi(r_1 r_2) = \overline{r_1 r_2} = \bar{r}_1 \bar{r}_2 = \varphi(r_1)\varphi(r_2).$$

This homomorphism is called the *natural homomorphism* of $\mathcal{R}$ onto $\mathcal{R}_m$.

Example 7.5.3
Let $\mathfrak{R}$ and $\mathfrak{S}$ both be the ring $\mathfrak{F}[\lambda]$ of polynomials with coefficients in a field $\mathfrak{F}$. Take $\alpha(\lambda)$ to be a fixed polynomial in $\mathfrak{F}[\lambda]$, and define a mapping φ_α of $\mathfrak{F}[\lambda]$ to itself by specifying its action on a polynomial $a(\lambda)$ to be

$$\varphi_\alpha(a(\lambda)) = a(\alpha(\lambda)).$$

Thus φ_α acts on a polynomial by evaluating it at α. The substitution rules

$$(a+b)(\alpha) = a(\alpha) + b(\alpha), \qquad (ab)(\alpha) = a(\alpha)b(\alpha),$$

proved in §3.2 for polynomials $a(\lambda)$, $b(\lambda) \in \mathfrak{F}[\lambda]$, show that φ_α is a homomorphism. Specifically,

$$\begin{aligned} \varphi_\alpha(a+b) &= (a+b)(\alpha) = a(\alpha) + b(\alpha) = \varphi_\alpha(a) + \varphi_\alpha(b), \\ \varphi_\alpha(ab) &= (ab)(\alpha) = a(\alpha)b(\alpha) = \varphi_\alpha(a)\varphi_\alpha(b). \end{aligned} \tag{2}$$

For each polynomial $\alpha(\lambda) \in \mathfrak{F}[\lambda]$, we obtain therefore a homomorphism φ_α from $\mathfrak{F}[\lambda]$ to itself. This homomorphism is called a *substitution homomorphism*, or an *evaluation homomorphism*. [Question: If polynomials $\alpha(\lambda)$, $\beta(\lambda)$ produce equal substitution homomorphisms φ_α, φ_β, are $\alpha(\lambda)$, $\beta(\lambda)$ equal?]

Example 7.5.4
Let $\mathfrak{F}$ be a field, and $\alpha(\lambda)$ a fixed *nonconstant* element of $\mathfrak{F}(\lambda)$. Define a mapping φ_α from $\mathfrak{F}(\lambda)$ to itself by specifying its action on a rational function $a(\lambda) \in \mathfrak{F}(\lambda)$ to be

$$\varphi_\alpha(a(\lambda)) = a(\alpha(\lambda)).$$

That is, φ_α acts on $a(\lambda)$ by evaluating it at α. We claim this evaluation is permissible, i.e., it does not produce a zero denominator. To prove this, let $g(\lambda) \in \mathfrak{F}[\lambda]$ be a nonzero polynomial, say

$$g(\lambda) = g_0 + g_1\lambda + \cdots + g_n\lambda^n, \qquad g_0, g_1, \ldots, g_n \in \mathfrak{F};\ g_n \neq 0,$$

and write $\alpha(\lambda)$ in lowest terms as

$$\alpha(\lambda) = \alpha_1(\lambda)/\alpha_2(\lambda), \qquad \alpha_1(\lambda), \alpha_2(\lambda) \in \mathfrak{F}[\lambda];\ (\alpha_1(\lambda), \alpha_2(\lambda)) = 1.$$

If $g(\alpha(\lambda)) = 0$, then $\alpha_2{}^n g(\alpha) = 0$, that is,

$$g_0\alpha_2{}^n + g_1\alpha_1\alpha_2{}^{n-1} + \cdots + g_{n-1}\alpha_1{}^{n-1}\alpha_2 + g_n\alpha_1{}^n = 0. \tag{3}$$

Let t be the smallest subscript such that $g_t \neq 0$; thus $g_0 = g_1 = \cdots = g_{t-1} = 0$. Since α is nonconstant, α_1 is nonzero, so that $\alpha_1{}^t$ may be cancelled from equation (3), giving

$$g_t\alpha_2{}^{n-t} + g_{t+1}\alpha_1\alpha_2{}^{n-t-1} + \cdots + g_{n-1}\alpha_1{}^{n-t-1}\alpha_2 + g_n\alpha_1{}^{n-t} = 0.$$

If $t = n$, this equation becomes $g_n = 0$, a contradiction. Therefore $t < n$. All terms in this equation, except perhaps $g_t\alpha_2{}^{n-t}$, are divisible by α_1, and all terms, except perhaps $g_n\alpha_1{}^{n-t}$, are divisible by α_2. Therefore $\alpha_1 \mid g_t\alpha_2{}^{n-t}$,

$\alpha_2 \mid g_n \alpha_1^{n-t}$. Since α_1, α_2 are relatively prime polynomials, Lemma 3.4.10 implies that $\alpha_1 \mid g_t$, $\alpha_2 \mid g_n$. Because g_t, g_n are nonzero constants, these conditions imply that both α_1, α_2 are constants, and therefore imply that $\alpha = \alpha_1/\alpha_2$ is a constant. This is contrary to hypothesis. We have proved that

evaluation of a nonzero polynomial at a nonconstant rational function always gives a nonzero result. (4)

In particular, the denominator of $a(\lambda)$ cannot become zero when evaluated at α. (The alert reader will notice that the proof just given is the proof of the rational and integral root theorem, 5.4.9, with rational functions in place of rational numbers.)

We claim that φ_α is a homomorphism. This follows immediately from (2) and the substitution rules

$$(a + b)(\alpha) = a(\alpha) + b(\alpha), \qquad (ab)(\alpha) = a(\alpha)b(\alpha) \tag{5}$$

proved in §6.4 for rational functions $a(\lambda), b(\lambda) \in \mathfrak{F}(\lambda)$.

As with polynomials, we call φ_α a *substitution homomorphism*, or an *evaluation homomorphism*.

Example 7.5.5

Let $\alpha \in \mathfrak{F}$ be a constant element of $\mathfrak{F}(\lambda)$, and consider the subset of $\mathfrak{F}(\lambda)$ consisting of the rational functions $a(\lambda)$ which do not have α as a *pole* (§6.5); i.e., the rational functions $a(\lambda)$ for which the denominator polynomial in the lowest terms representation

$$a(\lambda) = f(\lambda)/g(\lambda), \qquad f(\lambda), g(\lambda) \in \mathfrak{F}[\lambda]; \; (f(\lambda), g(\lambda)) = 1, \tag{6}$$

is not divisible by $\lambda - \alpha$ and therefore satisfies $g(\alpha) \neq 0$. Denote this subset of $\mathfrak{F}(\lambda)$ by $\mathfrak{F}(\lambda)^{\lambda-\alpha}$. Let also

$$b(\lambda) = f_1(\lambda)/g_1(\lambda), \qquad f_1(\lambda), g_1(\lambda) \in \mathfrak{F}[\lambda]; \; (f_1(\lambda), g_1(\lambda)) = 1,$$

be a member of $\mathfrak{F}(\lambda)^{\lambda-\alpha}$. Then $(\lambda - \alpha) \nmid g(\lambda)$, $(\lambda - \alpha) \nmid g_1(\lambda)$, and

$$a \pm b = (fg_1 \pm gf_1)/gg_1, \qquad ab = ff_1/gg_1.$$

Possibly the fractions on the right-hand sides here are not in lowest terms. After cancelling common factors from numerator and denominator, it follows that each denominator in the lowest terms representations of $a + b$, $a - b$, ab is a factor of gg_1 and hence cannot have the prime polynomial $\lambda - \alpha$ as a divisor. Thus $\mathfrak{F}(\lambda)^{\lambda-\alpha}$ is closed under addition, subtraction, multiplication, and therefore (Theorem 1.6.9) is a subring of $\mathfrak{F}(\lambda)$.

Define a mapping φ_α from $\mathfrak{F}(\lambda)^{\lambda-\alpha}$ to $\mathfrak{F}$ by specifying its action on a rational function $a(\lambda)$ in lowest terms to be

$$\varphi_\alpha(a(\lambda)) = a(\alpha).$$

Thus φ_α acts on a rational function by evaluating its lowest terms representation at α. This evaluation does not produce a zero denominator because $\mathfrak{F}(\lambda)^{\lambda-\alpha}$ contains only rational functions for which $\lambda - \alpha$ is not a factor of this denominator.

As in Example 7.5.4, the substitution rules (5) proved in §6.4 imply that φ_α is a homomorphism, in this case a homomorphism from $\mathfrak{F}(\lambda)^{\lambda-\alpha}$ to $\mathfrak{F}$. As before, we call φ_α a *substitution homomorphism*, or an *evaluation homomorphism*.

(We comment, in passing, that the ring $\mathfrak{F}(\lambda)^{\lambda-\alpha}$ of rational functions and its analogue for rational numbers will be examined from another point of view in §8.2.)

Another example of a homomorphism will be found in Exercise 16, where a homomorphic mapping from the polynomial ring $\mathfrak{F}[\lambda]$ to the ring of polynomial functions on $\mathfrak{F}$ is described.

The reader should carefully think through the meaning of the last paragraph and Examples 7.5.3, 7.5.4, 7.5.5, and clearly understand that it is *not* claimed that a polynomial function on a field is a homomorphism. For example, if $f(\lambda) = \lambda^2 \in \mathbf{R}[\lambda]$, it is not true that

$$f(r_1 + r_2) = f(r_1) + f(r_2)$$

for all r_1, r_2 in $\mathbf{R}$. Polynomial functions on $\mathfrak{F}$ obtained from polynomials in $\mathfrak{F}[\lambda]$ are therefore usually not homomorphisms.

Basic facts about homomorphisms

We give first two simple facts. Let $\mathfrak{R}$, $\mathfrak{S}$ be rings and φ a homomorphism from $\mathfrak{R}$ to $\mathfrak{S}$. Then:

$$\varphi(0) = 0, \tag{7}$$

$$\varphi(r_1 - r_2) = \varphi(r_1) - \varphi(r_2) \quad \text{for any } r_1, r_2 \in \mathfrak{R}. \tag{8}$$

PROOFS From $0 = 0 + 0$, we get $\varphi(0) = \varphi(0 + 0) = \varphi(0) + \varphi(0)$. Thus $\varphi(0) = \varphi(0) + \varphi(0)$; adding $-\varphi(0)$ gives $0 = \varphi(0)$. This proves (7). (Note that some of these 0's are the zero element of $\mathfrak{R}$, some the zero element of $\mathfrak{S}$.) From $(r_1 - r_2) + r_2 = r_1$, we get $\varphi(r_1 - r_2) + \varphi(r_2) = \varphi(r_1)$; adding $-\varphi(r_2)$ gives (8). □

There are two important sets associated with a homomorphism φ from $\mathfrak{R}$ to $\mathfrak{S}$: a subset of $\mathfrak{R}$, called the *kernel* of φ; and a subset of $\mathfrak{S}$, called the *range* of φ, or *the image of* $\mathfrak{R}$ *under* φ.

DEFINITION 7.5.6

(a) *The* kernel of φ *consists of all elements* r *of* $\mathfrak{R}$ *satisfying* $\varphi(r) = 0$.

(b) *The* range of φ *consists of all elements* s *of* $\mathfrak{S}$ *for which* $s = \varphi(r)$ *for at least one element* $r \in \mathfrak{R}$.

Example 7.5.2 (continued)
If $\mathcal{R}$ is **Z** or $\mathfrak{F}[\lambda]$, and $\mathcal{S}$ is the congruence class ring $\mathcal{R}_m$, the kernel of the natural homomorphism φ from $\mathcal{R}$ to $\mathcal{R}_m$ consists of all elements of $\mathcal{R}$ divisible by m, since $\bar{r} = \bar{0}$ if and only if $m \mid r$. The range of φ is all of $\mathcal{R}_m$.

Example 7.5.3 (continued)
The kernel of the evaluation homomorphism φ_α on the polynomial ring $\mathfrak{F}[\lambda]$ consists of all polynomials $f(\lambda)$ satisfying $f(\alpha(\lambda)) = 0$. If $\alpha(\lambda)$ is not a constant, (4) shows that $f(\alpha(\lambda)) \neq 0$ whenever $f(\lambda)$ is a nonzero polynomial. Therefore, if $\alpha(\lambda)$ is not a constant, the kernel of φ_α consists solely of the zero polynomial. If α is a constant, the kernel of φ_α consists of all polynomials in $\mathfrak{F}[\lambda]$ with α as a root, i.e., divisible by $\lambda - \alpha$. The range of φ_α will in general be a proper subset of $\mathfrak{F}[\lambda]$. For example, if $\alpha(\lambda) = \lambda^2$, the range of φ_α consists of all polynomials involving only even powers of λ. In §7.7 we shall determine when the range of φ_α is all of $\mathfrak{F}[\lambda]$.

Example 7.5.4 (continued)
If both $\mathcal{R}$ and $\mathcal{S}$ are $\mathfrak{F}(\lambda)$, and if $\alpha(\lambda)$ is a nonconstant element of $\mathfrak{F}(\lambda)$, the kernel of the evaluation homomorphism φ_α consists solely of 0, since $\varphi_\alpha(f(\lambda)/g(\lambda)) = 0$ implies $f(\alpha(\lambda)) = 0$, and by (4) this cannot happen if $f(\lambda)$ is a nonzero polynomial. The range of φ_α may be a proper subset of $\mathfrak{F}(\lambda)$; e.g., when $\alpha(\lambda) = \lambda^2$. In §7.7 we shall determine when the range of φ_α is all of $\mathfrak{F}(\lambda)$.

Example 7.5.5 (continued)
When $\alpha \in \mathfrak{F}$ is a constant, $\mathcal{R}$ is $\mathfrak{F}(\lambda)^{\lambda-\alpha}$, and $\mathcal{S}$ is $\mathfrak{F}$, the kernel of the evaluation homomorphism φ_α consists of the rational functions $f(\lambda)/g(\lambda)$ in lowest terms in which the numerator polynomial $f(\lambda)$ has $\lambda - \alpha$ as a factor and the denominator polynomial $g(\lambda)$ does not. The range of φ_α is all of $\mathfrak{F}$, since, if $\beta \in \mathfrak{F}$, then $\varphi_\alpha(\lambda - \alpha + \beta) = \beta$.

The basic properties of the kernel and range are given in the next theorem.

***THEOREM* 7.5.7**
Let $\mathcal{R}$, $\mathcal{S}$ be rings and φ a homomorphism from $\mathcal{R}$ to $\mathcal{S}$. Then:

(i) *The kernel of φ is a subring of $\mathcal{R}$ satisfying this strong multiplicative closure law: for any elements r in $\mathcal{R}$ and k in the kernel of φ, both rk and kr belong to the kernel of φ.*

(ii) *The range of φ is a subring of $\mathcal{S}$.*

PROOFS (i) Let k_1, k_2 be in the kernel of φ, and let $r \in \mathcal{R}$. Then $\varphi(k_1) = 0$, $\varphi(k_2) = 0$, so that $\varphi(k_1 + k_2) = \varphi(k_1) + \varphi(k_2) = 0 + 0 = 0$, $\varphi(k_1 - k_2) = \varphi(k_1) - \varphi(k_2) = 0 - 0 = 0$, and $\varphi(rk_2) = \varphi(r)\varphi(k_2) = \varphi(r)0 = 0$. Similarly $\varphi(k_1 r) = 0$. These equations imply that $k_1 + k_2$, $k_1 - k_2$, rk_2, $k_1 r$

all belong to the kernel of φ. In particular, k_1k_2 is in the kernel of φ. Therefore (Theorem 1.6.9), the kernel of φ is a subring of $\mathcal{R}$, and obviously it satisfies the strong multiplicative closure law. □

(ii) Let s_1, s_2 be elements of $\mathcal{S}$ in the range of φ, so that $s_1 = \varphi(r_1)$, $s_2 = \varphi(r_2)$ for suitable elements $r_1, r_2 \in \mathcal{R}$. Then $s_1 + s_2 = \varphi(r_1 + r_2)$, $s_1 - s_2 = \varphi(r_1 - r_2)$, $s_1s_2 = \varphi(r_1r_2)$, implying that $s_1 + s_2$, $s_1 - s_2$, s_1s_2 are all in the range of φ. Therefore, by Theorem 1.6.9, the range of φ is a subring of $\mathcal{S}$. □

Ideals

Subrings of $\mathcal{R}$ satisfying the strong multiplicative closure law described in Theorem 7.5.7 are given the special name *ideal.* Historically, ideals were introduced into algebra in order to circumvent difficulties caused by the failure of the unique factorization theorem (3.5.3) to be valid in many rings. (See Exercise 18.) The book by Pollard (page 455) is an introductory reference explaining the connection between ideals and the unique factorization theorem. We now state formally the definition of an ideal.

DEFINITION 7.5.8
A subring $\mathcal{I}$ of a ring $\mathcal{R}$ is an ideal *if it satisfies this strong multiplicative closure law:*

$$\text{For any elements } r \text{ of } \mathcal{R} \text{ and } i \text{ of } \mathcal{I}\text{, both } ri \text{ and } ir \text{ belong to } \mathcal{I}. \qquad (9)$$

The kernel of any homomorphism is therefore an ideal.

In any ring $\mathcal{R}$, the subset consisting solely on 0 is always an ideal, and $\mathcal{R}$ itself is also an ideal. These ideals are called, respectively, the *zero ideal* and the *unit ideal.* In some rings, these are the only ideals. Other rings, however, may have many additional ideals.

Example 7.5.9
Let $\mathcal{R}$ be a commutative ring, m an element of $\mathcal{R}$, and $\mathcal{I}$ the subset of $\mathcal{R}$ consisting of all elements divisible by m; that is,

$$i \in \mathcal{I} \quad \text{if and only if} \quad i = qm \quad \text{for some } q \in \mathcal{R}.$$

We claim that $\mathcal{I}$ is an ideal in $\mathcal{R}$. Observe that $\mathcal{I}$ is nonempty since setting $q = 0$ shows that $0 \in \mathcal{I}$. If $i_1, i_2 \in \mathcal{I}$, then $i_1 = q_1m$, $i_2 = q_2m$ for suitable $q_1, q_2 \in \mathcal{R}$, and hence

$$i_1 + i_2 = (q_1 + q_2)m, \qquad i_1 - i_2 = (q_1 - q_2)m,$$
$$ir = ri = (qr)m \qquad \text{for any } r \in \mathcal{R}.$$

These equations show that $\mathcal{I}$ is closed under addition, subtraction, and satisfies the strong closure law for multiplication. It therefore satisfies also the usual closure law for multiplication, hence is a subring of $\mathcal{R}$, and thus an ideal in $\mathcal{R}$. This type of ideal is called a *principal ideal,* and the element m giving rise to it is called a *generator* of the ideal. Different elements m may generate the same principal ideal; e.g., in $\mathbf{Z}$, the principal ideals generated by 2 and -2 both consist of all even integers.

The principal ideal generated by $m = 0$ is, of course, the zero ideal in $\mathcal{R}$. When $\mathcal{R}$ has a multiplicative identity 1, the principal ideal generated by $m = 1$ is the unit ideal.

The next theorem classifies all ideals in $\mathbf{Z}$, $\mathfrak{F}[\lambda]$, and in any field.

THEOREM 7.5.10

(i) *Any ideal in* $\mathbf{Z}$ *or* $\mathfrak{F}[\lambda]$ *is principal.*

(ii) *A field has only two ideals, the zero ideal and the unit ideal. Both are principal.*

PROOFS (i) Let $\mathcal{R}$ denote either $\mathbf{Z}$ or $\mathfrak{F}[\lambda]$, and let $\mathcal{I}$ be an ideal in $\mathcal{R}$. If $\mathcal{I}$ is the zero ideal, then $\mathcal{I}$ is principal, being generated by $m = 0$. Suppose that $\mathcal{I}$ is a nonzero ideal. Among all nonzero elements of $\mathcal{I}$, choose an element $m \in \mathcal{I}$ such that $|m|$ is least. If $i \in \mathcal{I}$, divide i by m to get elements q, r in $\mathcal{R}$ with

$$i = qm + r, \qquad |r| < |m|.$$

Since $m \in \mathcal{I}$, the strong closure law for multiplication implies that $qm \in \mathcal{I}$, and then, from $i \in \mathcal{I}$ and $qm \in \mathcal{I}$, we get $r = i - qm \in \mathcal{I}$. Thus $r \in \mathcal{I}$ and $|r| < |m|$. Since $|m|$ is least among all nonzero elements of $\mathcal{I}$, this implies $r = 0$. That is: if $i \in \mathcal{I}$, then $m \mid i$. Conversely, any element ρ of $\mathcal{R}$ divisible by m is in $\mathcal{I}$, since $m \mid \rho$ implies $\rho = qm$ for some $q \in \mathcal{R}$, and hence implies $\rho \in \mathcal{I}$ by the strong multiplicative closure law. Thus $\mathcal{I}$ consists precisely of the multiples of m and therefore is the principal ideal generated by m. □

(ii) Suppose $\mathcal{R}$ is a field and $\mathcal{I}$ is a nonzero ideal in $\mathcal{R}$; let β be a nonzero element of $\mathcal{I}$. If ρ is any element of $\mathcal{R}$, then $\rho = x\beta$ if $x = \rho\beta^{-1}$. Since $x \in \mathcal{R}$ and $\beta \in \mathcal{I}$, the strong multiplicative closure property implies that $\rho = x\beta \in \mathcal{I}$. Any element of $\mathcal{R}$ is therefore in $\mathcal{I}$. This means that $\mathcal{I} = \mathcal{R}$. □

Since the kernel of a homomorphism is always an ideal, and since all ideals in $\mathbf{Z}$, $\mathfrak{F}[\lambda]$, $\mathfrak{F}(\lambda)$ are principal (remember that $\mathfrak{F}(\lambda)$ is a field), the kernels of the homomorphisms in Examples 7.5.2, 7.5.3, 7.5.4 must be principal. In fact, if $\mathcal{R}$ is $\mathbf{Z}$ or $\mathfrak{F}[\lambda]$, the kernel of the natural homomorphism of $\mathcal{R}$ onto $\mathcal{R}_m$ is the principal ideal generated by m. The kernel of the evaluation homomorphism φ_α on $\mathfrak{F}[\lambda]$ is the zero ideal if α is not a constant, and the principal ideal generated by $\lambda - \alpha$ if α is a constant. For nonconstant $\alpha(\lambda) \in \mathfrak{F}(\lambda)$, the kernel of the evaluation homomorphism φ_α on $\mathfrak{F}(\lambda)$ is the zero ideal.

An example of a nonprincipal ideal is obtained by considering the ring $\mathbf{Z}[\lambda]$ of polynomials with integral coefficients. It is easily verified that the set $\mathcal{I}$ of all

$$2r_1 + \lambda r_2, \qquad \text{for } r_1, r_2 \in \mathbf{Z}[\lambda],$$

is an ideal in $\mathbf{Z}[\lambda]$. This ideal consists of all polynomials with integral coefficients and even constant term. It is not principal because a generator $m(\lambda)$, being a polynomial with even constant term and a divisor of 2, would have to be ± 2, but then could not satisfy $\lambda = g(\lambda)(\pm 2)$ for any polynomial $g(\lambda)$ with integral coefficients.

The ideals in the ring $\mathfrak{F}(\lambda)^{\lambda-\alpha}$ of Example 7.5.5 are also easily described. (See Exercise 12.) They are all principal. Further examples of rings with nonprincipal ideals are given in Exercises 18 and 19.

Ideals and the greatest common divisor

Let $\mathcal{R}$ be $\mathbf{Z}$ or $\mathfrak{F}[\lambda]$, and suppose $a_1, a_2, \ldots a_n \in \mathcal{R}$. We established, in Theorem 3.4.6, the existence and uniqueness of the greatest common divisor of $a_1, a_2, \ldots, a_n$. This is an element d of $\mathcal{R}$ which, simultaneously, is a common divisor of $a_1, \ldots, a_n$ and is divisible by any common divisor e of $a_1, \ldots, a_n$. It is also positive or monic when $a_1, \ldots, a_n$ are not all zero. We now repeat the existence part of the proof of Theorem 3.4.6, but casting it into the language of ideals, and using the fact that any ideal in $\mathcal{R}$ is principal.

Let $\mathcal{I}$ be the set of all elements of $\mathcal{R}$ of the form

$$i = x_1a_1 + x_2a_2 + \cdots + x_na_n, \qquad x_1, x_2, \ldots, x_n \in \mathcal{R}. \tag{10}$$

Adding or subtracting two elements of the form (10) clearly gives another element of this form. Multiplying (10) by any $r \in \mathcal{R}$ gives a result

$$ri = (rx_1)a_1 + (rx_2)a_2 + \cdots + (rx_n)a_n$$

of the form (10). Thus $\mathcal{I}$ is closed under addition, subtraction, and satisfies also the strong closure law for multiplication. Taking $x_1 = 1$, $x_2 = x_3 = \cdots = x_n = 0$ shows that $a_1 \in \mathcal{I}$, and similarly $a_2, a_3, \ldots, a_n \in \mathcal{I}$. Thus $\mathcal{I}$ is an ideal in $\mathcal{R}$ containing all of $a_1, a_2, \ldots a_n$. (We call $\mathcal{I}$ the ideal generated by $a_1, a_2, \ldots, a_n$.) By Theorem 7.5.10(i), any ideal in $\mathcal{R}$ is principal. Therefore $\mathcal{I}$ is principal and hence consists of all multiples of an element d of $\mathcal{R}$. If d is nonzero, we may multiply it by a unit in $\mathcal{R}$ to get it positive or monic, since (by Exercise 11 below) associated elements of $\mathcal{R}$ generate the same principal ideal. Since $a_1, a_2, \ldots, a_n$ all lie in the principal ideal $\mathcal{I}$ generated by d, each must be a multiple of d, implying that d is a common divisor of $a_1, a_2, \ldots, a_n$. Since $d \in \mathcal{I}$, it must have the form (10), i.e.,

$$d = y_1a_1 + y_2a_2 + \cdots + y_na_n \quad \text{for suitable } y_1, y_2, \ldots, y_n \in \mathcal{R}.$$

From this equation it is evident that any common divisor e of $a_1, a_2, \ldots, a_n$ is a divisor of d. This completes the proof that a greatest common divisor of $a_1, a_2, \ldots, a_n$ exists.

The proof that the greatest common divisor of $a_1, a_2, \ldots, a_n$ is unique is the same as in §3.4. □

Isomorphisms

We next study bijective homomorphisms.

DEFINITION 7.5.11
Let $\mathcal{R}$, $\mathcal{S}$ be rings. A homomorphism φ from $\mathcal{R}$ to $\mathcal{S}$ is an isomorphism *if it is bijective.*

We give a short discussion of isomorphisms before giving examples.

THEOREM 7.5.12
Let $\mathcal{R}$, $\mathcal{S}$ be rings. A homomorphism φ from $\mathcal{R}$ to $\mathcal{S}$ is injective if and only if its kernel is the zero ideal in $\mathcal{R}$, and surjective if and only if its range is $\mathcal{S}$. It is an isomorphism of $\mathcal{R}$ onto $\mathcal{S}$ if and only if it has both zero kernel and $\mathcal{S}$ as its range.

PROOF Suppose the homomorphism φ is injective. To prove that the kernel of φ is the zero ideal, we let r be an element of $\mathcal{R}$ in this kernel. Then $\varphi(r) = 0$ and, of course, $\varphi(0) = 0$. Thus $\varphi(r) = \varphi(0)$, implying, because φ is injective, that $r = 0$. The kernel of φ is therefore the zero ideal. Conversely, suppose the kernel of the homomorphism φ is the zero ideal, and let r_1, r_2 be elements of $\mathcal{R}$ with $\varphi(r_1) = \varphi(r_2)$. Then $\varphi(r_1 - r_2) = 0$, implying that $r_1 - r_2$ is in the kernel of φ and hence is zero. Thus $r_1 = r_2$, so that φ is injective. This proves the first assertion. The assertion that φ is surjective if and only if its range is $\mathcal{S}$ is nothing more than the definitions of surjectivity and range, and needs no proof. The last assertion is now evident since a homomorphism is an isomorphism if and only if it is both injective and surjective. □

If $\mathcal{R}$ and $\mathcal{S}$ are rings and there is at least one isomorphism from $\mathcal{R}$ to $\mathcal{S}$, we write

$$\mathcal{R} \cong \mathcal{S}; \tag{11}$$

read (11) as "$\mathcal{R}$ is isomorphic to $\mathcal{S}$" or as "$\mathcal{R}$ and $\mathcal{S}$ are isomorphic."

An example of isomorphic rings is obtained by noticing that the kernel of the natural homomorphism (Example 7.5.2) of $\mathbf{Z}$ or $\mathfrak{F}[\lambda]$ onto the ring of residue classes (mod m) is the zero ideal if $m = 0$. Being surjective, this mapping is therefore an isomorphism when $m = 0$, so that $\mathbf{Z} \cong \mathbf{Z}_0$ and $\mathfrak{F}[\lambda] \cong \mathfrak{F}[\lambda]_0$. For another example, let $\mathfrak{F}$ be an infinite field. Then the polynomial ring $\mathfrak{F}[\lambda]$ and the ring of polynomial functions on $\mathfrak{F}$ are isomorphic. (Exercise 16.) More examples will be given below.

We now show that isomorphism has the reflexive, symmetric, and transitive properties of an equivalence relation.

LEMMA 7.5.13
Let $\mathcal{R}$, $\mathcal{S}$, $\mathcal{T}$ *be rings. Then:*
(i) $\mathcal{R} \cong \mathcal{R}$;
(ii) *If* $\mathcal{R} \cong \mathcal{S}$, *then* $\mathcal{S} \cong \mathcal{R}$;
(iii) *If* $\mathcal{R} \cong \mathcal{S}$ *and* $\mathcal{S} \cong \mathcal{T}$, *then* $\mathcal{R} \cong \mathcal{T}$.

Proof of (i). The identity function $i_{\mathcal{R}}$ on $\mathcal{R}$ is an isomorphism of $\mathcal{R}$ to itself, since it is a bijection and

$$i_{\mathcal{R}}(r_1 + r_2) = r_1 + r_2 = i_{\mathcal{R}}(r_1) + i_{\mathcal{R}}(r_2), \qquad i_{\mathcal{R}}(r_1r_2) = r_1r_2 = i_{\mathcal{R}}(r_1)i_{\mathcal{R}}(r_2).$$

Therefore $\mathcal{R} \cong \mathcal{R}$. □
Proof of (ii). Let φ be an isomorphism from $\mathcal{R}$ to $\mathcal{S}$. Then φ is bijective, and therefore by Theorem 7.4.10 has an inverse bijection from $\mathcal{S}$ to $\mathcal{R}$, which we denote by θ, satisfying

$$\theta\varphi = i_{\mathcal{R}}, \qquad \varphi\theta = i_{\mathcal{S}}. \tag{12}$$

We claim that θ is a homomorphism. When proved, this will demonstrate that θ is an isomorphism and that $\mathcal{S} \cong \mathcal{R}$. If s_1, s_2 are elements of $\mathcal{S}$, let $r_1 = \theta(s_1)$, $r_2 = \theta(s_2)$. Then $s_1 = \varphi(r_1)$ and $s_2 = \varphi(r_2)$. Using (12) and the fact that φ is a homomorphism, we get

$$\theta(s_1 + s_2) = \theta(\varphi(r_1) + \varphi(r_2)) = \theta(\varphi(r_1 + r_2)) = r_1 + r_2 = \theta(s_1) + \theta(s_2),$$
$$\theta(s_1s_2) = \theta(\varphi(r_1)\varphi(r_2)) = \theta(\varphi(r_1r_2)) = r_1r_2 = \theta(s_1)\theta(s_2).$$

Therefore θ is an isomorphism, and $\mathcal{S} \cong \mathcal{R}$. □
This proof also shows that

the inverse of an isomorphism is an isomorphism.

Proof of (iii). Let φ_1 be a homomorphism from $\mathcal{R}$ to $\mathcal{S}$, and φ_2 a homomorphism from $\mathcal{S}$ to $\mathcal{T}$. Compose φ_1, φ_2 to get a mapping $\varphi_3 = \varphi_2\varphi_1$ from $\mathcal{R}$ to $\mathcal{T}$. We shall prove that φ_3 is a homomorphism. This will show that

composition of homomorphisms gives a homomorphism.

Taking φ_1 and φ_2 to be isomorphisms, and using the fact that composition of bijections gives a bijection, it will then be evident that φ_3 is an isomorphism. This will show that

composition of isomorphisms gives an isomorphism

and, in particular, that $\mathcal{R} \cong \mathcal{T}$. We therefore have only to prove that $\varphi_3 = \varphi_2\varphi_1$ is a homomorphism if φ_1, φ_2 are homomorphisms. Let $r_1, r_2 \in \mathcal{R}$. Then:

$$\begin{aligned}\varphi_3(r_1 + r_2) &= \varphi_2(\varphi_1(r_1 + r_2)) = \varphi_2(\varphi_1(r_1) + \varphi_1(r_2)) \\ &= \varphi_2(\varphi_1(r_1)) + \varphi_2(\varphi_1(r_2)) = \varphi_3(r_1) + \varphi_3(r_2),\end{aligned}$$

$$\begin{aligned}\varphi_3(r_1r_2) &= \varphi_2(\varphi_1(r_1r_2)) = \varphi_2(\varphi_1(r_1)\varphi_1(r_2)) = \varphi_2(\varphi_1(r_1))\varphi_2(\varphi_1(r_2)) \\ &= \varphi_3(r_1)\varphi_3(r_2).\end{aligned}$$

Thus $\varphi_3(r_1 + r_2) = \varphi_3(r_1) + \varphi_3(r_2)$ and $\varphi_3(r_1r_2) = \varphi_3(r_1)\varphi_3(r_2)$, completing the proof. □

REMARK Frequently a homomorphism φ from ring $\mathcal{R}$ to ring $\mathcal{S}$ is injective but not surjective. Let $\mathcal{S}_1$ be the range of φ. Then φ is a homomorphism from $\mathcal{R}$ to $\mathcal{S}_1$ which is both injective and surjective, and therefore is an isomorphism. That is,

an injective homomorphism from a ring $\mathcal{R}$ to a ring $\mathcal{S}$ is an isomorphism of $\mathcal{R}$ onto a subring of $\mathcal{S}$.

It is also important to note that

The image of an integral domain under an isomorphism is an integral domain, and that

the image of a field under an isomorphism is a field. (13)

PROOFS Let $\mathcal{R}$, $\mathcal{S}$ be rings and φ an isomorphism from $\mathcal{R}$ onto $\mathcal{S}$.

First suppose $\mathcal{R}$ is a ring with identity 1, and $1 \neq 0$. We claim that $\varphi(1)$ is an identity for $\mathcal{S}$, and $\varphi(1) \neq 0$. Let $s \in \mathcal{S}$. Then $s = \varphi(r)$ for a suitable element r in $\mathcal{R}$, and then $\varphi(1)s = \varphi(1)\varphi(r) = \varphi(1r) = \varphi(r) = s$. Similarly $s\varphi(1) = s$. Therefore $\varphi(1)$ is a multiplicative identity for $\mathcal{S}$. It is nonzero since $\varphi(1) = 0 = \varphi(0)$ implies (because φ is injective) that $1 = 0$, a contradiction.

Next, suppose $\mathcal{R}$ is commutative. We claim that $\mathcal{S}$ is also commutative. Let $s_1, s_2 \in \mathcal{S}$. Then $s_1 = \varphi(r_1)$, $s_2 = \varphi(r_2)$ for suitable $r_1, r_2 \in \mathcal{R}$, so that

$$s_1s_2 = \varphi(r_1)\varphi(r_2) = \varphi(r_1r_2) = \varphi(r_2r_1) = \varphi(r_2)\varphi(r_1) = s_2s_1 .$$

This shows that $\mathcal{S}$ is commutative.

Now suppose that $\mathcal{R}$ has no zero divisors. We claim that $\mathcal{S}$ also has none. Let $s_1, s_2 \in \mathcal{S}$ and suppose $s_1s_2 = 0$. With r_1, r_2 as in the last paragraph, we find that $\varphi(r_1r_2) = s_1s_2 = 0 = \varphi(0)$. Thus $r_1r_2 = 0$, and hence at least one of r_1, r_2 is 0. But then at least one of s_1, s_2 is 0, implying that $\mathcal{S}$ has no zero divisors.

Finally, suppose $\mathcal{R}$ contains a multiplicative inverse r^{-1} for any nonzero element r. We claim that $\mathcal{S}$ contains a multiplicative inverse for each of its nonzero elements. If s is a nonzero element of $\mathcal{S}$, with $s = \varphi(r)$, then $r \neq 0$, so that $r^{-1} \in \mathcal{R}$ and

$$s\varphi(r^{-1}) = \varphi(r)\varphi(r^{-1}) = \varphi(rr^{-1}) = \varphi(1).$$

Since $\varphi(1)$ is the multiplicative identity of $\mathcal{S}$, this equation shows that $\varphi(r^{-1})$ is a multiplicative inverse for $s = \varphi(r)$. Each nonzero element of $\mathcal{S}$ therefore has a multiplicative inverse. In particular,

$$\varphi(r)^{-1} = \varphi(r^{-1}). \tag{14}$$

If $\mathcal{R}$ is an integral domain, then it is a commutative ring with a nonzero multiplicative identity and without zero divisors. The above paragraphs show that $\mathcal{S}$ also has these properties and therefore it also is an integral domain (Theorem 1.6.4).

If $\mathcal{R}$ is a field, then it is a commutative ring with nonzero multiplicative identity and a multiplicative inverse for each of its nonzero elements. By the paragraphs above, $\mathcal{S}$ also has these properties, so that it also is a field (Theorem 1.6.7). □

Many rings have homomorphisms which are neither the zero function nor an isomorphism. For example, if the integer m is neither 0 nor ± 1, the natural homomorphism of $\mathbf{Z}$ onto $\mathbf{Z}_m$ is not the zero function and not an isomorphism. However, the very simple ideal structure of a field [Theorem 7.5.10(ii)] implies that a homomorphism of a field must be the zero function or an isomorphism (onto its range). We may therefore supplement (13) in this way:

A nonzero homomorphic image of a field is an isomorphic image. (15)

PROOF The kernel must be the zero ideal (in which case the homomorphism is an isomorphism—Theorem 7.5.12) or the unit ideal (and then the homomorphism is the zero function since every element of the field is in the kernel). □

Example 7.5.14 Embeddings

(i) When introducing complex numbers in §2.3, we first studied ordered pairs (x, y) in which $x, y \in \mathbf{R}$, then special ordered pairs $(x, 0)$ on the X axis. Define a mapping φ from $\mathbf{R}$ onto the ring of these special pairs by

$$\varphi(x) = (x, 0) \qquad \text{for all } x \in \mathbf{R}.$$

By (13) and (15), this mapping must be an isomorphism and the ring of special pairs a field. The mapping φ permits the notational change from $(x, 0)$ to x by providing a unique element x of $\mathbf{R}$ corresponding to each special pair $(x, 0)$, this correspondence being compatible with addition and multiplication. Substitution of x for $(x, 0)$ then *embeds* $\mathbf{R}$ within $\mathbf{C}$.

(ii) Let $\mathcal{R}$ be a commutative ring with identity. When introducing the polynomial ring $\mathcal{R}[\lambda]$ in §3.2, we first introduced infinite sequences $(r_0, r_1, r_2, \ldots)$ with entries from $\mathcal{R}$, almost all zero. Later, special infinite sequences $(r_0, 0, 0, 0, \ldots)$ were encountered, in which all entries after the initial entry are zero. Define a mapping φ from $\mathcal{R}$ to these special infinite sequences by specifying that

$$\varphi(r) = (r, 0, 0, 0, \ldots) \qquad \text{for all } r \in \mathcal{R}.$$

Trivially, φ is an isomorphism from $\mathcal{R}$ to the ring of special sequences. This mapping permits the notational change from $(r, 0, 0, 0, \ldots)$ to r by providing a unique element of $\mathcal{R}$ corresponding to each special sequence, this correspondence being compatible with addition and multiplication. Substitution of r for $(r, 0, 0, 0, \dots)$ then embeds $\mathcal{R}$ in $\mathcal{R}[\lambda]$.

(iii) Let $\mathcal{R}$ be a commutative ring with at least two elements and without zero divisors, and let ν be a nonzero element of $\mathcal{R}$. When introducing the field of fractions of $\mathcal{R}$ in §6.3, we introduced a mapping φ from $\mathcal{R}$ to certain equivalence classes of fractions by specifying $\varphi(m)$ to be the equivalence class of fractions containing $m\nu/\nu$, for each $m \in \mathcal{R}$. Equations (11), (12), (13) of §6.3 show that φ is an isomorphism from $\mathcal{R}$ to the ring of these special equivalence classes. This mapping permits the embedding of $\mathcal{R}$ within the field of fractions by providing a unique element of $\mathcal{R}$ corresponding to each special equivalence class of fractions, this correspondence being compatible with addition and multiplication. Substitution of m for the class of fractions $m\nu/\nu$ then embeds $\mathcal{R}$ in the field of fractions.

In general, a change of notation in which elements of a ring $\mathcal{S}$ are replaced with elements of another ring $\mathcal{R}$ is possible when the rings are isomorphic. The isomorphism allows the change from elements of $\mathcal{S}$ to elements of $\mathcal{R}$ to be made by providing a unique element $r \in \mathcal{R}$ corresponding to each element $s \in \mathcal{S}$, this correspondence being compatible with addition and multiplication. Replacing each element of $\mathcal{S}$ with the corresponding element of $\mathcal{R}$ accomplishes the change of notation. If it happens that $\mathcal{S}$ is a subring of a ring $\mathcal{T}$, this is the notational change that *embeds* $\mathcal{R}$ within $\mathcal{T}$.

Example 7.5.15

Consider the ring $\mathbf{R}[\lambda]_{m(\lambda)}$ of congruence classes of real polynomials modulo $m(\lambda)$, where $m(\lambda) = \lambda^2 + 1$. Since $m(\lambda)$ is a prime in $\mathbf{R}[\lambda]$, the ring $\mathbf{R}[\lambda]_{m(\lambda)}$ is a field (Theorem 7.3.1). What field is it? We shall show, to within change of notation, that this field is $\mathbf{C}$, the complex number field. The meaning of the last sentence is that $\mathbf{R}[\lambda]_{m(\lambda)} \cong \mathbf{C}$. Since $\deg m(\lambda) = 2$, the polynomials

$$a + b\lambda, \qquad a, b \in \mathbf{R} \tag{16}$$

constitute a system of distinct representatives for the congruence classes $(\text{mod}\ m(\lambda))$ [Theorem 7.2.3]. Addition, for these representatives, is the normal addition of polynomials, but multiplication is performed by first multiplying in the normal way for polynomials, then computing the remainder after division by $m(\lambda)$. Define a function φ on $\mathbf{R}[\lambda]_{m(\lambda)}$ to $\mathbf{C}$ in this way: If C is the congruence class represented by (16), set

$$\varphi(C) = a + bi,$$

i denoting here the complex number $(0, 1)$ in the Y axis. This mapping will be the required isomorphism. First we show that it is a homomorphism.

Let C_1, C_2 be congruence classes represented, respectively, by

$$a_1 + b_1\lambda, \qquad a_2 + b_2\lambda, \qquad a_1, a_2, b_1, b_2 \in \mathbf{R}.$$

Then

$$a_1 + a_2 + (b_1 + b_2)\lambda \quad \text{represents } C_1 + C_2,$$

$$a_1a_2 - b_1b_2 + (a_1b_2 + a_2b_1)\lambda \quad \text{represents } C_1C_2.$$

Thus $\varphi(C_1) = a_1 + b_1i, \qquad \varphi(C_2) = a_2 + b_2i,$

$$\varphi(C_1 + C_2) = a_1 + a_2 + (b_1 + b_2)i = (a_1 + b_1i) + (a_2 + b_2i) = \varphi(C_1) + \varphi(C_2),$$

$$\varphi(C_1C_2) = a_1a_2 - b_1b_2 + (a_1b_2 + a_2b_1)i = (a_1 + b_1i)(a_2 + b_2i) = \varphi(C_1)\varphi(C_2).$$

These equations show that φ is a homomorphism.

The mapping φ is therefore a nonzero homomorphism of a field $\mathbf{R}[\lambda]_{m(\lambda)}$ to $\mathbf{C}$, and hence, by (13) and (15), must be an isomorphism onto a field contained within $\mathbf{C}$. We now show that φ is surjective, in this way: Let $a + bi$ be a complex number, with $a, b \in \mathbf{R}$. If C is the congruence class represented by $a + b\lambda$, then $\varphi(C) = a + bi$. This demonstrates the surjectivity of φ and completes the proof that φ is an isomorphism of $\mathbf{R}[\lambda]_{m(\lambda)}$ onto $\mathbf{C}$. □

Endomorphisms and automorphisms

Homomorphisms and isomorphisms of a ring or field to itself are often encountered. Examples are the evaluation homomorphisms φ_α on $\mathfrak{F}[\lambda]$ or $\mathfrak{F}(\lambda)$ in 7.5.3, 7.5.4. When $\alpha(\lambda)$ is not a constant, φ_α has zero kernel, and therefore is an isomorphism onto its range, this range being a subring of $\mathfrak{F}[\lambda]$, or a subfield of $\mathfrak{F}(\lambda)$. In some instances, to be described in §7.7, the range of φ_α is all of $\mathfrak{F}[\lambda]$, or $\mathfrak{F}(\lambda)$, respectively, and in these instances φ_α is an isomorphism of a ring or field *onto itself*. Examples 7.5.16, 7.5.17 below give other occurrences of an isomorphism of a field onto itself.

Example 7.5.16

We claim that the mapping φ on $\mathbf{C}$ defined by

$$\varphi(z) = \bar{z}, \quad \text{the complex conjugate of } z, \quad \text{for } z \in \mathbf{C},$$

is an isomorphism of $\mathbf{C}$ onto itself. The rules

$$\overline{z_1 + z_2} = \bar{z}_1 + \bar{z}_2, \qquad \overline{z_1z_2} = \bar{z}_1\,\bar{z}_2,$$

proved in §2.4 for conjugation of complex numbers show that φ is a homomorphism. It is injective since any nonzero homomorphism on a field is injective [by (13) and (15)], and surjective since $\varphi(w) = z$ is satisfied by taking $w = \bar{z}$. □

Example 7.5.17
Let $\mathbf{Z}_2$ be the two element field described in §7.3, the two elements being denoted by 0 and 1. Take $\mathcal{R}$ to be $\mathbf{Z}_2[\lambda]_{m(\lambda)}$, where $m(\lambda) = \lambda^2 + \lambda + 1$. The polynomials 0, 1, λ, $\lambda + 1$ then are a system of distinct representatives for the elements of $\mathcal{R}$. A discussion of $\mathcal{R}$ and tables for addition and multiplication of these representatives may be found in §7.3. Since $m(\lambda)$ is a prime in $\mathbf{Z}_2[\lambda]$, this ring $\mathcal{R} = \mathbf{Z}_2[\lambda]_{m(\lambda)}$ is a field, so that any nonzero homomorphism on $\mathcal{R}$ must be an isomorphism onto a field. Define a mapping φ from $\mathcal{R}$ to itself by specifying that

$$\varphi(r) = r^2$$

for each element r in $\mathcal{R}$. If $r_1, r_2 \in \mathcal{R}$, then

$$\varphi(r_1 + r_2) = (r_1 + r_2)^2 = r_1^2 + 2r_1r_2 + r_2^2 = r_1^2 + r_2^2 = \varphi(r_1) + \varphi(r_2),$$

$$\varphi(r_1r_2) = (r_1r_2)^2 = r_1^2r_2^2 = \varphi(r_1)\varphi(r_2).$$

(We used here the fact that $2 = 0$ in this ring $\mathcal{R}$.) These equations show that φ is a homomorphism. Since $\varphi(1) = 1$, this mapping is a nonzero homomorphism of the field $\mathcal{R}$ and therefore is injective. Thus it must map the four elements of $\mathcal{R}$ to four distinct elements of $\mathcal{R}$, and hence it must also be surjective. Consequently, φ is an isomorphism of $\mathcal{R}$ onto itself. It is not the identity function since $\varphi(0) = 0$, $\varphi(1) = 1$, $\varphi(\lambda) = \lambda + 1$, $\varphi(\lambda + 1) = \lambda$. □

Homomorphisms of a ring to itself are often called *endomorphisms*, and isomorphisms of a ring *onto* itself are often called *automorphisms*. The evaluation homomorphisms are therefore examples of endomorphisms; and complex conjugation is an example of an automorphism, as is the mapping φ in Example 7.5.17. The next theorem gives basic facts about the endomorphisms and automorphisms of a ring.

THEOREM 7.5.18
Let $\mathcal{R}$ be a ring.

(i) *Under composition, the endomorphisms of $\mathcal{R}$ form a monoid, called the endomorphism monoid of $\mathcal{R}$.*

(ii) *Under composition, the automorphisms of $\mathcal{R}$ form a group, called the automorphism group of $\mathcal{R}$.*

PROOF Taking the rings $\mathcal{S}$ and $\mathcal{T}$ in the proof of Lemma 7.5.13 to be $\mathcal{R}$, we find that composition of endomorphisms gives an endomorphism, and composition of automorphisms gives an automorphism. Closure (G1) therefore holds. Associativity (G2) holds because composition of mappings is always associative (§7.4). There is an identity since the identity function $i_{\mathcal{R}}$ is an automorphism and satisfies $i_{\mathcal{R}}\varphi = \varphi i_{\mathcal{R}}$ for any mapping φ on $\mathcal{R}$. This verifies G3 and completes the verification that the endomorphisms of $\mathcal{R}$ form a monoid. Since any isomorphism has an inverse which is also an isomorphism, any automorphism has an inverse automorphism. The inverse axiom G4 therefore holds for automorphisms, completing the verification that the automorphisms form a group. □

A complete description of the endomorphism monoid and the automorphism group, when $\mathcal{R}$ is $\mathbf{Z}$, $\mathbf{Q}$, $\mathbf{Q}[\lambda]$, $\mathbf{R}[\lambda]$, $\mathbf{Q}(\lambda)$, $\mathbf{R}(\lambda)$ will be given in §7.7.

REMARK Although composition of endomorphisms gives an endomorphism, generally the sum or pointwise product does not. For example, $i_{\mathbf{Z}} + i_{\mathbf{Z}}$ is not an endomorphism of $\mathbf{Z}$, and the pointwise product of the evaluation endomorphism φ_0 with itself is not an endomorphism of $\mathbf{Q}[\lambda]$.

Prolongation

Let $\mathcal{R}$, $\mathcal{S}$ be rings, and suppose φ is a homomorphism from $\mathcal{R}$ to $\mathcal{S}$. Suppose further that $\mathcal{R}_1$, $\mathcal{S}_1$ are rings containing $\mathcal{R}$, $\mathcal{S}$, respectively, as subrings. (See Figure 7.5.1.) If there is a homomorphism Φ from $\mathcal{R}_1$ to $\mathcal{S}_1$ which satisfies

$$\Phi(r) = \varphi(r) \qquad \text{for all } r \text{ in } \mathcal{R},$$

we say that Φ is a *prolongation* of φ to $\mathcal{R}_1$. We then often use φ to denote both the original homomorphism and its prolongation, i.e., use φ in place of Φ. An important example of prolongation occurs when $\mathcal{R}_1$ is the polynomial ring over $\mathcal{R}$, and $\mathcal{S}_1$ the polynomial ring over $\mathcal{S}$. To avoid the confusion that might occur if we used the same symbol for two indeterminates, we denote the polynomial ring over $\mathcal{R}$ by $\mathcal{R}[\lambda]$ and that over $\mathcal{S}$ by $\mathcal{S}[\Lambda]$, λ being lower case "lambda", and Λ uppercase "lambda."

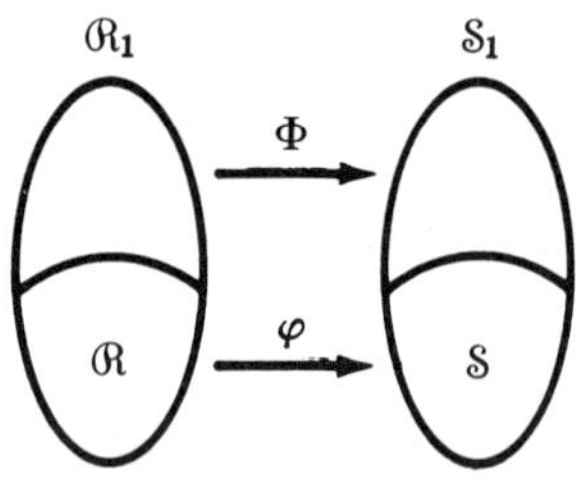

Figure 7.5.1

THEOREM 7.5.19
Let $\mathcal{R}$, $\mathcal{S}$ be commutative rings with identity, φ a homomorphism from $\mathcal{R}$ to $\mathcal{S}$. Suppose that $\varphi(1) = 1$, i.e., φ carries the identity of $\mathcal{R}$ to the identity of $\mathcal{S}$. Then φ may be prolonged to a homomorphism (also denoted by φ) of $\mathcal{R}[\lambda]$ to $\mathcal{S}[\Lambda]$, such that $\varphi(\lambda) = \Lambda$.

PROOF A typical element of $\mathcal{R}[\lambda]$ is a polynomial

$$a(\lambda) = a_0 + a_1\lambda + a_2\lambda^2 + \cdots = \sum_{j=0}^{\infty} a_j\lambda^j,$$

in which the coefficients a_j lie in $\mathcal{R}$ and are almost all zero. Define a function Φ by specifying its action on $a(\lambda)$ to be

$$\Phi(a(\lambda)) = \varphi(a_0) + \varphi(a_1)\Lambda + \varphi(a_2)\Lambda^2 + \cdots = \sum_{j=0}^{\infty} \varphi(a_j)\,\Lambda^j. \quad (17)$$

The symbols $\varphi(a_j)$ in this equation are meaningful since φ is given as a function defined on elements of $\mathcal{R}$. Since almost all a_j are zero and $\varphi(0) = 0$, almost all $\varphi(a_j)$ are zero; therefore the right-hand side of (17) is indeed a polynomial. Thus we obtain a mapping Φ from $\mathcal{R}[\lambda]$ to $\mathcal{S}[\Lambda]$. Applying Φ to the constant polynomial $a_0 + 0\lambda + 0\lambda^2 + \cdots$ shows that $\Phi(a_0) = \varphi(a_0)$. Therefore Φ is a prolongation of φ; henceforth use φ in place of Φ. Evidently

$$\varphi(\lambda) = \varphi(0 + 1\lambda + 0\lambda^2 + \cdots) = \varphi(0) + \varphi(1)\,\Lambda + \varphi(0)\,\Lambda^2 + \cdots = \Lambda.$$

Let

$$b(\lambda) = \sum_{j=0}^{\infty} b_j\lambda^j, \qquad b_j \in \mathcal{R};\ j = 0, 1, 2, \ldots,$$

also be a polynomial in $\mathcal{R}[\lambda]$. Then

$$a(\lambda) + b(\lambda) = \sum_{j=0}^{\infty} s_j\lambda^j, \qquad a(\lambda)b(\lambda) = \sum_{j=0}^{\infty} p_j\lambda^j,$$

where

$$s_j = a_j + b_j\,, \qquad p_j = a_0b_j + a_1b_{j-1} + \cdots + a_{j-1}b_1 + a_jb_0\,,$$

for each j. Applying φ to these elements of $\mathcal{R}$ yields

$$\varphi(s_j) = \varphi(a_j) + \varphi(b_j), \qquad \varphi(p_j) = \varphi(a_0)\varphi(b_j) + \cdots + \varphi(a_j)\varphi(b_0). \quad (18)$$

Using (18), we get

$$\varphi(a(\lambda) + b(\lambda)) = \sum_{j=0}^{\infty} \varphi(s_j)\,\Lambda^j = \sum_{j=0}^{\infty} \varphi(a_j)\,\Lambda^j + \sum_{j=0}^{\infty} \varphi(b_j)\,\Lambda^j$$
$$= \varphi(a(\lambda)) + \varphi(b(\lambda)).$$

Using (18) again, and recalling the manner in which the coefficients of a product of polynomials are obtained from the coefficients of the factors, we get

$$\varphi(a(\lambda)b(\lambda)) = \sum_{j=0}^{\infty} \varphi(p_j)\,\Lambda^j = \sum_{j=0}^{\infty} \{\varphi(a_0)\varphi(b_j) + \cdots + \varphi(a_j)\varphi(b_0)\}\,\Lambda^j$$

$$= \Big(\sum_{j=0}^{\infty} \varphi(a_j)\,\Lambda^j\Big)\Big(\sum_{j=0}^{\infty} \varphi(b_j)\,\Lambda^j\Big) = \varphi(a(\lambda))\varphi(b(\lambda)).$$

These calculations show that the prolongation of φ is indeed a homomorphism. □

If φ is an isomorphism of $\mathcal{R}$ onto $\mathcal{S}$, the reader will easily be able to prove that its prolongation to $\mathcal{R}[\lambda]$ is an isomorphism onto $\mathcal{S}[\Lambda]$.

Another example of prolongation occurs when we consider fields of fractions. Let $\mathcal{R}$ be a commutative ring with at least two elements and without zero divisors, and $\mathcal{K}$ the field of fractions of $\mathcal{R}$ described in §6.3. Let φ be an isomorphism of $\mathcal{R}$ onto a ring $\mathcal{S}$, and take $\mathcal{S}_1$ to be any field containing $\mathcal{S}$. We prolong φ to a mapping Φ of $\mathcal{K}$ to $\mathcal{S}_1$, in the following way. If $m, n \in \mathcal{R}$, with $n \neq 0$, then m/n is an element of $\mathcal{K}$, and we define Φ by

$$\Phi(m/n) = \varphi(m)\varphi(n)^{-1} \qquad (\in \mathcal{S}_1).$$

To show that Φ is compatible with equality, addition, multiplication of elements of $\mathcal{K}$, suppose also that $m_1, n_1 \in \mathcal{R}$, with n_1 nonzero. If $m/n = m_1/n_1$, then $mn_1 = nm_1$, hence $\varphi(mn_1) = \varphi(nm_1)$, yielding $\varphi(m)\varphi(n_1) = \varphi(n)\varphi(m_1)$, and thus $\varphi(m)\varphi(n)^{-1} = \varphi(m_1)\varphi(n_1)^{-1}$. That is,

$$\Phi(m/n) = \Phi(m_1/n_1) \qquad \text{if } m/n = m_1/n_1.$$

To show compatibility with addition and multiplication, we compute as follows:

$$\begin{aligned}
\Phi(m/n + m_1/n_1) &= \Phi((mn_1 + nm_1)/nn_1) = \varphi(mn_1 + nm_1)\varphi(nn_1)^{-1} \\
&= \{\varphi(m)\varphi(n_1) + \varphi(n)\varphi(m_1)\}\{\varphi(n)\varphi(n_1)\}^{-1} \\
&= \varphi(m)\varphi(n)^{-1} + \varphi(m_1)\varphi(n_1)^{-1} = \Phi(m/n) + \Phi(m_1/n_1); \\
\Phi(m/n \cdot m_1/n_1) &= \Phi(mm_1/nn_1) = \varphi(mm_1)\varphi(nn_1)^{-1} \\
&= \{\varphi(m)\varphi(n)^{-1}\}\{\varphi(m_1)\varphi(n_1)^{-1}\} \\
&= \Phi(m/n)\Phi(m_1/n_1).
\end{aligned}$$

From these equations, we infer that Φ is a homomorphism of $\mathcal{K}$ to $\mathcal{S}_1$. If $m/n = r$ is an element of $\mathcal{R}$, then $m = nr$, hence $\varphi(m) = \varphi(n)\varphi(r)$, and thus $\varphi(r) = \varphi(m)\varphi(n)^{-1} = \Phi(m/n) = \Phi(r)$. Thus Φ is indeed a prolongation of φ. In particular, Φ maps $\mathcal{R}$ onto $\mathcal{S}$ (since this is true of φ), and therefore Φ is a nonzero homomorphism on the field $\mathcal{K}$. Hence Φ is an isomorphism of $\mathcal{K}$ onto a subfield of $\mathcal{S}_1$ containing $\mathcal{S}$, since, by (15), Φ must be an isomorphism, and by (13), onto a field. This completes the proof of Theorem 7.5.20.

THEOREM 7.5.20
Let $\mathcal{R}$ be a commutative ring, with at least two elements and without zero divisors, and φ an isomorphism of $\mathcal{R}$ onto a ring $\mathcal{S}$. If $\mathcal{K}$ is the field of fractions of $\mathcal{R}$ described in §6.3, and $\mathcal{S}_1$ is any field containing $\mathcal{S}$ as a subring, then φ may be prolonged to an isomorphism of $\mathcal{K}$ to $\mathcal{S}_1$.

Conceivably, a commutative ring with at least two elements and without zero divisors may be embedded in a field in several ways, the discussion of fractions in §6.3 being just one way. The reader will find illustrations in Exercises 24 and 25, where two new embeddings of the polynomial ring in a field are described. We shall say that:

> A field $\mathcal{L}$ containing $\mathcal{R}$ as a subring is an abstract field of fractions for $\mathcal{R}$ if there is no proper subfield of $\mathcal{L}$ containing $\mathcal{R}$ (i.e., $\mathcal{L}$ is a "smallest" field containing $\mathcal{R}$).

From Theorem 7.5.20, we now infer the uniqueness (to within isomorphism) of the field of fractions of $\mathcal{R}$.

COROLLARY 7.5.21
If $\mathcal{R}$ is a commutative ring with at least two elements and without zero divisors, $\mathcal{K}$ its field of fractions (as in §6.3), *and $\mathcal{L}$ any field containing $\mathcal{R}$ as a subring, then $\mathcal{K}$ is isomorphic to a subfield of $\mathcal{L}$ containing $\mathcal{R}$. In particular, if $\mathcal{L}$ is an abstract field of fractions for $\mathcal{R}$, then $\mathcal{K} \cong \mathcal{L}$.*

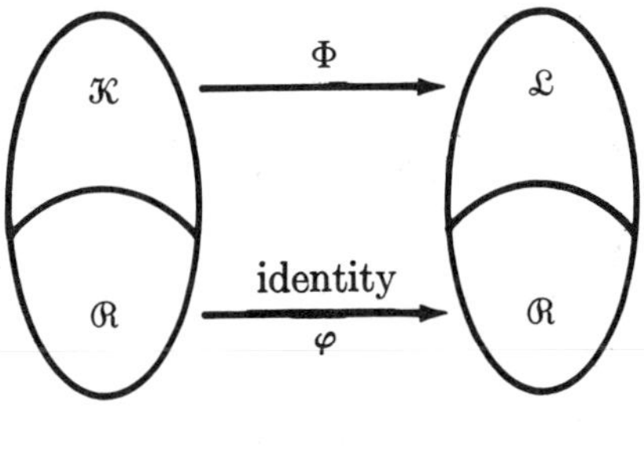

Figure 7.5.2

PROOF In the discussion above, take $\mathcal{S} = \mathcal{R}$ and φ to be the identity function. (See Figure 7.5.2.) Then, by Theorem 7.5.20, φ may be prolonged to an isomorphism Φ of $\mathcal{K}$ to a subfield of $\mathcal{L}$ containing $\mathcal{R}$. This proves the first assertion. If $\mathcal{L}$ is an abstract field of fractions of $\mathcal{R}$, it has no subfield containing $\mathcal{R}$ other than itself, so that Φ must be an isomorphism of $\mathcal{K}$ onto $\mathcal{L}$. Thus $\mathcal{K} \cong \mathcal{L}$. □

We shall use the prolongation Φ of the above discussion in §8.6 to obtain a field of fractions for the polynomial ring $\mathbf{R}[\lambda]$ with a distinctly different appearance. Other unusual fields of fractions will be found in Exercises 24, 25 below.

Abstract congruence class rings modulo an ideal (Optional)

Let $\mathfrak{R}$ be an arbitrary ring and $\mathfrak{I}$ an ideal in $\mathfrak{R}$. The construction in §7.2 of a ring of congruence classes may be generalized to an arbitrary ring if the notion of congruence is suitably generalized. If $a, b \in \mathfrak{R}$, we say that a, b are *congruent modulo* $\mathfrak{I}$ if $a - b \in \mathfrak{I}$; we write this as

$$a \equiv b \pmod{\mathfrak{I}}.$$

Notice that, when $\mathfrak{I}$ is the principal ideal generated by an element m (Example 7.5.9), congruence modulo $\mathfrak{I}$ means the same as congruence modulo m (§3.6). Congruence $(\text{mod}\ \mathfrak{I})$ is an equivalence relation on $\mathfrak{R}$. *Proof:* Let $r, r_1, r_2, r_3 \in \mathfrak{R}$. Then:

(i) $r \equiv r \pmod{\mathfrak{I}}$ since $r - r = 0 \in \mathfrak{I}$.

(ii) If $r_1 \equiv r_2 \pmod{\mathfrak{I}}$, then $r_1 - r_2 \in \mathfrak{I}$, so that $-(r_1 - r_2) \in \mathfrak{I}$; thus $r_2 - r_1 \in \mathfrak{I}$, implying that $r_2 \equiv r_1 \pmod{\mathfrak{I}}$.

(iii) If $r_1 \equiv r_2 \pmod{\mathfrak{I}}$ and $r_2 \equiv r_3 \pmod{\mathfrak{I}}$, then $r_1 - r_2 \in \mathfrak{I}$, $r_2 - r_3 \in \mathfrak{I}$. Adding gives $r_1 - r_3 \in \mathfrak{I}$, and thus $r_1 \equiv r_3 \pmod{\mathfrak{I}}$.

Since congruence $(\text{mod}\ \mathfrak{I})$ is an equivalence relation, $\mathfrak{R}$ partitions into equivalence classes, which, as before, are called *congruence* classes. Elements a, b of $\mathfrak{R}$ represent (belong to) the same congruence class if and only if $a \equiv b \pmod{\mathfrak{I}}$, i.e., $a - b \in \mathfrak{I}$.

We introduce addition and multiplication operations for congruence classes precisely as in §7.2, by using representatives. If A, B are congruence classes, with $a \in A$, $b \in B$, we define $A + B$ and AB to be the congruence classes represented, respectively, by $a + b$ and ab. This definition is independent of the representatives used, since if also $a' \in A$, $b' \in B$, then $a - a' \in \mathfrak{I}$, $b - b' \in \mathfrak{I}$, and then

$$(a + b) - (a' + b') = (a - a') + (b - b') \in \mathfrak{I},$$

$$ab - a'b' = a(b - b') + (a - a')b' \in \mathfrak{I},$$

so that $a + b$, $a' + b'$ represent the same congruence class, and ab, $a'b'$ represent the same congruence class. We used here the closure of $\mathfrak{I}$ under addition and strong multiplication.

It is now a simple matter to prove that under these operations, the congruence classes $(\text{mod}\ \mathfrak{I})$ form a ring. Letting C also be a congruence class, and $c \in C$, the proof is word for word the same as in §7.2.

We therefore obtain a ring of congruence classes. This ring is called the *quotient ring*, or the *difference ring*, or the *congruence class ring*, or the *residue class ring*, modulo $\mathfrak{I}$, and usually denoted by $\mathfrak{R}/\mathfrak{I}$. The mapping φ from $\mathfrak{R}$ to $\mathfrak{R}/\mathfrak{I}$ defined by

$$\varphi(a) = \text{the congruence class represented by } a, \qquad \text{for } a \in \mathfrak{R},$$

is called the natural homomorphism of $\mathfrak{R}$ onto $\mathfrak{R}/\mathfrak{I}$. It is a homomorphism since (using notation as above), a represents A, b represents B, $a + b$ represents $A + B$, ab represents AB, so that

$$\varphi(a + b) = A + B = \varphi(a) + \varphi(b), \qquad \varphi(ab) = AB = \varphi(a)\varphi(b).$$

It is easily seen, using the fact that the zero element of $\mathcal{R}/\mathcal{I}$ is the congruence class represented by 0, that the kernel of φ is $\mathcal{I}$.

When $\mathcal{R}$ is $\mathbf{Z}$ or $\mathfrak{F}[\lambda]$, and $\mathcal{I}$ is the principal ideal generated by m, the construction of $\mathcal{R}/\mathcal{I}$ just given specializes to the construction given in §7.2 of the ring $\mathcal{R}_m$, and the natural homomorphism of $\mathcal{R}$ onto $\mathcal{R}/\mathcal{I}$ specializes to the natural homomorphism of $\mathcal{R}$ onto $\mathcal{R}_m$ given in Example 7.5.2.

Ideals were introduced earlier in this section as a class of subrings of $\mathcal{R}$ including all kernels of homomorphisms. Since an arbitrary ideal $\mathcal{I}$ is the kernel of the natural homomorphism of $\mathcal{R}$ onto $\mathcal{R}/\mathcal{I}$, we now see that the ideals and the kernels of homomorphisms are, in fact, the same class of subrings of $\mathcal{R}$.

EXERCISES 7.5

I

1. When is the natural homomorphism of $\mathbf{Z}$ onto $\mathbf{Z}_m$ an isomorphism?
2. Give two examples of an automorphism of $\mathbf{C}[\lambda]$. Is the polynomial function on $\mathbf{C}$ obtained from the polynomial λ^n ever an endomorphism of $\mathbf{C}$?
3. What is the inverse of the complex conjugation automorphism of $\mathbf{C}$?
4. What is the inverse of the automorphism φ given in Example 7.5.17?
5. The rings $\mathbf{Z}_4$ and $\mathbf{Z}_2[\lambda]_{m(\lambda)}$ are both rings with four elements, if $m(\lambda) = \lambda^2 + \lambda + 1$. Are they isomorphic? What if $m(\lambda) = \lambda^2 + 1$?
6. Produce an example demonstrating that composition of evaluation endomorphisms is not commutative.
7. Find the kernel and range of the evaluation endomorphisms φ_α on $\mathbf{Q}[\lambda]$, when

   (i) $\alpha(\lambda) = 1$; (ii) $\alpha(\lambda) = \lambda + 1$; (iii) $\alpha(\lambda) = \lambda^2 + 1$.

   Use these examples to demonstrate that the range of a homomorphism need not be an ideal.

II

8. Let $\mathcal{R}$ consist of all $a + b\sqrt{2}$, for $a, b \in \mathbf{Z}$. Under the usual addition and multiplication of real numbers, prove that $\mathcal{R}$ is a ring, and that the mapping φ defined by

$$\varphi(a + b\sqrt{2}) = a - b\sqrt{2}, \qquad a, b \in \mathbf{Z},$$

   is an automorphism of $\mathcal{R}$.

9. If $\mathcal{R}$, $\mathcal{S}$ are rings, prove that the zero function (see Theorem 7.4.8) from $\mathcal{R}$ to $\mathcal{S}$ is a homomorphism. What is the kernel of this homomorphism?

10. Let $\mathcal{R}$, $\mathcal{S}$ be rings and φ a surjective homomorphism from $\mathcal{R}$ to $\mathcal{S}$. If $\mathcal{R}$ has a multiplicative identity 1, prove that $\varphi(1)$ is a multiplicative identity for $\mathcal{S}$. If $\mathcal{R}$ contains a multiplicative inverse r^{-1} for one of its elements r, prove that $\varphi(r)^{-1} = \varphi(r^{-1})$.

11. Let $\mathcal{R}$ be $\mathbf{Z}$ or $\mathfrak{F}[\lambda]$, and let $\mathcal{I}$ be a nonzero ideal in $\mathcal{R}$. Then $\mathcal{I}$ consists of all elements of $\mathcal{R}$ divisible by an element m (Theorem 7.5.10). Prove: Elements m, m_1 of $\mathcal{R}$ generate the same ideal $\mathcal{I}$ if and only if they are associates, i.e., $m = um_1$ with u a unit in $\mathcal{R}$.

12. Let $\alpha \in \mathfrak{F}$ and take $\mathfrak{F}(\lambda)^{\lambda-\alpha}$ to be the ring of Example 7.5.5.
   (i) If $f(\lambda)/g(\lambda)$ is in $\mathfrak{F}(\lambda)^{\lambda-\alpha}$, and $f(\lambda)/g(\lambda)$ is in lowest terms, prove that $g(\lambda)/f(\lambda)$ is also in $\mathfrak{F}(\lambda)^{\lambda-\alpha}$, provided $(\lambda - \alpha) \nmid f(\lambda)$.
   (ii) Let $\mathcal{I}$ be a nonzero ideal in $\mathfrak{F}(\lambda)^{\lambda-\alpha}$. Each nonzero fraction $f(\lambda)/g(\lambda)$ [in lowest terms] in $\mathcal{I}$ may be written as $f_1(\lambda)(\lambda - \alpha)^k/g(\lambda)$, where $f_1(\lambda)$ is a polynomial not divisible by $\lambda - \alpha$. Choose a nonzero fraction in $\mathcal{I}$ for which the exponent k is least. For this minimal k, prove that $\mathcal{I}$ consists of all elements of $\mathfrak{F}(\lambda)^{\lambda-\alpha}$ the form

   $$r(\lambda - \alpha)^k \qquad \text{for arbitrary } r \in \mathfrak{F}(\lambda)^{\lambda-\alpha}.$$

   Conclude that the nonzero ideals in $\mathfrak{F}(\lambda)^{\lambda-\alpha}$ are precisely the principal ideals generated by $(\lambda - \alpha)^k$, for $k = 0, 1, 2, \ldots$. Let $\mathcal{I}_k$ denote the principal ideal generated by $(\lambda - \alpha)^k$.
   (iii) Prove that $\mathcal{I}_0 = \mathfrak{F}(\lambda)^{\lambda-\alpha}$, and that $\mathcal{I}_0 \supset \mathcal{I}_1 \supset \mathcal{I}_2 \supset \cdots$, the inclusions being strict.
   (iv) Prove that an element β in $\mathfrak{F}(\lambda)^{\lambda-\alpha}$ has a multiplicative inverse in $\mathfrak{F}(\lambda)^{\lambda-\alpha}$ if and only if β is not in the ideal $\mathcal{I}_1$.
   (v) Show that $\mathcal{I}_1$ is the kernel of the evaluation homomorphism φ_α on $\mathfrak{F}(\lambda)^{\lambda-\alpha}$.

   A *local ring* is a ring in which there is a single "maximal" ideal, i.e., an ideal containing all ideals except the unit ideal. Prove that $\mathfrak{F}(\lambda)^{\lambda-\alpha}$ is a local ring and that $\mathcal{I}_1$ is its maximal ideal.

13. Let $\mathcal{R}$, $\mathcal{S}$ be fields and φ an isomorphism from $\mathcal{R}$ onto $\mathcal{S}$.
   (i) Show that φ may be prolonged to an isomorphism of $\mathcal{R}(\lambda)$ onto $\mathcal{S}(\Lambda)$ such that $\varphi(\lambda) = \Lambda$.
   (ii) Show that φ may also be prolonged to an isomorphism of $\mathcal{R}(\lambda)$ onto $\mathcal{S}(\Lambda)$ such that $\varphi(\lambda) = \Lambda^{-1}$.
   (iii) Apply (i), (ii) to discover four distinct automorphisms of $\mathbf{C}(\lambda)$.

14. Let $\mathcal{R}$, $\mathcal{S}$ be fields, and φ an isomorphism from $\mathcal{R}$ to $\mathcal{S}$. Compose the prolongation of φ given in Theorem 7.5.19 or Exercise 13 with an

evaluation endomorphism, and obtain a prolongation of φ to $\mathcal{R}[\lambda]$ or $\mathcal{R}(\lambda)$ which maps λ to any specified polynomial or nonconstant rational function in $\mathcal{S}[\Lambda]$, $\mathcal{S}(\Lambda)$ respectively.

15. Find all automorphisms of the field in Example 7.5.17.

16. Let $\mathfrak{F}$ be a field. Define a mapping φ from the polynomial ring $\mathfrak{F}[\lambda]$ to the ring of polynomial functions on $\mathfrak{F}$ (see §7.4) in this way: If $f(\lambda) \in \mathfrak{F}[\lambda]$, take $\varphi(f(\lambda))$ to be the polynomial function on $\mathfrak{F}$ determined by $f(\lambda)$; i.e.,

    $\varphi(f(\lambda))$ is the function on $\mathfrak{F}$ which, when evaluated at $\alpha \in \mathfrak{F}$, yields $f(\alpha)$.

    (i) Prove that φ is a surjective homomorphism of $\mathfrak{F}[\lambda]$ to the ring of polynomial functions on $\mathfrak{F}$.
    (ii) Show that the kernel of this homomorphism is the zero ideal if $\mathfrak{F}$ is an infinite field, and the principal ideal generated by the polynomial $m(\lambda)$ in Theorem 7.4.7, if $\mathfrak{F}$ is a finite field.
    (iii) Conclude that φ is an isomorphism if and only if $\mathfrak{F}$ is an infinite field.

III

17. Find an isomorphism between the rings in Exercise 3 of §7.3.

18. Let $\mathcal{R}$ be the ring of complex numbers of the form $a + b\sqrt{-5}$, where $a, b \in \mathbf{Z}$. Unique factorization into primes does not hold within this ring. See Exercise 3.5.16. Show that the subset of $\mathcal{R}$ consisting of all

$$3r_1 + (1 + 2\sqrt{-5})r_2 \qquad \text{for } r_1, r_2 \in \mathcal{R},$$

is a nonprincipal ideal in $\mathcal{R}$. (For number rings of this type, it can be proved that the unique factorization theorem is valid if and only if every ideal is principal. See the Pollard book listed on page 455.)

19. Consider the ring $\mathcal{R} = \mathbf{R}[\lambda, \mu]$ of polynomials in two indeterminates. (See Exercise 24, §3.2.) Show that all elements of $\mathcal{R}$ of the form

$$r_1\lambda + r_2\mu, \qquad r_1, r_2 \in \mathcal{R},$$

form an ideal in $\mathcal{R}$. Prove that this ideal is not principal. (In spite of the fact that this ring contains nonprincipal ideals, the unique factorization theorem does hold in this ring.)

20. Let $m(\lambda) \in \mathbf{Z}_2[\lambda]$, and take $\mathcal{R} = \mathbf{Z}_2[\lambda]_{m(\lambda)}$. Show that the function φ from $\mathcal{R}$ to itself defined by $\varphi(r) = r^2$, any $r \in \mathcal{R}$, is always an endomorphism of $\mathcal{R}$, and an automorphism when $m(\lambda)$ is a prime in $\mathbf{Z}_2[\lambda]$.

21. (For readers familiar with 2×2 matrices) Let 0, 1 denote the elements of $\mathbf{Z}_2$, and let $\mathcal{R}$ be the ring in Exercise 14, §1.5. Show that

the mapping φ from $\mathfrak{R}$ to 2×2 matrices defined by

$$\varphi(a) = \begin{bmatrix} 0 & 0 \\ 0 & 0 \end{bmatrix}, \quad \varphi(b) = \begin{bmatrix} 1 & 1 \\ 0 & 0 \end{bmatrix}, \quad \varphi(c) = \begin{bmatrix} 0 & 0 \\ 1 & 1 \end{bmatrix}, \quad \varphi(d) = \begin{bmatrix} 1 & 1 \\ 1 & 1 \end{bmatrix},$$

is an isomorphism.

22. Let $\mathcal{G}$ be a commutative group, the group operation being $+$. An endomorphism of $\mathcal{G}$ is a mapping φ of $\mathcal{G}$ to itself such that

$$\varphi(g_1 + g_2) = \varphi(g_1) + \varphi(g_2) \qquad \text{for all } g_1, g_2 \in \mathcal{G}.$$

Prove that, under function addition and composition, the endomorphisms of $\mathcal{G}$ form a ring. Produce an example in which this ring of endomorphisms is noncommutative.

23. Let $\mathcal{G}$ be a group, written multiplicatively, and $\mathfrak{N}$ a subgroup of $\mathcal{G}$ satisfying this condition: $gng^{-1} \in \mathfrak{N}$, for all $g \in \mathcal{G}$, $n \in \mathfrak{N}$. (This type of subgroup is called a *normal* subgroup.) Definition: Elements g_1, g_2 of $\mathcal{G}$ are congruent modulo $\mathfrak{N}$ if and only if $g_1g_2^{-1} \in \mathfrak{N}$.
   (i) Prove that congruence modulo $\mathfrak{N}$ is an equivalence relation on $\mathcal{G}$. We call the equivalence classes under this equivalence relation *congruence classes.* (They are often also called *cosets.*)
   (ii) Introduce an operation on the congruence classes, by using representatives, in the following way: If A, B are congruence classes, and $a \in A$, $b \in B$, let AB be the congruence class represented by ab. Prove: This definition of AB is independent of the representatives used of A, B.
   (iii) Prove that the congruence classes, under the operation just introduced, form a group. This group is called the congruence class group of $\mathcal{G}$, modulo $\mathfrak{N}$, and denoted by $\mathcal{G}/\mathfrak{N}$.
   (iv) Define a mapping φ from $\mathcal{G}$ to $\mathcal{G}/\mathfrak{N}$ by specifying that $\varphi(g)$ is the congruence class represented by g, for $g \in \mathcal{G}$. Prove that $\varphi(g_1g_2) = \varphi(g_1)\varphi(g_2)$, and that $\varphi(g)$ is the identity element of $\mathcal{G}/\mathfrak{N}$ if and only if $g \in \mathfrak{N}$. This mapping φ is called the *natural homomorphism* of $\mathcal{G}$ onto $\mathcal{G}/\mathfrak{N}$.

24. Let $\mathfrak{F}$ be a field, and take $\mathfrak{R} = \mathfrak{F}[[\lambda]]$, the ring of formal power series with coefficients in $\mathfrak{F}$ (see §3.2). Let $\mathfrak{L}$ consist of all *extended formal power series*

$$a(\lambda) = \sum_{i=-\infty}^{\infty} \alpha_i \lambda^i,$$

where each $\alpha_i \in \mathfrak{F}$ and almost all coefficients α_i with negative subscript i are 0, i.e., only finitely many powers λ^i with negative exponent i are actually present. Let also $b(\lambda) = \sum_{i=-\infty}^{\infty} \beta_i \lambda^i \in \mathfrak{L}$, where each $\beta_i \in \mathfrak{F}$. Endow $\mathfrak{L}$ with these operations:

$$a(\lambda) = b(\lambda) \quad \text{if and only if} \quad \alpha_i = \beta_i \text{ for every } i;$$

$$a(\lambda) + b(\lambda) = \sum_{i=-\infty}^{\infty} (\alpha_i + \beta_i)\lambda^i;$$

$$a(\lambda)b(\lambda) = \sum_{i=-\infty}^{\infty} \left(\sum_{j=-\infty}^{i} \alpha_j\beta_{i-j}\right)\lambda^i.$$

Notice that $\sum_{j=-\infty}^{i} \alpha_j\beta_{i-j}$ is actually a finite sum. Verify that these operations make $\mathfrak{L}$ into a field containing $\mathfrak{F}[[\lambda]]$ as a subring. (Suggestion: If $a(\lambda) = \sum_{i=t}^{\infty} \alpha_i\lambda^i$, with $\alpha_t \neq 0$, the multiplicative inverse of $a(\lambda)$ is $\sum_{i=-t}^{\infty} d_i\lambda^i$, where $d_{-t} = \alpha_t^{-1}$, and then the d_i for $i = -t+1, -t+2, \ldots$ are determined recursively from $\alpha_t d_i + \alpha_{t+1}d_{i-1} + \alpha_{t+2}d_{i-2} + \cdots = 0$.)

Remark: Because the polynomial ring $\mathfrak{F}[\lambda]$ is contained within the field $\mathfrak{L}$, its field of fractions $\mathfrak{F}(\lambda)$ is isomorphic to a subfield of $\mathfrak{L}$. This suggests that any rational function is expressible as a formal power series $\sum \alpha_i\lambda_i^i$ involving only finitely many powers of λ with negative exponent. This will be proved in §8.4.

25. Repeat the above exercise, but with this change: $\mathfrak{L}$ is to consist of extended formal power series involving only finitely many powers of λ with positive exponent. (The inner sum in the formula for $a(\lambda)b(\lambda)$ is now to be $\sum_{j=i}^{\infty} \alpha_j\beta_{i-j}$.) Again prove that $\mathfrak{L}$ is a field containing $\mathfrak{F}[\lambda]$ as a subring, and therefore containing a subfield isomorphic to $\mathfrak{F}(\lambda)$.

Remark: This suggests that a rational function is expressible as a power series in λ involving only finitely many powers of λ with positive exponent. This has already been seen in the Laurent series discussion in §6.5.

7.6 RATIONAL FACTORIZATIONS OF POLYNOMIALS WITH INTEGRAL COEFFICIENTS

Factorization, for the polynomial rings $\mathbf{R}[\lambda]$ and $\mathbf{C}[\lambda]$, was studied in some detail in §5.4, and studied in less detail for the ring $\mathbf{Q}[\lambda]$. We shall now apply the homomorphism theory of §7.5 to obtain further results concerning factorization within $\mathbf{Q}[\lambda]$. The basic result (Theorem 7.6.3) will be that, apart from insertion of constants, any factorization within $\mathbf{Q}[\lambda]$ of a polynomial with integral coefficients is achievable using factors with integral coefficients. This result will then be applied to exhibit primes of arbitrary degree in $\mathbf{Q}[\lambda]$.

We begin with a definition.

DEFINITION 7.6.1
The content *of a polynomial* $f(\lambda) = a_m\lambda^m + \cdots + a_0$ *with integral coefficients* $a_m, \ldots, a_0$ *is the greatest common divisor* $d = (a_m, \ldots, a_0)$ *of these coefficients. If the content* $d = 1$, *we say that* $f(\lambda)$ *is* primitive.

The following lemma is usually called the Lemma of Gauss, after its discoverer, C. F. Gauss (1777–1855).

LEMMA *7.6.2*
The product of primitive polynomials is primitive.

PROOF Let $f(\lambda) = a_m\lambda^m + \cdots + a_0$, $g(\lambda) = b_n\lambda^n + \cdots + b_0$ be polynomials with integral coefficients $a_m, \ldots, a_0, b_n, \ldots, b_0$. Set $h(\lambda) = f(\lambda)g(\lambda) = c_{m+n}\lambda^{m+n} + \cdots + c_0$, where $c_{m+n}, \ldots, c_0 \in \mathbf{Z}$. We assume that both $f(\lambda)$, $g(\lambda)$ are primitive, and wish to prove that $h(\lambda)$ is also primitive. If $h(\lambda)$ were not primitive, then $d = (c_{m+n}, \ldots, c_0) > 1$, so that a prime integral factor p of d exists. Thus $p \mid c_j$, for $j = 0, 1, \ldots, m + n$. Let φ be the natural homomorphism of $\mathbf{Z}$ onto $\mathbf{Z}_p$. (Example 7.5.2.) By Theorem 7.5.19, φ may be prolonged to a homomorphism of $\mathbf{Z}[\lambda]$ to $\mathbf{Z}_p[\Lambda]$ such that $\varphi(\lambda) = \Lambda$. Since $p \mid c_j$, the element $\varphi(c_j)$ of $\mathbf{Z}_p$ is the zero element, and hence

$$\varphi(h(\lambda)) = \varphi(c_{m+n})\Lambda^{m+n} + \cdots + \varphi(c_0)$$

is the zero polynomial in $\mathbf{Z}_p[\Lambda]$. Because $f(\lambda)$ is primitive, at least one coefficient $a_m, \ldots, a_0$ of $f(\lambda)$ is not divisible by p, and therefore at least one coefficient of

$$\varphi(f(\lambda)) = \varphi(a_m)\Lambda^m + \cdots + \varphi(a_0)$$

is a nonzero element of $\mathbf{Z}_p$. Thus $\varphi(f(\lambda))$ is a nonzero polynomial in $\mathbf{Z}_p[\Lambda]$; similarly, $\varphi(g(\lambda))$ is a nonzero polynomial in $\mathbf{Z}_p[\Lambda]$. From $h = fg$, we get

$$0 = \varphi(h(\lambda)) = \varphi(f(\lambda))\varphi(g(\lambda)).$$

This equation implies that the nonzero elements $\varphi(f(\lambda))$, $\varphi(g(\lambda))$ of $\mathbf{Z}_p[\Lambda]$ have a zero product. However, this is impossible because $\mathbf{Z}_p[\Lambda]$ is the polynomial ring over a field $\mathbf{Z}_p$ and therefore is an integral domain. (Theorems 1.6.4, 3.2.1, 7.3.1.) This contradiction shows that $h(\lambda)$ must be primitive. □

THEOREM ***7.6.3***
Let $F(\lambda)$ *be a nonzero polynomial with integral coefficients. If* $A(\lambda)$, $B(\lambda)$ *are polynomials with rational coefficients such that*

$$F(\lambda) = A(\lambda)B(\lambda), \tag{1}$$

then

$$F(\lambda) = da(\lambda)b(\lambda), \tag{2}$$

where (i) *d is the content of $F(\lambda)$;*
(ii) *the polynomials $a(\lambda)$, $b(\lambda)$ have integral coefficients and are primitive;*
(iii) *constants $\alpha, \beta \in \mathbf{Q}$ exist such that $a(\lambda) = \alpha A(\lambda)$, $b(\lambda) = \beta B(\lambda)$.*

PROOF Bringing all coefficients of $A(\lambda)$ to a common positive denominator, we may write $A(\lambda)$ as $A(\lambda) = (1/r_1)A_1(\lambda)$, where $A_1(\lambda)$ has integral coefficients and r_1 is a positive integer. If d_1 is the content of $A_1(\lambda)$, then $A_1(\lambda) = d_1 a(\lambda)$, where $a(\lambda)$ has integral coefficients and is primitive. Then $A(\lambda) = (d_1/r_1)a(\lambda)$. Similarly, $B(\lambda) = (d_2/r_2)b(\lambda)$, where d_2, r_2 are positive integers, and $b(\lambda)$ has integral coefficients and is primitive. Let d be the content of $F(\lambda)$, so that $F(\lambda) = df(\lambda)$, where $f(\lambda)$ has integral coefficients and is primitive. Then $F(\lambda) = A(\lambda)B(\lambda)$ becomes

$$dr_1r_2 f(\lambda) = d_1d_2 a(\lambda)b(\lambda). \tag{3}$$

By Lemma 7.6.2, the product $a(\lambda)b(\lambda)$ of primitive polynomials is primitive. The content of the right-hand side of (3) is therefore d_1d_2 ; and since $f(\lambda)$ is primitive, the content of the left-hand side is dr_1r_2. Therefore $d_1d_2 = fr_1r_2$, yielding

$$f(\lambda) = a(\lambda)b(\lambda).$$

Multiplying by d gives $F(\lambda) = da(\lambda)b(\lambda)$, as required; and taking $\alpha = r_1/d_1$, $\beta = r_2/d_2$ completes the proof. □

We call (2) a factorization of $F(\lambda)$ *equivalent* to the factorization (1).

We next use this result and the homomorphism theory to obtain a criterion often sufficient to prove that a polynomial with integral coefficients is a prime in $\mathbf{Q}[\lambda]$.

THEOREM 7.6.4 (Eisenstein's criterion)
Let $a(\lambda) = a_0 + a_1\lambda + \cdots + a_m\lambda^m$ be a polynomial with integral coefficients. If a prime integer p exists such that

$$p \mid a_0,\ p \mid a_1,\ \ldots,\ p \mid a_{m-1},\ p \nmid a_m,\ p^2 \nmid a_0, \tag{4}$$

then $a(\lambda)$ is a prime in $\mathbf{Q}[\lambda]$.

PROOF If $a(\lambda)$ is not a prime in $\mathbf{Q}[\lambda]$, polynomials

$$b(\lambda) = b_0 + b_1\lambda + \cdots + b_k\lambda^k, \qquad b_0, b_1, \ldots, b_k \in \mathbf{Q},$$
$$c(\lambda) = c_0 + c_1\lambda + \cdots + c_l\lambda^l \qquad c_0, c_1, \ldots, c_l \in \mathbf{Q},$$
$$k \geq 1,\ l \geq 1,\ k + l = m,$$

would exist such that $a(\lambda) = b(\lambda)c(\lambda)$. We may, by Theorem 7.6.3, assume that $b(\lambda)$, $c(\lambda)$ have integral coefficients. Clearly,

$$a_0 = b_0c_0.$$

Evidently, from (4), p divides one of b_0, c_0, but not both. Suppose $p \mid b_0$, $p \nmid c_0$.

Let φ be the natural homomorphism of $\mathbf{Z}$ onto $\mathbf{Z}_p$, and prolong φ to a homomorphism of $\mathbf{Z}[\lambda]$ to $\mathbf{Z}_p[\Lambda]$, with $\varphi(\lambda) = \Lambda$. Then $\varphi(a(\lambda)) = \varphi(a_0) + \cdots + \varphi(a_m)\Lambda^m$. By (4), $\varphi(a_0), \ldots, \varphi(a_{m-1})$, but not $\varphi(a_m)$, are the zero element of $\mathbf{Z}_p$. Thus $\varphi(a(\lambda))$ is a monomial of the type

$$\text{a nonzero constant} \times \Lambda^m.$$

From $a(\lambda) = b(\lambda)c(\lambda)$ we get $\varphi(a(\lambda)) = \varphi(b(\lambda))\varphi(c(\lambda))$. Since $\varphi(a(\lambda))$ is a nonzero constant multiple of Λ^m, both $\varphi(b(\lambda))$, $\varphi(c(\lambda))$ are divisors of Λ^m, thus must be monomials. However, $\varphi(c_0)$ is a nonzero element of $\mathbf{Z}_p$, since $p \nmid c_0$. From this fact and $\varphi(c(\lambda)) = \varphi(c_0) + \varphi(c_1)\Lambda + \cdots$, it is evident that the monomial $\varphi(c(\lambda))$ must be a constant. Consequently, the other factor $\varphi(b(\lambda))$ of $\varphi(a(\lambda))$ must be a nonzero constant multiple of Λ^m and hence must have degree m. However, the formula

$$\varphi(b(\lambda)) = \varphi(b_0) + \cdots + \varphi(b_k)\Lambda^k$$

shows that the degree of $\varphi(b(\lambda))$ is at most k. Thus $m \leq k$. But this contradicts $k \geq 1$, $l \geq 1$, $k + l = m$. Consequently, no factorization of $a(\lambda)$ is possible, implying that $a(\lambda)$ is a prime in $\mathbf{Q}[\lambda]$. □

We now apply Eisenstein's criterion to exhibit a prime element of $\mathbf{Q}[\lambda]$. In the exercises below, Eisenstein's criterion is used to find numerous other examples of primes in $\mathbf{Q}[\lambda]$.

COROLLARY 7.6.5

If n is a positive integer, the finite geometric series

$$a(\lambda) = 1 + \lambda + \lambda^2 + \cdots + \lambda^{n-1}$$

is a prime in $\mathbf{Q}[\lambda]$ if and only if n is a prime integer.

PROOF We give first a preliminary result needed in the proof. Let p be a positive prime integer. Recall (§4.2) that the binomial coefficient $\binom{p}{j}$ is given by

$$\binom{p}{j} = \frac{p!}{j!(p-j)!}.$$

Suppose $0 < j < p$. Multiplying by $j!(p-j)!$ produces the equation

$$1 \cdot 2 \cdots j \cdot 1 \cdot 2 \cdots (p-j) \cdot \binom{p}{j} = 1 \cdot 2 \cdots p,$$

and applying the natural homomorphism φ of $\mathbf{Z}$ to $\mathbf{Z}_p$ leads to

$$\varphi(1) \cdot \varphi(2) \cdots \varphi(j) \cdot \varphi(1) \cdot \varphi(2) \cdots \varphi(p-j) \cdot \varphi\left(\binom{p}{j}\right) = 0. \qquad (5)$$

Since $0 < j < p$, each of $1, 2, \ldots, j, 1, 2, \ldots, p - j$ is positive and smaller than p, and hence not divisible by p. This implies that $\varphi(1), \varphi(2), \ldots, \varphi(j), \varphi(1), \varphi(2), \ldots, \varphi(p - j)$ are nonzero elements in $\mathbf{Z}_p$. Since $\mathbf{Z}_p$ has no zero divisors, being a field, at least one factor on the left-hand side of (5) must be the zero element of $\mathbf{Z}_p$. Thus $\varphi(\binom{p}{j}) = 0$. Because the kernel of φ is the principal ideal generated by p, we see that

the integer $\binom{p}{j}$ *is divisible by* p *if* p *is a prime and* $0 < j < p$.

Let $n = p$ be a prime integer. From Lemma 2.2.3 we get

$$a(\lambda) = \frac{\lambda^p - 1}{\lambda - 1}.$$

Evaluating at $\lambda + 1$ yields

$$a(\lambda + 1) = \frac{(\lambda + 1)^p - 1}{\lambda}.$$

From the binomial theorem,

$$(\lambda + 1)^p = \sum_{j=0}^{p} \binom{p}{j} \lambda^j = 1 + p\lambda + \sum_{j=2}^{p-1} \binom{p}{j} \lambda^j + \lambda^p,$$

using $\binom{p}{0} = 1$, $\binom{p}{1} = p$, $\binom{p}{p} = 1$. Thus

$$a(\lambda + 1) = p + \sum_{j=2}^{p-1} \binom{p}{j} \lambda^{j-1} + \lambda^{p-1}.$$

Since $p \mid \binom{p}{j}$ for $j = 2, \ldots, p - 1$, evidently the coefficients of $a(\lambda + 1)$ meet Eisenstein's criterion. Therefore $a(\lambda + 1)$ is a prime in $\mathbf{Q}[\lambda]$.

If $a(\lambda)$ were not a prime in $\mathbf{Q}[\lambda]$, then $a(\lambda) = b(\lambda)c(\lambda)$, where $b(\lambda)$, $c(\lambda)$ are nonconstant polynomials with coefficients in $\mathbf{Q}$. Evaluating at $\lambda + 1$, we obtain

$$a(\lambda + 1) = b(\lambda + 1)c(\lambda + 1).$$

Since $b(\lambda + 1)$, $c(\lambda + 1)$ have coefficients in $\mathbf{Q}$ and are not constants, this factorization of $a(\lambda + 1)$ contradicts the fact that $a(\lambda + 1)$ is a prime in $\mathbf{Q}[\lambda]$. Therefore $a(\lambda)$ is a prime in $\mathbf{Q}[\lambda]$ when $n = p$ is a prime integer.

To complete the proof, we show that $a(\lambda)$ is not a prime polynomial if n is not a prime integer. If n is not prime, it is either 1 or composite. If $n = 1$, then $a(\lambda) = 1$, obviously not a prime in $\mathbf{Q}[\lambda]$. If n is composite, say $n = kl$ with $k > 1$, $l > 1$, then

$$a(\lambda) = \frac{\lambda^n - 1}{\lambda - 1} = \frac{\lambda^{kl} - 1}{\lambda^k - 1} \cdot \frac{\lambda^k - 1}{\lambda - 1}$$

$$= (\lambda^{k(l-1)} + \lambda^{k(l-2)} + \cdots + \lambda^k + 1)(\lambda^{k-1} + \lambda^{k-2} + \cdots + \lambda + 1).$$

Thus $a(\lambda)$ is composite. □

EXERCISES 7.6

I

1. Find integral factorizations equivalent to the following rational factorizations:

$$2\lambda^3 - 3\lambda^2 + 3\lambda - 1 = (\tfrac{2}{3}\lambda - \tfrac{1}{3})(3\lambda^2 - 3\lambda + 3),$$
$$2\lambda^2 - 2 = (\tfrac{4}{3}\lambda - \tfrac{4}{3})(\tfrac{3}{2}\lambda + \tfrac{3}{2}).$$

2. What is the content of $(8\lambda^4 - 6\lambda^2 + 10)(6\lambda^3 - 8)$?

3. Find the factorization of $\lambda^3 - p$ into primes in $\mathbf{Q}[\lambda]$. Here p is a prime integer.

4. Let p be a prime integer, n an integer with $n > 1$. Which of the following are primes in $\mathbf{Q}[\lambda]$?

$\lambda^n + p, \quad \lambda^n - p, \quad \lambda^n + p\lambda + p, \quad \lambda^n + p\lambda^{n-1} + p,$

$\lambda^n + \lambda, \quad \lambda^n - 1, \quad (\lambda + 2)^n + p\lambda + 3p$

5. Which of the following polynomials are primes in $\mathbf{Q}[\lambda]$?

$\lambda^2 + 5\lambda + 6, \quad \lambda^3 + 8\lambda + 6, \quad \lambda^{11} + 15\lambda^5 + 30,$

$2\lambda^4 + 3\lambda + 12, \quad \lambda^4 - 2\lambda^2 + 1, \quad \lambda^6 - 5\lambda^3 + 6.$

Factor each of these polynomials into a product of primes in $\mathbf{Q}[\lambda]$.

II

6. Use Theorem 7.6.3 to give a new proof of the rational and integral root theorem, Theorem 5.4.9. (Hint: If $f(r/s) = 0$, then $f(\lambda) = (\lambda - r/s)g(\lambda)$ where g has rational coefficients. Obtain an equivalent factorization in which only integral coefficients appear.

7. Let p be a prime integer. Eisenstein's criterion may be expressed this way: If the digits of the polynomial $a(\lambda)$ in base λ are integers, all but the highest divisible by p, and the lowest not divisible by p^2, then $a(\lambda)$ is a prime in $\mathbf{Q}[\lambda]$. Is the same true for integers; i.e., must an integer, which in a base b has all digits except the highest divisible by p and the lowest not divisible by p^2, necessarily be a prime integer?

8. Let $\alpha(\lambda)$ be a polynomial in $\mathbf{Q}[\lambda]$. Does the substitution homomorphism φ_α, for which $\varphi_\alpha(a(\lambda)) = a(\alpha(\lambda))$, necessarily carry a prime polynomial $a(\lambda)$ to another prime polynomial? What if $\deg \alpha(\lambda) = 1$? or $\deg \alpha(\lambda) > 1$?

9. Twice a prime in $\mathbf{Z}$ is never a prime in $\mathbf{Z}$. Twice a prime in $\mathbf{Q}[\lambda]$ is always a prime in $\mathbf{Q}[\lambda]$. Prove these statements.

10. Let p be a prime integer and φ the natural homomorphism of $\mathbf{Z}$ to $\mathbf{Z}_p$. Prolong φ to a homomorphism of $\mathbf{Z}[\lambda]$ to $\mathbf{Z}_p[\Lambda]$, with $\varphi(\lambda) = \Lambda$.

If the polynomial $a(\lambda)$ has integral coefficients and is a prime in $\mathbf{Q}[\lambda]$, must $\varphi(a(\lambda))$ be a prime in $\mathbf{Z}_p[\Lambda]$?

III

11. Let p be a positive prime integer. Is $1 + \lambda + \lambda^2 + \cdots + \lambda^{p-1}$ a prime in $\mathbf{R}[\lambda]$ or $\mathbf{C}[\lambda]$? If not, factor it into primes in these rings.

12. Let $c(f)$ denote the content of a polynomial f having integral coefficients. If f, g are polynomials with integral coefficients, show that $c(fg) = c(f)c(g)$.

13. Formulate and prove a "unique factorization into primes" theorem for the ring $\mathbf{Z}[\lambda]$. Your theorem must take into account both prime polynomials and prime integers.

14. Gauss's lemma and Theorem 7.6.3 were proved for polynomials with coefficients in $\mathbf{Z}$. Let $\mathcal{F}$ be a field, and take $\mathcal{R}$ to be the ring $\mathcal{F}[\mu]$. Consider elements of $\mathcal{R}[\lambda]$. If $f(\lambda) \in \mathcal{R}[\lambda]$, define the content of $f(\lambda)$ to be the greatest common divisor of the coefficients of $f(\lambda)$—these coefficients being elements of $\mathcal{F}[\mu]$. Then prove Gauss's lemma, the analogue of Theorem 7.6.3, and Eisenstein's criterion, all for polynomials with coefficients in $\mathcal{R}$.

7.7 THE CLASSIFICATION OF HOMOMORPHISMS OF SOME IMPORTANT RINGS (OPTIONAL)

In this section we give theorems classifying the homomorphic images, the endomorphism monoid, and the automorphism group, of some important rings. We begin with a preliminary result.

***LEMMA* 7.7.1**

Let $\mathcal{R}$, $\mathcal{S}$, $\mathcal{T}$ be rings, Φ a surjective homomorphism of $\mathcal{R}$ to $\mathcal{S}$, and φ a surjective homomorphism of $\mathcal{R}$ to $\mathcal{T}$. If Φ, φ have the same kernel, then $\mathcal{S} \cong \mathcal{T}$.

PROOF The proof begins by defining a function θ from $\mathcal{S}$ to $\mathcal{T}$, proving that it is a homomorphism, and finally, that it is a bijection. The mapping θ will then be known to be an isomorphism of $\mathcal{S}$ onto $\mathcal{T}$, and the proof will be complete. (See Figure 7.7.1.)

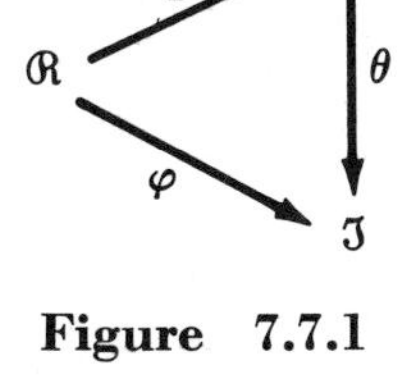

Figure 7.7.1

Let $\mathcal{J}$ be the common kernel of Φ, φ. If $r \in \mathcal{R}$, the equation $\Phi(r) = 0$ then holds if and only if $\varphi(r) = 0$, since either holds if and only if $r \in \mathcal{J}$.

Let $s \in \mathcal{S}$. Then $s = \Phi(r)$ for some $r \in \mathcal{R}$, since Φ is surjective. Define θ by specifying $\theta(s)$ to be $\varphi(r)$; i.e.,

$$\theta(s) = \varphi(r) \qquad \text{if } \Phi(r) = s. \tag{1}$$

This definition of $\theta(s)$ is independent of the particular element r selected from $\mathcal{R}$ to satisfy $\Phi(r) = s$, since if $r' \in \mathcal{R}$ and $\Phi(r') = s$, then $\Phi(r - r') = 0$, implying that $\varphi(r - r') = 0$, and hence that $\varphi(r) = \varphi(r')$.

We now show that θ is a homomorphism. Let s, $s' \in \mathcal{S}$, and select elements r, $r' \in \mathcal{R}$ to satisfy

$$\Phi(r) = s, \qquad \Phi(r') = s'.$$

Since Φ is a homomorphism, from these equations we obtain

$$\Phi(r + r') = s + s', \qquad \Phi(rr') = ss'.$$

The definition of θ then implies that

$$\theta(s) = \varphi(r), \quad \theta(s') = \varphi(r'), \quad \theta(s + s') = \varphi(r + r'), \quad \theta(ss') = \varphi(rr').$$

Since φ is a homomorphism, we then get

$$\theta(s + s') = \varphi(r) + \varphi(r') = \theta(s) + \theta(s'),$$
$$\theta(ss') = \varphi(r)\varphi(r') = \theta(s)\theta(s').$$

It is now clear that θ is a homomorphism.

We next prove that θ is injective. With s, s' as above, we assume that $\theta(s) = \theta(s')$, and wish to prove that $s = s'$. With r, r' as before, we have $\theta(s) = \varphi(r), \theta(s') = \varphi(r')$, so that $\varphi(r) = \varphi(r')$, and hence $\varphi(r - r') = 0$. This gives $\Phi(r - r') = 0$, and therefore $\Phi(r) = \Phi(r')$. Since $s = \Phi(r)$, $s' = \Phi(r')$, we have $s = s'$, as desired. The homomorphism θ is therefore injective.

Now we prove that θ is surjective. Let t be an element of $\mathfrak{T}$. We wish to find an element s in $\mathcal{S}$ such that $\theta(s) = t$. Since φ is surjective, there is an element r in $\mathcal{R}$ such that $\varphi(r) = t$. Let $s = \Phi(r)$. Then we have

$$\Phi(r) = s, \qquad \varphi(r) = t,$$

and therefore, by (1), $\theta(s) = \varphi(r) = t$. Consequently, θ is surjective.

Being injective and surjective, the homomorphism θ is bijective and therefore is an isomorphism. Thus $\mathcal{S} \cong \mathfrak{T}$. □

We now apply the lemma just proved to classify the homomorphic images of the integer and polynomial rings.

THEOREM 7.7.2

If $\mathcal{R}$ is the integer ring $\mathbf{Z}$ or the polynomial ring $\mathfrak{F}[\lambda]$, a homomorphic image of $\mathcal{R}$ is isomorphic to a congruence class ring $\mathcal{R}_m$ for a suitable element m of $\mathcal{R}$.

PROOF Let the ring $\mathcal{S}$ be a homomorphic image of $\mathcal{R}$ and Φ a surjective homomorphism of $\mathcal{R}$ to $\mathcal{S}$. Denote the kernel of Φ by $\mathcal{I}$. Then $\mathcal{I}$ is an ideal in $\mathcal{R}$, and since every ideal in $\mathcal{R}$ is principal, $\mathcal{I}$ is generated by an element m of $\mathcal{R}$. For this m, let φ be the natural homomorphism of $\mathcal{R}$ onto the congruence class ring $\mathcal{R}_m$. Then $\mathcal{I}$ is also the kernel of φ. Thus Φ, φ are homomorphisms from $\mathcal{R}$ onto $\mathcal{S}$, $\mathcal{R}_m$ respectively, with the same kernel, implying by Lemma 7.7.1 that $\mathcal{S} \cong \mathcal{R}_m$. □

REMARK Because the rings of primary interest in this book have been **Z** and $\mathfrak{F}[\lambda]$, we have stated Theorem 7.7.2 for these rings only. This theorem, however, is a special case of a result valid for any ring. Let $\mathcal{R}$ be an arbitrary ring, and $\mathcal{I}$ an ideal in $\mathcal{R}$. We described the ring of congruence classes of $\mathcal{R}$ modulo $\mathcal{I}$ in the last subsection of §7.5. This ring is denoted by $\mathcal{R}/\mathcal{I}$, and there is a natural homomorphism φ from $\mathcal{R}$ onto $\mathcal{R}/\mathcal{I}$, for which the kernel is $\mathcal{I}$. When $\mathcal{R}$ is **Z** or $\mathfrak{F}[\lambda]$, and $\mathcal{I}$ is the principal ideal in $\mathcal{R}$ generated by m, the congruence class ring $\mathcal{R}/\mathcal{I}$ is simply the ring $\mathcal{R}_m$ described in §7.2, and the natural homomorphism φ is the mapping given in Example 7.5.2. We now have the following generalization of Theorem 7.7.2.

A homomorphic image of a ring $\mathcal{R}$ is isomorphic to a congruence class ring $\mathcal{R}/\mathcal{I}$ for a suitably chosen ideal $\mathcal{I}$ in $\mathcal{R}$.

The proof is the same as before. If Φ is a surjective homomorphism of $\mathcal{R}$ onto a ring $\mathcal{S}$, let $\mathcal{I}$ be the kernel of Φ, and then let φ be the natural homomorphism of $\mathcal{R}$ onto $\mathcal{R}/\mathcal{I}$. Then $\mathcal{I}$ is also the kernel of φ, and therefore, by Lemma 7.7.1, $\mathcal{S} \cong \mathcal{R}/\mathcal{I}$. □

We now turn to the classification of the endomorphisms of some of the rings frequently appearing in this book. The symbol φ will now be used to denote an endomorphism. We comment that the structure of the monoid of endomorphisms or the group of automorphisms of a ring depends very much on the ring being considered, as the results below will show. Any ring $\mathcal{R}$, however, always has at least two endomorphisms: the zero endomorphism for which $\varphi(r) = 0$ for all $r \in \mathcal{R}$, and the identity automorphism $i_{\mathcal{R}}$.

THEOREM 7.7.3
*The endomorphism monoids of **Z**, **Q**, **R** consist solely of the zero and identity functions; and the automorphism groups solely of the identity function.*

PROOF (i) Let φ be an endomorphism of **Z**. Applying φ to the equation $1 = 1\cdot 1$ gives $\varphi(1) = \varphi(1)\varphi(1)$; hence $\varphi(1)(\varphi(1) - 1) = 0$, and therefore $\varphi(1) = 0$ or $\varphi(1) = 1$. If $\varphi(1) = 0$, then, for any x, we have $\varphi(x) = \varphi(x\cdot 1) = \varphi(x)\varphi(1) = 0$. In this case, φ is the zero endomorphism. Suppose $\varphi(1) = 1$. If m is a positive integer, then

$$m = 1 + 1 + \cdots + 1 \qquad (m \text{ terms})$$

so that

$$\begin{aligned}\varphi(m) &= \varphi(1) + \varphi(1) + \cdots + \varphi(1) &\qquad (m \text{ terms})\\ &= 1 + 1 + \cdots + 1 &\qquad (m \text{ terms})\\ &= m.\end{aligned}$$

Thus $\varphi(m) = m$ for any integer $m > 0$. Then $\varphi(-m) = -\varphi(m) = -m$, and we also have $\varphi(0) = 0$. Consequently, $\varphi(k) = k$ for every integer k, so that φ is the identity function. This completes the proof for **Z**.

(ii) Now let φ be an endomorphism of **Q**. By the proof in part (i), either $\varphi(1) = 0$ or $\varphi(1) = 1$. If $\varphi(1) = 0$, then, as in (i), $\varphi(x) = 0$ for every x. In this case φ is the zero endomorphism. If $\varphi(1) = 1$, then, as in part (i), $\varphi(k) = k$ for any integer k. Let $q = m/n$ be a rational number, with $m, n \in$ **Z**. Then $nq = m$, so that $\varphi(n)\varphi(q) = \varphi(m)$. Taking k to be m or n gives $\varphi(m) = m$, $\varphi(n) = n$, and therefore $n\varphi(q) = m$; i.e., $\varphi(q) = q$. Thus $\varphi(q) = q$ for all $q \in$ **Q**, implying that φ is the identity function. This completes the proof for **Q**.

(iii) Let φ be an endomorphism of **R**. Then, as in part (i), $\varphi(1) = 0$ or $\varphi(1) = 1$. If $\varphi(1) = 0$, then, as in (i), $\varphi(x) = 0$ for any x, so that φ is the zero endomorphism. If $\varphi(1) = 1$, then, as in parts (i) and (ii), $\varphi(q) = q$ for every rational number q. Let r be a nonrational real number. If $\varphi(r) \neq r$, then either $\varphi(r) < r$ or $\varphi(r) > r$. Consider the first possibility. Let q be a rational number with $\varphi(r) < q < r$. Because $r - q$ is positive, $r - q = s^2$ for some real number s. Then $\varphi(r - q) = \varphi(s)^2$ and therefore $\varphi(r - q) \geq 0$. Thus $\varphi(r) - \varphi(q) \geq 0$, so that $\varphi(r) \geq \varphi(q)$. This contradicts $\varphi(r) < q = \varphi(q)$, implying that $\varphi(r) < r$ is impossible. Similarly, $\varphi(r) > r$ is impossible. Therefore $\varphi(r) = r$, for every real number r. This means that φ is the identity function. □

We turn next to the corresponding results for polynomials and rational functions. First, we investigate the evaluation endomorphism φ_α on $\mathfrak{F}[\lambda]$, where the polynomial $\alpha(\lambda) \in \mathfrak{F}[\lambda]$.

THEOREM 7.7.4

The evaluation endomorphism φ_α on the polynomial ring $\mathfrak{F}[\lambda]$ is injective if and only if $\alpha(\lambda)$ is not a constant, and surjective if and only if $\deg \alpha(\lambda) = 1$.

PROOF The endomorphism φ_α is injective if and only if its kernel is the zero ideal, by Theorem 7.5.12. The kernel of φ_α was found in Example 7.5.3 to be the principal ideal generated by $\lambda - \alpha$ when α is a constant, and to be the zero ideal when α is not a constant. This proves the first assertion.

Now suppose φ_α is surjective. Let $\alpha(\lambda) = a_0 + a_1\lambda + \cdots + a_m\lambda^m$, where $a_0, a_1, \ldots, a_m \in \mathfrak{F}$. Since φ_α is surjective, there is a polynomial

$$f(\lambda) = f_0 + f_1\lambda + \cdots + f_n\lambda^n, \qquad f_0, f_1, \ldots, f_n \in \mathfrak{F},$$

such that $\varphi_\alpha(f(\lambda)) = \lambda$; i.e.,

$$f_0 + f_1\alpha(\lambda) + f_2\alpha(\lambda)^2 + \cdots + f_n\alpha(\lambda)^n = \lambda. \tag{2}$$

If $m = 0$ or $n = 0$, the left-hand side of (2) is a constant, contradicting the fact that the right-hand side is not. Therefore $m > 0$, $n > 0$, and we may take $a_m \neq 0$, $f_n \neq 0$. The degree of the left-hand side of (2) is mn, since the term $f_n\alpha(\lambda)^n$ has degree mn and all other terms have degree not exceeding $m(n-1)$. Since the degree of the right-hand side is 1, we obtain $mn = 1$, implying that $m = 1$. Thus $\deg \alpha(\lambda) = m = 1$.

Now suppose $\deg \alpha(\lambda) = 1$, say $\alpha(\lambda) = u\lambda + v$, where $u, v \in \mathfrak{F}$, $u \neq 0$. Let $\beta(\lambda) = u^{-1}(\lambda - v)$. Then $\varphi_\alpha(\beta(\lambda)) = \beta(\alpha(\lambda)) = \lambda$. Letting $h(\lambda)$ be any polynomial, to show that φ_α is surjective, we must find a polynomial $H(\lambda)$ such that $\varphi_\alpha(H(\lambda)) = h(\lambda)$. Taking $H(\lambda) = h(\beta(\lambda))$, we get

$$\varphi_\alpha(H(\lambda)) = \varphi_\alpha(h(\beta(\lambda)) = h(\beta(\alpha(\lambda))) = h(\lambda).$$

Thus φ_α is surjective if $\deg \alpha(\lambda) = 1$. □

COROLLARY 7.7.5

If $\alpha(\lambda) \in \mathfrak{F}[\lambda]$, the evaluation endomorphism φ_α on $\mathfrak{F}[\lambda]$ is an automorphism if and only if $\deg \alpha(\lambda) = 1$.

PROOF An endomorphism is an automorphism if and only if it is injective and surjective. □

Now we give the corresponding results for evaluation endomorphisms on $\mathfrak{F}(\lambda)$. Let $\alpha(\lambda)$ be a nonconstant element of the rational function field $\mathfrak{F}(\lambda)$.

THEOREM 7.7.6

The evaluation endomorphism φ_α on $\mathfrak{F}(\lambda)$ is always injective. It is surjective if and only if $\alpha(\lambda)$ has the form

$$\alpha(\lambda) = \frac{u\lambda + v}{w\lambda + x}, \qquad u, v, w, x \in \mathfrak{F}, \tag{3}$$

with

$$ux - vw \neq 0. \tag{4}$$

PROOF The kernel of φ_α, being an ideal in a field, must be either the zero ideal or the unit ideal. Since φ_α is not the zero function, the kernel cannot be the unit ideal and therefore must be the zero ideal. (This kernel was also computed directly in Example 7.5.4.) Therefore (Theorem 7.5.12) φ_α is injective.

Suppose now that φ_α is surjective. Let

$$\alpha(\lambda) = \alpha_1(\lambda)/\alpha_2(\lambda),$$

where $\alpha_1(\lambda)$, $\alpha_2(\lambda)$ are relatively prime polynomials. Since φ_α is surjective, there is a rational function $h(\lambda) \in \mathfrak{F}(\lambda)$ for which $\varphi_\alpha(h(\lambda)) = \lambda$. Since $\varphi_\alpha(h(\lambda))$ is not a constant, $h(\lambda)$ cannot be a constant and therefore cannot be 0. Write $h(\lambda)$ as a quotient of polynomials, in lowest terms, as $h(\lambda) = f(\lambda)/g(\lambda)$, where

$$f(\lambda) = f_0 + f_1\lambda + \cdots + f_m\lambda^m, \qquad g(\lambda) = g_0 + g_1\lambda + \cdots + g_n\lambda^n,$$

$$f_0, f_1, \ldots, f_m, g_0, g_1, \ldots, g_n \in \mathfrak{F}; \quad f_m g_n \neq 0; \quad (f(\lambda), g(\lambda)) = 1.$$

Inserting zero terms into one of these polynomials, we may take $m = n$; however, we can then only say that at least one of f_m, g_m is nonzero. At least one of f_0, g_0 is also nonzero, since if both were zero, λ would be a common factor of the relatively prime polynomials $f(\lambda)$, $g(\lambda)$.

The equation $\varphi_\alpha(h(\lambda)) = \lambda$ takes the form

$$\frac{f_0 + f_1(\alpha_1/\alpha_2) + \cdots + f_m(\alpha_1/\alpha_2)^m}{g_0 + g_1(\alpha_1/\alpha_2) + \cdots + g_m(\alpha_1/\alpha_2)^m} = \lambda,$$

yielding

$$f_0\alpha_2^m + f_1\alpha_1\alpha_2^{m-1} + \cdots + f_m\alpha_1^m = \lambda(g_0\alpha_2^m + g_1\alpha_1\alpha_2^{m-1} + \cdots + g_m\alpha_1^m).$$

Rearranging terms produces the equations

$$(\lambda g_0 - f_0)\alpha_2^m = \alpha_1 \times \text{a polynomial},$$

$$(\lambda g_m - f_m)\alpha_1^m = \alpha_2 \times \text{a polynomial}.$$

Thus $\alpha_1 \mid (\lambda g_0 - f_0)\alpha_2^m$ and $\alpha_2 \mid (\lambda g_m - f_m)\alpha_1^m$. Because α_1, α_2 are relatively prime, we obtain (Lemma 3.4.10)

$$\alpha_1 \mid (\lambda g_0 - f_0), \qquad \alpha_2 \mid (\lambda g_m - f_m). \tag{5}$$

Since g_0, f_0 are constants, not both 0, and g_m, f_m are constants, not both 0, both $\lambda g_0 - f_0$, $\lambda g_m - f_m$ are nonzero polynomials of degree at most 1. Conditions (5) therefore imply that $\deg \alpha_1(\lambda) \leq 1$, $\deg \alpha_2(\lambda) \leq 1$, so that $\alpha(\lambda) = \alpha_1(\lambda)/\alpha_2(\lambda)$ has the form displayed in (3). If $ux - vw$ were zero, we would obtain

$$\alpha(\lambda) = u/w \quad \text{if } x = 0, \qquad \alpha(\lambda) = v/x \quad \text{if } x \neq 0,$$

using the fact that w, x cannot both be 0. But this implies that $\alpha(\lambda)$ is a constant, contrary to hypothesis. Therefore $ux - vw \neq 0$.

Now we suppose that $\alpha(\lambda)$ is given by (3) and (4), and we wish to prove that φ_α is surjective. When (4) holds, it is easily checked that $\alpha(\lambda)$ is not a constant. Let

$$A = x/(ux - vw), \quad B = -v/(ux - vw),$$

$$C = -w/(ux - vw), \quad D = u/(ux - vw).$$

A simple calculation shows that

$$(AD - BC)(ux - vw) = 1.$$

Therefore $AD - BC \neq 0$, so that C, D are not both zero, and the rational function $\beta(\lambda)$ defined by

$$\beta(\lambda) = \frac{A\lambda + B}{C\lambda + D}$$

is not a constant. Straightforward calculating shows that

$$\varphi_\alpha(\beta(\lambda)) = \beta(\alpha(\lambda)) = \lambda.$$

To prove that φ_α is surjective, we let $h(\lambda)$ be any rational function, and we wish to find a rational function $H(\lambda)$ such that $\varphi_\alpha(H(\lambda)) = h(\lambda)$. We claim that

$$H(\lambda) = h(\beta(\lambda))$$

will work. Using $\beta(\alpha(\lambda)) = \lambda$, this is immediate, since

$$\varphi_\alpha(H(\lambda)) = \varphi_\alpha(h(\beta(\lambda)) = h(\beta(\alpha(\lambda))) = h(\lambda).$$

Thus φ_α is indeed surjective if $\alpha(\lambda)$ has the form given by (3) and (4). □

The rational functions of the form (3), (4) are called *Mobius functions*, or *linear fractional transformations*. They occur in several other parts of mathematics, including the theory of a complex variable and non-Euclidean geometry.

COROLLARY *7.7.7*

If $\alpha(\lambda)$ is a nonconstant element of the function field $\mathfrak{F}(\lambda)$, the evaluation endomorphism φ_α is an automorphism if and only if $\alpha(\lambda)$ is a Mobius function.

PROOF Same as the proof of Corollary 7.7.5. □

Notice that composition of evaluation endomorphisms gives an evaluation endomorphism, since if $\alpha(\lambda), \beta(\lambda) \in \mathfrak{F}[\lambda]$, then, for any $f(\lambda) \in \mathfrak{F}[\lambda]$,

$$(\varphi_\alpha\varphi_\beta)(f(\lambda)) = \varphi_\alpha(f(\beta(\lambda))) = f(\beta(\alpha(\lambda))) = \varphi_{\beta(\alpha(\lambda))}(f(\lambda)),$$

implying that

$$\varphi_{\alpha(\lambda)}\varphi_{\beta(\lambda)} = \varphi_{\beta(\alpha(\lambda))}\,.$$

The same calculation holds for evaluation endomorphisms on $\mathfrak{F}(\lambda)$. The identity function is also an evaluation endomorphism, namely, φ_λ. The evaluation endomorphisms on $\mathfrak{F}[\lambda]$, $\mathfrak{F}(\lambda)$ therefore form a monoid, a submonoid of the monoid of all endomorphisms. This submonoid is never the full monoid since the zero endomorphism is never an evaluation endomorphism.

The evaluation automorphisms on $\mathfrak{F}[\lambda]$, $\mathfrak{F}(\lambda)$ form a group, a subgroup of the group of all automorphisms, since (applying Theorem 1.6.10): (i) The composite of evaluation automorphisms is both an automorphism and an evaluation endomorphism, therefore an evaluation automorphism; (ii) Each evaluation automorphism φ_α has an inverse evaluation automorphism, namely, φ_β, where $\beta(\lambda)$ is the degree one polynomial or the Mobius function found in the proofs of Theorems 7.7.4, 7.7.6, respectively. These groups of evaluation automorphisms are usually noncommutative since, if c is an element of $\mathfrak{F}$ other that 0 or 1, then $\varphi_{c\lambda}\varphi_{\lambda+1}$, $\varphi_{\lambda+1}\varphi_{c\lambda}$ are unequal. This element c exists for all fields except $\mathbf{Z}_2$. For the case $\mathfrak{F} = \mathbf{Z}_2$, see Exercises 2 and 3.

For some fields $\mathfrak{F}$, the evaluation endomorphisms constitute all nonzero endomorphisms of $\mathfrak{F}[\lambda]$, $\mathfrak{F}(\lambda)$, and the evaluation automorphisms constitute all automorphisms. We now show this to be the case when $\mathfrak{F}$ is $\mathbf{Q}$ or $\mathbf{R}$. The exercises below show this not to be the case when $\mathfrak{F}$ is $\mathbf{C}$.

THEOREM 7.7.8

Let the field $\mathfrak{F}$ be $\mathbf{Q}$ or $\mathbf{R}$. Then:

(i) *Any nonzero endomorphism of $\mathfrak{F}[\lambda]$ is an evaluation endomorphism φ_α for a polynomial $\alpha(\lambda) \in \mathfrak{F}[\lambda]$;*

(ii) *Any nonzero endomorphism of $\mathfrak{F}(\lambda)$ is an evaluation endomorphism φ_α for a nonconstant rational function $\alpha(\lambda) \in \mathfrak{F}(\lambda)$.*

PROOF Let φ be a nonzero endomorphism of $\mathfrak{F}[\lambda]$ or $\mathfrak{F}(\lambda)$. From the proof of Theorem 7.7.3, we get $\varphi(1) = 0$ or $\varphi(1) = 1$. If $\varphi(1) = 0$, the endomorphism φ is the zero function. Therefore $\varphi(1) = 1$. We now prove, by examining cases, that $\varphi(x) = x$ for every constant $x \in \mathfrak{F}$.

Let φ be a nonzero endomorphism of $\mathbf{Q}[\lambda]$. Exactly as in parts (i), (ii) of the proof of Theorem 7.7.3, we deduce that $\varphi(q) = q$ for all $q \in \mathbf{Q}$.

Let φ be a nonzero endomorphism of $\mathbf{Q}(\lambda)$. In the same way we deduce that $\varphi(q) = q$ for all $q \in \mathbf{Q}$.

Next, let φ be a nonzero endomorphism of $\mathbf{R}[\lambda]$. Let ρ be a positive real number. Then $\varphi(\rho)$ is an element of $\mathbf{R}[\lambda]$, and we wish to show that it is a constant. Let n be any positive integer and take $s = \rho^{1/n}$ to be the real nth root of ρ. Then

$$\varphi(\rho) = \varphi(s^n) = \varphi(s)^n.$$

The degrees of the elements $\varphi(\rho)$, $\varphi(s)$ of $\mathbf{R}[\lambda]$ therefore satisfy

$$\deg \varphi(\rho) = n \deg \varphi(s).$$

Thus $n \mid \deg \varphi(\rho)$. Because n is arbitrary, $\deg \varphi(\rho)$ cannot be a positive integer, implying that $\varphi(\rho)$ is a constant. Thus $\varphi(\rho) \in \mathbf{R}$ for each positive real number ρ. Since $\varphi(-\rho) = -\varphi(\rho) \in \mathbf{R}$, $\varphi(0) = 0 \in \mathbf{R}$, we see that $\varphi(r) \in \mathbf{R}$ for every real number r. Thus φ, when acting on real numbers (only), gives an endomorphism of $\mathbf{R}$. Being a nonzero endomorphism (because $\varphi(1) = 1$), it must, by Theorem 7.7.3, be the identity. This means that $\varphi(r) = r$ for all $r \in \mathbf{R}$.

Now let φ be a nonzero endomorphism of $\mathbf{R}(\lambda)$. Let ρ be a positive real number. Then $\varphi(\rho)$, being a rational function, may be written in lowest terms as

$$\varphi(\rho) = f(\lambda)/F(\lambda),$$

where $f(\lambda)$, $F(\lambda)$ are relatively prime polynomials in $\mathbf{R}[\lambda]$. Let n be a positive integer, and take $s = \rho^{1/n}$ to be a real nth root of ρ. Then $\varphi(s)$, also being a rational function, may be written in lowest terms as

$$\varphi(s) = g(\lambda)/G(\lambda),$$

where $g(\lambda)$, $G(\lambda)$ are relatively prime polynomials. Then we have

$$f/F = \varphi(\rho) = \varphi(s^n) = \varphi(s)^n = g^n/G^n.$$

Since both f/F and g^n/G^n are in lowest terms, we get $f = g^n$, $F = G^n$, implying that $\deg f = n \deg g$, $\deg F = n \deg G$. Hence $n \mid \deg f$, $n \mid \deg F$. Because n is an arbitrary positive integer, neither $\deg f$ nor $\deg F$ can be a positive integer. Hence f, F are both constants, so that $\varphi(\rho) = f/F$ is also a constant. Thus $\varphi(\rho) \in \mathbf{R}$. From this, as in the last case, we deduce $\varphi(r) = r$ for all $r \in \mathbf{R}$.

In each case, therefore, we have $\varphi(x) = x$ for every $x \in \mathfrak{F}$.

We now finish the proof for the polynomial ring $\mathfrak{F}[\lambda]$. Certainly $\varphi(\lambda)$, as an element of $\mathfrak{F}[\lambda]$, is a polynomial. Denote it by $\alpha(\lambda)$. Then $\alpha(\lambda) \in \mathfrak{F}[\lambda]$ and $\varphi(\lambda) = \alpha(\lambda)$. We claim that φ equals the evaluation endomorphism φ_α. To show this, let

$$f(\lambda) = f_0 + f_1\lambda + \cdots + f_m\lambda^m, \qquad f_0, f_1, \ldots, f_m \in \mathfrak{F},$$

be any element of $\mathfrak{F}[\lambda]$. Using $\varphi(x) = x$ for all x in $\mathfrak{F}$ and $\varphi(\lambda) = \alpha$, we get

$$\begin{aligned}\varphi(f(\lambda)) &= \varphi(f_0) + \varphi(f_1)\varphi(\lambda) + \cdots + \varphi(f_m)\varphi(\lambda)^m \\ &= f_0 + f_1\alpha(\lambda) + \cdots + f_m\alpha(\lambda)^m \\ &= \varphi_\alpha(f(\lambda)).\end{aligned}$$

Thus $\varphi(f(\lambda)) = \varphi_\alpha(f(\lambda))$, for every $f(\lambda) \in \mathfrak{F}[\lambda]$, implying that $\varphi = \varphi_\alpha$. This completes the proof in the polynomial case.

Finally, we complete the proof in the rational function case. Here $\varphi(\lambda)$ is a rational function; denote it by $\alpha(\lambda)$, so that $\alpha(\lambda) \in \mathfrak{F}(\lambda)$ and $\varphi(\lambda) = \alpha(\lambda)$. If $\alpha(\lambda)$ were a constant, then $\varphi(\alpha) = \alpha$ (by the result proved in paragraphs above), and $\varphi(\lambda) = \alpha$. This implies that φ is not injective (since $\alpha \neq \lambda$, as α is a constant), a contradiction, because a nonzero homomorphism on any field must be injective. Therefore $\alpha(\lambda)$ is not a constant. We claim that φ equals the evaluation endomorphism φ_α. If $f(\lambda)$ is a polynomial, the calculation in the previous paragraph shows that $\varphi(f(\lambda)) = \varphi_\alpha(f(\lambda))$. We use this to show that $\varphi(h(\lambda)) = \varphi_\alpha(h(\lambda))$ for any rational function $h(\lambda)$. Let $h(\lambda) = f(\lambda)/g(\lambda)$, where $f(\lambda)$, $g(\lambda)$ are polynomials. Then

$$\begin{aligned}\varphi(h(\lambda)) &= \varphi(f(\lambda)g(\lambda)^{-1}) = \varphi(f(\lambda))\varphi(g(\lambda))^{-1}\\ &= \varphi_\alpha(f(\lambda))\varphi_\alpha(g(\lambda))^{-1} = \varphi_\alpha(f(\lambda)g(\lambda)^{-1})\\ &= \varphi_\alpha(h(\lambda)).\end{aligned}$$

[We used here $\varphi(g^{-1}) = \varphi(g)^{-1}$, $\varphi_\alpha(g^{-1}) = \varphi_\alpha(g)^{-1}$; see Exercise 10 and equation (14) of §7.5.] Thus $\varphi(h(\lambda)) = \varphi_\alpha(h(\lambda))$ for every $h(\lambda) \in \mathfrak{F}(\lambda)$, implying that $\varphi = \varphi_\alpha$. □

COROLLARY 7.7.9

Let the field $\mathfrak{F}$ be **Q** *or* **R**.

(i) *The endomorphism monoid of $\mathfrak{F}[\lambda]$ consists of the zero function and the evaluation endomorphisms $\varphi_{\alpha(\lambda)}$, for $\alpha(\lambda) \in \mathfrak{F}[\lambda]$. The automorphism group consists of the $\varphi_{\alpha(\lambda)}$ for which $\alpha(\lambda) \in \mathfrak{F}[\lambda]$ and* $\deg \alpha(\lambda) = 1$.

(ii) *The endomorphism monoid of $\mathfrak{F}(\lambda)$ consists of the zero function and the evaluation endomorphisms $\varphi_{\alpha(\lambda)}$ for $\alpha(\lambda)$ a nonconstant element of $\mathfrak{F}(\lambda)$. The automorphism group consists of the $\varphi_{\alpha(\lambda)}$ for which $\alpha(\lambda)$ is a Mobius function in $\mathfrak{F}(\lambda)$.*

These monoids and groups are noncommutative.

PROOF Every nonzero endomorphism in these cases is an evaluation endomorphism, and Corollaries 7.7.5 and 7.7.7 give the conditions for an evaluation endomorphism to be an automorphism. The noncommutativity was established immediately above the statement of Theorem 7.7.8. □

We have not yet discussed the endomorphisms of **C**. We already know two automorphisms: the identity, and complex conjugation. In view of the paucity of endomorphisms of **Q** and **R**, a natural conjecture is that these are the only nonzero endomorphisms of **C**. Surprisingly, this conjecture is false. Although it can be proved that any nonzero endomorphism of **C** must be an automorphism, it can also be proved that **C** has infinitely many automorphisms. These proofs require techniques too advanced to be included in this book. The proof that a nonzero endomorphism of **C** is an automorphism uses the fundamental theorem of algebra, and the proof of the existence of automorphisms of **C** other than the identity and complex conjugation requires an advanced result known as Zorn's Lemma. Although **C** has infinitely many automorphisms, no explicit formulas are known for any except the identity and complex conjugation.

We omit the discussion of the endomorphisms of **C**[λ] and **C**(λ). See the exercises below.

We conclude by drawing attention to the interaction of the various kinds of multiplication, confining the discussion to polynomials with real coefficients. When introducing polynomial multiplication in §3.2, we combined the coefficients of the factors to get the coefficients of the product by the "convolution" method. When reinterpreting polynomials as polynomial functions in §7.4, the multiplication became the "pointwise product." But when evaluating polynomials at other polynomials, we obtained the evaluation endomorphisms for which "composition" is the multiplication. Convolution, the pointwise product, and composition therefore all have a role to play in polynomial theory. For further insight into these multiplications, see §8.6.

EXERCISES 7.7

I

1. Evaluate $\varphi_{2\lambda}\varphi_{\lambda+1}$, $\varphi_{\lambda+1}\varphi_{2\lambda}$ at λ, $\lambda + 1$, $(\lambda + 2)/(2\lambda + 1)$. Verify that $\varphi_{2\lambda}\varphi_{\lambda+1} \neq \varphi_{\lambda+1}\varphi_{2\lambda}$. The ring here is $\mathbf{R}(\lambda)$.

2. Consider these evaluation endomorphisms on $\mathbf{Z}_2[\lambda]$: $\varphi_{\alpha(\lambda)}$, for $\alpha(\lambda) = \lambda$ and $\lambda + 1$. Are these endomorphisms automorphisms? Under composition, do they form a group? A commutative group? Are they the entire automorphism group of $\mathbf{Z}_2[\lambda]$?

3. Consider these evaluation endomorphisms on $\mathbf{Z}_2(\lambda)$: $\varphi_{\alpha(\lambda)}$ for $\alpha(\lambda) = (u\lambda + v)/(w\lambda + x)$, where $u, v, w, x \in \mathbf{Z}_2$ and $ux - vw \neq 0$. Show that this set consists of six endomorphisms. Are these endomorphisms automorphisms? Under composition, do they form a group? A commutative group? Are they the entire automorphism group of $\mathbf{Z}_2(\lambda)$?

4. If the evaluation endomorphism φ_α on $\mathbf{Q}[\lambda]$ satisfies $\varphi_\alpha{}^n$ = the identity function, find α and φ_α. The multiplication here is composition and the integer superscript n denotes nth power under this multiplication.

5. Let $\alpha(\lambda) = 2\lambda + 3 \in \mathbf{Q}[\lambda]$. Solve for the polynomial $f(\lambda)$ if $\varphi_\alpha(f(\lambda)) = \lambda^3 + \lambda^2 + \lambda + 1$. If instead $\alpha(\lambda) = (\lambda + 1)/(\lambda + 2)$, solve for a rational function $f(\lambda)$.

II

6. Use your knowledge of the kernels of evaluation endomorphisms (§7.5), together with Lemma 7.7.1, to determine when the ranges of a pair of evaluation endomorphisms φ_α, φ_β are isomorphic rings. Use your results to give examples of proper subrings of $\mathfrak{F}[\lambda]$ isomorphic to $\mathfrak{F}[\lambda]$, and proper subfields of $\mathfrak{F}(\lambda)$ isomorphic to $\mathfrak{F}(\lambda)$.

7. (i) Let $\mathcal{R}$, $\mathcal{S}$, $\mathfrak{I}$ be rings. Suppose Φ is a homomorphism of $\mathcal{R}$ onto $\mathcal{S}$, and φ a homomorphism of $\mathcal{R}$ onto $\mathfrak{I}$, with kernels $\mathcal{I}_\Phi$, $\mathcal{I}_\varphi$, respectively. If $\mathcal{I}_\Phi \subseteq \mathcal{I}_\varphi$, show that $\mathfrak{I}$ is a homomorphic image of $\mathcal{S}$. Imitate the proof of Lemma 7.7.1.
(ii) Let $\mathcal{R}$ be $\mathbf{Z}$ or $\mathfrak{F}[\lambda]$, and m_1, m_2 elements of $\mathcal{R}$. If $m_1 \mid m_2$,

apply (i) to prove that $\mathcal{R}_{m_1}$ is a homomorphic image of $\mathcal{R}_{m_2}$.
(iii) As a consequence, show that $\mathbf{Z}_2$ is a homomorphic image of $\mathbf{Z}_4$. Give this homomorphism explicitly.

8. Imitate the discussion in the text and find all endomorphisms of $\mathbf{Z}[\lambda]$. Show that the automorphism group consists of evaluation endomorphisms φ_α with $\alpha(\lambda) = u\lambda + v$, where $u = \pm 1$ and v any integer. (Compare Example 1.4.8.)

9. Let p be a positive prime integer. Imitate the discussion in the text pertaining to $\mathbf{Q}[\lambda]$, $\mathbf{Q}(\lambda)$, and find all endomorphisms of $\mathbf{Z}_p[\lambda]$, $\mathbf{Z}_p(\lambda)$. Find all automorphisms. Show that the endomorphism monoids and automorphism groups are finite.

10. The zero, identity, and complex conjugation endomorphisms of $\mathbf{C}$ have the property that they map real numbers to real numbers. Prove the converse: Any endomorphism of $\mathbf{C}$ which maps real numbers to real numbers must be the zero, identity, or conjugation mappings. (Hint: From $i^2 + 1 = 0$ follows $\varphi(i)^2 + \varphi(1) = 0$.)

11. (a) Let $\mathfrak{F}$ be a field and φ an isomorphism of $\mathfrak{F}$ to itself. Using Theorem 7.5.19, prolong φ to an isomorphism of $\mathfrak{F}[\lambda]$ to itself for which $\varphi(\lambda) = \lambda$. If φ is not the identity function on $\mathfrak{F}$, prove that the prolongation of φ cannot be an evaluation endomorphism of $\mathfrak{F}[\lambda]$.
(b) Let γ be the complex conjugation automorphism of $\mathbf{C}$. Prolong γ to an automorphism of $\mathbf{C}[\lambda]$ which acts on any polynomial by replacing each coefficient with its complex conjugate. Show that γ is not an evaluation endomorphism on $\mathbf{C}[\lambda]$.
(c) With φ as in (a), show that φ may be prolonged to an isomorphism of $\mathfrak{F}(\lambda)$ to itself for which $\varphi(\lambda) = \lambda$.

12. (i) If φ is an endomorphism of $\mathbf{C}[\lambda]$ or $\mathbf{C}(\lambda)$, prove that it must map elements of $\mathbf{C}$ to elements of $\mathbf{C}$. Imitate the proofs in the text.
(ii) If φ is a nonzero endomorphism of $\mathbf{C}[\lambda]$ or $\mathbf{C}(\lambda)$, and if φ maps real numbers to real numbers, prove that φ is either an evaluation endomorphism or the composite of an evaluation endomorphism and the conjugation endomorphism γ in Exercise 11. Imitate proofs in the text.
(iii) Conclude that not every endomorphism of $\mathbf{C}[\lambda]$ or $\mathbf{C}(\lambda)$ is an evaluation endomorphism.

13. Imitate proofs in the text, and find all endomorphisms and automorphisms of the Gaussian integers and Gaussian rationals (Exercise 13, §2.4).

14. [Continuation of Exercise 16 of §7.5] Let $\mathfrak{F}$ be a finite field. Apply Lemma 7.7.1 and deduce that the ring of polynomial functions on $\mathfrak{F}$

is isomorphic to the congruence class ring $\mathfrak{F}[\lambda]_{m(\lambda)}$, where $m(\lambda) = \lambda^n - \lambda$ is the polynomial described in Theorem 7.4.7.

III

15. Let the field $\mathfrak{F}$ be $\mathbf{Q}$ or $\mathbf{R}$, let $\alpha \in \mathfrak{F}$, and let $\mathfrak{F}(\lambda)^{\lambda-\alpha}$ ring of rational functions in Example 7.5.5, for which $\lambda - \alpha$ is not a factor of the denominator polynomial. Imitate the methods used in the text and prove that the nonzero endomorphisms of $\mathfrak{F}(\lambda)^{\lambda-\alpha}$ consist precisely of the evaluation endomorphisms $\varphi_{\beta(\lambda)}$ for which the rational function $\beta(\lambda) \in \mathfrak{F}(\lambda)^{\lambda-\alpha}$ and has α as a *fixed point*, i.e., $\beta(\alpha) = \alpha$. Find the automorphisms.

16. Let $\mathfrak{R}$ be $\mathbf{Z}$ or $\mathfrak{F}[\lambda]$, and let $\mathfrak{S}$, $\mathfrak{I}$ be homomorphic images of $\mathfrak{R}$ under homomorphisms Φ, φ, respectively. If $\mathfrak{S} \cong \mathfrak{I}$, must Φ, φ have the same kernel?

17. Let $\mathfrak{g}$ be an ideal in the ring $\mathfrak{F}(\lambda)^{\lambda-\alpha}$ of Example 7.5.5. If the congruence class ring $\mathfrak{F}(\lambda)^{\lambda-\alpha}/\mathfrak{g}$ is an integral domain or a field, find $\mathfrak{g}$.

18. The endomorphisms and automorphisms of $\mathfrak{F}[\lambda]$ or $\mathfrak{F}(\lambda)$ heretofore discussed always carry constants to constants. Show that this need not always be the case, as follows: Let $\mathfrak{F}_1$ be a field, and take $\mathfrak{F} = \mathfrak{F}_1(\mu)$. Find an automorphism of $\mathfrak{F}(\lambda)$ which does not map elements of $\mathfrak{F}$ to elements of $\mathfrak{F}$.

7.8 REVIEW

The study of homomorphic mappings of rings, the kernels and ranges of these mappings (ideals, congruence class rings), and applications of these mappings to factorization theory in $\mathbf{Q}[\lambda]$, constituted the central theme of this chapter. The homomorphic images of $\mathbf{Z}$ and $\mathfrak{F}[\lambda]$, namely, the congruence class rings modulo m, were investigated in some detail in §§7.2, 7.3. These congruence class rings provide many new examples of rings and fields, including finite rings and fields. A very useful finite field is $\mathbf{Z}_p$, where p is a positive prime integer. In general, the congruence class rings $\mathbf{Z}_m$ and $\mathfrak{F}[\lambda]_m$ are fields if and only if m is a prime, and integral domains if and only if m is zero or prime. The next section, §7.4, gave a short review of functions in general, with a discussion of surjectivity, injectivity, bijectivity, and inverses of functions. Homomorphisms of rings were introduced in §7.5, particular attention being given to the natural homomorphism of $\mathbf{Z}$ or $\mathfrak{F}[\lambda]$ onto a congruence class ring, and to the evaluation homomorphisms (endomorphisms) on a polynomial ring or rational function field. An isomorphism is a bijective homomorphism; if an isomorphic mapping exists between a pair of rings, their algebraic properties will be the same, and they may be regarded as one ring in different notations. An important result governing factorization of polynomials with integral coefficients, proved in §7.6, is this: a polynomial with integral coefficients, factorable using polynomials with rational coefficients, is factorable in an equivalent way using polynomials with integral coefficients. The last section, §7.7, classified congruence class rings as the most general (to within isomorphism) homomorphic image **Z** or $\mathfrak{F}[\lambda]$. Finally, the endomorphism monoids and automorphism groups of **Z**, **Q**, **R**, **Q**$[\lambda]$, **Q**(λ), **R**$[\lambda]$, **R**(λ), were studied in close detail in §7.7.

REVIEW EXERCISES 7.8

1. If p is a prime integer, what field is the field of fractions of $\mathbf{Z}_p$?

2. Is there a field of fractions for $\mathbf{Z}_{21}$?

3. The complex numbers $x + yi$, for $x, y \in \mathbf{Q}$, form a field (the Gaussian rationals) isomorphic to $\mathbf{Q}[\lambda]_{\lambda^2+1}$. Prove this.

4. If $m(\lambda) = a_n\lambda^n + \cdots + a_1\lambda + a_0$, where $a_n, \ldots, a_1, a_0 \in \mathfrak{F}$, is a nonzero polynomial in $\mathfrak{F}[\lambda]$, show that the congruence class ring $\mathfrak{F}[\lambda]_{m(\lambda)}$ contains an element x satisfying $f(x) = a_nx^n + \cdots + a_1x + a_0 = 0$.

5. Factor $\lambda^3 + 2\lambda + 2$ into primes in $\mathbf{Z}_3[\lambda]$, $\mathbf{Z}_5[\lambda]$, $\mathbf{Z}_7[\lambda]$.

6. Let φ be a homomorphism of $\mathbf{Q}(\lambda)$ to another field satisfying $\varphi(\lambda^3 + \lambda) = 0$. Prove that φ is the zero function. If $\varphi(\lambda^3 + \lambda) \neq 0$, prove that φ is an isomorphism.

7. Can distinct evaluation endomorphisms on $\mathbf{R}[\lambda]$ have the same kernel? What if neither is an isomorphism?

8

SPECIAL TOPICS (Optional)

8.1 INTRODUCTION

A variety of special topics will be studied in this chapter, all having direct application to ideas encountered earlier in this book and in other parts of introductory college mathematics.

We know, from §§2.2 and 6.4, that the fields of rational numbers and rational functions are endowed with absolute value functions measuring the size of their elements. In §8.2 we shall introduce infinitely many new ways, called *valuations*, of measuring the size of rational numbers and functions. Several applications of valuation theory will be given in this chapter. The first, still in §8.2, will be to show how valuations provide a mechanism for describing many subrings of $\mathbf{Q}$ and $\mathfrak{F}(\lambda)$, and ideals in these subrings, in terms of inequalities. Then we go on to §8.3, where a new type of representation of a rational number or function, called the *partial fraction representation*, will be explained. Valuation theory will be used in the proof of the uniqueness of this representation. An important application of partial fractions occurs in calculus, when evaluating integrals of rational functions. This application will be outlined in the exercises. As another application of partial fractions, we give a complete proof of a theorem (6.5.6) about convergence of Laurent series proved earlier, but under a restrictive hypothesis. In §8.4 valuation theory will be applied to develop additional representations, called *b-adic representations*, or rational numbers and functions. These representations generalize the base b representations of integers and polynomials described in Chapter 4. In the rational function case, they give *Taylor series*, a type of infinite series encountered in calculus. In the rational number case, they provide an explanation for the behaviour of *complements*, previously encountered in Chapter 4 when attempting to apply to a negative integer the procedure for calculating the base b representation of a positive integer.

Another type of representation of a rational number, called a *continued fraction*, will be given in §8.5. This is simply Euclid's procedure for calculating the greatest common divisor of a pair of integers, but with attention focussed on the quotients appearing in the procedure, rather than on the remainders. The continued fraction representation may be generalized to

apply to real numbers, and there it permits the calculation of rational numbers closely approximating real numbers. When applied to π, it produces the familiar 22/7 and other interesting, sharper approximations. The material in §8.5 uses nothing from the other sections of Chapter 8, and may be read without having read these other sections.

Finally, in §8.6, which also is independent of the other parts of Chapter 8, we shall reexamine the ring of polynomials with real coefficients, and by introducing a new type of multiplication on it called *convolution*, we shall discover how differentiation may be interpreted as a fraction.

8.2 VALUATIONS

The first portion of this section, to page 386, is prerequisite to part of §8.3 and all of §8.4. Throughout this section, the ring $\mathcal{R}$ is **Z** and the field $\mathcal{K}$ is **Q**, or the ring $\mathcal{R}$ is $\mathcal{F}[\lambda]$ and the field $\mathcal{K}$ is $\mathcal{F}(\lambda)$. As always, $\mathcal{F}$ denotes a field, which may usually be taken to be **Q**, **R**, or **C**. In some instances, always explicitly indicated, statements about $\mathcal{K}$ may also apply to **R** and **C**, and possibly to any field.

We first give the definition of a valuation on $\mathcal{K}$.

DEFINITION 8.2.1

A valuation on $\mathcal{K}$ is a function $\|\ \|$ from $\mathcal{K}$ to the real numbers **R**, *such that:*

V1 *For each element a of $\mathcal{K}$, $\|a\|$ is a nonnegative real number;*

V2 $\|a\| = 0$ *if and only if* $a = 0$;

V3 $\|ab\| = \|a\|\,\|b\|$ *for all a, b in $\mathcal{K}$;*

V4 $\|a + b\| \leq \|a\| + \|b\|$ *for all a, b in $\mathcal{K}$.*

We read the symbol $\|a\|$ as "the valuation of a." The inequality in V4 is usually called the *triangle inequality.*

We are already familiar with two examples of valuations. First, if $\mathcal{K}$ is **Q** (or **R**, or **C**), the usual absolute value $|\ |$ of rational, real, or complex numbers described in Chapter 2 satisfies axioms V1–V4 and therefore is a valuation. In fact, the above axioms V1–V4 are the same as the properties V1–V4 listed in Chapter 2 (and also in §1.2). Second, if $\mathcal{K}$ is $\mathcal{F}(\lambda)$, the absolute value $|\ |$ for rational functions described in §6.4 is a valuation, the axioms V1–V4 above being the same as the properties V1–V4 listed in §6.4. In both of these cases, we may regard $|a|$ as a measurement of the size of a. A good way, therefore, of thinking of $\|a\|$, when $\|\ \|$ is an arbitrary valuation, is as a measurement of the size of a. The reader may be surprised to learn that other ways exist, besides the two just mentioned, of measuring the size of a rational number or function. We now describe these.

Let p be a prime in $\mathcal{R}$, with p positive when $\mathcal{R}$ is $\mathbf{Z}$, and monic when $\mathcal{R}$ is $\mathfrak{F}[\lambda]$. If a is a nonzero element of $\mathcal{K}$, the prime p may appear as a factor of the numerator or the denominator of a. In all cases, we may write a as

$$a = p^e \frac{M}{N},$$

where M, N are elements of $\mathcal{R}$, neither divisible by p, and e is an integer which may be positive, negative, or zero. We now describe a function $|\ |_p$, called the *p-adic valuation.* The definition differs slightly according as $\mathcal{K}$ is $\mathbf{Q}$ or $\mathfrak{F}(\lambda)$. We let ρ be the fixed real number satisfying $0 < \rho < 1$ already used in the definition (6.4.3) of the absolute value of a rational function.

DEFINITION 8.2.2
The p-adic valuation $|\ |_p$ *on* $\mathcal{K}$ *is defined by:*

$$\textit{If } a = 0, \quad \textit{then } |a|_p = 0;$$

If $a = p^e \dfrac{M}{N}$, *where* M, N *are elements of* $\mathcal{R}$ *relatively prime to* p, *then*

$$|a|_p = \begin{cases} (1/p)^e & \text{when } \mathcal{K} \text{ is } \mathbf{Q}, \\ \rho^e & \text{when } \mathcal{K} \text{ is } \mathfrak{F}(\lambda). \end{cases}$$

We first justify calling $|\ |_p$ a valuation by showing that it satisfies axioms V1–V4.

THEOREM 8.2.3 *The p-adic valuation* $|\ |_p$ *is a valuation.*

PROOF We give, one by one, the verifications of V1–V4.

Verification of V1 *and* V2. The definition clearly shows that $|a|_p$ is a nonnegative real number, for any $a \in \mathcal{K}$, and $|a|_p = 0$ if and only if $a = 0$. □

Verification of V3. We are to prove that

$$|ab|_p = |a|_p\, |b|_p\,.$$

This equation is certainly valid if a or b is zero, since each side is then zero. Assume $a \neq 0$, $b \neq 0$. Let

$$a = p^e M/N, \qquad b = p^f M'/N', \tag{1}$$

where M, N, M', N' are elements of $\mathcal{R}$ relatively prime to p. Then

$$ab = p^{e+f} MM'/NN'.$$

Since MM', NN' are also relatively prime to p, from these equations we get

$$\begin{aligned} &|a|_p = (1/p)^e, \quad |b|_p = (1/p)^f, \quad |ab|_p = (1/p)^{e+f}, \quad \text{when } \mathcal{K} \text{ is } \mathbf{Q}, \\ &|a|_p = \rho^e, \qquad\ \ |b|_p = \rho^f, \qquad\ \ |ab|_p = \rho^{e+f}, \qquad\ \ \text{when } \mathcal{K} \text{ is } \mathfrak{F}(\lambda). \end{aligned} \tag{2}$$

These equations clearly show that $|ab|_p = |a|_p\, |b|_p$. □

Before verifying V4, we establish the following sharper property V4′, called the *strong triangle law*, or the *strong triangle inequality:*

V4′ *For all a, b in $\mathcal{K}$,* $|a + b|_p \leq \max\{|a|_p, |b|_p\}$. (3)

(The symbol max denotes the larger of the quantities enclosed in braces { }.) Notice that this strong triangle law is directly analogous to the strong triangle law described in §6.4 for the absolute value of rational functions.

Proof of the strong triangle law. The inequality (3) is valid when $a = 0$ since both sides then equal $|b|_p$. Similarly, it holds when $b = 0$. It also holds when $a \neq 0$, $b \neq 0$, but $a + b = 0$, since the left-hand side then equals zero, and the right-hand side the positive number $\max\{|a|_p, |b|_p\}$. So assume $a \neq 0$, $b \neq 0$, $a + b \neq 0$. Take a, b as in (1). Then $|a|_p$, $|b|_p$ are given in (2). One of the quantities $|a|_p$, $|b|_p$ is the larger, say $|a|_p \geq |b|_p$. Then $(1/p)^e \geq (1/p)^f$, or $\rho^e \geq \rho^f$, implying that $f \geq e$. Adding equations (1), we get

$$a + b = p^e \frac{MN' + p^{f-e}M'N}{NN'}. \tag{4}$$

Because $f \geq e$, $MN' + p^{f-e}M'N$ is an element of $\mathcal{R}$. The denominator NN' in (4) is relatively prime to p, but the numerator $MN' + p^{f-e}M'N$ (which is nonzero because $a + b \neq 0$) may not be. If p^k is the highest power of p dividing this numerator, then

$$a + b = p^{e+k}M''/NN',$$

where $k \geq 0$, and M'' is relatively prime to p. But then

$$|a + b|_p = \begin{cases}(1/p)^{e+k} \\ \\ \rho^{e+k}\end{cases} \leq \begin{cases}(1/p)^e \\ \\ \rho^e\end{cases} = |a|_p = \max\{|a|_p, |b|_p\}.$$

The strong triangle inequality is now established. □

Verification of V4. Using the strong triangle inequality, we get

$$|a + b|_p \leq \max\{|a|_p, |b|_p\} = |a|_p \quad \text{or} \quad |b|_p \leq |a|_p + |b|_p. \qquad \square$$

This completes the proof of Theorem 8.2.3.

Parallel to terminology used in §6.4, the inequality

$$|a + b|_p \leq |a|_p + |b|_p$$

is sometimes called the *weak triangle inequality*; usually, however, it is simply called the *triangle inequality*.

The p-adic valuation just introduced plays a major role in advanced algebra and number theory. It is a relatively new algebraic concept, being introduced since 1900 and with many of its applications coming more recently, e.g., since 1930.

Example 8.2.4

(i) Find the 2-adic and 3-adic valuations of $a = 81/56$.

SOLUTION Since $a = 2^{-3}(81/7)$, we get $|a|_2 = (1/2)^{-3} = 2^3$. We also have $a = 3^4(1/56)$, so that $|a|_3 = (1/3)^4 = 3^{-4}$. □

(ii) Find the λ-adic valuation of the polynomial $\lambda^3 + \lambda^2 + \lambda$. (Note: λ is a prime polynomial.)

SOLUTION Since $\lambda^3 + \lambda^2 + \lambda = \lambda(\lambda^2 + \lambda + 1)$, we get $|\lambda^3 + \lambda^2 + \lambda|_\lambda = \rho$. □

(iii) Which of $a_1 = -127/2$, $a_2 = 31/64$ is, in your opinion, closer to $a = 1/2$, when distances are measured 2-adically?

SOLUTION We say that a_1 is closer than a_2 to a, in the 2-adic sense, if $|a_1 - a|_2 < |a_2 - a|_2$. This is correct since $|a_1 - a|_2 = |-64|_2 = 2^{-6}$, whereas $|a_2 - a|_2 = |-1/64|_2 = 2^6$. □

Figure 8.2.1

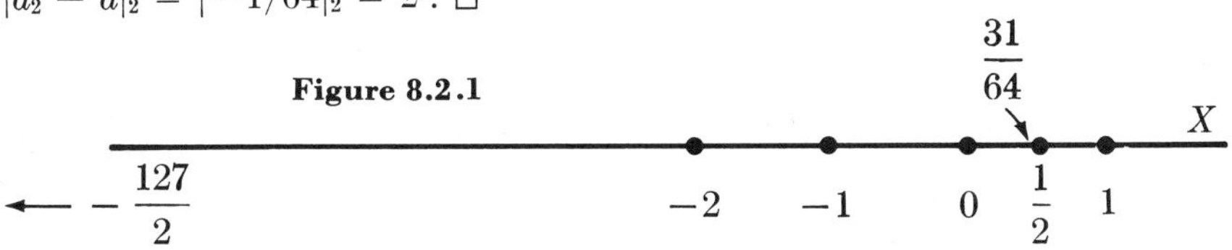

Part (iii) above and Figure 8.2.1 show that the 2-adic way of measuring distances is quite distinct from the normal way using the ordinary absolute value, since, when plotted as points on an X axis, a_1 is far off the left-hand edge of the page, yet is closer 2-adically to a than is a_2, which in the figure seems almost coincident with a.

Many authors employing valuations use the symbol $|a|$ for an arbitrary valuation, i.e., instead of $||a||$, they write $|a|$. When this notation is used, another notation is needed for the usual absolute value, and the symbol often adopted is $|a|_\infty$. One then has a valuation for each monic or positive prime p in $\mathfrak{R}$, and a valuation for ∞, and it is sometimes convenient to think of ∞ also as a prime. In more advanced discussions of valuations, this is the notation customarily followed. We shall not follow this notation. For us, $||a||$ will denote an arbitrary valuation, and $|a|$ will be the usual absolute value described in Chapter 2 and §6.4.

Valuations may be defined for any field, not just for $\mathcal{K}$. If $\mathcal{L}$ is an arbitrary field, the definition of a valuation $||\ ||$ on $\mathcal{L}$ is just Definition 8.2.1, with $\mathcal{L}$ substituted throughout for $\mathcal{K}$. For example, if $\mathcal{L}$ is any field, the function $||\ ||$ from $\mathcal{L}$ to $\mathbf{R}$ defined by

$$||0|| = 0, \qquad ||a|| = 1 \quad \text{for any nonzero element } a \text{ of } \mathcal{L},$$

is a valuation, as is easily checked. This valuation is called the *trivial valuation,* and it always satisfies the strong triangle law $||a + b|| \leq \max\{||a||, ||b||\}$. If $\mathcal{L}$ is a finite field, it is not difficult to prove that the trivial valuation is the only valuation on $\mathcal{L}$. (Exercise 21.)

The following simple properties hold for any valuation $|| \ ||$ and so, in particular, hold for the p-adic valuation:

(i) $||1|| = 1$;
(ii) $||-1|| = 1$;
(iii) $||-\beta|| = ||\beta||$;
(iv) $||\alpha/\beta|| = ||\alpha||/||\beta||$ if $\beta \neq 0$.

PROOF (i) From $1 = 1^2$, we get, by V3, $||1|| = ||1||^2$, and hence $||1|| = 1$ (since $||1|| \neq 0$, by V2).

(ii) From $1 = (-1)^2$, we get $1 = ||1|| = ||-1||^2$, and hence $||-1|| = 1$ since $||-1|| \geq 0$, by V1.

(iii) $||-\beta|| = ||(-1)\cdot\beta|| = ||-1|| \, ||\beta|| = ||\beta||$.

(iv) From $\beta\cdot\alpha/\beta = \alpha$, we get $||\beta|| \, ||\alpha/\beta|| = ||\alpha||$, giving $||\alpha/\beta|| = ||\alpha||/||\beta||$. □

The next property, called the *principle of domination*, has already been encountered for the absolute value of rational functions in §6.4. It holds for any valuation satisfying the strong triangle law. We shall use it, for the p-adic valuations, in §§8.3, 8.4. Let a, b, a_1, a_2, ..., a_n be elements of the field under consideration.

THEOREM 8.2.5

If the valuation $|| \ ||$ *on a field satisfies the strong triangle law*

$$||a + b|| \leq \max \{||a||, ||b||\},$$

then it satisfies this principle of domination:

$$||a_1 + a_2 + \cdots + a_n|| = ||a_1|| \quad \text{if } ||a_1|| > ||a_j|| \text{ for } j = 2, 3, \ldots, n. \tag{5}$$

In particular, if the field is $\mathcal{K}$,

$$|a_1 + a_2 + \cdots + a_n|_p = |a_1|_p \quad \text{if } |a_1|_p > |a_j|_p \text{ for } j = 2, 3, \ldots, n.$$

The principle of domination asserts that if a term a_1 in the sum $a_1 + a_2 + \cdots + a_n$ is strictly larger than the other terms a_2, a_3, ..., a_n, then the valuation of the sum is the valuation of the biggest term.

PROOF We first show that $||a + b|| = ||a||$ if $||a|| > ||b||$. From the strong triangle inequality, we get

$$||a + b|| \leq \max \{||a||, ||b||\} = ||a||.$$

However, applying the strong triangle inequality again, we obtain

$$||a|| = ||(a + b) - b|| \leq \max \{||a + b||, ||-b||\} = \max \{||a + b||, ||b||\}.$$

If the maximum indicated here were $||b||$, it would follow that $||a|| \leq ||b||$, contrary to hypothesis. Therefore the maximum is $||a + b||$, implying that $||a|| \leq ||a + b||$. Since we just proved that $||a + b|| \leq ||a||$, we obtain $||a + b|| = ||a||$, as desired.

We now establish (5) by induction on n. For $n = 2$, (5) follows from the preceding paragraph by setting $a = a_1$ and $b = a_2$. Assuming the validity of the principle of domination when there are $n - 1$ terms, we get

$$||a_1 + a_2 + \cdots + a_{n-1}|| = ||a_1||.$$

Putting $a = a_1 + a_2 + \cdots + a_{n-1}$ and $b = a_n$, we therefore get $||a|| = ||a_1|| > ||a_n|| = ||b||$, so that, by the result of the preceding paragraph, $||a + b|| = ||a||$, i.e.,

$$||a_1 + a_2 + \cdots + a_n|| = ||a_1||.$$

Thus the principle of domination holds for n terms, completing the induction step. □

The λ-adic valuation of the polynomial $\lambda^3 + \lambda^2 + \lambda$, already calculated in Example 8.2.4(ii), may also be calculated this way, using the principle of domination: We have $|\lambda|_\lambda = \rho$, $|\lambda^2|_\lambda = \rho^2$, $|\lambda^3|_\lambda = \rho^3$. Therefore λ is strictly larger (in the λ-adic sense) than λ^2 and λ^3, so that $|\lambda^3 + \lambda^2 + \lambda|_\lambda = |\lambda|_\lambda = \rho$.

Employing the p-adic valuation in an algebraic investigation means that a single prime, p, has been selected for a preferred role, with all other primes being cast into a secondary role. Often one wishes to give the preferred role to several primes simultaneously, perhaps even to infinitely many primes. This may be achieved by employing the p-adic valuations for all these primes. We give illustrations of this remark in the next two paragraphs, where we show how algebraic concepts may sometimes be reexpressed in terms of inequalities involving p-adic valuations for a family of primes p. Further illustrations will be given in the discussion of valuation rings at the end of this section, in the exercises below, and in §8.4

Observe that a nonzero element a of $\mathcal{K}$ satisfies $|a|_p = 1$ for all primes p, apart from a finite number of exceptions (Exercise 7); i.e., $|a|_p = 1$ for almost all p. Furthermore, if $a \in \mathcal{R}$, then $|a|_p \leq 1$ for every prime p. Conversely, if $a \in \mathcal{K}$ and $|a|_p \leq 1$ for every prime p, then a must be an element of $\mathcal{R}$ (Exercise 13). We may use these properties to describe ideals in $\mathcal{R}$ in terms of inequalities. If $\mathcal{I}$ is a nonzero ideal in $\mathcal{R}$, generated by b, then an element β of $\mathcal{R}$ is in $\mathcal{I}$ if and only if

$$|\beta|_p \leq |b|_p \qquad \text{for every prime } p,$$

since β/b is an element of $\mathcal{R}$ if and only if $|\beta/b|_p \leq 1$ for every p. Since $|\beta|_p \leq 1$ for every p, and $|b|_p = 1$ for almost all p, the inequality $|\beta|_p \leq |b|_p$ is automatically satisfied for almost all p, and therefore we describe the elements of $\mathcal{I}$ by requiring $|\beta|_p \leq |b|_p$ only for the finitely many p for which $|b|_p < 1$. For example, the ideal in $\mathbf{Z}$ generated by 600 consists of all integers β for which

$$|\beta|_2 \leq 1/8, \quad |\beta|_3 \leq 1/3, \quad |\beta|_5 \leq 1/25,$$

and the ideal in $\mathbf{R}[\lambda]$ generated by $\lambda^3 - 1$ consists of all polynomials $\beta(\lambda)$ for which

$$|\beta(\lambda)|_{\lambda-1} \leq \rho, \qquad |\beta(\lambda)|_{\lambda^2+\lambda+1} \leq \rho.$$

Conversely, if S is any set of primes from $\mathcal{R}$, and if real numbers r_p are given, one for each p in S, with $0 \leq r_p \leq 1$ for each p, then the set of all elements β of $\mathcal{R}$ satisfying

$$|\beta|_p \leq r_p, \qquad \text{for every } p \text{ in } S,$$

is always an ideal in $\mathcal{R}$. *Proof:* This set is nonempty (it contains 0), and it is easily shown to be closed under addition, subtraction, and strong multiplication, by using the strong triangle inequality and V3. □

In summary, then, for the rational numbers we have a p-adic valuation $| \ |_p$ for each positive prime integer p, together with the usual absolute value $| \ |$. The p-adic valuations all satisfy the strong triangle law, but the usual absolute does not. For rational functions, we have a p-adic valuation $| \ |_p$ for each monic prime polynomial $p(\lambda)$ in $\mathcal{F}[\lambda]$, together with the usual absolute value $| \ |$, all of which satisfy the strong triangle law.

The remainder of this section gives p-adic generalizations of some of the concepts encountered in earlier chapters, but is not needed for reading subsequent sections of this chapter.

Archimedean property for valuations

This was described for the absolute value $| \ |$ in §6.4. Generalizing that discussion, we say that a valuation $\| \ \|$ on an arbitrary field is *archimedean* if, whenever q is an element of the field, there is a natural number k (dependent on q) such that

$$\|\underbrace{1 + 1 + \cdots + 1}_{k \text{ terms}}\| > \|q\|.$$

A valuation for which this is not the case is said to be *nonarchimedean*.

THEOREM 8.2.6
A valuation $\| \ \|$ satisfying the strong triangle law is always nonarchimedean. In particular, the p-adic valuation $| \ |_p$ on $\mathcal{K}$ is nonarchimedean, as is the usual absolute value on $\mathcal{F}(\lambda)$.

PROOF Take $q = 1$. Then $||q|| = ||1|| = 1$. However, by the strong triangle law,

$$||\underbrace{1 + 1 + \cdots + 1}_{k \text{ terms}}|| \leq \max \{||1||, ||1||, \ldots, ||1||\} = 1.$$

Therefore $||1 + 1 + \cdots + 1|| > ||1||$ is false for any number k of terms. □

Exercise 20 gives the converse of Theorem 8.2.6, namely, that a nonarchimedean valuation must satisfy the strong triangle law.

Discreteness

Since the p-adic valuation of a nonzero element of $\mathcal{K}$ has the form $(1/p)^e$ or ρ^e, for e an integer, there is no element a in $\mathcal{K}$ for which

$$1/p < |a|_p < 1, \qquad \text{or} \qquad \rho < |a|_p < 1.$$

We express this by saying that the p-adic valuation is *discrete*. The usual absolute value, for **Q**, **R**, or **C**, is not discrete in the sense that, if ϵ is any real number with $0 < \epsilon < 1$, there is always an element a for which $\epsilon < |a| < 1$. However, the usual absolute value for $\mathfrak{F}(\lambda)$ is discrete. (See Exercise 8.)

Valuation subrings of $\mathcal{K}$

Let p be a fixed positive or monic prime in $\mathcal{R}$, and denote by $\mathcal{K}^p$ the subset of $\mathcal{K}$ consisting of all

$$a \in \mathcal{K} \quad \text{with } |a|_p \leq 1.$$

We claim that $\mathcal{K}^p$ is a subring of $\mathcal{K}$. *Proof:* Let $a_1, a_2 \in \mathcal{K}^p$. Then $|a_1|_p \leq 1$, $|a_2|_p \leq 1$; by the strong triangle law and V3, we get

$$|a_1 + a_2|_p \leq \max \{|a_1|_p, |a_2|_p\} \leq 1,$$

$$|a_1 - a_2|_p \leq \max \{|a_1|_p, |-a_2|_p\} \leq 1,$$

$$|a_1a_2|_p = |a_1|_p |a_2|_p \leq 1.$$

These inequalities show that $a_1 + a_2$, $a_1 - a_2$, a_1a_2 all belong to $\mathcal{K}^p$, implying by Theorem 1.6.9 that $\mathcal{K}^p$ is a subring of $\mathcal{K}$. We call $\mathcal{K}^p$ the *valuation subring of $\mathcal{K}$ belonging to the prime p*, or the *valuation subring of $\mathcal{K}$ belonging to the valuation* $| \ |_p$.

When the element a of $\mathcal{K}$ is written in lowest terms as

$$a = p^e \frac{M}{N},$$

where M, N are elements of $\mathcal{R}$, relatively prime to each other and to p, then $|a|_p = (1/p)^e$ or ρ^e, so that $|a|_p \leq 1$ if and only if $e \geq 0$. Thus $a \in \mathcal{K}^p$ if and only if the factor p^e belongs to the numerator of a. Therefore, $\mathcal{K}^p$

may be described as *all fractions in $\mathcal{K}$ which, when written in lowest terms, have denominator relatively prime to p.* In particular, when $\mathcal{K}$ is $\mathfrak{F}(\lambda)$ and p is the prime $\lambda - \alpha$, where α is a constant, the valuation ring $\mathcal{K}^p = \mathfrak{F}(\lambda)^{\lambda-\alpha}$ is precisely the ring $\mathfrak{F}(\lambda)^{\lambda-\alpha}$ described in Example 7.5.5.

This ring $\mathcal{K}^p$ has interesting properties. We describe (i) the unique factorization theorem for $\mathcal{K}^p$, and (ii) the ideals in $\mathcal{K}^p$.

(i) A nonzero element u in $\mathcal{K}^p$ is said to be a *unit* for $\mathcal{K}^p$ if u^{-1} is also in $\mathcal{K}^p$. We claim that u is a unit for $\mathcal{K}^p$ if and only if $|u|_p = 1$. *Proof:* Let both $u, u^{-1} \in \mathcal{K}^p$, so that $|u|_p \leq 1$, $|u^{-1}|_p \leq 1$. If we had $|u|_p < 1$, then $1 = |1|_p = |uu^{-1}|_p = |u|_p\,|u^{-1}|_p < 1$. This contradiction proves that $|u|_p = 1$ if u is a unit for $\mathcal{K}^p$. Conversely, if $|u|_p = 1$, then $|u^{-1}|_p = 1/|u|_p = 1$, implying that both u, u^{-1} belong to $\mathcal{K}^p$, and thus that u is a unit for $\mathcal{K}^p$.

A unit for $\mathcal{K}^p$ may alternatively be described as an element of $\mathcal{K}$ which, when expressed in lowest terms, has both numerator and denominator relatively prime to p.

A nonzero element of $\mathcal{K}^p$ is a *prime* for $\mathcal{K}^p$ if it is not a unit and can be factored as a product of two elements of $\mathcal{K}^p$ only when one of the factors is a unit. We claim that p is a prime for $\mathcal{K}^p$. *Proof:* Let $p = ab$, where $a, b \in \mathcal{K}^p$, so that $|a|_p \leq 1$, $|b|_p \leq 1$. If neither a nor b is a unit for $\mathcal{K}^p$, then $|a|_p < 1$, $|b|_p < 1$, so that

$$|a|_p \leq 1/p, \quad |b|_p \leq 1/p, \qquad \text{or} \qquad |a|_p \leq \rho, \quad |b|_p \leq \rho.$$

But then

$$\left.\begin{matrix} 1/p \\ \\ \rho \end{matrix}\right\} = |p|_p = |ab|_p = |a|_p\,|b|_p \leq \left\{\begin{matrix} 1/p \cdot 1/p, \\ \\ \rho \cdot \rho. \end{matrix}\right.$$

This is impossible because $1/p > (1/p)^2$ and $\rho > \rho^2$. Therefore, p is a prime for $\mathcal{K}^p$.

We may now state the *unique factorization theorem* for $\mathcal{K}^p$:

THEOREM 8.2.7

If a is a nonzero element of $\mathcal{K}^p$, it is uniquely factorable as

$$a = up^e,$$

where u is a unit for $\mathcal{K}^p$, and e is a nonnegative integer. (As proved above, p is a prime for $\mathcal{K}^p$.)

PROOF Let the p-adic valuation of a be $|a|_p = (1/p)^e$ or ρ^e. Then $|p^e|_p = |a|_p$. Set $u = a/p^e$. Then $|u|_p = |a|_p/|p^e|_p = 1$, so that u is a unit for $\mathcal{K}^p$. Since $a = up^e$, this completes the existence part of the proof. To prove uniqueness, suppose also $a = vp^f$, where v is a unit for $\mathcal{K}^p$. Thus $|v|_p = 1$, so that $|a|_p = (1/p)^f$ or ρ^f. Since we already have $|a|_p = (1/p)^e$ or ρ^e, we get $f = e$. Then $a = vp^e = up^e$, so $v = u$. □

The theorem of unique factorization into primes is therefore valid for $\mathcal{K}^p$, but in the simplest possible way, namely: there essentially is only one prime for $\mathcal{K}^p$. In fact, every prime for $\mathcal{K}^p$ is given by the formula up, where u is a unit for $\mathcal{K}^p$. See Exercise 10. (We call up an *associate* of p.) Exercise 16 gives a class of rings defined by valuations, this class including both $\mathcal{R}$ and $\mathcal{K}^p$ as special cases, for which the unique factorization theorem holds. In some of these rings the number of (nonassociated) primes is more than one, but still finite.

(ii) If $\mathcal{I}$ is a nonzero ideal in $\mathcal{K}^p$, select an element a in $\mathcal{I}$ for which $|a|_p$ is largest. There is such an element a since $|a|_p \leq 1$ (because $a \in \mathcal{K}^p$) and the p-adic valuation is discrete. Let the valuation of this maximal element a of $\mathcal{I}$ be $|a|_p = (1/p)^k$ or ρ^k. Then $k \geq 0$, and a simple proof (which we leave as Exercise 12) shows that $\mathcal{I}$ is the principal ideal generated by p^k. If we let $\mathcal{I}_k$ denote this principal ideal, we see that $\mathcal{I}_0, \mathcal{I}_1, \mathcal{I}_2, \ldots$ constitute all nonzero ideals in $\mathcal{K}^p$. In fact,

$$\mathcal{K}^p = \mathcal{I}_0 \supset \mathcal{I}_1 \supset \mathcal{I}_2 \supset \cdots,$$

the inclusions being strict. Referring to the definition of a *local ring* in Exercise 12 of §7.5, we therefore see that $\mathcal{K}^p$ is a local ring, and that the principal ideal $\mathcal{I}_1$ generated by p is its maximal ideal.

EXERCISES 8.2

I

1. Calculate $|a|_p$ for $p = 2, 3, 5$, when $a = -72/25$ and when $a = 40/9$.

2. Calculate $|a|_p$ for $p = \lambda, \lambda + 1, \lambda - 1$, when $a = \lambda^2/(\lambda^2 - 1)$.

3. Let $\mathcal{K} = \mathbf{C}(\lambda)$. If $f(\lambda)$ is a nonzero polynomial in $\mathbf{C}[\lambda]$, with a root α of multiplicity e, calculate $|f(\lambda)|_p$ when $p(\lambda) = \lambda - \alpha$.

4. Let q be a prime in $\mathbf{Z}$. Calculate $|q|_p$ and $|1/q|_p$, for each positive prime p in $\mathbf{Z}$.

5. Let a be a positive integer. If the three lowest digits of a in base 7 are $d_0 = 0$, $d_1 = 0$, $d_2 = 4$, compute $|a|_7$. If, instead, the four lowest digits of a in base ten are $d_0 = d_1 = d_2 = 0$, $d_3 = 7$, compute $|a|_2$ and $|a|_5$.

6. Describe the elements of the principal ideal in $\mathbf{Z}$ generated by 1225 in terms of inequalities involving p-adic valuations. Do the same for the ideal in $\mathbf{Q}[\lambda]$ generated by $\lambda^4 - 1$. Find a generator for the ideal in $\mathbf{Z}$ consisting of all integers β with $|\beta|_3 \leq 1/4$, $|\beta|_5 \leq 1/10$, $|\beta|_7 \leq 1/30$.

II

7. Let a be a nonzero element of $\mathcal{K}$, and suppose its factorization into primes in $\mathcal{R}$ is $a = up_1^{e_1} \cdots p_r^{e_r}$, where u is a unit in $\mathcal{R}$, $p_1, \ldots, p_r$ are distinct positive or monic primes in $\mathcal{R}$, and $e_1, \ldots, e_r$ are integers.
(i) If the prime p is not any of $p_1, \ldots, p_r$, prove that $|a|_p = 1$.
(ii) Prove that the product of all the numbers $\|a\|$, as $\|\ \|$ ranges over the absolute value valuation and all the p-adic valuations, is 1. In this product, there are only finitely many factors different from 1. Why?
This result is called the *product formula for valuations.*

8. Prove that no rational function $q(\lambda) \in \mathcal{F}(\lambda)$ exists such that $\rho < |q(\lambda)| < 1$.

9. (i) Show that the field of fractions of $\mathcal{K}^p$ is $\mathcal{K}$.
(ii) Give two other proofs for the existence part of Theorem 8.2.7, in this way: For the first, write a as a product of primes in $\mathcal{R}$ and a unit in $\mathcal{R}$, as in (1) of §6.4, and take p to be p_1. For the second, write $a = p^e M/N$, as in the text above, where M, N are relatively prime to p. Show that M/N is a unit in $\mathcal{K}^p$.
(iii) Show that the unique factorization theorem for $\mathcal{K}^p$ extends to $\mathcal{K}$, parallel to the manner in which the unique factorization theorem 3.5.3 for $\mathcal{R}$ extends to $\mathcal{K}$ as 6.4.2, by permitting the exponent e in Theorem 8.2.7 to be negative as well as nonnegative.

10. Prove: Every prime for $\mathcal{K}^p$ is an associate of p, i.e., has the form up with u a unit for $\mathcal{K}^p$. Conversely, if u is a unit for $\mathcal{K}^p$, prove that up is always a prime for $\mathcal{K}^p$.

11. For each positive prime integer p, let r_p be a real number satisfying $0 \leq r_p \leq 1$. If $r_p < 1$ for infinitely many p, find the ideal in $\mathbf{Z}$ consisting of all integers β with

$$|\beta|_p \leq r_p \qquad \text{for all } p.$$

12. If $\mathcal{I}$ is an ideal in $\mathcal{K}^p$, and a is an element of $\mathcal{I}$ for which $|a|_p$ is maximal, prove that $\mathcal{I}$ is the principal ideal generated by p^k, where k is the integer given by $|a|_p = (1/p)^k$ or ρ^k. Show also that an element β of $\mathcal{K}$ is in $\mathcal{I}$ if and only if $|\beta|_p \leq |p^k|_p$.

13. (i) Show that an element a of $\mathcal{K}$ is in $\mathcal{R}$ if and only if $|a|_p \leq 1$ for every prime p in $\mathcal{R}$.
(ii) If $\mathcal{I}$ is an ideal in $\mathcal{R}$, generated by a, show that an element β of $\mathcal{K}$ is in $\mathcal{I}$ if and only if $|\beta|_p \leq |a|_p$ for every prime p in $\mathcal{R}$.
(iii) If $\beta \in \mathcal{K}$, prove that no finite set of inequalities of the form $|\beta|_p \leq |a|_p$ will guarantee that β is in $\mathcal{R}$, let alone in $\mathcal{I}$.

14. A sequence $a_1, a_2, a_3, \ldots$ *converges* to a in a valuation $\|\ \|$ if, whenever a tolerance ϵ is given (a positive real number), there is a

natural number K (dependent on ϵ) such that $||a - a_k|| < \epsilon$ for all $k > K$.
(i) Prove that the sequence $-1 + 2^k$, for $k = 0, 1, 2, \ldots$, converges to -1 in the 2-adic valuation, but not in the 3-adic valuation, and not in the usual absolute value.
(ii) Prove that p^k, for $k = 0, 1, 2, \ldots$, converges to 0 in the p-adic valuation, but not in the q-adic, for any prime $q \neq p$, and not in the usual absolute value.

15. Let $p_1, p_2, \ldots, p_r$ be a finite set of distinct positive or monic primes in $\mathcal{R}$. Use the Chinese remainder theorem (§3.6) to prove this simultaneous approximation property: If $a_1, a_2, \ldots, a_r \in \mathcal{R}$, there is an element x in $\mathcal{R}$ such that

$$|x - a_1|_{p_1} < 1, \quad |x - a_2|_{p_2} < 1, \ldots, \quad |x - a_r|_{p_r} < 1.$$

Generalize, letting $a_1, \ldots, a_r \in \mathcal{K}$, and ϵ be an arbitrary positive real number, and show that an element x in $\mathcal{K}$ exists such that

$$|x - a_1|_{p_1} < \epsilon, \quad |x - a_2|_{p_2} < \epsilon, \ldots, \quad |x - a_r|_{p_r} < \epsilon.$$

(This is called the *weak approximation theorem.*)

III Exercises 16, 17, 18 form a connected set and describe a class of rings containing $\mathcal{R}$ and $\mathcal{K}^p$ as special cases.

16. Let S be a (finite or infinite) set of primes in $\mathcal{R}$, and take $\mathcal{K}^S$ to consist of all elements a of $\mathcal{K}$ for which

$$|a|_p \leq 1 \qquad \text{for every prime } p \text{ in } S.$$

(i) Prove that $\mathcal{K}^S$ is a subring of $\mathcal{K}$. When S is all primes in $\mathcal{R}$, show that $\mathcal{K}^S$ is $\mathcal{R}$, and when S is a single prime p, that $\mathcal{K}^S$ is $\mathcal{K}^p$.
(ii) Prove: An element of $\mathcal{K}$ is in $\mathcal{K}^S$ if and only if its lowest terms representation has denominator relatively prime to every p in S.
(iii) Show that a nonzero element u of $\mathcal{K}^S$ is a *unit for* $\mathcal{K}^S$ (i.e., $u^{-1} \in \mathcal{K}^S$) if and only if $|u|_p = 1$ for every $p \in S$.
(iv) A nonzero element of $\mathcal{K}^S$ is a *prime for* $\mathcal{K}^S$ if it is not a unit for $\mathcal{K}^S$ and can be factored within $\mathcal{K}^S$ only when one of the factors is a unit for $\mathcal{K}^S$. Prove: Each $p \in S$ is a prime for $\mathcal{K}^S$.
(v) Prove this *unique factorization theorem for* $\mathcal{K}^S$: A nonzero element of $\mathcal{K}^S$ is uniquely factorable as a product of a unit for $\mathcal{K}^S$ and powers of finitely p in S.
(vi) Prove that any prime for $\mathcal{K}^S$ has the form up, with u a unit for $\mathcal{K}^S$ and $p \in S$, and that any element of this form is a prime for $\mathcal{K}^S$.
(vii) Primes p_1, p_2 for $\mathcal{K}^S$ are said to be *associated* if $p_1 = up_2$ with u a unit for $\mathcal{K}^S$. Prove that association is an equivalence relation on the set of primes for $\mathcal{K}^S$, and that the elements of S are a system of distinct representatives for the equivalence classes.

17. Let $\mathcal{I}$ be an ideal in the ring $\mathcal{K}^S$ of Exercise 16.
(i) Prove that $\mathcal{I}$ is principal. (Hint: For each $p \in S$, choose an element $a_p \in \mathcal{I}$ for which $|a_p|_p$ is maximal. Adjust a_p by units for $\mathcal{K}^S$ to get $a_p \in \mathcal{R}$. If a is the greatest common divisor in $\mathcal{R}$ of these a_p, prove that a generates $\mathcal{I}$.)
(ii) [Generalization of Exercises 12 and 13] Prove that an ideal in $\mathcal{K}^S$ may always be described in terms of inequalities in this way: There are nonnegative real numbers $r_p \leq 1$, for each p in S, such that $\mathcal{I}$ consists of all $\beta \in \mathcal{K}$ with

$$|\beta|_p \leq r_p, \qquad \text{for each } p \text{ in } S.$$

(Hint: Take $r_p = |a|_p$ if a is the generator of $\mathcal{I}$.)
(iii) [Converse to (ii)] If arbitrary nonnegative real numbers $r_p \leq 1$ are given, one for each p in S, show that the set of all $\beta \in \mathcal{K}$ with $|\beta|_p \leq r_p$, for each p in S, forms an ideal in $\mathcal{K}^S$. If S has infinitely many elements, prove that this ideal $\mathcal{I}$ is nonzero if and only if no r_p is zero and almost all r_p equal 1.

18. If $\alpha, \beta \in \mathcal{K}^S$, we say that β divides α in $\mathcal{K}^S$ if $\alpha = \gamma\beta$ for some γ in $\mathcal{K}^S$. Let $\alpha_1, \alpha_2, \ldots, \alpha_k$ be elements of $\mathcal{K}^S$. Show that the set of all $x_1\alpha_1 + x_2\alpha_2 + \cdots + x_k\alpha_k$, for $x_1, x_2, \ldots, x_k$ in $\mathcal{K}^S$, form an ideal in $\mathcal{K}^S$. This ideal being principal, let it be generated by d. Show that d has the properties expected of a greatest common divisor of α_1, $\alpha_2, \ldots, \alpha_k$, namely, it divides all of $\alpha_1, \alpha_2, \ldots, \alpha_k$ and is divisible by any common divisor of $\alpha_1, \alpha_2, \ldots, \alpha_k$.

19. Let $\mathcal{S}$ consist of all elements a of $\mathcal{F}(\lambda)$ for which $|a| \leq 1$. Prove that $\mathcal{S}$ is a subring of $\mathcal{F}(\lambda)$, and give a unique factorization theorem for it, analogous to that for $\mathcal{K}^p$. Give a verbal description of the elements of $\mathcal{S}$ and your prime for $\mathcal{S}$.

20. Let $\|\ \|$ be a nonarchimedean valuation on an arbitrary field $\mathcal{L}$. Prove that it satisfies the strong triangle law, as follows.
(i) Show that a fixed real number c exists with $\|k\| \leq c$, for every natural number k, where k denotes $1 + 1 + \cdots + 1$ (k terms).
(ii) Let $a, b \in \mathcal{L}$, with $\|a\| \leq \|b\|$. Assume $b \neq 0$. Let n be a positive integer. Use the binomial theorem, the weak triangle law, and part (i), to show that

$$\|a + b\|^n = \|(a + b)^n\| = \|b\|^n\,\|(1 + a/b)^n\| \leq \|b\|^n\Big(\sum_{j=0}^{n-1} c\,\|a/b\|^j\Big)$$
$$\leq \|b\|^n \cdot cn.$$

(iii) Deduce that $\|a + b\| \leq \|b\|\,(cn)^{1/n}$.
(iv) Let $n \to \infty$, and deduce that $\|a + b\| \leq \|b\| = \max\{\|a\|, \|b\|\}$.

21. The only valuation on a finite field is the trivial valuation. Prove this. Use the polynomial $\lambda^n - \lambda$ in Theorem 7.4.7.

8.3 PARTIAL FRACTIONS

This section is independent of other sections of this chapter, except that the uniqueness proof for Theorem 8.3.1 will use the p-adic valuations described in §8.2.

Integrals of the form

$$\int \frac{f(x)}{g(x)}\,dx,$$

involving a quotient of polynomials f, g with real coefficients, are often encountered in calculus. The partial fraction decomposition for rational functions, which we are about to describe, provides a method for evaluating this type of integral. Since this decomposition can be obtained both for rational numbers and rational functions, throughout this section we continue our frequently used convention that $\mathcal{R}$ denotes the ring **Z** and $\mathcal{K}$ the field **Q**, or $\mathcal{R}$ denotes the ring $\mathfrak{F}[\lambda]$ and $\mathcal{K}$ the field $\mathfrak{F}(\lambda)$. The reader, if he wishes, may take the field $\mathfrak{F}$ to be **Q**, **R**, or **C**, although our discussion will be valid for any $\mathfrak{F}$.

Let $q = m/n$ be an element of $\mathcal{K}$, in lowest terms, where $m, n \in \mathcal{R}$. We are about to write the fraction m/n in terms of other fractions. Suppose

$$n = p_1^{e1}p_2^{e2}\cdots p_T^{eT}$$

is the factorization of n into powers of distinct positive or monic primes $p_1, p_2, \ldots, p_T$. Set

$$n_j = n/p_j^{ej}, \qquad \text{for } j = 1, 2, \ldots, T.$$

Then

$$n_j = p_1^{e1}\cdots p_{j-1}^{ej-1}p_{j+1}^{ej+1}\cdots p_T^{eT},$$

so that n_j is an element of $\mathcal{R}$ having no prime factors other that $p_1, \ldots, p_{j-1}, p_{j+1}, \ldots, p_T$. Clearly, $n_1, \ldots, n_T$ cannot have any common prime factor, implying that the greatest common divisor of $n_1, \ldots, n_T$ must be 1. Elements $f_1, \ldots, f_T$ of $\mathcal{R}$ therefore exist (by 3.4.7) such that

$$1 = f_1n_1 + f_2n_2 + \cdots + f_Tn_T.$$

Multiplying by m/n gives

$$m/n = mf_1/p_1^{e1} + \cdots + mf_T/p_T^{eT}. \tag{1}$$

Applying the division algorithm (§3.3) to mf_j, we get elements g_j, r_j in $\mathcal{R}$ with

$$mf_j = g_jp_j^{ej} + r_j, \qquad \begin{cases} 0 \le r_j < p_j^{ej}, \\ \\ \deg r_j < \deg p_j^{ej}, \end{cases} \qquad j = 1, 2, \ldots, T.$$

Substituting into (1) and setting $g = g_1 + \cdots + g_T$, we then obtain

$$m/n = g + r_1/p_1^{e1} + \cdots + r_T/p_T^{eT}. \tag{2}$$

This proves that *any fraction $m/n \in \mathcal{K}$ may be written as an element g of $\mathcal{R}$ plus proper fractions $r_j/p_j^{e_j}$ in which each denominator is a prime power.*

Focus attention, for a moment, on r_j for a fixed j. We may apply results from §4.2 and express r_j in any base that we wish. Expressing r_j in base p_j, we get digits $d_{j0}, d_{j1}, d_{j2}, \ldots \in \mathcal{R}$ such that

$$r_j = d_{j0} + d_{j1}p_j + d_{j2}p_j^2 + \cdots,$$
$$0 \leq d_{jt} < p_j, \qquad \text{or} \qquad \deg d_{jt} < \deg p_j, \quad \text{for all } t.$$

Because $r_j < p_j^{e_j}$, or $\deg r_j < \deg p_j^{e_j}$, no power of p_j as high as $p_j^{e_j}$ can appear with nonzero coefficient in this representation of r_j, so that we may write

$$r_j = d_{j0} + d_{j1}p_j + \cdots + d_{j,e_j-1}p_j^{e_j-1}.$$

Inserting this expression for r_j into (2), and renaming the digits by setting $D_{j1} = d_{j,e_j-1}$, $D_{j2} = d_{j,e_j-2}$, $\ldots D_{j,e_j} = d_{j0}$, we get:

$$m/n = g + \sum_{j=1}^{T} (D_{j1}/p_j + D_{j2}/p_j^2 + \cdots + D_{je_j}/p_j^{e_j}), \tag{3}$$

where

$$g \in \mathcal{R}, \quad D_{jt} \in \mathcal{R}, \quad \text{and} \quad \begin{cases} 0 \leq D_{jt} < p_j, & \text{when } \mathcal{R} \text{ is } \mathbf{Z}, \\ \deg D_{jt} < \deg p_j & \text{when } \mathcal{R} \text{ is } \mathcal{F}[\lambda], \end{cases} \tag{4}$$

for all j, t.

We call (3), subject to the conditions listed in (4), the *decomposition of m/n into partial fractions.*

We next prove the uniqueness of this partial fraction decomposition, by applying the p-adic valuations, the strong triangle inequality, and the principle of domination from §8.2. Readers who have not read §8.2 may omit this proof, if they wish; an alternative proof not using these tools is outlined in Exercise 10.

Suppose we had a second partial fraction representation of m/n. We may assume the primes p_j and positive integers e_j in the two representations are the same, since we may put additional terms into either or both of these representations using zero digits D_{jt}. So suppose

$$g + \sum_{j=1}^{T} (D_{j1}/p_j + \cdots + D_{je_j}/p_j^{e_j}) = g' + \sum_{j=1}^{T} (D'_{j1}/p_j + \cdots + D'_{je_j}/p_j^{e_j}), \tag{5}$$

where both (4) and the following conditions are satisfied:

$$g' \in \mathcal{R}, \quad D'_{jt} \in \mathcal{R}, \quad 0 \leq D'_{jt} < p_j \quad \text{or} \quad \deg D'_{jt} < \deg p_j,$$

for all j, t. We shall deduce that $D_{jt} = D'_{jt}$ for all j, t. It is sufficient to give this argument for $j = 1$, the argument for other values of j being similar.

If we do not have $D_{1t} = D'_{1t}$ for each t, let τ be the highest value of t for which this is false. Then $\tau \geq 1$. Using $D_{1t} = D'_{1t}$ for $t > \tau$, from (5) we get

$$(D_{11} - D'_{11})/p_1 + \cdots (D_{1\tau} - D'_{1\tau})/p_1^{\tau}$$
$$= g' - g + \sum_{j=2}^{T} ((D'_{j1} - D_{j1})/p_j + \cdots + (D'_{je_j} - D_{je_j})/p_j^{e_j}). \quad (6)$$

We compute the p_1-adic valuation of each side of (6). Since the p_1-adic valuation of an element of $\mathcal{R}$ is at most 1, we have

$$|g' - g|_{p_1} \leq 1 \qquad \text{and} \qquad |D'_{jt} - D_{jt}|_{p_1} \leq 1 \qquad \text{for all } j, t.$$

Because p_j, p_1 are distinct if $j \neq 1$, we also have

$$|p_j|_{p_1} = 1 \qquad \text{if } j \neq 1.$$

The p_1-adic valuation of each term on the right-hand side of (6) is therefore ≤ 1, implying, by the strong triangle inequality, that the p_1-adic valuation of the entire right-hand side of (6) is ≤ 1. We also have

$$|D_{1t} - D'_{1t}|_{p_1} \leq 1 \quad \text{for } t = 1, \ldots, \tau - 1, \qquad \text{and} \qquad |D_{1\tau} - D'_{1\tau}|_{p_1} = 1,$$

the last because the nonzero element $D_{1\tau} - D'_{1\tau}$ cannot be divisible by p_1 since it satisfies

$$-p_1 < D_{1\tau} - D'_{1\tau} < p_1, \qquad \text{or} \qquad \deg (D_{1\tau} - D'_{1\tau}) < \deg p_1 .$$

Thus

$$|(D_{1t} - D'_{1t})/p_1^t|_{p_1} \leq |1/p_1^t|_{p_1} < |1/p_1^{\tau}|_{p_1}, \quad \text{for } t = 1, 2, \ldots, \tau - 1,$$
$$|(D_{1\tau} - D'_{1\tau})/p_1^{\tau}|_{p_1} = |1/p_1^{\tau}|_{p_1} > 1. \quad (7)$$

The term $(D_{1\tau} - D'_{1\tau})/p_1^{\tau}$ is therefore, p_1-adically, strictly larger that all other terms on the left-hand side of (6), implying by the principle of domination that the p_1-adic valuation of the entire left-hand side of (6) equals the p_1-adic valuation of this term, and therefore by (7), is >1. But this means that the two sides of (6) have unequal p_1-adic size, a contradiction. Therefore τ cannot exist, establishing the uniqueness of the partial fraction decomposition. □

THEOREM 8.3.1
A rational number or a rational function is uniquely representable in terms of partial fractions, as in (3) *and* (4).

The proof above showed that the primes p_j and exponents e_j occurring in the partial fraction representation of m/n come from the factorization of the denominator n into a product of powers of primes.

The most common applications of partial fractions are in elementary integration theory and occur when $\mathcal{R}$ is $\mathbf{R}[\lambda]$. Here (see §5.4) a prime p_j is given by

$$p_j(\lambda) = \lambda - \alpha_j, \qquad \text{or by} \qquad p_j(\lambda) = (\lambda - \alpha_j)^2 + \beta_j^2,$$

with $\alpha_j, \beta_j \in \mathbf{R}$ and β_j nonzero. The description of the partial fraction decomposition in this case takes the following form:

(i) Belonging to a multiplicity e_j factor $\lambda - \alpha_j$ of $n(\lambda)$, i.e., belonging to a real root α_j of $n(\lambda)$ of multiplicity e_j, there is a set of partial fractions of the type

$$D_{j1}/(\lambda - \alpha_j),\ D_{j2}/(\lambda - \alpha_j)^2,\ \ldots,\ D_{je_j}/(\lambda - \alpha_j)^{e_j},$$

where [by (4)] each D_{jt} is a real constant; and,

(ii) Belonging to a multiplicity e_j prime factor $(\lambda - \alpha_j)^2 + \beta_j^2$ of $n(\lambda)$, i.e., belonging to a pair $\alpha_j \pm \beta_j i$ of conjugate nonreal roots of $n(\lambda)$ of multiplicity e_j, there is a set of partial fractions of the form

$$D_{j1}/((\lambda - \alpha_j)^2 + \beta_j^2),\ D_{j2}/((\lambda - \alpha_j)^2 + \beta_j^2)^2,\ \ldots,\ D_{je_j}/((\lambda - \alpha_j)^2 + \beta_j^2)^{e_j},$$

where (by (4)) each D_{jt} is a polynomial of the form $a + b\lambda$ with $a, b \in \mathcal{R}$.

The manner in which these partial fractions are used in integration is outlined in Exercise 6.

Application: Convergence of a Laurent series

Let $m(\lambda)$, $n(\lambda)$ be polynomials with complex coefficients, and suppose the rational function $q(\lambda) = m(\lambda)/n(\lambda)$ is in lowest terms. Let α be a fixed complex number. In §6.5 we encountered this *Laurent series* for $q(\lambda)$ about α:

$$q(\lambda) = d_l(\lambda - \alpha)^l + d_{l-1}(\lambda - \alpha)^{l-1} + \cdots + d_0 + d_{-1}(\lambda - \alpha)^{-1} + \cdots, \tag{8}$$

where $d_l, d_{l-1}, \ldots, d_0, d_{-1}, \ldots$ are complex constants. This series was the rational function counterpart of the base b representation of a rational number. In §6.5 we evaluated both sides of (8) at a complex number z, and then asked whether

$$q(z) = d_l(z - \alpha)^l + d_{l-1}(z - \alpha)^{l-1} + \cdots + d_0 + d_{-1}(z - \alpha)^{-1} + \cdots. \tag{9}$$

The meaning to be given to (9) is that, if ϵ is a tolerance, i.e., a positive real number, there is an integer K (dependent on ϵ) such that truncating the right-hand side at the term $d_{-k}(z - \alpha)^{-k}$, for any $k \geq K$, gives an approximation of the left-hand side within this tolerance. In symbols,

$$|q(z) - \{d_l(z - \alpha)^l + \cdots + d_{-k}(z - \alpha)^{-k}\}| < \epsilon$$

for all $k \geq K$, the bars $|\ |$ denoting the usual absolute value for complex numbers. If this condition is satisfied for every tolerance ϵ, the series of complex numbers on the right-hand side of (9) is said to *converge* to $q(z)$. We claimed, in §6.5, that this convergence holds if z lies outside the smallest circle C in the complex plane centered at α enclosing all roots of $n(\lambda)$. We proved this claim in §6.5 under the additional assumption that $n(\lambda)$ has simple roots only. We shall now apply the complex case of the partial fraction decomposition and prove this result, Theorem 6.5.6, in full generality.

Suppose $$n(\lambda) = (\lambda - z_1)^{e_1} \cdots (\lambda - z_T)^{e_T},$$

where $z_1, \ldots, z_T$ are distinct complex numbers and $e_1, \ldots, e_T$ are positive integers. The distinct prime factors of $n(\lambda)$ within $\mathbf{C}[\lambda]$ are then $\lambda - z_1, \ldots, \lambda - z_T$, so that the partial fraction decomposition of $m(\lambda)/n(\lambda)$ becomes

$$q(\lambda) = m(\lambda)/n(\lambda) = g(\lambda) + \sum_{j=1}^{T} \left(\sum_{t=1}^{e_j} D_{jt}/(\lambda - z_j)^t\right), \tag{10}$$

where the digits D_{jt} are complex constants. The proof depends on the following specific representation of a typical fraction $1/(\lambda - z_j)^t$ in this decomposition.

***LEMMA* 8.3.2**
If $k \geq t \geq 1$, polynomials $P_{jtk}(\lambda)$, $Q_{jtk}(\lambda)$ exist such that

$$1/(\lambda - z_j)^t = P_{jtk}(\lambda)/(\lambda - \alpha)^k + Q_{jtk}(\lambda)/\{(\lambda - z_j)^t(\lambda - \alpha)^k\}, \tag{11}$$

where $$Q_{jtk}(\lambda) = \sum_{s=0}^{t-1} \binom{k}{s} (z_j - \alpha)^{k-s}(\lambda - z_j)^s. \tag{12}$$

PROOF Using the binomial theorem (§4.2), we obtain

$$\begin{aligned}(\lambda - \alpha)^k &= ((\lambda - z_j) + (z_j - \alpha))^k \\ &= \sum_{s=0}^{k} \binom{k}{s} (\lambda - z_j)^s(z_j - \alpha)^{k-s} \\ &= \sum_{s=0}^{t-1} \binom{k}{s} (\lambda - z_j)^s(z_j - \alpha)^{k-s} \\ &\qquad + (\lambda - z_j)^t \left\{\sum_{s=t}^{k} \binom{k}{s} (\lambda - z_j)^{s-t}(z_j - \alpha)^{k-s}\right\}.\end{aligned}$$

Letting $P_{jtk}(\lambda)$ denote the polynomial enclosed in braces here, we find that

$$(\lambda - \alpha)^k = Q_{jtk}(\lambda) + (\lambda - z_j)^t P_{jtk}(\lambda).$$

Multiplying by $1/\{(\lambda - \alpha)^k(\lambda - z_j)^t\}$ gives the result. □

Observe that (12) shows that $\deg Q_{jtk}(\lambda) \leq t - 1$.

Let $k \geq \deg n(\lambda)$. Substituting (11) into (10), then multiplying by $n(\lambda)(\lambda - \alpha)^k$, we find that

$$m(\lambda)(\lambda - \alpha)^k = [g(\lambda)(\lambda - \alpha)^k + \sum_{j=1}^{T} \sum_{t=1}^{e_j} D_{jt}P_{jtk}(\lambda)]n(\lambda)$$

$$+ \{\sum_{j=1}^{T} \sum_{t=1}^{e_j} D_{jt}Q_{jtk}(\lambda)n(\lambda)/(\lambda - z_j)^t\}. \quad (13)$$

Since $\deg Q_{jtk}(\lambda) \leq t - 1$, and since $n(\lambda)/(\lambda - z_j)^t$ is a polynomial of degree $\deg n(\lambda) - t$, the expression in braces in (13), namely,

$$\sum_{j=1}^{T} \sum_{t=1}^{e_j} D_{jt}Q_{jtk}(\lambda)n(\lambda)/(\lambda - z_j)^t, \quad (14)$$

is a polynomial, and of degree at most $\deg n(\lambda) - 1$.

Continuing to let $k \geq \deg n(\lambda)$, from Theorem 6.5.1 we get

$$q(\lambda) = m(\lambda)/n(\lambda)$$

$$= d_l(\lambda - \alpha)^l + \cdots + d_{-k}(\lambda - \alpha)^{-k} + m_{-k}(\lambda)/\{(\lambda - \alpha)^k n(\lambda)\}, \quad (15)$$

where $m_{-k}(\lambda)$ is a polynomial with $\deg m_{-k}(\lambda) < \deg n(\lambda)$. Multiplying by $n(\lambda)(\lambda - \alpha)^k$ produces

$$m(\lambda)(\lambda - \alpha)^k = \{d_l(\lambda - \alpha)^{k+l} + \cdots + d_{-k}\}n(\lambda) + m_{-k}(\lambda). \quad (16)$$

Evidently, from (13) and (16), the polynomials (14) and $m_{-k}(\lambda)$ are remainders after division of $m(\lambda)(\lambda - \alpha)^k$ by $n(\lambda)$. This remainder being unique when its degree is at most $\deg n(\lambda) - 1$, we deduce that

$$m_{-k}(\lambda) = \sum_{j=1}^{T} \sum_{t=1}^{e_j} D_{jt}Q_{jtk}(\lambda)n(\lambda)/(\lambda - z_j)^t.$$

Thus

$$m_{-k}(\lambda)/\{(\lambda - \alpha)^k n(\lambda)\} = \sum_{j=1}^{T} \sum_{t=1}^{e_j} D_{jt}Q_{jtk}(\lambda)/\{(\lambda - \alpha)^k(\lambda - z_j)^t\}. \quad (17)$$

Let z be a complex number outside the smallest circle C centered at α enclosing all of $z_1, \ldots, z_T$. Then $|z - \alpha| > |z_j - \alpha|$, for $j = 1, \ldots, T$, the bars $|\ |$ still signifying the usual absolute value for complex numbers. Using (15), (17), and the triangle inequality, we get

$$|q(z) - \{d_l(z - \alpha)^l + \cdots + d_{-k}(z - \alpha)^{-k}\}|$$

$$= |m_{-k}(z)/\{(z - \alpha)^k n(z)\}|$$

$$\leq \sum_{j=1}^{T} \sum_{t=1}^{e_j} |D_{jt}|\ |Q_{jtk}(z)/\{(z - \alpha)^k(z - z_j)^t\}|. \quad (18)$$

The definition (12) of $Q_{jtk}(\lambda)$ gives

$$|Q_{jtk}(z)/\{(z-\alpha)^k(z-z_j)^t\}|$$
$$\leq \sum_{s=0}^{t-1}\binom{k}{s}|z_j-\alpha|^{k-s}|z-z_j|^s/\{|z-\alpha|^k|z-z_j|^t\}$$
$$=\sum_{s=0}^{t-1}\binom{k}{s}\left|\frac{z_j-\alpha}{z-\alpha}\right|^k|z_j-\alpha|^{-s}|z-z_j|^{s-t}. \tag{19}$$

From $|z-\alpha|>|z_j-\alpha|$, we get

$$|(z_j-\alpha)/(z-\alpha)|<1,$$

and therefore, by Exercise 9 with $x=|z_j-\alpha|/|z-\alpha|$, the quantity

$$\binom{k}{s}|(z_j-\alpha)/(z-\alpha)|^k$$

approaches zero as k grows large. Therefore, from (19),

$$|Q_{jtk}(z)/\{(z-\alpha)^k(z-z_j)^t\}|$$

approaches zero as k grows large. By (18), this implies that

$$|q(z)-\{d_l(z-\alpha)^l+\cdots+d_{-k}(z-\alpha)^{-k}\}|$$

is bounded above by a fixed number of terms, each of which approaches zero as k increases. If the tolerance ϵ is given, there will therefore be a natural number K, dependent on ϵ, such that for any $k\geq K$, we have

$$|q(z)-\{d_l(z-\alpha)^l+\cdots+d_{-k}(z-\alpha)^{-k}\}|<\epsilon.$$

This completes the proof of Theorem 6.5.6. □

EXERCISES 8.3

I

1. Decompose the following rational functions into partial fractions, within $\mathbf{R}(\lambda)$, and within $\mathbf{C}(\lambda)$.

(a) $$\frac{3\lambda^3-\lambda^2-\lambda+3}{(\lambda-1)^2(\lambda^2+1)}$$ (b) $$\frac{2\lambda^5+5\lambda^3+2\lambda^2+3\lambda+3}{(\lambda^2+1)^2(\lambda^2+2)}$$

(c) $$\frac{\lambda^3-\lambda+2}{(\lambda-1)^2(\lambda+1)}$$

2. Evaluate the following indefinite integrals, using partial fractions:

$$\int\frac{1+x^2}{1-x^2}\,dx,\qquad \int\frac{(1-2x^2)}{(1-x^2)^2}\,dx.$$

3. Decompose $-599/3600$ and $82/63$ into partial fractions within $\mathbf{Q}$.

4. If the decomposition of $\alpha \in \mathbf{Q}$ into partial fractions within $\mathbf{Q}$ is

$$\alpha = -13 + 1/3 + 2/3^2 + 1/2 + 1/4 + 8/11 + 10/11^2,$$

compute the p-adic valuation of α for $p = 2, 3, 11$.

5. Find the partial fraction decomposition within $\mathbf{Z}_2(\lambda)$ of $1/\{\lambda^2(\lambda^2 + \lambda + 1)^2\}$.

II

6. (Integration theory for rational functions) After decomposing a rational function $q(\lambda) \in \mathbf{R}(\lambda)$ into partial fractions within $\mathbf{R}(\lambda)$, the evaluation of $\int q(x)\,dx$ reduces to the evaluation of these three types of integrals:

$$\int x^t\,dx, \quad \int \frac{1}{(x-\alpha)^t}\,dx, \quad \int \frac{ax+b}{(x-\alpha)^2+\beta^2}\,dx, \qquad t > 0.$$

The first is trivial, and so is the second, its value being $-\{(t-1)(x-\alpha)^{t-1}\}^{-1}$ if $t > 1$, and $\log|x-\alpha|$ if $t = 1$. The third type reduces to these two integrals:

$$\int \frac{x}{(x^2+1)^t}\,dx, \qquad \int \frac{1}{(x^2+1)^t}\,dx.$$

Prove this. The first of these is also simple, its value being

$$\{-2(t-1)(x^2+1)^{t-1}\}^{-1} \text{ if } t > 1, \text{ and } \tfrac{1}{2}\log(x^2+1) \text{ if } t = 1.$$

Verify all these statements.

Consider the last type, and denote it by $I_t = \int (x^2+1)^{-t}\,dt$. If $t = 1$, evidently $I_1 = \tan^{-1} x$. Let $t > 1$. By writing I_t as

$$\int \frac{x^2+1-x^2}{(x^2+1)^t}\,dx,$$

show that $I_t = I_{t-1} - \int x^2(x^2+1)^{-t}\,dx$. Integrate by parts, taking $u = x$ and $dv/dx = x/(x^2+1)^t$ in the standard integration-by-parts formula, and obtain an expression for this last integral in terms of I_{t-1}. Then deduce that

$$I_t = \frac{2t-3}{2t-2} I_{t-1} + \frac{x}{2(t-1)(x^2+1)^{t-1}}.$$

This formula may be used to calculate $I_2, I_3, I_4, \ldots$ recursively in terms of I_1.

Conclude that an integral of a rational function may always be evaluated in terms of other rational functions, logarithms, and inverse tangents.

7. Assume that $(m, n) = 1$, and consider formula (2) in the text, in which the r_j satisfy the conditions listed above (2).

(i) Prove that r_j is not divisible by p_j. (Hint: If $p_j \mid r_j$, then $p_j^{e_j}$ cannot appear in the denominator of m/n. Or else compute the p_j-adic valuation.)
(ii) Show that $|r_1/p_1^{e_1} + \cdots + r_T/p_T^{e_T}| \leq 1$ if $\mathcal{K}$ is $\mathfrak{F}(\lambda)$. Is this always true if $\mathcal{K}$ is $\mathbf{Q}$?
(iii) For rational functions only, if m/n is a proper fraction, show that $g = 0$. Must this necessarily hold for rational numbers?

8. Write $1/\{(\lambda - \alpha)^2 + \beta^2\}$ in terms of partial fractions in $\mathbf{C}(\lambda)$. If $q(\lambda) = m(\lambda)/n(\lambda)$, and $n(\lambda)$ has simple roots only, explain the relation between the partial fraction decomposition of $q(\lambda)$ in $\mathbf{R}(\lambda)$ and that in $\mathbf{C}(\lambda)$.

9 This exercise outlines a proof of the following fact needed in the text: If x is a real number with $0 < x < 1$, and s is a positive integer, then $\binom{k}{s}x^k$ approaches 0 as k grows large. Let $y = 1/x$. Then $y > 1$. Define h by setting $y = 1 + h$. Then h is positive.
(a) Take $k > s$ and apply the binomial theorem to obtain

$$y^k = (1 + h)^k = \cdots + \binom{k}{s+1} h^{s+1} + \cdots > \binom{k}{s+1} h^{s+1}.$$

(b) Show that $\displaystyle\binom{k}{s+1} = \binom{k}{s}\frac{k - s}{s + 1}.$

(c) Hence show that

$$\binom{k}{s} x^k = \frac{\binom{k}{s}}{y^k} < \frac{\binom{k}{s}(s + 1)}{\binom{k}{s}(k - s)h^{s+1}} = \frac{(s + 1)}{(k - s)h^{s+1}} \to 0$$

as k grows large.

10. (Alternative uniqueness proof for Theorem 8.3.1) Multiply (6) by $p_1^{\tau-1}p_2^{e_2}p_3^{e_3}\cdots p_T^{e_T}$. Observe that all terms are then elements of $\mathcal{R}$, except the term arising from $(D_{1\tau} - D'_{1\tau})/p_1^{\tau}$. Hence deduce the uniqueness of the partial fraction decomposition.

III

11. Let $m(\lambda)$, $n(\lambda)$ be monic polynomials in $\mathbf{C}[\lambda]$. If the roots $\alpha_1, \ldots, \alpha_T$ of $n(\lambda)$ are simple, and if $m(\lambda)$ is monic with $\deg m(\lambda) = T - 1$, use the partial fraction decomposition of m/n to prove that

$$\sum_{s=1}^{T} \frac{m(\alpha_s)}{n'(\alpha_s)} = 1,$$

$n'(\lambda)$ denoting the derivative of $n(\lambda)$.

12. Let $q(\lambda) = 1/(\lambda - w)^e$, where w is a complex number and e a positive integer. Prove that the Laurent series for $q(\lambda)$ in decreasing powers of $\lambda - \alpha$, when evaluated at z, does not converge to $q(z)$ if z lies inside the circle centered at α and passing through w.
13. Find a simpler formula for

$$\sum_{j=0}^{T} \frac{1}{(\lambda + j)(\lambda + j + 1)}.$$

14. If α, β are real numbers, with β nonzero, give methods for calculating the complex partial fraction decomposition of $1/\{(\lambda - \alpha)^2 + \beta^2\}^t$.

8.4 *b*-adic REPRESENTATIONS

The discussion of valuations in §8.2 is prerequisite to an understanding of this section. As in previous sections of this chapter, we let $\mathcal{R}$ and $\mathcal{K}$ be either $\mathbf{Z}$ and $\mathbf{Q}$ or $\mathfrak{F}[\lambda]$ and $\mathfrak{F}(\lambda)$.

Assume throughout this section that b is a fixed element of $\mathcal{R}$, with $b > 1$ when $\mathcal{R}$ is $\mathbf{Z}$, and monic of positive degree when $\mathcal{R}$ is $\mathfrak{F}[\lambda]$. In both cases, $|b| > 1$. Suppose that

$$b = p_1^{e_1} p_2^{e_2} \cdots p_r^{e_r},$$

where $p_1, p_2, \ldots, p_r$ are distinct positive or monic primes, and $e_1, e_2, \ldots, e_r$ are positive integers.

We now introduce a new concept of approximation which depends on a family of p-adic valuations.* If α and β are elements of $\mathcal{K}$, and if ϵ is a tolerance (a positive real number), we say that

β is an approximation of α, within tolerance ϵ, in the b-adic sense, if

$$|\alpha - \beta|_{p_1} < \epsilon, \ |\alpha - \beta|_{p_2} < \epsilon, \ \ldots, \ |\alpha - \beta|_{p_r} < \epsilon.$$

If $r = 1$, so that b has a single prime factor, p_1, approximation in the b-adic sense becomes *approximation in the p_1-adic sense.* However, if b has two or more distinct prime factors, approximation in the b-adic sense is a requirement more stringent than approximation in the p-adic sense for any single prime p. For example, when $\mathcal{K}$ is $\mathbf{Q}$ and b is 10 (thus $p_1 = 2$ and $p_2 = 5$), then 10^t is a close 10-adic approximation to zero when t is large since both

$$|10^t|_2 = 2^{-t}, \qquad |10^t|_5 = 5^{-t}$$

are small real numbers. However, 2^t will not become a close 10-adic approximation to zero, no matter how large t is taken, because $|2^t|_5 = 1$ for every t, even though $|2^t|_2 = 2^{-t}$ is a small real number for large t.

---

* By generalizing the concept of valuation, this family of inequalities may be written as a single inequality. See Exercise 12(iv).

Let q be an element of $\mathcal{K}$. When $\mathcal{K}$ is $\mathbf{Q}$, we permit q to be negative. Indeed, some of the most interesting facets of the results about to be described occur when q is negative. When q is written in lowest terms, its denominator may not be relatively prime to b. However, for a sufficiently large nonnegative integer l, the fraction $b^l q$ will have a denominator relatively prime to b since any common prime factors with b will cancel. Let $q_0 = b^l q$, and set

$$q_0 = m_0/n,$$

where m_0, n are elements of $\mathcal{R}$, with n positive or monic and relatively prime to b. We give, in the next two lemmas, a recursive process that begins with q_0, then progressively calculates new fractions

$$q_1 = m_1/n,\ q_2 = m_2/n,\ q_3 = m_3/n,\ \ldots,$$

where m_1, m_2, m_3, ... are all in $\mathcal{R}$, and calculates also elements δ_0, δ_1, δ_2, ... of $\mathcal{R}$ such that, for each t,

$$0 \le \delta_t < b \quad \text{when } \mathcal{R} \text{ is } \mathbf{Z}, \qquad \deg \delta_t < \deg b \quad \text{when } \mathcal{R} \text{ is } \mathcal{F}[\lambda]. \tag{1}$$

LEMMA 8.4.1
If $q_0 = m_0/n$, where m_0, n are elements of $\mathcal{R}$ and n is relatively prime to b, then elements m_1, δ_0 of $\mathcal{R}$ exist such that

$$q_0 = (m_1/n)\cdot b + \delta_0, \tag{2}$$

and δ_0 satisfies (1).

PROOF Since n and b are relatively prime, elements u, v of $\mathcal{R}$ exist such that $1 = un + vb$. Thus $m_0 = (m_0u)n + (m_0v)b$. Dividing m_0u by b, we obtain elements U, δ_0 of $\mathcal{R}$ such that $m_0u = Ub + \delta_0$, where δ_0 satisfies (1). Then $m_0 = (Ub + \delta_0)n + (m_0v)b = (Un + m_0v)b + \delta_0 n$. If we let $m_1 = Un + m_0v$, then $m_0 = m_1b + \delta_0 n$, and multiplying by $1/n$ gives (2). □

LEMMA 8.4.2
With q_0 as in Lemma 8.4.1, *there are elements δ_0, δ_1, δ_2, ... of $\mathcal{R}$, all satisfying* (1), *and further elements m_1, m_2, m_3, ... of $\mathcal{R}$, such that*

$$q_0 = (m_{k+1}/n)\cdot b^{k+1} + \delta_k b^k + \delta_{k-1}b^{k-1} + \cdots + \delta_1 b + \delta_0, \tag{3}$$

for each nonnegative integer k.

PROOF We certainly have (3) for $k = 0$, since it then is the same as (2). If we have obtained (3) for a particular value of k, we apply Lemma 8.4.1 to m_{k+1}/n, obtaining thereby elements m_{k+2} and δ_{k+1} of $\mathcal{R}$, with δ_{k+1} satisfying (1), such that

$$m_{k+1}/n = (m_{k+2}/n)\cdot b + \delta_{k+1}.$$

Substituting this into (3) produces

$$q_0 = (m_{k+2}/n)\cdot b^{k+2} + \delta_{k+1}b^{k+2} + \delta_k b^k + \cdots + \delta_0 .$$

Thus k may be increased, so that, beginning with $k = 0$, we progressively obtain (3) for each nonnegative integer k. □

Writing down (3) with $k + l$ in place of k, substituting $q_0 = b^l q$, and changing notation by setting

$$d_{-l} = \delta_0 ,\ d_{-(l-1)} = \delta_1 ,\ \ldots,\ d_k = \delta_{k+l} ,\ M_{k+1} = m_{k+l+1} ,$$

we get

$$q = E_{k+1}b^{k+1} + d_k b^k + d_{k-1}b^{k-1} + \cdots + d_{-(l-1)}b^{-(l-1)} + d_{-l}b^{-l}, \quad (4)$$

where

$$E_{k+1} = M_{k+1}/n, \qquad M_{k+1} \in \mathcal{R}, \quad (5)$$

and

$$d_t \in \mathcal{R},\quad 0 \le d_t < b \quad \text{or} \quad \deg d_t < \deg b, \quad \text{for all } t. \quad (6)$$

We call the term $E_{k+1}b^{k+1}$ in (4) the *error term*. Since $E_{k+1} = M_{k+1}/n$, with $M_{k+1} \in \mathcal{R}$ and n relatively prime to b, we have

$$|E_{k+1}|_{p_j} \le 1, \qquad \text{for each prime divisor } p_j \text{ of } b.$$

Thus

$$|E_{k+1}b^{k+1}|_{p_j} \le |b|_{p_j}{}^{k+1}.$$

Since $|b|_{p_j} < 1$, the p_j-adic size of the error term approaches zero as k increases, for each j, implying that the error term becomes close to zero in the b-adic sense when k grows large. This means that the finite series

$$d_k b^k + \cdots + d_{-l}b^{-l}, \quad (7)$$

obtained from the right-hand side of (4) by omitting the error term, closely approximates q in the b-adic sense for large k. Parallel to earlier discussions of infinite series (in §§2.2 and 6.5), we write

$$q = \cdots + d_k b^k + \cdots + d_{-l}b^{-l} \quad (8)$$

as an infinite series (read most conviently from right to left). The meaning we wish to give to this series is parallel to the meanings given to infinite series encountered earlier, except that precision of approximation is now measured in the b-adic sense: *If a tolerance ϵ is given (a positive real number), there is to be an integer K (dependent on ϵ) such that truncating* (8) *at the term $d_k b^k$, for any $k \ge K$, gives an approximation of q in the b-adic sense within this tolerance.* In symbols: *For all $k \ge K$,*

$$|q - \{d_k b^k + \cdots d_{-l}b^{-l}\}|_{p_j} < \epsilon,$$

for each prime divisor p_j of b. If this approximation property holds for every tolerance ϵ, we say that the infinite series on the right-hand side of (8) is *convergent to q in the b-adic sense.*

LEMMA 8.4.3
The infinite series (8) obtained above is convergent to q in the b-adic sense.

PROOF Although this should be almost evident, we write out a formal proof. Let ϵ be a tolerance. Since $|b|_{p_j} < 1$, for each prime divisor p_j of b, for sufficiently large K we have $|b^K|_{p_j} < \epsilon$ for all j. For any $k \geq K$ and all j, we then have

$$|q - \{d_k b^k + \cdots + d_{-l} b^{-l}\}|_{p_j} = |E_{k+1} b^{k+1}|_{p_j} \leq |b^{k+1}|_{p_j} \leq |b^K|_{p_j} < \epsilon. \qquad \square$$

The representations of q obtained in (4), (8) above are wholly analogous to the representations in base b discussed in §6.5. Specifically, in §6.5 we had

$$q = d_l b^l + \cdots + d_{-k} b^{-k} + q_{-k} b^{-k}, \quad \text{with } |q_{-k}| < 1, \qquad \text{(7) of §6.5,}$$

$$q = d_l b^l + \cdots + d_{-k} b^{-k} + \cdots, \qquad \text{(10) of §6.5,}$$

with the digits d_t satisfying (6) above. The meaning given to (10) of §6.5 is that, if ϵ is a tolerance, there is an integer K (dependent on ϵ) such that truncating at the term $d_{-k}b^{-k}$, for any $k \geq K$, gives an approximation of q within this tolerance, *when the usual absolute value* $|\ |$ *is used to measure the closeness of the approximation.* In symbols,

$$|q - \{d_l b^l + \cdots + d_{-k} b^{-k}\}| < \epsilon, \qquad \text{for all } k \geq K,$$

and this follows immediately from (7) of §6.5 since the error term $q_{-k}b^{-k}$ grows small (in the sense of the usual absolute value) for large k because $|q_{-k}| < 1$, $|b| > 1$. The difference between the discussion of §6.5 and that above is therefore solely the different method of measurement.

Parallel to the discussion in Chapters 4 and 6, we give the name *b-adic digits* to the elements

$$d_k, d_{k-1}, \ldots, d_{-(l-1)}, d_{-l}$$

of $\mathcal{R}$ appearing in (4) and (8) and satisfying (6). We now show that these digits are unique. No generality is lost when comparing two sets of digits if we take l to be the same for both, since extra terms may be inserted into (4) or (8) by using zero digits.

LEMMA 8.4.4
Suppose, in addition to the representation of q in (4), (5), (6), *we also have*

$$q = E_{k+1}' b^{k+1} + d_k' b^k + \cdots + d_{-l}' b^{-l},$$

with $d_t' \in \mathcal{R}$, $0 \leq d_t' < b$ or deg $d_t' <$ deg b, *for each t, and $E_{k+1}' \in \mathcal{K}$ with $|E_{k+1}'|_{p_j} \leq 1$ for each prime divisor p_j of b. Then*

$$d_t = d_t'$$

for $t = -l, \ldots, k$.

PROOF Equating the two representations of q, then multiplying by b^{l-1}, we get

$$(d_{-l} - d_{-l}')/b = Eb^{k+l} + D_k b^{k+l-1} + \cdots + D_{-(l-1)}, \tag{9}$$

where

$$E = E_{k+1}' - E_{k+1} \quad \text{and} \quad D_t = d_t' - d_t \quad \text{for each } t.$$

Then

$$|E|_{p_j} \le \max\{|E_{k+1}'|_{p_j}, |E_{k+1}|_{p_j}\} \le 1,$$
$$|D_t|_{p_j} \le 1, \qquad |b|_{p_j} \le 1, \qquad \text{since } D_t, b \in \mathcal{R}.$$

These inequalities imply, by the strong triangle inequality, that the p_j-adic size of the right-hand side of (9) is at most 1. Hence

$$|(d_{-l} - d_{-l}')/b|_{p_j} \le 1, \qquad \text{for each prime divisor } p_j \text{ of } b.$$

However, if p is a prime element of $\mathcal{R}$, but not a divisor of b, then $|b|_p = 1$ and $|d_{-l} - d_{-l}'|_p \le 1$, the latter because $d_{-l} - d_{-l}' \in \mathcal{R}$. Thus we also have

$$|(d_{-l} - d_{-l}')/b|_p \le 1 \qquad \text{for each prime } p \text{ not dividing } b.$$

Consequently, we have $|(d_{-l} - d_{-l}')/b|_p \le 1$ for every p, whether dividing b or not. Since an element of $\mathcal{K}$ whose p-adic valuation is ≤ 1 for every p is an element of $\mathcal{R}$, we find that $(d_{-l} - d_{-l}')/b \in \mathcal{R}$, and hence that $b \mid (d_{-l} - d_{-l}')$. However, $d_{-l} - d_{-l}'$ satisfies

$$-b < d_{-l} - d_{-l}' < b, \qquad \text{or} \qquad \deg(d_{-l} - d_{-l}') < \deg b,$$

so that it can be divisible by b only if it is zero. Therefore $d_{-l} = d_{-l}'$.

Applying the result of the last paragraph to $q - d_{-l}b^{-l} = \cdots + d_{-(l-1)}b^{-(l-1)} = \cdots + d_{-(l-1)}'b^{-(l-1)}$, we get $d_{-(l-1)} = d_{-(l-1)}'$. Applying it again to $q - d_{-l}b^{-l} - d_{-(l-1)}b^{-(l-1)}$, we get $d_{-(l-2)} = d_{-(l-2)}'$. Continuing in this way, we get $d_t = d_t'$ for every t. □

THEOREM 8.4.5

An element q of $\mathcal{K}$ is uniquely representable as an infinite series (8) with digits d_t satisfying (6), such that the initial finite segments

$$d_k b^k + \cdots + d_{-l} b^{-l}$$

satisfy

$$|q - \{d_k b^k + \cdots + d_{-l} b^{-l}\}|_p \le |b^{k+1}|_p$$

for each prime divisor p of b. In particular, if

$$q = E_{k+1}b^{k+1} + d_k b^k + \cdots + d_{-l} b^{-l} \tag{4}$$

and $|E_{k+1}|_p \le 1$ for each p dividing b, then $d_k, \ldots, d_{-l}$ are the initial (lowest) b-adic digits of q.

We call the infinite series (8) *the b-adic representation of q*, and parallel to the base b representations in §6.5, we write it more compactly as

$$q = \ldots d_3 d_2 d_1 d_0 . d_{-1} d_{-2} \ldots d_{-l}\,, \tag{10}$$

using a *b-adic point* to separate digits belonging to negative powers of b from those belonging to nonnegative powers. Here, there may be infinitely many digits to the *left* of the b-point. Of course, (10) is nothing more than a convenient abbreviation for the infinite series (8). We call b the *base* in (8) or (10); since it does not explicitly appear in (10), the particular base to be used must be clear from context, or may be attached as a subscript, as in §§4.2 and 6.5.

If q is a rational number or function for which the base b representation obtained in §6.5 is terminating, say

$$q = d_u b^u + \cdots + d_{-v} b^{-v}, \quad \text{with } u \geq -v \text{ and digits } d_t \text{ satisfying (6)},$$

then this is also the b-adic representation of q, by Theorem 8.4.5, since we may then write q in the form (4), for any $k \geq u$, with error $E_{k+1} = 0$ and digits $d_t = 0$, for $t > u$.

We may perform arithmetic using the b-adic representations (10), just as we performed arithmetic in Chapters 4 and 6. We give this discussion next, for the case $\mathcal{K} = \mathbf{Q}$.

Addition

Let a_1, a_2 be rational numbers, represented in b-adic form as

$$\begin{aligned} a_1 &= \cdots + d_{-(l-1)}^{(1)} b^{-(l-1)} + d_{-l}^{(1)} b^{-l}, \\ a_2 &= \cdots + d_{-(l-1)}^{(2)} b^{-(l-1)} + d_{-l}^{(2)} b^{-l}, \end{aligned} \tag{11}$$

with integral digits $d_t^{(1)}$, $d_t^{(2)}$ satisfying $0 \leq d_t^{(1)}, d_t^{(2)} < b$. No generality is lost by taking l to be the same for both a_1, a_2 since we may insert extra terms using zero digits. To compute the b-adic representation for $a_1 + a_2$, we first add equations (11), getting

$$a_1 + a_2 = \cdots + D_{-(l-1)} b^{-(l-1)} + D_{-l} b^{-l},$$

where $D_t = d_t^{(1)} + D_t^{(2)}$ for each t. That is, we add the digits, position by position. Now we *reduce and carry*, as follows. Divide D_{-l} by b to get $D_{-l} = c_{-(l-1)} b + r_{-l}$, where $0 \leq r_{-l} < b$, and then we have

$$a_1 + a_2 = \cdots + D_{-(l-2)} b^{-(l-2)} + (c_{-(l-1)} + D_{-(l-1)}) b^{-(l-1)} + r_{-l} b^{-l}.$$

Here $c_{-(l-1)}$ is a *carry integer*. Next, divide $c_{-(l-1)} + D_{-(l-1)}$ by b to get $c_{-(l-1)} + D_{-(l-1)} = c_{-(l-2)} b + r_{-(l-2)}$, where $0 \leq r_{-(l-1)} < b$ and $c_{-(l-2)}$ is a new carry integer, and then

$$\begin{aligned} a_1 + a_2 = \cdots + D_{-(l-3)} b^{-(l-3)} &+ (c_{-(l-2)} + D_{-(l-2)}) b^{-(l-2)} \\ &+ r_{-(l-1)} b^{-(l-1)} + r_{-l} b^{-l}. \end{aligned}$$

Plainly this may be continued. The process may be justified by writing (11) as

$$\begin{aligned} a_1 &= E_{k+1}'b^{k+1} + d_k^{(1)}b^k + \cdots + d_{-l}^{(1)}b^{-l}, \quad |E_{k+1}'|_p \leq 1 \text{ for all } p \mid b, \\ a_2 &= E_{k+1}''b^{k+1} + d_k^{(2)}b^k + \cdots + d_{-l}^{(2)}b^{-l}, \quad |E_{k+1}''|_p \leq 1 \text{ for all } p \mid b. \end{aligned} \tag{12}$$

Adding these equations, then reducing and carrying, as above, we obtain $a_1 + a_2 = (c_{k+1} + E_{k+1}' + E_{k+1}'')b^{k+1} + r_k b^k + \cdots + r_{-l}b^{-l}$. Since $|c_{k+1} + E_{k+1}' + E_{k+1}''|_p \leq \max\{|c_{k+1}|_p, |E_{k+1}'|_p, |E_{k+1}''|_p\} \leq 1$, for each prime divisor p of b, it follows from Theorem 8.4.5 that $\ldots, r_k, \ldots, r_{-l}$ are indeed the b-adic digits of $a_1 + a_2$.

The thoughtful reader will observe that the process of adding digits, reducing and carrying, applied to the b-adic approximations

$$d_k^{(1)}b^k + \cdots + d_{-l}^{(1)}b^{-l}, \qquad d_k^{(2)}b^k + \cdots + d_{-l}^{(2)}b^{-l}$$

of a_1 and a_2, is the same as the process of adding digits, reducing and carrying, applied to these rational numbers A_1, A_2:

$$A_1 = d_k^{(1)}b^k + \cdots + d_{-l}^{(1)}b^{-l}, \qquad A_2 = d_k^{(2)}b^k + \cdots + d_{-l}^{(2)}b^{-l}.$$

Because of this, calculating machines may be used to perform 10-adic arithmetic, and the properties of 10-adic arithmetic may be used to explain phenomena seen when performing ordinary arithmetic on a desk calculating machine.*

Example: On the ten-digit base ten adding machine described in the footnote, if 1 is added to 9 999 999 999, the result is 0 000 000 000. What is the explanation of this? If we regard 1 and 9 999 999 999 as integers, the result of adding digits, reducing and carrying, should be 10, 000 000 000. However, the machine cannot handle eleven digits, the eleventh digit (the 1) is an overflow, and the machine simply fails to record it. A second interpretation of this calculation is obtained by noticing that the 10-adic representation of -1 is

$$-1 = \cdots + 9 \cdot 10^2 + 9 \cdot 10 + 9$$

with every digit a 9. This can be seen by noticing that $-1 = (-1)10 + 9$, so that (using this equation repeatedly) we get

$$\begin{aligned} -1 &= (-1)10 + 9 = ((-1)10 + 9)10 + 9 = (-1)10^2 + 9 \cdot 10 + 9 \\ &= ((-1)10 + 9)10^2 + 9 \cdot 10 + 9 = (-1)10^3 + 9 \cdot 10^2 + 9 \cdot 10 + 9 \\ &= \cdots \text{ etc.} \end{aligned}$$

* Our desk calculating machine is understood to be one built to use complements. (See footnote, pages 162–3, §4.2.) An example of such a machine is the Marchant, Model AB10FA. This machine is capable of handling ten-digit numbers in additions and subtractions and can also record all twenty digits appearing in a product when two ten-digit numbers are multiplied.

We may therefore regard 9,999,999,999.0 as a close 10-adic approximation to -1. In fact $9{,}999{,}999{,}999 = 10^{10} - 1$ and 10^{10} is small in both the 2-adic and 5-adic valuations. So if we add 1 and 9,999,999,999, the result should be a close approximation to zero. And it is! The sum in fact is 10^{10}, and this is too *small* a number for the machine to recognize—it is off the left-hand end of the dial.

Subtraction

The discussion of subtraction in §4.3 easily adapts to the present situation. Let

$$a_1 = E_{k+1}^{(1)}b^{k+1} + d_k^{(1)}b^k + \cdots + d_0^{(1)} = \cdots + d_k^{(1)}b^k + \cdots + d_0^{(1)},$$

$$a_2 = E_{k+1}^{(2)}b^{k+1} + d_k^{(2)}b^k + \cdots + d_0^{(2)} = \cdots + d_k^{(2)}b^k + \cdots + d_0^{(2)},$$

be two numbers in b-adic form, where $|E_{k+1}^{(1)}|_p \le 1$, $|E_{k+1}^{(2)}|_p \le 1$ for each prime divisor p of b. (For simplicity we have taken $l = 0$.) Let $S = a_1 - a_2$. The first step in calculating S is to subtract the digits, position by position, precisely as described in §4.3, thereby obtaining

$$S = E_{k+1}b^{k+1} + D_kb^k + \cdots + D_0 = \cdots + D_kb^k + \cdots + D_0$$

where $E_{k+1} = E_{k+1}^{(1)} - E_{k+1}^{(2)}$ and $D_t = d_t^{(1)} - d_t^{(2)}$ for each t. The integers D_t may not lie in the proper range, 0 to $b - 1$, so starting from the right-hand end we bring them to this range by borrowing (i.e., reducing and carrying). Several applications of this process give us digits $r_0, r_1, \ldots, r_k$, and a carry integer c_{k+1}, such that

$$S = (E_{k+1} + c_{k+1})b^{k+1} + r_kb^k + \cdots + r_1b + r_0 .$$

Since $|E_{k+1} + c_{k+1}|_p \le 1$ for each prime divisor p of b, we deduce from Theorem 8.4.5 that $\ldots, r_1, r_0$ are indeed the correct b-adic digits of S. Thus

$$S = \cdots + r_kb^k + \cdots + r_1b + r_0 .$$

What happens if we try to subtract a positive integer a from 0? Modifying the notation used above slightly to make it conform to notation used in §4.2, we write a as

$$a = d_lb^l + \cdots + d_1b + d_0 ,$$

where $0 \le d_t < b$, for each t. The first step in calculating $0 - a$ is to obtain

$$-a = (-d_l)b^l + \cdots + (-d_1)b + (-d_0).$$

Referring to equations (19) and (21) in §4.2, we see that the reducing and carrying process to bring $-d_0, -d_1, \ldots$ to the proper range, 0 to $b - 1$, amounts to the steps shown in the following equations:

$$\begin{aligned} -a &= (-d_l)b^l + \cdots + (-d_2)b^2 + (-d_1 - c_1)b + \delta_0 \\ &= (-d_l)b^l + \cdots + (-d_3)b^3 + (-d_2 - c_2)b^2 + \delta_1 b + \delta_0 \\ &= \cdots\cdots\cdots\cdots\cdots \\ &= -c_{l+1}b^{l+1} + \delta_l b^l + \cdots + \delta_1 b + \delta_0 \\ &= -c_{l+2}b^{l+2} + \delta_{l+1}b^{l+1} + \cdots + \delta_1 b + \delta_0 \\ &= \cdots\cdots\cdots\cdots\cdots \\ &= -b^{t+1} + (b-1)b^t + \cdots + (b-1)b^{l+1} + \delta_l b^l + \cdots + \delta_0, \\ &\qquad \text{for any } t \geq l. \qquad (13) \end{aligned}$$

This is exactly the *complement* of a discussed in §4.2—see equation (22) of §4.2. The error term in (13) is $-b^{t+1}$, and for large t it is small in the p-adic valuation, for each prime divisor p of b.

Example: On a ten-digit base ten adding machine using complements, subtract 1 from 0. The result is 9,999,999,999. What does this mean? It is simply the series (13) for $a = -1$, $b = 10$; $l = 0$, and $t = 9$. Each digit $\delta_0, \delta_1, \ldots, \delta_9$ equals nine, and (13) gives

$$-1 = -10^{10} + 9\cdot 10^9 + 9\cdot 10^8 + \cdots + 9\cdot 10 + 9.$$

The error term here is -10^{10}, and (10-adically) it is small; so small, in fact, that it is too small for the machine to recognize. (It's off the left end of the dial.) So, when we subtract 1 from 0, the adding machine gives a close 10-adic approximation to -1.

Multiplication

It is convenient when discussing multiplication to write a b-adic number q as $q = b^{-l}q_0$, with l chosen so that q_0 has no digits to the right of the b-adic point. Thus

$$q = b^{-l}(E_{k+1}b^{k+1} + \delta_k b^k + \cdots + \delta_1 b + \delta_0)$$

where $|E_{k+1}|_p \leq 1$ for each prime divisor p of b. The quantities $\delta_0, \delta_1, \ldots, \delta_k$ are the initial (i.e. lowest) b-adic digits of q, renumbered so that the lowest digit carries 0 as its subscript. Let a_1, a_2 be numbers to be multiplied, written in this form as

$$a_1 = b^{-l'}(E_{k+1}'b^{k+1} + \delta_k'b^k + \cdots + \delta_1'b + \delta_0') = b^{-l'}(E_{k+1}'b^{k+1} + A_1),$$
$$a_2 = b^{-l''}(E_{k+1}''b^{k+1} + \delta_k''b^k + \cdots + \delta_1''b + \delta_0'') = b^{-l''}(E_{k+1}''b^{k+1} + A_2).$$

Here $A_1 = \delta_k'b^k + \cdots + \delta_0'$, $A_2 = \delta_k''b^k + \cdots + \delta_0''$. Then

$$a_1a_2 = b^{-l'-l''}((E_{k+1}'E_{k+1}''b^{k+1} + E_{k+1}'A_2 + E_{k+1}''A_1)b^{k+1} + A_1A_2). \qquad (14)$$

Multiplying A_1 and A_2 by the ordinary rules for multiplying base b numbers, as explained in §4.3, we get

$$A_1A_2 = \cdots + D_2b^2 + D_1b + D_0$$

where D_0, D_1, ... are the usual base b digits of the product A_1A_2. Collecting all terms in this product involving powers of b with exponent higher than k, we get

$$A_1A_2 = Eb^{k+1} + D_kb^k + \cdots + D_0 \tag{15}$$

where E is an integer. Inserting (15) into (14), we obtain

$$a_1a_2 = b^{-l'-l''}(\epsilon b^{k+1} + D_kb^k + \cdots + D_0),$$

with $\epsilon = E + E_{k+1}'E_{k+1}''b^{k+1} + E_{k+1}'A_2 + E_{k+1}''A_1$. Since E, A_1, A_2 are integers, we have $|E|_p \leq 1$, $|A_1|_p \leq 1$, $|A_2|_p \leq 1$. Since also $|E_{k+1}'|_p \leq 1$, $|E_{k+1}''|_p \leq 1$ for each $p \mid b$, it follows that $|\epsilon|_p \leq 1$ for each $p \mid b$. From Theorem 8.4.5, we then deduce that D_k, ..., D_0 must be lowest b-adic digits of a_1a_2. In other words: to multiply the b-adic representations of a_1 and a_2, truncate by retaining only the $k + 1$ lowest b-adic digits in each number, multiply these truncated numbers by the rules for base b multiplication, then truncate again by retaining only the lowest $k + 1$ digits in the product. These are then the initial $k + 1$ digits in the b-adic representation of a_1a_2. Furthermore, the power of b belonging to the lowest digit in a_1a_2 is $b^{-l'-l''}$ when $b^{-l'}$, $b^{-l''}$ are the lowest powers of b in a_1 and a_2. That is, if a_1 has l' digits to the right of the b-point, and a_2 has l'' digits to the right of the b-point, then a_1a_2 has $l' + l''$ digits to the right of the b-point.

Example: In our ten-digit base ten adding machine, we know that $a = 9{,}999{,}999{,}999 = 10^{10} - 1$ is a close 10-adic approximation to -1. Our adding machine is capable of recording all 20 digits obtained when two 10-digit numbers are multiplied.* The machine thus shows

$$a^2 = 99{,}999{,}999{,}980{,}000{,}000{,}001 = 10^{20} - 2 \cdot 10^{10} + 1.$$

Since the number 9,999,999,999 is the correct ten-digit 10-adic approximation to -1, the theory above shows that the lowest ten digits of its square must be the correct ten-digit 10-adic approximation to $(-1)^2 = 1$. Since the lowest ten digits in the product are 0,000,000,001, this is in fact the case.

The above discussion of addition, subtraction, and multiplication of b-adic series is easily adapted to apply to b-adic series of rational functions. We omit writing out the details. Thus term-by-term addition or subtraction of the b-adic series for rational functions $a_1(\lambda)$, $a_2(\lambda)$ gives the b-adic series for $a_1(\lambda) + a_2(\lambda)$ and $a_1(\lambda) - a_2(\lambda)$. These operations are simpler than for rational numbers in the sense that there never will be a "carry." Multiplication of the b-adic series for $a_1(\lambda)$, $a_2(\lambda)$ also gives the series for $a_1(\lambda)a_2(\lambda)$, and here there will again be no "carries," provided $\deg b(\lambda) = 1$.

* Many of the inexpensive calculating machines coming onto the market in recent years do not have this very useful feature.

Periodicity

We proved in §6.5 that the representation of a rational number in decreasing powers of b ultimately becomes periodic. We shall now prove that the representation in increasing powers of b—the b-adic representation above—is also ultimately periodic. In broad outline, this proof will be parallel to the proof in §6.5, but the technical details differ somewhat. Adopting notation used at the beginning of this section, let q be a rational number, and set $q_0 = b^l q = m_0/n$, where m_0, n are integers, and n is positive and relatively prime to b. Since the b-adic digits of q are those of q_0 shifted l places to the right, it will be sufficient to prove that the digits of q_0 are ultimately periodic.

From the proof of Lemma 8.4.2, we know that the b-adic representation of q_0 is obtained from m_0 by recursively calculating integers $m_1, m_2, m_3, \ldots$ and integral digits $\delta_0, \delta_1, \delta_2, \ldots$ in this way:

$$\begin{aligned} m_k/n &= (m_{k+1}/n)\cdot b + \delta_k,\ 0 \le \delta_k < b;\ k = 0, 1, 2, \ldots . \\ m_k &= m_{k+1}b + \delta_k n, \end{aligned} \tag{16}$$

We now estimate the size of the integers $m_1, m_2, \ldots.$ appearing in this process, using the usual absolute value $|\ |$ for rational numbers. Let M be the larger of $|m_0|$, n; and set $c = bM/(b-1)$. By induction on k, we prove that $|m_k| \le c$ for $k = 0, 1, 2, \ldots$. This clearly holds for $k = 0$, by the choice of M. Suppose that $|m_k| \le c$ has been proved. Using (16), the triangle inequality, $|\delta_k| < b$, $|m_k| \le c$, $c = Mb/(b-1)$, and $n \le M$, we get

$$\begin{aligned} |m_{k+1}| &= |m_k/b - \delta_k n/b| \le |m_k|/b + |\delta_k|n/b \\ &\le c/b + n \le M/(b-1) + M = c. \end{aligned}$$

The induction step is completed; thus $|m_k| \le c$ holds for $k = 0, 1, 2, \ldots$.

Since $m_0, m_1, m_2, \ldots$ are integers, restricted by $|m_k| \le c$ for all k, distinct subscripts r, s exist, with $r < s$, such that $m_r = m_s$. To simplify notation, set $q_k = m_k/n$, for $k = 0, 1, 2, \ldots$. Thus $q_k = q_{k+1}b + \delta_k$. Using this equation for $k = r, r+1, \ldots. r+s-1$, then $q_r = q_s$. we get

$$\begin{aligned} q_r &= q_s b^{s-r} + \delta_{s-1}b^{s-r-1} + \cdots + \delta_{r+1}b + \delta_r \\ &= q_r b^{s-r} + \delta_{s-1}b^{s-r-1} + \cdots + \delta_{r+1}b + \delta_r . \end{aligned} \tag{17}$$

Setting $P = s - r$ and using (17) repeatedly, we get

$$\begin{aligned} q_0 &= q_r b^r + (\delta_{r-1}b^{r-1} + \cdots + \delta_0) \\ &= q_r b^{r+P} + (\delta_{r+P-1}b^{r+P-1} + \cdots + \delta_r b^r) + (\delta_{r-1}b^{r-1} + \cdots + \delta_0) \\ &= q_r b^{r+2P} + (\delta_{r+P-1}b^{r+2P-1} + \cdots + \delta_r b^{r+P}) \\ &\qquad + (\delta_{r+P-1}b^{r+P-1} + \cdots + \delta_r b^r) + (\delta_{r-1}b^{r-1} + \cdots + \delta_0) \\ &= \cdots . \end{aligned}$$

Since $|q_r|_p = |m_r/n|_p \leq 1$. for each prime divisor p of b, the uniqueness part of Theorem 8.4.5 enables us to conclude that the b-adic representation of q_0 is

$$q_0 = \ldots \underbrace{\delta_{r+P-1} \ldots \delta_r}_{P \text{ digits}} \underbrace{\delta_{r+P-1} \ldots \delta_r}_{P \text{ digits}} \underbrace{\delta_{r-1} \ldots \delta_0}_{r \text{ digits}}, \tag{18}$$

and thus is ultimately periodic.

A convenient notation for the ultimately periodic b-adic series (18) is

$$\overline{\delta_{r+P-1} \ldots \delta_r}\, \delta_{r-1} \ldots \delta_0 , \tag{19}$$

the horizontal bar being placed over the periodic digits.

Suppose now that we are given an ultimately periodic b-adic series (19). The rational number giving rise to it may be found by summing a geometric series, as in Corollary 2.2.4, in the following way:

$$\begin{aligned} &\overline{\delta_{r+P-1} \ldots \delta_r}\, \delta_{r-1} \ldots \delta_0 \\ &= \cdots + (\delta_{r+P-1} b^{r+2P-1} + \cdots + \delta_r b^{r+P}) \\ &\qquad + (\delta_{r+P-1} b^{r+P-1} + \cdots + \delta_r b^r) + (\delta_{r-1} b^{r-1} + \cdots + \delta_0) \\ &= (\cdots + b^{2P} + b^P + 1)(\delta_{r+P-1} b^{r+P-1} + \cdots + \delta_r b^r) + (\delta_{r-1} b^{r-1} + \cdots + \delta_0) \\ &= \frac{1}{1 - b^P} A + B, \end{aligned}$$

$$\text{where} \quad A = \delta_{r+P-1} b^{r+P-1} + \cdots + \delta_r b^r , \qquad B = \delta_{r-1} b^{r-1} + \cdots + \delta_0 . \tag{20}$$

However, we must be cautious here since application of Corollary 2.2.4 to sum $\cdots + b^{2P} + b^P + 1$ as $1/(1 - b^P)$ is permissible only if $|b| < 1$. Since $|b| > 1$, justification for this procedure is required. The first step in this justification is to show that the b-adic series for $1/(1 - b^P)$ is indeed $\cdots + b^{2P} + b^P + 1$. For any integral $t \geq 0$, the summation formula (Lemma 2.2.3) for a finite geometric series gives

$$\frac{1 - b^{tP}}{1 - b^P} = b^{(t-1)P} + \cdots + b^P + 1,$$

so that
$$\begin{aligned} \frac{1}{1 - b^P} &= \frac{b^{tP}}{1 - b^P} + \{b^{(t-1)P} + \cdots + b^P + 1\} \\ &= \text{error} \times b^{(t-1)P+1} + \{b^{(t-1)P} + \cdots + b^P + 1\}, \end{aligned}$$

where the error $b^{P-1}/(1 - b^P)$ satisfies $|b^{P-1}/(1 - b^P)|_p \leq 1$, for each prime divisor p of b (because $|1 - b^P|_p = 1$). Invoking the uniqueness assertion of Theorem 8.4.5, we conclude that $b^{(t-1)P} + \cdots + b^P + 1$ is an initial segment of the b-adic series for $1/(1 - b^P)$, for any $t \geq 0$, and therefore $\cdots + b^{2P} + b^P + 1$ is indeed the b-adic series for $1/(1 - b^P)$. Using this, we now show that (19) is the b-adic series for $(1/(1 - b^P))A + B$.

From the discussion of addition and multiplication of b-adic series, we know that the b-adic series for $(1/(1 - b^P))A + B$ is obtained by multiplying the series for $1/(1 - b^P)$ and the series for A, then adding the series for B. Multiplying $\cdots + b^{2P} + b^P + 1$ by the (finite) series (20) for A, then adding the (finite) series (20) for B, produces the series (19), as direct computation shows. Our application of the summation formula for geometric series is therefore justified since it leads to the number $(1/(1 - b^P))A + B$ whose b-adic series is the given series (19).

We have thus proved that *every rational number has an ultimately periodic b-adic series, and any ultimately periodic b-adic series may be summed as as above to yield a rational number.*

In §6.5, we proved that the length P of the period in the base b representation of m_0/n satisfies $P \leq n$. This is also true of the b-adic representation of m_0/n. The proof is outlined in Exercise 17, and amounts to estimating more precisely than above the size of the integers $m_1, m_2, m_3, \ldots$.

When q_0 is a nonnegative integer, its b-adic series is finite, and when q_0 is a negative integer, all digits in its b-adic series to the left of some fixed digit equal $b - 1$. This is evident from the discussion of complements in §4.2. In both of these cases, $P = 1$ and the periodic digit is 0 or $b - 1$. The converses of these statements are also true, namely,

$$\overline{0}\delta_{r-1}\ldots\delta_0 \quad \text{is a nonnegative integer,}$$

$$\overline{(b-1)}\delta_{r-1}\ldots\delta_0 \quad \text{is a negative integer.}$$

These are clear since $P = 1$ and the integer A above is either 0 or $(b - 1)b^r$; in the latter case $(1/(1 - b))A + B = -b^r + B < 0$. The general ultimately periodic b-adic series may, of course, have $P > 1$ or have periodic digits other than 0 or $b - 1$, and we may regard these b-adic series as the extension to rational numbers of the concept of complement developed in §4.2 for integers.

This discussion of the periodicity of the b-adic series applies also to elements of $\mathfrak{F}(\lambda)$, provided that $\mathfrak{F}$ is a finite field, if obvious minor changes are made, such as replacing "integer" with "polynomial," and replacing the absolute value for rational numbers with the absolute value for rational functions. We take M to be the larger of $|m_0|$, $|n|$, and set $c = |b|M/|b - 1| = M$. The inequality $|m_k| \leq c$, for $k = 0, 1, 2, \ldots$, together with the finiteness of $\mathfrak{F}$, permits the conclusion that $m_r = m_s$ for distinct subscripts r, s, since the polynomials $m_k(\lambda)$ then have bounded degree and only finitely many distinct values possible for their coefficients. However, when $\mathfrak{F}$ is an infinite field, the b-adic series for a rational function need not be ultimately periodic, as is shown by this example from $\mathbf{C}(\lambda)$:

$$1/(1 - \lambda)^2 = \cdots + 3\lambda^2 + 2\lambda + 1,$$

with digit $d_t = t + 1$, for each $t \geq 0$. The analogy between rational numbers and rational functions, observed in §6.5, continues therefore to be sharpest when the rational functions have their coefficients in a finite field.

Confining our attention once more to rational numbers, now assume that $b = p$ is a positive prime integer. In §6.5, we saw that rational numbers (and only rational numbers) have ultimately periodic expansions in base b (decreasing powers of b), and that a larger number system exists—the real numbers—consisting of both ultimately periodic and not ultimately periodic expansions in base b. In this section, with $b = p$, we have seen that rational numbers have ultimately periodic p-adic expansions (increasing powers of p), and that any ultimately periodic p-adic expansion may be summed to yield a rational number. Parallel to the situation in which the real numbers are a number system containing the rational numbers, there is another number system also containing the rational numbers and consisting of all p-adic series which are not necessarily ultimately periodic. Just as the real numbers form a field, this number system also forms a field, called the *field of p-adic numbers*. Its construction, very much like that of the real numbers, uses $| \ |_p$ instead of $| \ |$. Some of its striking properties may be found in the book *Introduction to p-adic Numbers and Valuation Theory*, by G. Bachman (Academic Press, 1964, reprinted 1972). For emphasis, and to make quite clear the relation of the p-adic and real numbers to the rational numbers, we state the following facts:

Let b be a positive integer

I(a) The rational numbers are all numbers representable as an ultimately periodic series

$$\pm(d_l b^l + d_{l-1} b^{l-1} + \cdots) \tag{21}$$

in which the digits d_t are integers with $0 \leq d_t < b$.

I(b) The real numbers are all numbers representable as a not necessarily ultimately periodic series (21), in which the digits d_t are integers with $0 \leq d_t < b$.

A series (21) is convergent when the usual absolute value is used to measure the precision of an approximation.

Let $b = p$ be a prime

II(a) The rational numbers are all numbers representable as an ultimately periodic series

$$\cdots + d_{-(l-1)} b^{-(l-1)} + d_{-l} b^{-l} \tag{22}$$

in which the digits d_t are integers with $0 \leq d_t < b$.

II(b) The p-adic numbers are those numbers representable as a not necessarily ultimately periodic series (22) in which the digits d_t are integers with $0 \leq d_t < b$.

A series (22) is convergent when the p-adic valuation is used to measure the precision of an approximation.

For the past 40 years, and especially for the past 20 years, p-adic numbers have occupied a central position in abstract algebra. They generally play this role: When solving an equation or a system of equations, p-adic solutions may sometimes be found relatively easily. Therefore, the equation or system of equations is solved p-adically, for every p, and from these solutions a solution in "ordinary" numbers (rational numbers) is obtained. It always turns out that p-adic solutions are needed for only finitely many primes p. Recently, this method has been adapted for use on large electronic computing machines to solve systems of linear equations of the form

$$ax + by = \alpha,$$

$$cx + dy = \beta,$$

where a, b, c, d, α, β are given, and x, y are unknowns. This illustrative system has two equations and two unknowns. In applications, systems of this type occur involving n equations and unknowns, where n may be 100 or larger. (See the article by R. T. Gregory cited in page 455.) In practice, this method amounts to calculating solutions in the finite field $\mathbf{Z}_p$ or the finite ring $\mathbf{Z}_{p^k}$ for several primes p and suitable exponents k, then using the Chinese remainder theorem (3.6.3) to calculate the solution in ordinary numbers.

Taylor and Laurent series

We now return to the case $\mathcal{R} = \mathfrak{F}[\lambda]$, $\mathcal{K} = \mathfrak{F}(\lambda)$, and we take

$$b(\lambda) = \lambda - \alpha,$$

where α is an element of $\mathfrak{F}$; i.e., a constant. Then $b(\lambda)$ is a prime in $\mathcal{R}$ Applying Theorem 8.4.5 to a rational function $q(\lambda)$, but now writing series in the customary left-to-right fashion, instead of right-to-left, we get

$$q(\lambda) = d_{-l}(\lambda - \alpha)^{-l} + \cdots + d_k(\lambda - \alpha)^k + E_{k+1}(\lambda)(\lambda - \alpha)^{k+1}. \quad (23)$$

Each digit d_t here, being a polynomial with $|d_t| < |\lambda - \alpha|$, is a constant. The integer l is the multiplicity of α as a pole (§6.5) of $q(\lambda)$, i.e., as a root of the denominator polynomial when $q(\lambda)$ is written in lowest terms, so that

$$(\lambda - \alpha)^l q(\lambda) = m_0(\lambda)/n(\lambda), \qquad l \geq 0,$$

where $m_0(\lambda)$, $n(\lambda)$ are relatively prime polynomials, and $n(\lambda)$ is monic and relatively prime to $\lambda - \alpha$. The error factor $E_{k+1}(\lambda)$ in the last term is given by

$$E_{k+1}(\lambda) = m_{k+1}(\lambda)/n(\lambda),$$

with $m_{k+1}(\lambda)$ also a polynomial. As in (8), we let k grow arbitrarily large, and write $q(\lambda)$ as an infinite series

$$q(\lambda) = d_{-l}(\lambda - \alpha)^{-l} + \cdots + d_k(\lambda - \alpha)^k + \cdots. \quad (24)$$

The meaning conveyed by this infinite series is that sufficiently many terms provide close approximations to $q(\lambda)$ in the $(\lambda - \alpha)$-adic sense. More precisely, if ϵ is a tolerance (a positive real number), there is an integer K, dependent on ϵ, such that

$$|q(\lambda) - \{d_{-l}(\lambda - \alpha)^{-l} + \cdots + d_k(\lambda - \alpha)^k\}|_{\lambda-\alpha} < \epsilon,$$

for all $k \geq K$. Only one valuation is employed in this convergence criterion since $\lambda - \alpha$ is the only prime factor of $b(\lambda)$.

We now take $\mathfrak{F}$ to be the complex number field **C**. In this case, the series on the right-hand side of (24) is called a *Laurent series* or a *Laurent expansion of* $q(\lambda)$ *about* α. This Laurent series expresses $q(\lambda)$ in terms of ascending powers of $\lambda - \alpha$, whereas the Laurent series previously encountered in §6.5 expressed $q(\lambda)$ in descending powers of $\lambda - \alpha$. The special case $l = 0$ of the Laurent series (24), which arises when α is not a pole of $q(\lambda)$, is of considerable importance, and is called the *Taylor series*. It is often encountered in calculus courses; for many types of functions (but usually in calculus for functions involving real numbers only). If $q(\lambda)$ is a polynomial, its Taylor series is the finite series given in Theorem 4.2.7. A detailed study of Laurent and Taylor series properly belongs to courses and books in the *theory of a complex variable*. We discuss here only the evaluation of a Laurent series at a complex number z.

If we evaluate both sides of (24) at z, where $z \neq \alpha$, do we obtain

$$q(z) = d_{-l}(z - \alpha)^{-l} + \cdots + d_k(z - \alpha)^k + \cdots? \tag{25}$$

Since the right-hand side of this formula is an infinite series of complex numbers, we adopt the convergence criterion previously employed for infinite series, but now measure the precision of an approximation using the *usual absolute value* $|\ |$ *for complex numbers*, and not using a p-adic valuation. Thus we say that the infinite series in (25) is convergent to $q(z)$ if this criterion is satisfied: Whenever ϵ is a tolerance (a positive real real number), there is an integer K, dependent on ϵ, such that

$$|q(z) - \{d_{-l}(z - \alpha)^{-l} + \cdots + d_k(z - \alpha)^k\}| < \epsilon,$$

for all $k \geq K$. (See Figure 8.4.1.) Is this convergence criterion satisfied?

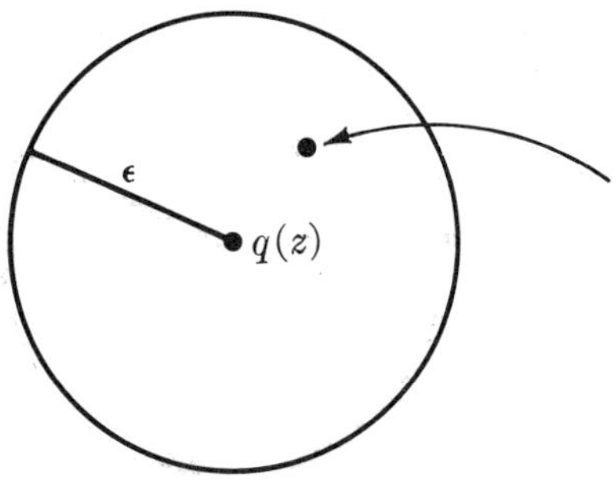

$d_{-l}(z - \alpha)^{-l} + \cdots + d_k(z - \alpha)^k$ lies within a circle of radius ϵ centered at $q(z)$, for all $k \geq K$.

Figure 8.4.1

Not always. If C is the largest circle centered at α in the complex plane and containing no root of $n(\lambda)$ in its interior, then the series on the right-hand side of (25) converges to $q(z)$ in the sense just described whenever z belongs to the interior of C, and does not converge in this sense (to anything) whenever z belongs to the exterior of C. Convergence may or may not occur if z is on the circumference of C. This circle C is illustrated in Figure 8.4.2, where $z_1, z_2, \ldots, z_T$ denote the roots of $n(\lambda)$.

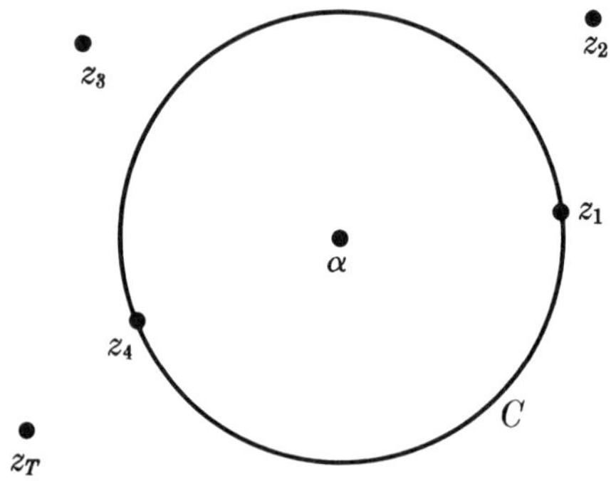

If $n(\lambda)$ is not a constant, at least one root of $n(\lambda)$ is on the circumference of C

Figure 8.4.2

The proof that convergence does not occur if z is outside C involves more complex variable theory than can be included in this book, and must therefore be omitted. However, a proof of convergence if z is inside C may be based on the convergence results obtained in §§6.5, 8.3 for Laurent series in descending powers of $\lambda - \alpha$. We give this proof in the next subsection.

The convergence proof

Let z belong to the interior of C, with $z \neq \alpha$. Our proof employs three valuations: the usual absolute value $|\;|$ for rational functions (§6.4), the $(\lambda - \alpha)$-adic valuation for rational functions (§8.2), and the usual absolute value $|\;|$ for complex numbers (§2.4). Because distinct absolute values are denoted by one symbol $|\;|$, care must be taken to understand which absolute value is intended whenever these vertical bars are encountered. We first prove a lemma involving the usual absolute value $|\;|$ for rational functions and the $(\lambda - \alpha)$-adic valuation $|\;|_{\lambda-\alpha}$.

LEMMA 8.4.6
Let $\Phi(\lambda)$ be a rational function with $|\Phi(\lambda)| < 1$. Then

$$\left| \frac{\Phi(\alpha + (\lambda - \alpha)^{-1})}{\lambda - \alpha} \right|_{\lambda-\alpha} \leq 1.$$

PROOF The result is clearly correct if $\Phi(\lambda) = 0$. Assume that Φ is a nonzero rational function, and write its numerator and denominator in base $\lambda - \alpha$, so that

$$\Phi(\lambda) = \frac{a_0 + a_1(\lambda - \alpha) + \cdots + a_s(\lambda - \alpha)^s}{b_0 + b_1(\lambda - \alpha) + \cdots + b_S(\lambda - \alpha)^S},$$

where $a_0, \ldots, a_s, b_0, \ldots, b_S$ are constants with $a_s \neq 0$, $b_S \neq 0$. Then $|\Phi(\lambda)| = \rho^{S-s}$. Since $|\Phi(\lambda)| < 1$, we get $S > s$. Notice that substituting $\alpha + (\lambda - \alpha)^{-1}$ for λ converts $\lambda - \alpha$ to $(\lambda - \alpha)^{-1}$. Using this fact to evaluate $\Phi(\lambda)$ at $\alpha + (\lambda - \alpha)^{-1}$, then multiplying the resulting numerator and denominator by $(\lambda - \alpha)^S$, we obtain

$$\frac{\Phi(\alpha + (\lambda - \alpha)^{-1})}{\lambda - \alpha} = (\lambda - \alpha)^{S-s-1} \frac{(a_0(\lambda - \alpha)^s + \cdots + a_s)}{(b_0(\lambda - \alpha)^S + \cdots + b_S)}.$$

Neither numerator nor denominator of the fraction appearing on the right-hand side of this formula contains $\lambda - \alpha$ as a factor, since $a_s \neq 0$, $b_S \neq 0$, so that

$$\left|\frac{\Phi(\alpha + (\lambda - \alpha)^{-1})}{(\lambda - \alpha)}\right|_{\lambda-\alpha} = \rho^{S-s-1} \leq 1. \qquad \square$$

We now state the convergence theorem that we shall prove.

THEOREM 8.4.7

Let $q(\lambda)$ be a rational function with complex coefficients, α a complex number, and C the largest circle centered at α in the complex plane not containing any pole (root of the denominator) of $q(\lambda)$ other than α in its interior. If z is a complex number strictly inside C, $z \neq \alpha$, then the Laurent series in ascending powers of $q(\lambda)$, when evaluated at z, becomes a series of complex numbers converging to $q(z)$.

PROOF Until the change explicitly mentioned later in this proof, the symbols $|\ |$, $|\ |_{\lambda-\alpha}$ denote the usual absolute value for rational functions and the $(\lambda - \alpha)$-adic valuation, respectively.

Let $Q(\lambda) = q(\alpha + (\lambda - \alpha)^{-1})$. Applying the discussion of Laurent series in §§6.5 and 8.3 to $Q(\lambda)$, we get an integer $L \geq 0$, digits (complex constants) $\delta_L, \delta_{L-1}, \ldots,$ and error factors (rational functions) $Q_{-1}(\lambda)$, $Q_{-2}(\lambda), \ldots,$ such that

$$Q(\lambda) = \delta_L(\lambda - \alpha)^L + \cdots + \delta_{-k}(\lambda - \alpha)^{-k} + (\lambda - \alpha)^{-k}Q_{-k}(\lambda),$$

for any $k \geq 0$, with $|Q_{-k}(\lambda)| < 1$. Evaluating at $\alpha + (\lambda - \alpha)^{-1}$ produces

$$\begin{aligned} q(\lambda) &= Q(\alpha + (\lambda - \alpha)^{-1}) \\ &= \delta_L(\lambda - \alpha)^{-L} + \cdots + \delta_{-k}(\lambda - \alpha)^k \\ &\qquad + (\lambda - \alpha)^{k+1} \frac{Q_{-k}(\alpha + (\lambda - \alpha)^{-1})}{\lambda - \alpha}. \end{aligned} \tag{26}$$

Setting $\Phi(\lambda) = Q_{-k}(\lambda)$, Lemma 8.4.6 and $|Q_{-k}(\lambda)| < 1$ imply that the factor in the last term satisfies

$$\left|\frac{Q_{-k}(\alpha + (\lambda - \alpha)^{-1})}{\lambda - \alpha}\right|_{\lambda-\alpha} \leq 1.$$

We also have

$$q(\lambda) = d_{-l}(\lambda - \alpha)^{-l} + \cdots + d_k(\lambda - \alpha)^k + (\lambda - \alpha)^{k+1}E_{k+1}(\lambda), \quad (4)$$

where the error factor $E_{k+1}(\lambda)$ satisfies

$$|E_{k+1}(\lambda)|_{\lambda-\alpha} \leq 1.$$

Comparing (4) and (26), and noticing the estimates just given of the $(\lambda - \alpha)$-adic size of the error factors in the last terms, we infer from the uniqueness part of Theorem 8.4.5 that the digits in these two expressions for $q(\lambda)$ must coincide. Thus $L = -l$, $d_{-l} = \delta_L, \ldots, d_k = \delta_{-k}$, so that

$$q(\lambda) = d_{-l}(\lambda - \alpha)^{-l} + \cdots + d_k(\lambda - \alpha)^k + (\lambda - \alpha)^k Q_{-k}(\alpha + (\lambda - \alpha)^{-1}). \quad (27)$$

Throughout the remainder of this proof, the vertical bars $|\ |$ are to denote the usual absolute value for complex numbers. Let D be the smallest circle centered at α in the complex plane enclosing all poles of $Q(\lambda)$. If w is a complex number strictly outside D, then the Laurent series for $Q(\lambda)$ in descending powers of $\lambda - \alpha$, namely, $\delta_L(\lambda - \alpha)^L + \delta_{L-1}(\lambda - \alpha)^{L-1} + \cdots$, becomes a series of complex numbers converging to $Q(w)$ when evaluated at w. This is Theorem 6.5.6. In particular, by the proof in §8.3, the error term

$$(w - \alpha)^{-k}Q_{-k}(w)$$

grows small for large k, when its size is measured in the sense of the usual absolute value for complex numbers. Specifically, if ϵ is a tolerance, there is an integer K, dependent on ϵ, such that

$$\left|\frac{Q_{-k}(w)}{(w - \alpha)^k}\right| < \epsilon, \quad (28)$$

for all $k \geq K$.

What are the poles of $Q(\lambda)$? To answer this, let

$$n(\lambda) = (\lambda - z_1)\cdots(\lambda - z_T),$$

where $z_1, \ldots, z_T$ are the not necessarily distinct roots of $n(\lambda)$. None of these roots equals α, by hypothesis. Set $w_s = \alpha + (z_s - \alpha)^{-1}$, for $s = 1, 2, \ldots, T$. Plainly $w_s \neq \alpha$, and we claim that the poles of $Q(\lambda)$ lie among the numbers $\alpha, w_1, \ldots, w_T$. To see this, write the numerator of $q(\lambda)$ in the base $\lambda - \alpha$, so that

$$q(\lambda) = \frac{\theta_0 + \theta_1(\lambda - \alpha) + \cdots + \theta_S(\lambda - \alpha)^S}{(\lambda - \alpha)^l(\lambda - z_1)\cdots(\lambda - z_T)},$$

where $\theta_0, \theta_1, \ldots, \theta_S$ are complex constants. Evaluating at $\alpha + (\lambda - \alpha)^{-1}$, and noticing that $\lambda - z_s$ becomes

$$\alpha + (\lambda - \alpha)^{-1} - z_s = (\lambda - \alpha)^{-1} - (w_s - \alpha)^{-1}$$
$$= (\lambda - w_s)/\{(\lambda - \alpha)(\alpha - w_s)\},$$

we get

$$Q(\lambda) = (\lambda - \alpha)^{T-S+l} \frac{\theta_0(\lambda - \alpha)^S + \cdots + \theta_S}{(\lambda - w_1)\cdots(\lambda - w_T) \times \text{nonzero constant}}.$$

This formula clearly shows that only α, $w_1, \ldots, w_T$ can be poles (roots of the denominator) of $Q(\lambda)$. (We do not claim that all these are actually poles since we have not shown that this representation of $Q(\lambda)$ is in lowest terms.)

We therefore know that the circle D mentioned above is the smallest circle centered at α in the complex plane enclosing certain of $w_1, \ldots, w_T$. If w is a complex number satisfying

$$|w - \alpha| > |w_s - \alpha|, \qquad \text{for } s = 1, 2, \ldots, T,$$

then w certainly lies outside D, so that (28) holds for all $k \geq K$.

Let $w = \alpha + (z - \alpha)^{-1}$. Then

$$\frac{w - \alpha}{w_s - \alpha} = \frac{z_s - \alpha}{z - \alpha}. \tag{29}$$

Because z lies inside C, we have $|z - \alpha| < |z_s - \alpha|$, implying by (29) that $|w - \alpha| > |w_s - \alpha|$; $s = 1, \ldots, T$. Thus w lies outside D and therefore (28) holds for this w.

Evaluating both sides of (27) at z, and using $z - \alpha = (w - \alpha)^{-1}$, $\alpha + (z - \alpha)^{-1} = w$, we get

$$q(z) - \{d_{-l}(z - \alpha)^{-l} + \cdots + d_k(z - \alpha)^k\} = \frac{Q_{-k}(w)}{(w - \alpha)^k}.$$

From (28), we infer that

$$|q(z) - \{d_{-l}(z - \alpha)^{-l} + \cdots + d_k(z - \alpha)^k\}| < \epsilon,$$

for all $k \geq K$. Convergence of $d_{-l}(z - \alpha)^{-l} + \cdots + d_k(z - \alpha)^k + \cdots$ to $q(z)$ is therefore established. □

Let C' be the largest circle centered at α enclosing no pole of $q(\lambda)$ other than α in its interior, and let C'' be the smallest circle centered at α enclosing all poles of $q(\lambda)$. Thus C'' encloses C'. For any complex number z outside C'', $q(z)$ may be written as a Laurent series in decreasing powers of $z - \alpha$, by §§6.5, 8.3; and for any z strictly inside C' (other than α), $q(z)$ may be written as a Laurent series in increasing powers of $z - \alpha$, by Theorem 8.4.7. What about the complex numbers in the "gap" between C' and C''? For many of these complex numbers z, we may obtain Laurent series that converge to $q(z)$, but which involving both increasing and decreasing powers of $z - \alpha$. This is proved rather easily, by separating $q(\lambda)$ into a sum of two terms (by using partial fractions), such that the Laurent series of §6.5 applies to one of these terms and the Laurent series of Theorem 8.4.6 to the other. (See Figure 8.4.3 and Exercise 16.)

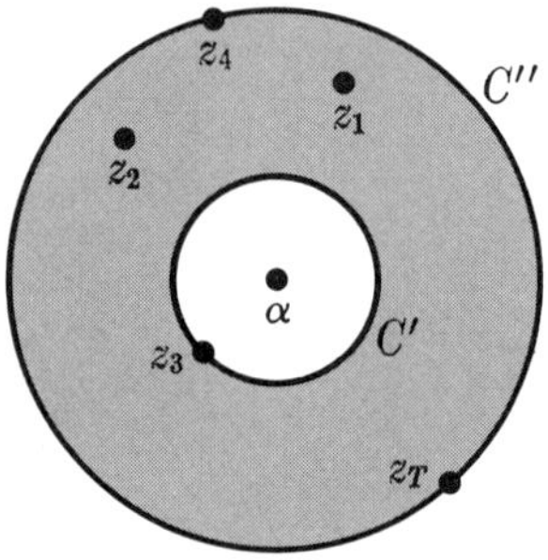

Figure 8.4.3

EXERCISES 8.4

I

1. Find the 2-adic, 3-adic and 6-adic expansions of -1. In each case square the result and check that the correct representations of 1 are obtained.

2. (i) Find the rational number q if $q = \overline{4}231.13$ in 10-adic numbers.
   (ii) Repeat, if $q = \overline{4}231.13$ in 6-adic numbers.

3. Find the (base ten) complements of 2 and 3 (i.e. find -2 and -3 in 10-adic numbers). Add, subtract and multiply these complements, and check that the correct 10-adic representations of 5, -1, 6 are obtained.

4. The method described in the proofs of Lemmas 8.4.1, 8.4.2 is the general procedure for writing a fraction M/N in b-adic form. When N is relatively prime to b, the calculation of this b-adic form may be carried out using a long division procedure which works from right to left.

   Example: Divide $M = 441$ by $N = 21$, in 10-adic numbers.
   Solution:

	441(21
1	21
	420
20	420
	000

The procedure is to use N and multiples of it to eliminate the digits of M, beginning at the right. Here we subtracted $1 \times N$ from M to eliminate the lowest digit in M, then subtracted $20 \times N$ to eliminate the lowest remaining digit. The division has then terminated, with remainder 0 and quotient $1 + 20 = 21$. Often the division does not terminate, and it may be continued until the remainder is sufficiently small (b-adically!) or until periodicity becomes evident.

(i) Use long division, as above, to calculate the 10-adic form of 1599/13.

(ii) Use long division, as above, to obtain the 10-adic form of 674/9, continuing until periodicty is evident.

(iii) Repeat (ii) for 679/11.

5. Divide $\alpha = \bar{9}\,1$ by $\beta = \bar{9}\,7$, these being 10-adic numbers.

6. Find the Laurent series of $1/\lambda$ in increasing powers of $\lambda - 1$. For which complex numbers z does evaluating at z yields a convergent series of complex numbers?

II

7. The twenty-digit number $\alpha = 99{,}999{,}999{,}999{,}999{,}999{,}991$ is a close 10-adic approximation to -9. The ten-digit number $\beta = 9{,}999{,}999{,}997$ is a close 10-adic approximation to -3. The quotient α/β should therefore be a close 10-adic approximation to 3. Show that this is so by finding the quotient exactly.

8. Let p be a prime, and set $b = p^2$. Find the b-adic digits of p.

9. In which base b is $\bar{1}$ the b-adic representation of $-\frac{1}{3}$.

10. If in b-adic numbers $a = \ldots d_k d_{k-1} \ldots d_1 d_0.$, show that

$$\frac{1}{b}\, a = \ldots d_k d_{k-1} \ldots d_1 . d_0 \,.$$

11. An element q of $\mathcal{K}$ is called a b-adic integer if it has zero digits only to the right of the b-adic point. Prove that q is a b-adic integer if and only if

$$|q|_{p1} \leq 1,\ |q|_{p2} \leq 1,\ \ldots,\ |q|_{pr} \leq 1,$$

and that the b-adic integers form a subring of $\mathcal{K}$ containing $\mathcal{R}$. (Compare Exercise 16 of §8.2.)

12. Let $b = p_1^{e_1} \ldots p_r^{e_r}$ be an element of $\mathcal{R}$, as in the text. Define the *b-adic norm* $\|\ \|_b$, also called the *b-adic pseudo valuation,* as follows: If $\alpha \in \mathcal{K}$, then

$$\|\alpha\|_b = \max\{|\alpha|_{p1},\ \ldots,\ |\alpha|_{pr}\},$$

"max" denoting the largest of the real numbers enclosed in braces.

(i) Prove that $\|\ \|_b$ satisfies these four axioms:

V1: $||\alpha||_b$ is a nonnegative real number, for all $\alpha \in \mathcal{K}$;
V2: $||\alpha||_b = 0$ if and only if $\alpha = 0$;
V3′: $||\alpha\beta||_b \leq ||\alpha||_b||\beta||_b$, for all $\alpha, \beta \in \mathcal{K}$;
V4′: $||\alpha + \beta||_b \leq \max\{||\alpha||_b, ||\beta||_b\}$, for all $\alpha, \beta \in \mathcal{K}$.

Outline: If the largest of $|\alpha\beta|_{p_1}, \ldots, |\alpha\beta|_{p_r}$ is $|\alpha\beta|_{p_i}$, then $||\alpha\beta||_b = |\alpha\beta|_{p_i} = |\alpha|_{p_i}|\beta|_{p_i} \leq ||\alpha||_b||\beta||_b$. This establishes V3′ and a similar calculation establishes V4′.

(ii) From V4′, deduce the triangle inequality

V4: $||\alpha + \beta||_b \leq ||\alpha||_b + ||\beta||_b$.

Thus $|| \; ||_b$ satisfies all axioms of a valuation, except that V3 is replaced by the weaker property V3′. For this reason, $|| \; ||_b$ is called a *pseudo valuation.*

(iii) Verify, by an example, that $||\alpha\beta||_b = ||\alpha||_b||\beta||_b$ is not generally true.

(iv) Verify that β is a b-adic approximation of α within tolerance ϵ if and only if $||\alpha - \beta||_b < \epsilon$.

(v) Rework the discussion of b-adic approximation given in the text, replacing all statements like "$|E|_p < 1$ for all prime divisors p of b" with statements of the form "$||E||_b < 1$."

13. Let $q(\lambda) = m(\lambda)/n(\lambda)$ be a rational function in $\mathbf{C}(\lambda)$, α a complex constant, and $n(\alpha) \neq 0$. Then formula (4) becomes

$$q(\lambda) = d_0 + d_1(\lambda - \alpha) + \cdots + d_k(\lambda - \alpha)^k + (\lambda - \alpha)^{k+1}\frac{m_{k+1}(\lambda)}{n(\lambda)}.$$

Prove: $\quad d_t = D^t q(\lambda)|_\alpha, \qquad t = 0, 1, \ldots, k,$

That is, d_t is the tth derivative of $q(\lambda)$ evaluated at α.
Hint: differentiate each side t times, then evaluate at α. [Thus the Laurent series in ascending powers of $\lambda - \alpha$ is the usual Taylor series taught in calculus, if α is not a pole of $q(\lambda)$.]

III

14. Let $q = \ldots d_k \ldots d_1d_0.$ be in b-adic form,where the lowest digit d_0 satisfies $(d_0, b) = 1$. We wish to find $x = \ldots x_k \ldots x_1x_0.$, a b-adic number with digits $x_0, x_1, \ldots,$ such that $qx = 1$. Show that the following procedure will work. Choose x_0 $(=0, 1, 2, \ldots,$ or $b - 1)$ such that $d_0x_0 \equiv 1 \pmod{b}$. Set $d_0x_0 = 1 + t_1b$. Then choose x_1 $(=0, 1, \ldots,$ or $b - 1)$ such that $d_0x_1 + d_1x_0 + t_1 \equiv 0 \pmod{b}$, say $d_0x_1 + d_1x_0 + t_1 = t_2b$. Then choose x_2 $(=0, 1, \ldots,$ or $b - 1)$ such that $d_0x_2 + d_1x_1 + d_2x_0 + t_2 \equiv 0 \pmod{b}$, say $d_0x_2 + d_1x_1 + d_2x_0 + t_2 = t_3b$.

Continue in this way. Show that at each stage it is possible to choose x_j as claimed, and that the product will, as j increases, become an arbitrarily close b-adic approximation to 1. In other words: $x = q^{-1}$.

15. Show how to modify the above construction for finding q^{-1} in b-adic numbers if

$$q = \dots d_k \dots d_0.d_{-1} \dots d_{-l} \neq 0,$$

with $(d_{-l}, b) = 1$. Observe the similarity with the calculation of the inverse power series in Exercises 24, 25 of §7.5.

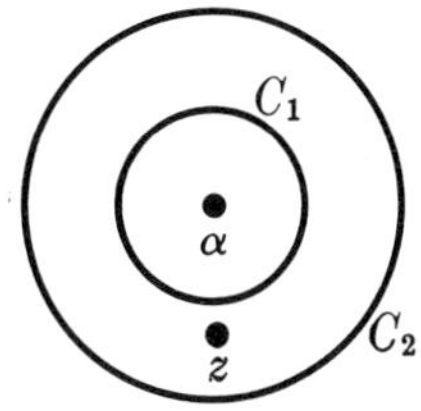

16. Let $q(\lambda)$ be a rational function with complex coeficients and α a complex number. Let C_1 and C_2 be circles centered at α such that no pole of $q(\lambda)$ lies in the annulus between C_1 and C_2. That is, a pole of $q(\lambda)$ is on C_1, inside C_1, on C_2, or outside C_2. Let $z_1, \dots, z_S$ be the poles of $q(\lambda)$ on or inside C_1, and $z_{S+1}, \dots, z_T$ the poles of $q(\lambda)$ on or outside C_2. (α may be one of $z_1, \dots, z_S$.) Let z be a complex number in the annulus strictly between C_1 and C_2.

(a) Apply the partial fraction decomposition to $q(\lambda)$ and show that it may be written as $q(\lambda) = q_1(\lambda) + q_2(\lambda)$ where $q_1(\lambda)$, $q_2(\lambda)$ are rational functions such that

$$z_1, \dots, z_S \text{ (only) are poles of } q_1(\lambda);$$

$$z_{S+1}, \dots, z_T \text{ (only) are poles of } q_2(\lambda).$$

Furthermore, $q_1(\lambda)$ is a proper fraction.

(b) Apply the theory of §6.5 to show that $q_1(z)$ may be expressed as a convergent infinite series of complex numbers:

$$q_1(z) = d_{-1}(z-\alpha)^{-1} + d_{-2}(z-\alpha)^{-2} + \cdots$$

where $d_{-1}, d_{-2}, \dots$ are constants, independent of z.

(c) Apply the theory of §8.4 to show that $q_2(z)$ may be expressed as a convergent infinite series of complex numbers

$$q_2(z) = d_0 + d_1(z-\alpha) + d_2(z-\alpha)^2 + \cdots$$

where $d_0, d_1, d_2, \dots$ are also constants, independent of z.

(d) Conclude that, for any z in the annulus between C_1 and C_2 (but on neither C_1 nor C_2), we have

$$q(z) = \sum_{j=-\infty}^{+\infty} d_j(z-\alpha)^j$$

Remark: This is the general type of Laurent series mentioned in the text. It is often encountered in complex variable theory, not only for rational functions, but also for more general types of functions. This expression for $q(z)$ is to be understood as the sum

$$\sum_{j=-\infty}^{-1} d_j(z-\alpha)^j + \sum_{j=0}^{+\infty} d_j(z-\alpha)^j$$

of two infinite series, each of which is convergent, the first to $q_1(z)$, the second to $q_2(z)$.

This type of Laurent series arises only in the function case and does not have a counterpart in the theory of rational numbers.

(e) Calculate the general Laurent series for $q(\lambda) = (\lambda - 1)^{-1} + (\lambda - 2)^{-1}$, if

(i) $z = 1/2$; (ii) $z = 3/2$; (iii) $z = 5/2$.

17. Let $m_0, n \in \mathbf{Z}$, with n positive and relatively prime to b. The purpose of this exercise is to prove that the length of the period of the b-adic representation of $q_0 = m_0/n$ is at most n. As in the text, we have $m_k = m_{k+1}b + \delta_k n$, for $k = 0, 1, 2, \ldots$

(i) Writing this as $m_{k+1} = m_k/b - \delta_k n/b$, and using $0 \le \delta_k < b$, prove by induction on k that $m_k \le m_0/b^k$, for $k = 0, 1, 2, \ldots$.

(ii) Similarly, prove by induction on k that $m_k \ge (m_0 + 1)/b^k - n$, for $k = 0, 1, 2, \ldots$.

(iii) If $m_0 \ge 0$, use these estimates and the fact that m_k is integral to show that $-n + 1 \le m_k \le 0$ holds for all sufficiently large k.

(iv) If $m_0 < 0$, show in a similar way that $-n \le m_k \le -1$ holds for all sufficiently large k.

(v) In either case, conclude that all m_k from some suffix onwards lie in an interval containing only n integers, and hence that $m_r = m_s$ for a pair of subscripts r, s with $0 < s - r \le n$. Then deduce, as in the text, that the b-adic representation of m_0/n is ultimately periodic with period $\le s - r \le n$.

8.5 CONTINUED FRACTIONS. THE NUMBER π

In this section, which may be read independently of other sections of this chapter, we shall work exclusively with integers, rational numbers, and real numbers.

Let $\alpha = a/b$ be a fraction, where a, b are integers with b positive. Suppose α is not an integer, so that $b \nmid a$. When computing the greatest common divisor of a and b using least nonnegative remainders in Euclid's algorithm (§3.4), we obtain the following system of equations in which $a = r_0$, $b = r_1$ and in which $q_1, \ldots, q_m, r_0, r_1, \ldots, r_m$ are integers, all positive except possibly q_1 and r_0 :

$$\begin{aligned}
r_0 &= q_1 r_1 + r_2\,, && 0 < r_2 < r_1\,,\\
r_1 &= q_2 r_2 + r_3\,, && 0 < r_3 < r_2\,,\\
&\cdots\cdots\cdots\cdots && \cdots\cdots\cdots\\
r_j &= q_{j+1} r_{j+1} + r_{j+2}\,, && 0 < r_{j+2} < r_{j+1}\,,\\
&\cdots\cdots\cdots\cdots && \cdots\cdots\cdots\\
r_{m-2} &= q_{m-1} r_{m-1} + r_m\,, && 0 < r_m < r_{m-1}\,,\\
r_{m-1} &= q_m r_m\,.
\end{aligned}$$

We rewrite these equations in the following way:

$$\begin{aligned}
\frac{r_0}{r_1} &= q_1 + \frac{r_2}{r_1} = q_1 + \frac{1}{r_1/r_2}\,, && 0 < \frac{r_2}{r_1} < 1,\\
\frac{r_1}{r_2} &= q_2 + \frac{r_3}{r_2} = q_2 + \frac{1}{r_2/r_3}\,, && 0 < \frac{r_3}{r_2} < 1,\\
&\cdots\cdots\cdots\cdots && \cdots\cdots\cdots\\
\frac{r_j}{r_{j+1}} &= q_{j+1} + \frac{r_{j+2}}{r_{j+1}} = q_{j+1} + \frac{1}{r_{j+1}/r_{j+2}}\,, && 0 < \frac{r_{j+2}}{r_{j+1}} < 1,\\
&\cdots\cdots\cdots\cdots && \cdots\cdots\cdots\\
\frac{r_{m-2}}{r_{m-1}} &= q_{m-1} + \frac{r_m}{r_{m-1}} = q_{m-1} + \frac{1}{r_m/r_{m-1}}\,, && 0 < \frac{r_m}{r_{m-1}} < 1,\\
\frac{r_{m-1}}{r_m} &= q_m\,.
\end{aligned}$$

Setting $\alpha_j = r_j/r_{j+1}$, for $j = 0, 1, \ldots, m-1$, we then have

$$\begin{aligned}
\alpha = \alpha_0 &= q_1 + 1/\alpha_1\,, && 0 < 1/\alpha_1 < 1,\\
\alpha_1 &= q_2 + 1/\alpha_2\,, && 0 < 1/\alpha_2 < 1,\\
&\cdots\cdots\cdots\cdots && \cdots\cdots\cdots\\
\alpha_j &= q_{j+1} + 1/\alpha_{j+1}\,, && 0 < 1/\alpha_{j+1} < 1,\\
&\cdots\cdots\cdots\cdots && \cdots\cdots\cdots
\end{aligned} \tag{1}$$

the process eventually reaching

$$\alpha_{m-2} = q_{m-1} + 1/\alpha_{m-1}\,, \qquad 0 < 1/\alpha_{m-1} < 1, \tag{1'}$$

and terminating with

$$\alpha_{m-1} = q_m\,. \tag{2}$$

Using equations (1) in succession, we obtain

$$\alpha = q_1 + \frac{1}{\alpha_1} = q_1 + \cfrac{1}{q_2 + \cfrac{1}{\alpha_2}} = q_1 + \cfrac{1}{q_2 + \cfrac{1}{q_3 + \cfrac{1}{\alpha_3}}}$$

$$= q_1 + \cfrac{1}{q_2 + \cfrac{1}{q_3 + \cfrac{1}{q_4 + \cfrac{1}{\alpha_4}}}} = \cdots, \text{ etc.}$$

Eventually we reach equation (1′), and then (2), and so obtain

$$\alpha = q_1 + \cfrac{1}{q_2 + \cfrac{1}{q_3 + \cdots + \cfrac{1}{q_{m-1} + \cfrac{1}{q_m}}}}$$

We call the right-hand side of this equation a *finite continued fraction.* In it we see the quotients q_j in Euclid's algorithm for computing the greatest common divisor of the numerator and denominator of α. Notice that (1) and (1′) show that q_{j+1} *is the largest integer not exceeding* α_j ; $j = 0, 1, 2, \ldots, m - 1$.

Now let α be a real number, but *not* a rational number. The recursive calculations shown in (1) may be applied to this number, as follows. We start by setting $\alpha_0 = \alpha$. If we have obtained an irrational number α_j, we let q_{j+1} be the largest integer not exceeding α_j, so that $0 < \alpha_j - q_{j+1} < 1$, then set $\alpha_{j+1} = 1/(\alpha_j - q_{j+1})$. Thus $\alpha_{j+1} > 1$, $\alpha_j = q_{j+1} + 1/\alpha_{j+1}$, and α_{j+1} is irrational (if it were rational, α_j would be also). Consequently, we obtain equations (1), with $j = 0, 1, 2, \ldots$, and from these equations we deduce exactly as before that

$$\alpha = q_1 + \cfrac{1}{q_2 + \cfrac{1}{q_3 + \cdots + \cfrac{1}{q_{n-1} + \cfrac{1}{q_n + \cfrac{1}{\alpha_n}}}}} \tag{3}$$

for any $n \geq 0$. Since $\alpha_{j+1} > 1$, the largest integer q_{j+2} not exceeding α_{j+1} must satisfy $q_{j+2} \geq 1$. This holds for any $j \geq 0$, so that q_1, q_2, q_3, q_4, ... are integers, all positive except possibly q_1, and α_0, α_1, α_2, α_3, ... are irrational numbers, all except possibly α_0 strictly greater than 1.

We wish to study the properties, as n increases, of the finite continued fraction

$$C_n = q_1 + \cfrac{1}{q_2 + \cfrac{\ddots}{\ddots + \cfrac{1}{q_{n-1} + \cfrac{1}{q_n}}}}$$

obtained from (3) by omitting the term $1/\alpha_n$. To save space, if x_1, x_2, ..., x_k are real numbers, all positive except perhaps x_1, we let

$$\langle x_1, x_2, \ldots, x_k \rangle = x_1 + \cfrac{1}{x_2 + \cfrac{1}{x_3 + \cfrac{\ddots}{\ddots + \cfrac{1}{x_{k-1} + \cfrac{1}{x_k}}}}} \tag{4}$$

Equation (3) may then be written as $\alpha = \langle q_1, q_2, \ldots, q_n, \alpha_n \rangle$, and this holds for any $n \geq 0$. (Note that angular parentheses $\langle\ \rangle$ were previously used in §3.4 for the least common multiple. That use of these parentheses has no relation to the present use.)

Observe that

$$\langle x_1, x_2, \ldots, x_{k-2}, x_{k-1} + 1/x_k \rangle = \langle x_1, x_2, \ldots, x_{k-1}, x_k \rangle, \tag{5}$$

as is evident from an inspection of (4).

In order to study the finite continued fraction C_n and its relationship to the irrational number α, we introduce integers N_0, N_1, N_2, N_3, ... and D_0, D_1, D_2, D_3, ... by defining them recursively according to these rules (compare the last part of the greatest common divisor discussion in §3.4):

$$\left.\begin{aligned} N_0 &= 0, \quad N_1 = 1, \quad \text{then} \quad N_{j+2} = q_{j+1}N_{j+1} + N_j, \\ D_0 &= 1, \quad D_1 = 0, \quad \text{then} \quad D_{j+2} = q_{j+1}D_{j+1} + D_j, \end{aligned}\right\} \quad \text{for all } j \geq 0. \tag{6}$$

Here q_1, q_2, q_3, ... are the integers appearing in (1) and (3), so that q_{j+1} is the largest integer not exceeding α_j; $j = 0, 1, 2, \ldots$. Notice that

$D_2 = 1$, $D_3 = q_2 \geq 1$, and then (6) and an easy induction on j show that $D_2, D_3, D_4, \ldots$ are all positive. Consequently, $D_{j+2} > q_{j+1}D_{j+1} \geq D_{j+1}$, if $j \geq 2$, so that

$$1 = D_2 \leq D_3 < D_4 < D_5 < \cdots. \tag{7}$$

The next lemma provides a basic formula in the theory of continued fractions.

LEMMA 8.5.1
For any positive real number x, and any integer $n \geq 0$, we have

$$\langle q_1, q_2, \ldots, q_n, x\rangle = \frac{xN_{n+1} + N_n}{xD_{n+1} + D_n}. \tag{8}$$

PROOF The proof is by induction on n. For $n = 0$, we interpret the left-hand side of (8) as $\langle x\rangle = x$, and by (6) this is also the value of the right-hand side. For $n = 1$, the left-hand side of (8) is $\langle q_1, x\rangle = q_1 + 1/x$, and by (6) this is again the value of the right-hand side. Suppose we have proved that

$$\langle q_1, q_2, \ldots, q_{j-1}, x\rangle = \frac{xN_j + N_{j-1}}{xD_j + D_{j-1}}.$$

for any positive x. Using (5), then this formula with $q_j + 1/x$ in place of x, and then (6), we get

$$\begin{aligned}\langle q_1, q_2, \ldots, q_{j-1}, q_j, x\rangle &= \left\langle q_1, q_2, \ldots, q_{j-1}, q_j + \frac{1}{x}\right\rangle \\ &= \frac{(q_j + x^{-1})N_j + N_{j-1}}{(q_j + x^{-1})D_j + D_{j-1}} \\ &= \frac{x(q_jN_j + N_{j-1}) + N_j}{x(q_jD_j + D_{j-1}) + D_j} \\ &= \frac{xN_{j+1} + N_j}{xD_{j+1} + D_j}.\end{aligned}$$

The inductive step being complete, formula (8) is established. □

Putting $x = q_{n+1}$ in (8), then using (6), we get $\langle q_1, \ldots, q_n, q_{n+1}\rangle = N_{n+2}/D_{n+2}$. Replacing n with $n - 1$, we then have

$$C_n = \langle q_1, \ldots, q_n\rangle = N_{n+1}/D_{n+1}, \quad \text{for all } n \geq 1.$$

Thus the integers N_{n+1} and D_{n+1} are numerator and denominator of the rational number C_n. Conceivably, N_{n+1}/D_{n+1} is a fraction not in lowest terms. We show that this fraction is in lowest terms by proving the identity

$$N_{n+1}D_n - D_{n+1}N_n = (-1)^n, \quad \text{for } n \geq 0. \tag{9}$$

This equation shows that N_{n+1}, D_{n+1} are relatively prime since any common divisor must divide $(-1)^n$. To verify (9), substitute $N_{n+1} = q_nN_n + N_{n-1}$, $D_{n+1} = q_nD_n + D_{n-1}$ into the left-hand side, thereby obtaining

$$N_{n+1}D_n - D_{n+1}N_n = -(N_nD_{n-1} - D_nN_{n-1}).$$

Repeated use of this formula, for steadily decreasing n, leads to

$$N_{n+1}D_n - D_{n+1}N_n = (-1)^n(N_1D_0 - D_1N_0) = (-1)^n,$$

thus verifying (9).

We now calculate an expression for the difference between α and $C_n = N_{n+1}/D_{n+1}$.

***LEMMA* 8.5.2** *For $n \geq 1$,*

$$\alpha - C_n = \alpha - \frac{N_{n+1}}{D_{n+1}} = \frac{(-1)^{n+1}}{D_{n+1}(\alpha_nD_{n+1} + D_n)}. \tag{10}$$

PROOF Using $\alpha = \langle q_1, \ldots, q_n, \alpha_n\rangle$, that is, (3), together with (8) for $x = \alpha_n$, we get

$$\alpha - \frac{N_{n+1}}{D_{n+1}} = \langle q_1, \ldots, q_n, \alpha_n\rangle - \frac{N_{n+1}}{D_{n+1}} = \frac{\alpha_nN_{n+1} + N_n}{\alpha_nD_{n+1} + D_n} - \frac{N_{n+1}}{D_{n+1}}$$

$$= \frac{N_nD_{n+1} - D_nN_{n+1}}{D_{n+1}(\alpha_nD_{n+1} + D_n)}$$

$$= \frac{(-1)^{n+1}}{D_{n+1}(\alpha_nD_{n+1} + D_n)}.$$

The last step uses (9). □

We use the formula just proved to show that, as n increases, the rational numbers C_n yield close approximations to α. Specifically, we have:

***THEOREM* 8.5.3**
Let α be an irrational number. Set $\alpha_0 = \alpha$ and recursively define integers $q_1, q_2, q_3, \ldots$ and irrational numbers $\alpha_1, \alpha_2, \ldots$ by (1), where for each j, q_{j+1} is the largest integer not exceeding α_j. Then recursively define integers $N_0, N_1, N_2, N_3, \ldots, D_0, D_1, D_2, D_3, \ldots$ by (6). If $C_n = \langle q_1, q_2, \ldots, q_n\rangle$, then

$$C_n = \frac{N_{n+1}}{D_{n+1}}, \qquad n = 1, 2, 3, \ldots,$$

and N_{n+1}/D_{n+1} is in lowest terms. Furthermore,

$$\lim_{n\to\infty} C_n = \lim_{n\to\infty} \frac{N_{n+1}}{D_{n+1}} = \alpha. \tag{11}$$

PROOF That N_{n+1}/D_{n+1} is in lowest terms was observed above. Because $\alpha_n > 1$ for $n \geq 1$, the denominator $D_{n+1}(\alpha_n D_{n+1} + D_n)$ in (10) is at least $D_{n+1}(D_{n+1} + D_n)$, and thus, by (7), can be made arbitrarily large for all sufficiently large n. If ϵ is a tolerance (a positive real number), an integer K dependent on ϵ will therefore exist such that

$$\left| \alpha - \frac{N_{n+1}}{D_{n+1}} \right| < \epsilon \qquad \text{for all } n \geq K.$$

This is precisely the meaning conveyed by $\lim_{n \to \infty} N_{n+1}/D_{n+1} = \alpha$. □

(The bars $|\ |$ are the usual absolute value of real numbers.)

Because of (11), we say that α equals the *infinite continued fraction*

$$q_1 + \cfrac{1}{q_2 + \cfrac{1}{q_3 + \cfrac{1}{q_4 + \cdots}}}$$

in which there is no last entry. The finite continued fraction $\langle q_1, q_2, \ldots, q_n \rangle = C_n$ obtained by truncating this infinite continued fraction at q_n is called the nth *convergent to* α, or sometimes just *a partial convergent to* α. Theorem 8.5.3 shows, therefore, that the nth convergent to α approximates α within a prescribed tolerance ϵ for all sufficiently large n.

The manner in which the fractions C_n approach α is of considerable interest and is described in the next theorem.

THEOREM 8.5.4
The following facts are valid:

$$C_1 < C_3 < C_5 < \cdots < \alpha < \cdots < C_5 < C_4 < C_2, \tag{12}$$

$$|\alpha - C_n| < \frac{1}{D_{n+1}D_{n+2}} \qquad \textit{for } n \geq 1, \tag{13}$$

$$|\alpha - C_{n+1}| < |\alpha - C_n| \qquad \textit{for } n = 1, 2, 3, \ldots. \tag{14}$$

Inequalities (12) assert that the C_j with *odd* subscripts j approach α from *below*, while those with *even* subscripts approach α from *above*. Inequality (13) shows that C_n is indeed a very good approximation to α since the error is bounded above by a fraction $1/D_{n+1}D_{n+2}$ with denominator $D_{n+1}D_{n+2}$ larger (and usually much larger) than the denominator D_{n+1} of C_n, and inequality (14) asserts that the C_n grow steadily nearer α as n increases.

PROOF Setting $n = 2j + 1$ in (10) shows that $\alpha - C_{2j+1} > 0$, and setting $n = 2j$ shows that $\alpha - C_{2j} < 0$. Thus every C_n with odd subscript n lies below α; and every C_n with even subscript, above α. If we prove (14), it will then be clear that (12) holds.

Before proving (14), we must first prove (13). Taking the absolute value of each side of (10), and using $\alpha_n > q_{n+1}$, we obtain

$$|\alpha - C_n| = \frac{1}{D_{n+1}(\alpha_n D_{n+1} + D_n)} < \frac{1}{D_{n+1}(q_{n+1}D_{n+1} + D_n)} = \frac{1}{D_{n+1}D_{n+2}},$$

where, in the last step, we used (6) with $j = n$. This establishes (13), and multiplying each side by D_{n+1} yields

$$|\alpha D_{n+1} - N_{n+1}| < \frac{1}{D_{n+2}}, \qquad \text{for } n \geq 1. \tag{15}$$

The next step in proving (14) is to show that

$$|\alpha D_{n+2} - N_{n+2}| < |\alpha D_{n+1} - N_{n+1}|, \qquad \text{for } n \geq 1. \tag{16}$$

First notice that $q_{n+1} + 1 > \alpha_n$ since q_{n+1} is the *largest* integer not exceeding α_n. For $n \geq 1$ we then have:

$$\begin{aligned} \alpha_n D_{n+1} + D_n &< (q_{n+1} + 1)D_{n+1} + D_n = (q_{n+1}D_{n+1} + D_n) + D_{n+1} \\ &= D_{n+2} + D_{n+1} \\ &\leq q_{n+2}D_{n+2} + D_{n+1} = D_{n+3}\,. \end{aligned}$$

[In order, we used here $\alpha_n < q_{n+1} + 1$, (6) with $j = n$, $1 \leq q_{n+2}$, and (6) with $j = n + 1$.] Thus $\alpha_n D_{n+1} + D_n < D_{n+3}$ if $n \geq 1$, so that using (10), we get

$$|\alpha - C_n| = \frac{1}{D_{n+1}(\alpha_n D_{n+1} + D_n)} > \frac{1}{D_{n+1}D_{n+3}} \qquad \text{for } n \geq 1.$$

Multiplying this inequality by D_{n+1} and using (15) with $n + 1$ in place of n, we obtain

$$|\alpha D_{n+1} - N_{n+1}| > \frac{1}{D_{n+3}} > |\alpha D_{n+2} - N_{n+2}|.$$

Therefore, (16) is established.

We complete the proof of (14) by deducing it from (16) and (7). For $n \geq 1$ we have

$$\begin{aligned} |\alpha - C_{n+1}| &= \left|\alpha - \frac{N_{n+2}}{D_{n+2}}\right| = \frac{1}{D_{n+2}}|\alpha D_{n+2} - N_{n+2}| < \frac{1}{D_{n+2}}|\alpha D_{n+1} - N_{n+1}| \\ &\leq \frac{1}{D_{n+1}}|\alpha D_{n+1} - N_{n+1}| = \left|\alpha - \frac{N_{n+1}}{D_{n+1}}\right| = |\alpha - C_n|. \qquad \square \end{aligned}$$

Application: Rational approximations to π

The circumference π of the circle of diameter 1 is an extremely important number in mathematics. Often the rational number 22/7 is used as a substitute for it. This is not strictly correct because π is irrational; see the first Niven book cited on page 455. The use of 22/7 is justified on the basis that it is a rational number approximately equal to π and having small integers in both numerator and denominator. We shall now use continued fraction theory to derive this and some other sharper approximations to π.

In calculus, it is shown that, rounded to the nearest digit in the tenth decimal place,

$$\pi = 3.14159\,26536. \qquad \text{This means that} \qquad \pi = 3.14159\,26536 + \epsilon_0,$$

where ϵ_0 is an error satisfying $|\epsilon_0| \leq 5 \times 10^{-11}$. We shall now apply the recursive procedure outlined in equations (1) to π. Our calculations necessarily will be approximate, since only a finite number of decimal places are being used, so that we must keep track of the size of the error. For this, the following formula will be useful: If a, b are real numbers, with $a > 0$ and $|b| < a$, then

$$\frac{1}{a+b} = \frac{1}{a} + E, \qquad \text{where } |E| \leq \frac{|b|}{a(a-|b|)}. \tag{17}$$

The proof is easy: Using the triangle inequality in the form $|a+b| \geq |a| - |b|$, we have

$$|E| = \left|\frac{1}{a+b} - \frac{1}{a}\right| = \left|\frac{-b}{(a+b)a}\right| \leq \frac{|b|}{(a-|b|)a}.$$

Let $\alpha_0 = \pi$. The largest integer not exceeding α_0 is $q_1 = 3$, so that

$$1/\alpha_1 = .14159\,26536 + \epsilon_0 = \delta_0 + \epsilon_0\,,$$

where $\delta_0 = .14159\,26536$. Then $\alpha_1 = 1/(\delta_0 + \epsilon_0)$. By direct calculation $1/\delta_0 = 7.06251\,3305 + \epsilon_1'$, the error ϵ_1' satisfying $|\epsilon_1'| < 5 \times 10^{-10}$. By (17), the difference $\epsilon_1'' = 1/(\delta_0 + \epsilon_0) - 1/\delta_0$ satisfies $|\epsilon_1''| \leq |\epsilon_0|/\delta_0(\delta_0 - |\epsilon_0|) \leq 5 \times 10^{-11}/(.14)^2 < 2.6 \times 10^{-9}$. Thus $\alpha_1 = 1/\delta_0 + \epsilon_1'' = 7.06251\,3305 + \epsilon_1$, with $\epsilon_1 = \epsilon_1' + \epsilon_1''$, and $|\epsilon_1| \leq |\epsilon_1'| + |\epsilon_1''| \leq 3.1 \times 10^{-9}$.

The largest integer not exceeding α_1 is $q_2 = 7$, so that

$$1/\alpha_2 = .06251\,3305 + \epsilon_1 = \delta_1 + \epsilon_1\,,$$

where $\delta_1 = .06251\,3305$. Then $\alpha_2 = 1/(\delta_1 + \epsilon_1)$. By direct calculation, $1/\delta_1 = 15.99659\,46 + \epsilon_2'$, the error ϵ_2' satisfying $|\epsilon_2'| < 5 \times 10^{-8}$. By (17), the difference $\epsilon_2'' = 1/(\delta_1 + \epsilon_1) - 1/\delta_1$ satisfies $|\epsilon_2''| \leq |\epsilon_1|/\delta_1(\delta_1 - |\epsilon_1|) \leq 3.1 \times 10^{-9}/(.06)^2 < 8.7 \times 10^{-7}$. Thus $\alpha_2 = 1/\delta_1 + \epsilon_2'' = 15.99659\,46 + \epsilon_2$, with $\epsilon_2 = \epsilon_2' + \epsilon_2''$, and $|\epsilon_2| \leq |\epsilon_2'| + |\epsilon_2''| \leq 9.2 \times 10^{-7}$.

The largest integer not exceeding α_2 is $q_3 = 15$, so that

$$1/\alpha_3 = .99659\,46 + \epsilon_2 = \delta_2 + \epsilon_2\,,$$

where $\delta_2 = .99659\,46$. Then $\alpha_3 = 1/(\delta_2 + \epsilon_2)$. By direct calculation, $1/\delta_2 = 1.00341\,70 + \epsilon_3'$, the error ϵ_3' satisfying $|\epsilon_3'| \leq 5 \times 10^{-8}$. By (17), the difference $\epsilon_3'' = 1/(\delta_2 + \epsilon_2) - 1/\delta_2$ satisfies $|\epsilon_3''| \leq |\epsilon_2|/\delta_2(\delta_2 - |\epsilon_2|) \leq 9.2 \times 10^{-7}/(.99)^2 < 9.4 \times 10^{-7}$. Thus $\alpha_3 = 1/\delta_3 + \epsilon_3'' = 1.00341\,70 + \epsilon_3$, with $\epsilon_3 = \epsilon_3' + \epsilon_3''$, and $|\epsilon_3| \leq |\epsilon_3'| + |\epsilon_3''| < 10^{-6}$.

The largest integer not exceeding α_3 is $q_4 = 1$, so that

$$1/\alpha_4 = .00341\,70 + \epsilon_3 = \delta_3 + \epsilon_3\,,$$

where $\delta_3 = .00341\,70$. Then $\alpha_4 = 1/(\delta_3 + \epsilon_3)$. By direct calculation, $1/\delta_3 = 292.654 + \epsilon_4'$, the error ϵ_4' satisfying $|\epsilon_4'| \leq 5 \times 10^{-4}$. By (17), the difference $\epsilon_4'' = 1/(\delta_3 + \epsilon_3) - 1/\delta_3$ satisfies $|\epsilon_4''| \leq |\epsilon_3|/\delta_3(\delta_3 - |\epsilon_3|) < 10^{-6}/(.0034)^2 < 8.7 \times 10^{-2}$. Thus $\alpha_4 = 1/\delta_3 + \epsilon_3'' = 292.654 + \epsilon_4$, with $\epsilon_4 = \epsilon_4' + \epsilon_4''$, and $|\epsilon_4| \leq |\epsilon_4'| + |\epsilon_4''| \leq .088$.

From the estimate of the size of ϵ_4, it is clear that α_4 lies between 292.566 and 292.742. Hence $q_5 = 292$, and $1/\alpha_5$ lies between .566 and .742. Thus α_5 lies between 1/.742 and 1/.566, hence between 1.34 and 1.77. Therefore, $q_6 = 1$, and $1/\alpha_6$ lies between .34 and .77. Consequently, because we can only say that α_6 lies between 1.29 and 2.95, we are only able to say that q_7 is 1 or 2.

We therefore know that $q_1 = 3$, $q_2 = 7$, $q_3 = 15$, $q_4 = 1$, $q_5 = 292$, $q_6 = 1$. From (6), we then get

$$N_2 = 3,\ N_3 = 22,\ N_4 = 333,\ N_5 = 355,\ N_6 = 103{,}993,\ N_7 = 104{,}348,$$
$$D_2 = 1,\ D_3 = 7,\ D_4 = 106,\ D_5 = 113,\ D_6 = 33{,}102,\ D_7 = 33{,}215,$$

and therefore, by Theorem 8.5.4,

$$3 < \frac{333}{106} < \frac{103993}{33102} < \pi < \frac{104348}{33215} < \frac{355}{113} < \frac{22}{7}\,.$$

We see here the familiar number 22/7. How close is it to π? According, to (12) and (13),

$$0 < \frac{22}{7} - \pi < \frac{1}{D_3D_4} = \frac{1}{742} < .0014.$$

Since $742 = 2 \times 7 \times 53$, this inequality may be written as

$$0 < \frac{22 \times 53}{7 \times 53} - \pi < \frac{1}{2}\,\frac{1}{7 \times 53}\,,$$

implying that among all fractions with denominator $7 \times 53 = 371$, the closest to π is the one with numerator $22 \times 53 = 1166$. In other words, 22/7 is a very good approximation for π.

Figure 8.5.1

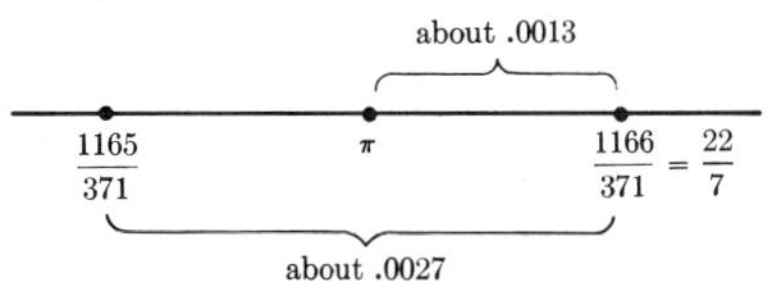

The fraction $C_4 = 355/113$ produces an exceptionally satisfactory approximation to π. By (13), it satisfies

$$0 < \frac{355}{113} - \pi < \frac{1}{D_5 D_6} < 3 \times 10^{-7}.$$

(The actual error is about 2.7×10^{-7}.) So 355/113, which still has a relatively small denominator, is an approximation to π accurate to within three units in the seventh decimal place.

We may summarize the theory this way: Continued fractions yield rational numbers with reasonably small denominators, which very accurately approximate an irrational number α.

EXERCISES 8.5

I

1. Compute the finite continued fraction expansions of 37/10, −3/10, 111/71, 22/7, 73/28.

2. Compute the continued fraction expansions for $\sqrt{2}$, $\sqrt{3}$, $\sqrt{5}$, $\sqrt{7}$. (*Remark:* For each of these numbers there will be integers $r < s$ with $\alpha_r = \alpha_s$. Using this you will be able to show that the q_j are ultimately periodic, and hence be able to find the entire continued fraction expansion. Use your continued fractions to some rational numbers closely approximating $\sqrt{2}$, $\sqrt{3}$, $\sqrt{5}$.)

3. Let $\alpha = \langle 1, 6, 1, 6, 1, 6, \ldots \rangle$ be an infinite continued fraction in which the entries are alternately 1 and 6. Evaluate α. (Hint: $\alpha = \langle 1, 6, \alpha \rangle$. Solve this equation for α. Then reverse your step and, using your value of α, compute as in the text the continued fraction for α.)

4. Find a rational number which approximates $\sqrt{5}$ to an accuracy of about 10^{-6}. Make the denominators of your rational numbers as small as you can.

5. Why is it theoretically impossible to calculate q_5 knowing only an eight decimal place approximation to π?

II

6. Rounded at the tenth decimal place, $e = 2.7182818284$. Find the first few entries in the continued fraction expansion of e. As you make your calculations, simultaneously perform an error analysis establishing the accuracy of your results. Find rational numbers closely approximating e. Estimate the accuracy of your approximations.

7. If $q_n > 1$, show that $\langle q_1, q_2, \ldots, q_{n-1}, q_n\rangle = \langle q_1, q_2, \ldots, q_{n-1}, q_n - 1, 1\rangle$. (The entries are integers, all positive except perhaps q_1.) Hence deduce that last entry q_n in a finite continued fraction may always be assumed to satisfy $q_n > 1$ (by eliminating the last entry if it is 1, and taking an altered next-to-last entry as the new last entry).

8. If $\langle q_1, q_2, \ldots, q_m\rangle = \langle q_1', q_2', \ldots, q_n'\rangle$, where $q_m > 1$ and $q_n' > 1$, show that $m = n$ and $q_j = q_j'$ for $j = 1, 2, \ldots, m$. (The entries are integers, all positive except perhaps q_1, q_1'.) [Hint: Show that the largest integer not exceeding $\langle q_1, q_2, \ldots, q_m\rangle$ is q_1.]

9. Let $\xi = \langle q_1, q_2, \ldots, q_n, \alpha_n\rangle = \langle q_1', q_2', \ldots, q_n', \beta_n\rangle$ where α_n, β_n are real numbers with $\alpha_n > 1$, $\beta_n > 1$, and where the q_j, q_j' are integers, all positive except perhaps for q_1, q_1'. Show that $q_j = q_j'$ for all j. Then deduce the uniqueness of the infinite continued fraction expansion of an irrational number ξ.

10. Let α be an irrational number. Show that there are infinitely many rational numbers N/D in lowest terms such that $|\alpha - N/D| < 1/D^2$.

11. Let α be an irrational number, and suppose N, D are relatively prime integers, with D positive. Prove, according to the following outline, that if

$$\left|\alpha - \frac{N}{D}\right| < \left|\alpha - \frac{N_n}{D_n}\right|$$

then $D > D_n$. (*Remark:* This shows that among *all* fractions with denominator $\leq D_n$, the continued fraction is the one which best approximates α.)

(i) Use the fact that N_n/D_n and N_{n-1}/D_{n-1} lie on opposite sides of α, and that N_n/D_n is closer to α than N_{n-1}/D_{n-1}, to show that N/D lies between N_{n-1}/D_{n-1} and N_n/D_n, hence that

$$\left|\frac{N_n}{D_n} - \frac{N_{n-1}}{D_{n-1}}\right| > \left|\frac{N}{D} - \frac{N_{n-1}}{D_{n-1}}\right|.$$

(Draw a picture.)

(ii) Use $N_nD_{n-1} - D_nN_{n-1} = \pm 1$ to obtain

$$\frac{1}{D_nD_{n-1}} > \left|\frac{ND_{n-1} - DN_{n-1}}{DD_{n-1}}\right|.$$

(iii) Deduce: $1/D_n > |ND_{n-1} - DN_{n-1}|/D$.

(iv) Deduce: $D > D_n\,|ND_{n-1} - DN_{n-1}|$.

(v) Show that $ND_{n-1} - DN_{n-1} \neq 0$.

(vi) Deduce from (iv) and (v) that $D > D_n$. Use the fact that $ND_{n-1} - DN_{n-1}$ is an integer.

III

12. If irrational numbers α and β have the same convergents $C_1, C_2, \ldots, C_n$, prove that their continued fraction expansions up to q_n are the same, i.e., if $\alpha = \langle q_1, q_2, q_3, \ldots \rangle$, $\beta = \langle q_1', q_2', q_3', \ldots \rangle$ then $q_t = q_t'$ for all $t \leq n$.

13. Let α, β, γ be irrational numbers with $\alpha < \beta < \gamma$. Suppose that α and γ have the same convergents $C_1, C_2, \ldots, C_n$. Prove that β also has these convergents.

8.6 THE OPERATIONAL CALCULUS

In this section, which is independent of other sections of Chapter 8, we shall study polynomials from a new point of view, by introducing a type of multiplication called *convolution*. This will be similar to but not precisely the same as the convolution method in §3.2 for calculating the coefficients of a product of polynomials.

The discussion of polynomials in §3.2, when restricted to polynomials with real coefficients, began with sequences

$$a = (a_0, a_1, a_2, \ldots), \qquad b = (b_0, b_1, b_2, \ldots),$$
$$s = (s_0, s_1, s_2, \ldots), \qquad p = (p_0, p_1, p_2, \ldots),$$

etc., in which the entries were real numbers, almost all zero. The sum $s = a + b$ and product $p = ab$ of sequences a, b were specified by

$$s_i = a_i + b_i, \; p_i = \sum_j a_j b_{i-j}, \qquad i = 0, 1, 2, \ldots, \tag{1}$$

the summation in the second part of (1) extending over all j for which both j and $i - j$ are nonnegative, i.e., $j = 0, 1, \ldots, i$. Eventually, in §3.2, we came to write sequences such as a in the usual form of a polynomial,

$$a(\lambda) = \sum_{i=0}^{\infty} a_i \lambda^i,$$

and then, in the terminology of Chapter 4, $a_0, a_1, a_2, \ldots$ are interpreted as the digits of $a(\lambda)$ in base λ. A polynomial $a(\lambda)$ therefore has a digit a_i for each nonnegative integer i, the digits are almost all zero, and their totality completely specifies the polynomial. Formulas (1) give the method for computing the digits of a sum or product, "convolution" being the name given to the method for the product.

Later, in Chapter 6, we formed the field of fractions of the polynomial ring $\mathbf{R}[\lambda]$, obtaining thereby the rational function field $\mathbf{R}(\lambda)$. In Chapter 7 (Example 7.4.4), we went in a different direction and examined the polynomial function $a(x)$ defined for real numbers x by $a(x) = \sum_i a_i x^i$, i.e.,

$$a(x) = \sum (i\text{th digit of } a) \times x^i, \tag{2}$$

the summation extending over all digits. This sum is finite since almost all digits are zero. It turned out that polynomial functions add and multiply in this way:

$$s(x) = a(x) + b(x), \qquad p(x) = a(x)b(x), \qquad \text{for all real } x. \tag{3}$$

Thus, by changing from polynomial digits to polynomial functions, the multiplication changed from convolution to the point wise product.

We now modify our approach to polynomials somewhat by regarding $a(x)$, for x an arbitrary nonnegative real number, as a new type of digit. We know, from Example 7.4.4, that the totality of values $a(x)$ for nonnegative real x specifies the polynomial completely. Thus we are now to have a digit $a(x)$ for each real $x \geq 0$, instead of a digit a_i for each integral $i \geq 0$. We wish to imitate the program above and use these new digits to introduce the addition and multiplication operations of a ring, the new type of convolution mentioned above being the method for calculating the digits of a product from those of the factors. We verify (in the next subsection) that the ring obtained in this way has no zero divisors, and then study its field of fractions. In the final subsection, we ask whether the convolution method of multiplication may be converted to a pointwise product.

Parallel to the first part of (1), we require that the digits of the sum $s = a + b$ are given by

$$s(x) = a(x) + b(x), \qquad \text{for all real nonnegative } x.$$

Thus digit addition is now to be the usual rule for function addition (pointwise addition). Parallel to the second part of (1), the rule for computing the digits of the product $p = ab$ should be that $p(x)$ is the sum of all products of the type

$$(\text{the } t\text{th digit of } a) \times (\text{the } (x - t)\text{th digit of } b),$$

the sum extending over all t for which both t and $x - t$ are nonnegative. However, we have a continuum of values of t, so that summation is not possible. Since integration is possible, we take the product definition to be

$$p(x) = \int_0^x a(t)b(x - t)\, dt, \qquad \text{for each real } x \geq 0.$$

This method of multiplication is the continuous analogue of the discrete type of convolution used earlier and is also called "convolution." We write this product as $p = a \circ b$.

The polynomial ring and its field of fractions

Motivated by the discussion above, we now study polynomial functions of a real nonnegative number x. Let $f(\lambda)$, $g(\lambda)$ be polynomials with real coefficients. Then $f(x)$, $g(x)$, for $x \geq 0$, are the associated polynomial functions.

Definition of equality: Polynomial functions f, g are equal, $f = g$, if and only if $f(x) = g(x)$ for all $x \geq 0$. By Example 7.4.4, this definition implies that $f = g$ if and only if the coefficients of like powers of x in $f(x)$, $g(x)$ are the same.

Addition: The sum of polynomial functions f, g is the function $f + g$ defined by

$$(f + g)(x) = f(x) + g(x), \qquad \text{for all } x \geq 0.$$

Clearly, $f + g$ is again a polynomial function.

Multiplication: The product of polynomial functions f, g is the function $f \circ g$ specified by

$$(f \circ g)(x) = \int_0^x f(t)g(x - t)\, dt, \qquad \text{for all real } x \geq 0.$$

Actually, this definition, as well as the definition of function addition, is applicable to more general types of functions, and yields functions still denoted by $f + g$ and $f \circ g$. To see that convolution of polynomial functions produces a polynomial function, observe that the individual powers of $x - t$ in $g(x - t)$ may be multiplied out and terms involving like powers of t collected, so that $g(x - t)$ takes the form

$$A_0(x) + A_1(x)t + A_2(x)t^2 + \cdots + A_n(x)t^n,$$

where $A_0(x)$, $A_1(x)$, ..., $A_n(x)$ are polynomials in x. Thus

$$\int_0^x f(t)g(x - t)\, dt$$
$$= A_0(x) \int_0^x f(t)\, dt + A_1(x) \int_0^x f(t)t\, dt + \cdots + A_n(x) \int_0^x f(t)t^n\, dt,$$

showing that $f \circ g$ is a polynomial function since each integral

$$\int_0^x f(t)t^j\, dt, \qquad j = 0, 1, \ldots, n,$$

is plainly a polynomial in x.

We must verify that these definitions of addition and multiplication do produce a commutative ring.

Closure under addition: Obvious, since the sum of polynomial functions is a polynomial function.

Associativity of addition: Since function addition is associative (§7.4), addition of polynomial functions is associative.

Existence of zero: The zero polynomial function, arising from the zero polynomial, plays the role of $\mathbb{O}$.

Existence of additive inverses: The additive inverse of the polynomial function f is the polynomial function $-f$ for which $(-f)(x) = -f(x)$ for each x.

Closure under multiplication: Evident, since $f \circ g$ is a polynomial function, as was checked above.

We verify the remaining axioms of a commutative ring in the following order: first the commutative law for multiplication, then the distributive law, finally, the associative law.

Commutative law for multiplication: Making the variable change $s = x - t$ in the integral

$$(f \circ g)(x) = \int_0^x f(t)g(x - t)\, dt$$

yields

$$(f \circ g)(x) = \int_0^x f(x - s)g(s)\, ds = (g \circ f)(x).$$

This holds for all $x \geq 0$, so that $f \circ g = g \circ f$.

Distributivity: Let f, g, h be polynomial functions. To verify $f \circ (g + h) = f \circ g + f \circ h$, observe that

$$\begin{aligned}(f \circ (g + h))(x) &= \int_0^x f(t)(g + h)(x - t)\, dt \\ &= \int_0^x f(t)\{g(x - t) + h(x - t)\}dt \\ &= \int_0^x f(t)g(x - t)dt + \int_0^x f(t)h(x - t)\, dt \\ &= (f \circ g)(x) + (f \circ h)(x) \\ &= (f \circ g + f \circ h)(x).\end{aligned}$$

Since this holds for all $x \geq 0$, we obtain $f \circ (g + h) = f \circ g + f \circ h$.

Associativity of multiplication: Two verifications will be given. The first is conceptually simple and is applicable to functions more general than polynomials, but involves ideas normally encountered in a second calculus course. The second is a straightforward computation, but is applicable only to polynomials. We describe the first proof here; the second is outlined in Exercises 10, 11, 12.

The proof about to be given requires a knowledge of *double integration.* Let f, g, h be polynomial functions.

First observe, for $0 \leq x \leq t$, that

$$(h \circ g)(t - x) = \int_0^{t-x} h(y)g(t - x - y)dy.$$

Hence $(f\circ(h\circ g))\ (t) = \int_0^t f(x)\left\{\int_0^{t-x} h(y)g(t-x-y)dy\right\} dx$

$$= \int_0^t \left\{\int_0^{t-x} f(x)h(y)g(t-x-y)dy\right\} dx.$$

Referring to Figure 8.6.1, for each fixed x we let the point (x, y) range along the line l, so that y goes from 0 to $t - x$, then let x range from 0 to t. Thus the point (x, y) covers the triangle T, so that

$$(f\circ(h\circ g))(t) = \iint_T f(x)h(y)g(t-x-y)dx\,dy$$

is the double integral of $f(x)h(y)g(t - x - y)$ over this triangle. Referring now to Figure 8.6.2, we can also evaluate this double integral by first fixing y and letting (x, y) run along L, so that x goes from 0 to $t - y$, afterwards letting y vary from 0 to t. Thus

$$(f\circ(h\circ g))(t) = \int_0^t h(y)\left\{\int_0^{t-y} f(x)g(t-x-y)dx\right\} dy$$

$$= \int_0^t h(y)\{(f\circ g)(t-y)\}dy$$

$$= (h\circ(f\circ g))(t).$$

Since this holds for every $t \geq 0$, we obtain $f\circ(h\circ g) = h\circ(f\circ g)$. From this, using the commutative law for multiplication proved above, we get

$$f\circ(g\circ h) = f\circ(h\circ g) = h\circ(f\circ g) = (f\circ g)\circ h.$$

Thus $f\circ(g\circ h) = (f\circ g)\circ h$, verifying associativity.

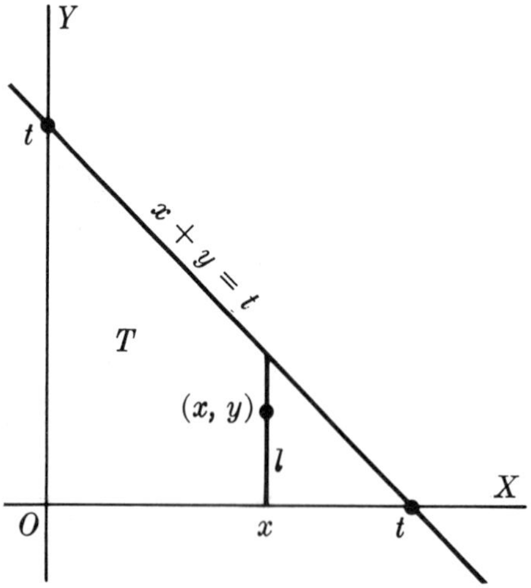

Figure 8.6.1

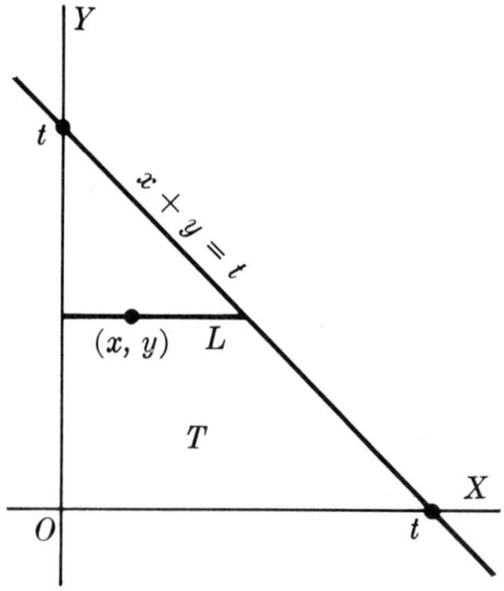

Figure 8.6.2

We now know that $\mathcal{R}$ is a commutative ring. We next show that it has no zero divisors. Assume that $f \circ g = 0$, so that $(f \circ g)(x) = 0$ for all $x \geq 0$, and hence that

$$\int_0^x f(t)g(x-t)\,dt = 0, \qquad \text{for all } x \geq 0. \tag{4}$$

If neither f nor g is the zero polynomial function, then neither is the zero polynomial, and hence only finitely many real numbers are roots of f or g. If r is the smallest positive root of f or g (or an arbitrary positive number if there is no positive root), then neither f nor g can become zero when evaluated at any t satisfying $0 < t < r$. Consequently, neither f nor g can undergo a sign change for these values of t (by Theorem 5.4.7). Take x to satisfy $0 < x < r$. Then, if t ranges from 0 to x, we have $0 \leq t \leq r$ and $0 \leq x - t < r$, so that $f(t)$, $g(x - t)$ have fixed nonzero signs (except, possibly, $f(0) = 0$ or $g(0) = 0$). In any case,

$$\int_0^x f(t)g(x-t)\,dt$$

is the integral of a quantity $f(t)g(x - t)$ of fixed nonzero sign except possibly at the endpoints $t = 0$ or x. This integral therefore cannot be zero, contradicting (4). Thus, if $f \circ g = 0$, at least one of f or g must be the zero polynomial function, implying that there are no zero divisors in the ring of polynomial functions.

We summarize these results in the following theorem.

THEOREM 8.6.1

Under pointwise addition and convolution, the set of polynomial functions with real coefficients forms a commutative ring without zero divisors.

Let us denote this ring by $\mathcal{R}$. Because $\mathcal{R}$ is commutative and has no zero divisors, it has a field of fractions, by Theorem 6.3.1, which we denote by $\mathcal{K}$. Beware! An element of $\mathcal{K}$, although written as a quotient f/g of polynomial functions f, g, is *not* a rational function in the sense of §6.4. This is because the multiplication here is different from that in §6.4. In fact, if f_1, f_2, g_1, g_2 are polynomial functions, with g_1, g_2 nonzero, the fractional equality $f_1/g_1 = f_2/g_2$ holds if and only if $f_1 \circ g_2 = g_1 \circ f_2$. The rules for fraction addition and multiplication are now

$$f_1/g_1 + f_2/g_2 = (f_1 \circ g_2 + g_1 \circ f_2)/g_1 \circ g_2,$$

$$f_1/g_1 \circ f_2/g_2 = (f_1 \circ f_2)/(g_1 \circ g_2).$$

Because $\mathcal{K}$ is a field, it has a multiplicative identity element, i.e., an identity for convolution. This element is usually denoted by δ and called the *delta function*. (Caution: The terminology is misleading because δ is not a function in the usual sense, as will be shown below.) When working with elements of $\mathcal{K}$, it is *absolutely necessary* to remember that δ, and not 1, is the multiplicative identity.

Let I denote the constant polynomial function 1. We use I for a reason to become clear in a moment. For any polynomial function f, we have

$$(I \circ f)(x) = (f \circ I)(x) = \int_0^x f(t)I(x-t)\,dt = \int_0^x f(t)\,dt.$$

Thus I acts on a polynomial by integrating it. For this reason, we call I an *integration operator*. Specifically, $I \circ f$ is that integral of f having zero constant term.

Let $F(x) = a_1x + a_2x^2 + \cdots$ be a polynomial with zero constant term, and take $f(x) = a_1 + 2a_2x + 3a_3x^2 + \cdots$ to be its derivative. Then $I \circ f = F$. Since $\mathcal{K}$ is a field, this equation may be solved for f to obtain $f = I^{-1} \circ F$. Thus the element I^{-1} of $\mathcal{K}$ acts as a *differentiation operator* when applied to any polynomial F with zero constant term. In other words: $(\text{integration})^{-1} = \text{differentiation}$. It is useful to denote I^{-1} by D. Then $D \circ I = \delta$ since δ is the identity of $\mathcal{K}$.

The delta function seems to have been invented by physicists, and is encountered in both physics and engineering. It satisfies $\delta \circ f = f$ for any polynomial function f. Could δ be a polynomial function? This would imply that

$$\int_0^x \delta(t)\,f(x-t)\,dt = f(x), \tag{5}$$

for every $x \geq 0$ and every polynomial function f. Taking $f(x) = 1$ gives

$$\int_0^x \delta(t)\,dt = 1 \tag{6}$$

for all x, and differentiation with respect to x then yields $\delta(x) = 0$ for all $x \geq 0$. This plainly contradicts (6). Thus δ cannot be a polynomial function, and the same argument shows that it cannot be any type of function with reasonable properties.

The δ function is often presented in physics and engineering as a "function" with the following (impossible) properties:

$$\delta(x) = 0 \text{ for all } x > 0,\ \delta(0) = \infty,\quad \int_0^x \delta(t)\,dt = 1 \text{ for all } x > 0. \tag{7}$$

Although functions cannot be found with properties (5) and (7), functions with approximately these properties are easily obtained, by considering a graph—see Figure 8.6.3—with a high narrow "spike" near the origin covering one unit of area. If we take $\delta(x)$ to be a function with the graph shown in Figure 8.6.3, equation (5) is then approximately correct, if the width ϵ of the spike is very small and $x \geq \epsilon$, since the integrand is nonzero only for $t \leq \epsilon$ and $f(x-t)$ is approximately constant over the range $0 \leq t \leq \epsilon$, so that we have the following:

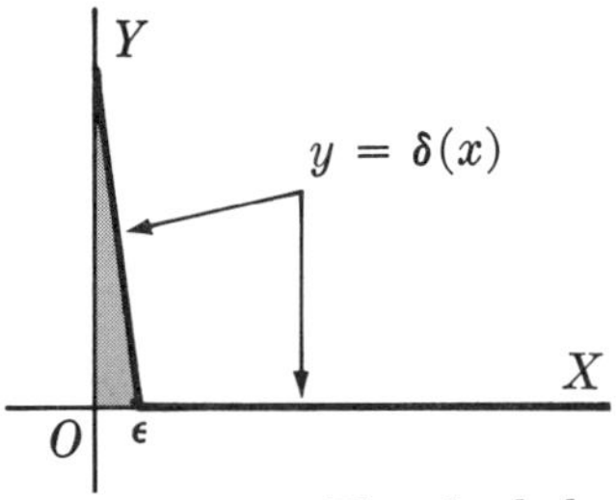

Figure 8.6.3 The shaded area under the graph equals 1

$$\int_0^x \delta(t)\, f(x-t)\, dt = \int_0^\epsilon \delta(t)\, f(x-t)\, dt$$
$$\approx f(x) \int_0^\epsilon \delta(t)\, dt = f(x).$$

(The symbol $\approx$ denotes approximate equality.)

In addition to the δ function, the field $\mathcal{K}$ contains a number of (genuine) functions that the reader may not expect. For example, the exponential and trigonometric functions e^x, $\cos x$, $\sin x$ are all elements of $\mathcal{K}$. In fact,

$$e^x = \frac{x}{1-x}, \qquad \cos x = \frac{x}{1+\frac{1}{2}x^2}, \qquad \sin x = \frac{\frac{1}{2}x^2}{1+\frac{1}{2}x^2}. \tag{8}$$

Of course, the right-hand sides of these equations are elements of $\mathcal{K}$ and not rational functions in the sense of §6.4. The proof of (8) is sketched in Exercise 21, and amounts to justifying multiplication by $(1-x)^{-1}$ and $(1+\frac{1}{2}x^2)^{-1}$ in the equations

$$(1-x)\circ e^x = x, \qquad (1+\tfrac{1}{2}x^2)\circ \cos x = x, \qquad (1+\tfrac{1}{2}x^2)\circ \sin x = \tfrac{1}{2}x^2. \tag{9}$$

The left-hand sides of (9) are convolution integrals of the indicated functions and these equations may be verified in a straightforward manner using integration by parts.

The discussion above is a brief introduction to *operational calculus.* This subject, originally developed by engineers, is extensively employed in parts of physics and electrical and control engineering. One of its attractive features is that it provides a precise set of rules for working with differentiation and integration much as one works with fractions, e.g., if a differentiation appears on one side of an equation, and you do not wish it there, divide through by it, whereupon it appears on the other side as an integration.

Note that precise and fully justifiable rules are needed if differentiation and integration are to be treated in this manner, since otherwise contradictions are easily obtained. For example, if c is a nonzero constant, and if we had $D\circ c = 0$, multiplying by D^{-1} would give the false result $c = 0$. This contradiction is prevented from arising in the operational calculus by the fact that $D\circ c$ is not zero. (Exercise 6.) In fact, $D\circ f(x)$ is not the usual derivative of $f(x)$ when the polynomial $f(x)$ has nonzero constant term. (See Exercise 17.)

The algebraic approach to the operational calculus sketched above is a quite recent development, discovered in the late 1940's. An introduction to the subject is given in the book of Berg listed on page 455. It is interesting to note that Berg's book, which is intended for use by engineers (among others), begins with a discussion of algebraic ideas such as rings, homomorphisms, ideals, residue classes, and the field of fractions, all of which have been encountered earlier in this book.

The associated pointwise product

As noted above, the digits (coefficients) of a product of polynomials are obtained from the digits of the factors by convolution, but the multiplication of the polynomial functions themselves is the pointwise product. Is there a similar shift, from convolution to a pointwise product, when we have a continuum of digits $a(x)$ [for each real $x \geq 0$] instead of a discrete set a_i [for each integer $i \geq 0$]? We now show that this is the case. Starting with digits (functional values) $a(x)$, for all real $x \geq 0$, we wish to create from these digits a function A, with the intent that multiplication for the functions A will be the pointwise product. Parallel to (2), we expect that A, on evaluation at y, should give the sum of all products of the type

$$(x\text{th digit of } a) \times y^x,$$

the sum extending over all $x \geq 0$. Because we have a continuum of values of x, summation is not possible. However, integration is possible, so we consider

$$\int a(x) y^x \, dx,$$

the integration extending over all digits, i.e., over all real $x \geq 0$. Therefore, we take

$$A(y) = \int_0^\infty a(x) y^x \, dx. \tag{10}$$

Owing to the infinite range of integration, there is no guarantee that this integral exists. To obtain an integral that does exist, we restrict y to the range $0 < y < 1$. It is customary to make the variable change $y = e^{-s}$, with $s > 0$. The integral (10) is then a function of s, namely, $\int_0^\infty a(x) e^{-sx} \, dx$, but also dependent on a. To indicate both constituents of this integral, we agree to denote it by $L(a)$, instead of A, with $L(a)$ a function of s given by

$$L(a)(s) = \int_0^\infty a(x) e^{-sx} \, dx, \qquad \text{for all } s > 0. \tag{11}$$

It turns out—see Exercises 8 and 9—that the integral

$$\int_0^\infty x^j e^{-sx} \, dx, \qquad s > 0; j = 0, 1, 2, \ldots,$$

exists and has value $j!/s^{j+1}$. Thus, if

$$a(x) = a_0 + a_1 x + \cdots + a_m x^m, \qquad a_0, \ldots, a_m \text{ real constants,}$$

then (11) becomes

$$L(a)(s) = \sum_{t=0}^{m} j!a_j/s^{j+1}, \qquad s > 0. \tag{12}$$

(If we had retained y as a variable, (12) would be more complicated, the powers of s being replaced with powers of the natural logarithm of y. The substitution $y = e^{-s}$ may therefore be justified on the grounds of the greater simplicity it affords.)

The function $L(a)$ just introduced occupies a prominent position in applied mathematics, and is called the *Laplace transform* of a. It can be studied for classes of functions $a(x)$ much wider than polynomials, (11) being the definition for general functions. We now give a discussion of its properties, confining our attention to polynomial functions.

If a, b are polynomial functions, then

$$\left.\begin{aligned} L(a + b)(s) &= L(a)(s) + L(b)(s), \\ L(a \circ b)(s) &= L(a)(s) \times L(b)(s), \end{aligned}\right\} \quad \text{for all } s > 0, \tag{13}$$

the symbol $\times$ denoting the pointwise product. The first of these formulas is proved in a straightforward way. The second, which shows that convolution is indeed converted by L to the pointwise product, has a more delicate proof. A calculational proof, valid for polynomials only, is given in Exercise 13 below. We describe here a more conceptual proof, which requires a knowledge of double integration, but which is applicable to a broader class of functions. From the definitions of $a \circ b$ and $L(a)(s)$, we get

$$L(a \circ b)(s) = \int_0^\infty e^{-sx} \left\{ \int_0^x a(t)b(x - t)\,dt \right\} dx.$$

The inside integration here is along the line l_1, and the outside integration shifts l_1 to the right from 0, so that the point (x, t) covers the full region T between the nonnegative X axis and the line $t = x$. Thus

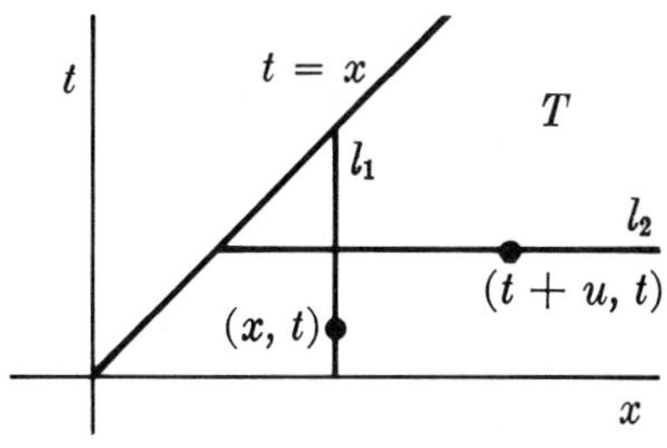

$$L(a \circ b)(s) = \iint_T e^{-sx} a(t)b(x - t)\,dx\,dt.$$

Figure 8.6.4

We may also cover this region by setting $x = t + u$, with u increasing from 0, so that (x, t) moves along the line l_2, then shifting l_2 by varying t from 0 to ∞. Making this change of variable (the Jacobian involved is 1), we get $$L(a \circ b)(s) = \int_0^\infty \left\{\int_0^\infty e^{-st}e^{-su}a(t)b(u)\,du\right\}dt$$

$$= \int_0^\infty e^{-st}a(t)\,dt \int_0^\infty e^{-su}b(u)\,du$$

$$= L(a)(s) \times L(b)(s).$$

A good many details must be supplied to justify the steps in this argument. Providing these details is beyond the scope of this book, and the reader may find them in the book *Modern Operational Mathematics in Engineering*, by R. V. Churchill, published by McGraw-Hill.

Rewriting (13) solely in terms of functions by omitting the variable s, we get

$$L(a + b) = L(a) + L(b), \quad L(a \circ b) = L(a)L(b), \quad \text{for } a, b \in \mathcal{R}. \tag{14}$$

These formulas show that L is a homomorphism on the ring $\mathcal{R}$. Furthermore, (12) shows that L maps a nonzero polynomial function (in variable x) to a nonzero polynomial function (in variable $1/s$) having no constant term. In particular, L must have zero kernel, so that:

The Laplace transform is an isomorphism from the ring $\mathcal{R}$ of polynomial functions in x (multiplication being convolution) to the ring of polynomial functions in $1/s$ without constant term (multiplication being the pointwise product).

Denote this latter ring by $\mathcal{S}$. Since the actual numerical value of s plays no role in the theory, we may regard $\mathcal{S}$ as the subring of $\mathbf{R}[1/s]$ consisting of polynomials in the symbol $1/s$ without constant term. (More precisely, $\mathcal{S}$ is isomorphic to a subring of $\mathbf{R}[\lambda]$; for simplicity, we retain $1/s$ and avoid introducing a new indeterminate.) The isomorphism from $\mathcal{R}$ to $\mathcal{S}$ allows algebraic properties of $\mathcal{R}$ to be deduced from corresponding algebraic properties of $\mathcal{S}$. For example, since an element of $\mathcal{S}$ is factorable into a product of irreducibles (primes for $\mathcal{S}$), but usually not uniquely (Exercise 15), the same assertion applies to elements of $\mathcal{R}$. In fact, an element of $\mathcal{S}$ is irreducible (within $\mathcal{S}$) if and only if the coefficient of the term $1/s$ is nonzero, so that $a(x) = a_0 + a_1x + \cdots + a_mx^m$ is irreducible in $\mathcal{R}$ if and only if $a_0 \neq 0$.

The isomorphism L of $\mathcal{R}$ to $\mathcal{S}$ may be prolonged to become an isomorphism of the field of fractions of $\mathcal{R}$ to the field of fractions of $\mathcal{S}$, i.e., from $\mathcal{K}$ to the rational function field $\mathbf{R}(1/s) = \mathbf{R}(s)$, using the method given in the proof of Theorem 7.5.20. If $a(x)$, $b(x)$ are polynomial functions (elements of $\mathcal{R}$), with b nonzero, we define L on a/b by

$$L(a/b) = L(a)/L(b),$$

with a/b here being an element of $\mathcal{K}$ and $L(a)/L(b)$ being a rational function in the usual sense, i.e., an element of $\mathbf{R}(s)$.

If $a_1(x)$, $b_1(x)$ are also elements of $\mathfrak{R}$, with b_1 nonzero, the proof of Theorem 7.5.20 shows that:

(i) If $a/b = a_1/b_1$, then $L(a/b) = L(a_1/b_1)$;

(ii) $L(a/b + a_1/b_1) = L(a/b) + L(a_1/b_1)$;

(iii) $L(a/b \circ a_1/b_1) = L(a/b) \cdot L(a_1/b_1)$.

Furthermore, L is an isomorphism from $\mathcal{K}$ to the field of fractions of $\mathcal{S}$. Since any rational function in $\mathbf{R}(s)$ may be expressed as a quotient of polynomials in $1/s$ without constant term, the field of fractions of $\mathcal{S}$ is just $\mathbf{R}(s)$. Thus:

L is an isomorphism between $\mathcal{K}$ and $\mathbf{R}(s)$.

Because an isomorphism must carry the multiplicative identity to the multiplicative identity, we see that

the Laplace transform of the delta function is the number 1.

We now derive some properties of the Laplace transform.

(i) Let α be an element of $\mathcal{K}$. Then

$$L(I \circ \alpha) = \frac{1}{s} L(\alpha).$$

Thus integration corresponds to multiplication of the transform by $1/s$. The proof is immediate, because I is the constant 1, and $L(1) = 1/s$.

(ii) If $a_0, a_1, \ldots, a_m$ are constants and $a(x) = a_0 + a_1 x + \cdots + a_m x^m \in \mathfrak{R}$, then the (usual) derivative of s is $a'(x) = a_1 + 2a_2 x + \cdots + m a_m x^{m-1}$, and applying (12) to both a and a' yields

$$\begin{aligned} L(a')(s) &= \sum_{t=1}^{m} j a_j \cdot (j-1)!/s^j = s \sum_{t=0}^{m} a_j \cdot j!/s^{j+1} - a_0 \\ &= sL(a)(s) - a(0). \end{aligned}$$

Thus the Laplace transform of $a(x)$ and its usual derivative are connected by

$$L(a') = sL(a) - a(0).$$

(iii) Another property is

$$L(xa(x)) = -\frac{d}{ds} L(a)(s),$$

as may be verified by calculating that each side equals $\sum_{j=0}^{m} a_j (j+1)!/s^{j+2}$.

Numerous other formulas may be developed. For a lengthy list, see the Churchill book cited above.

As a simple illustration of the manner in which these formulas are used, we solve the differential equation

$$y' - 2x = 0$$

for the unknown polynomial function $y(x)$. Taking the Laplace transform of each side yields the algebraic equation

$$sL(y) - y(0) - 2/s^2 = 0,$$

where $L(y)$ is the Laplace transform of the unknown function y. Solving this algebraic equation for $L(y)$ gives

$$L(y) = c/s + 2/s^3,$$

where $c = y(0)$ is a constant, so that

$$y = L^{-1}(c/s + 2/s^3) = c + x^2.$$

The solution therefore is $y(x) = c + x^2$.

If we wish to use the operator D, we must remember that differentiation of y coincides with $D \circ y$ only if y has zero constant term. If c is the constant term of y, then $y' = (y - c)' = D \circ (y - c)$, so that $y' - 2x = 0$ takes the form

$$D \circ (y - c) = 2x.$$

Multiplying by D^{-1} gives

$$y - c = D^{-1} \circ 2x = I \circ 2x = x^2.$$

Therefore $y = c + x^2$, as before.

In general, the Laplace transform provides a method for solving many types of differential and integral equations by converting them to simpler types of equations, often to algebraic equations. In order to be useful, Laplace transform theory must be developed for a class of functions much wider than polynomials. This has been done, and the reader may consult the Berg or Churchill books for the details. Caution: Berg's book uses a somewhat different convolution product (called the *Duhamel product*) defined to be

$$\frac{d}{dx}\int_0^x a(t)b(x - t)\,dx = \int_0^x a(t)b'(x - t)\,dt + a(x)b(0),$$

and uses also a corresponding multiplication for polynomials defined in terms of the usual multiplication to be $(1 - \lambda)a(\lambda)b(\lambda)$.

8.6. EXERCISES.

All polynomials in these exercises have real coefficients. Throughout, we denote by the symbols $\mathcal{R}$, $\mathcal{K}$, $\mathcal{S}$, and $\mathbf{R}(s)$ the rings and fields described in the text.

I

1. Use the operators D, I to solve for the polynomial function $y = f(x)$, if (a) $dy/dx = 2x^2 - x$; (c) $d^2y/dx^2 = 1 + x$; (b) $dy/dx = x^3 - x^2$; (d) $\int_0^x y\,dx = x^2 - x$. Repeat, using L.

2. Express $x \circ x$, $x \circ x \circ x$, $x \circ x \circ x \circ x$, etc., in terms of ordinary powers of x.

3. Solve the integral equation $x \circ y = x^3$ for the polynomial function $y = f(x)$.

4. Let a be a constant. Does $(a \circ f)(x) = af(x)$ for an arbitrary polynomial function $f(x)$?

5. Is $D \circ (x^2) = D \circ x \circ x$? Is $I \circ (x^2) = I \circ x \circ x$? Is $L(x^2) = L(x \circ x)$?

II

6. If a is a constant, show that $D \circ a$ is not zero if a is nonzero. (Hint: $\mathcal{K}$ is a field.) Conclude that the derivative of a constant is *not* zero in the operational calculus. If b is also a constant, verify that $D \circ a = D \circ b$ if and only if $a = b$.

7. If f, g are polynomial functions with zero constant term, show that $D \circ (f + g) = D \circ f + D \circ g$, $D \circ (fg) = f(D \circ g) + (D \circ f)g$, the multiplications without the symbol $\circ$ being pointwise products. Show, by an example, that $D \circ (f \circ g)$ is not equal to $(D \circ f) \circ g + f \circ (D \circ g)$. [Hint: for polynomials with zero constant term, $D \circ f$ is the ordinary derivative f'.]

8. Let k be a positive integer, s a positive real number. Verify that $x^k e^{-sx}$ approaches 0 as x grows large, in the following way.
   (a) Show that $e^x \geq x$ for all $x \geq 0$, by observing that the function $f(x) = e^x - x$ satisfies $f(0) = 1$ and $f'(x) \geq 0$.
   (b) Applying (a) to sx, obtain $e^{sx} \geq sx$ for all $x \geq 0$.
   (c) From (b), obtain $x/e^{sx} \leq 1/s$ for all $x \geq 0$.
   (d) Using (c) with $s/(k+1)$ in place of s, show that

$$x^{k+1}e^{-sx} = (xe^{-s/(k+1)})^{k+1} \leq ((k+1)/s)^{k+1} \qquad \text{for } x \geq 0.$$

   (e) From (d), conclude that

$$x^k/e^{sx} \leq ((k+1)/s)^{k+1}/x,$$

   for $x > 0$, and hence that x^k/e^{sx} approaches 0 as x grows arbitrarily large.

9. Let s be a positive real number and k a nonnegative integer. By induction on k, prove that

$$\int_0^\infty x^k e^{-sx}\,dx = k!/s^{k+1}.$$

Use integration by parts, taking $u = x^k$ and $v' = e^{-sx}$ in $\int uv'dx = uv - \int vu'dx$, together with the fact that $x^k e^{-sx}$ approaches 0 as x increases. (Exercise 8.)

10. Let m, n be positive integers. By taking $u = (x - t)^n$, $v' = t^m$ in the integration by parts formula, verify that

$$\int_0^x t^m(x-t)^n dt = n(m+1)^{-1}\int_0^x t^{m+1}(x-t)^{n-1}dt.$$

Applying this formula repeatedly, verify that

$$\int_0^x t^m(x-t)^n dt = n!m!x^{m+n+1}/(m+n+1)!\,.$$

Hence conclude that

$$x^m \circ x^n = n!m!x^{m+n+1}/(m+n+1)!\,.$$

11. Applying four times the formula proved in Exercise 10, verify that both $x^p \circ (x^q \circ x^r)$ and $(x^p \circ x^q) \circ x^r$ equal $p!q!r!x^{p+q+r+2}/(p+q+r+2)!$, and hence that

$$x^p \circ (x^q \circ x^r) = (x^p \circ x^q) \circ x^r$$

for all nonnegative integers p, q, r.

12. Let $f(x) = f_0 + f_1x + f_2x^2 + \cdots$, $g(x) = g_0 + g_1x + g_2x^2 + \cdots$, $h(x) = h_0 + h_1x + h_2x^2 + \cdots$, where $f_0, f_1, \ldots, g_0, g_1, \ldots, h_0, h_1, \ldots$ are constants. Multiply the formula proved in Exercise 11 by f_p and sum over p to get

$$f(x) \circ (x^q \circ x^r) = (f(x) \circ x^q) \circ x^r.$$

Then multiply by g_q and sum over q to get

$$f(x) \circ (g(x) \circ x^r) = (f(x) \circ g(x)) \circ x^r.$$

Finally, multiply by h_r and sum over r to obtain

$$f(x) \circ (g(x) \circ h(x)) = (f(x) \circ g(x)) \circ h(x).$$

Conclude that convolution of polynomial functions is associative.

13. Let p, q be nonnegative integers.
    (a) Applying first the formula proved in Exercise 10, then the formula proved in Exercise 9, verify that

$$L(x^p \circ x^q) = L(p!q!x^{p+q+1}/(p+q+1)!) = p!q!/s^{p+q+2}.$$

Using twice the formula proved in Exercise 9, show also that $L(x^p)L(x^q) = p!q!/s^{p+q+2}$, and hence conclude that

$$L(x^p \circ x^q) = L(x^p)L(x^q).$$

(b) Take $f(x)$, $g(x)$ as in Exercise 12. Multiply the formula proved in part (a) by f_p and sum over p to obtain

$$L(f(x) \circ x^q) = L(f(x))L(x^q),$$

then multiply by g_q and sum over q to obtain

$$L(f(x) \circ g(x)) = L(f(x))L(g(x)).$$

Conclude that the Laplace transform converts convolution of polynomials to a pointwise product.

14. Let $f(x)$, $g(x)$ be polynomial functions. Verify that the fraction $f(x)/g(x)$ in $\mathcal{K}$ equals 1 if and only if $f(x) = \int_0^x g(t)\,dt$, and equals δ if and only if $f(x) = g(x)$.

15. Verify that an element of $\mathcal{S}$ is prime (nonfactorable) within $\mathcal{S}$ if and only if the coefficient of the term $1/s$ is nonzero. Produce an example demonstrating that unique factorization into primes does not hold in $\mathcal{S}$. (Hint: See Exercise 12 of §3.5.)

16. An element of $\mathcal{K}$ has the form $f(x)/g(x)$, where $f(x)$, $g(x)$ are polynomial functions, with g nonzero. A nonzero polynomial function g may still satisfy $g(x) = 0$ for finitely many values of x. Why is it *not* necessary to require that the x in $f(x)/g(x)$ satisfy $g(x) \neq 0$?

17. If $a(x) = a_0 + a_1x + \cdots + a_mx^m$, where $a_0, a_1, \ldots, a_m$ are constants, verify that

$$D \circ a = a' + D \circ a_0,$$

where a' is the usual derivative of a, and $D \circ a_0$ is a nonzero element of $\mathcal{K}$, but not an element of $\mathcal{R}$.

III

18. Let a_0 be a real constant. Show that a fraction in $\mathcal{K}$ equals $D \circ a_0$ if and only if it has the form $a_0g(x)/g(x)$, for $g(x)$ a polynomial function in $\mathcal{R}$.

19. The quantity $\delta' = D \circ \delta$ is sometimes called the "derivative of the δ function." If $f(x)$ is a polynomial with zero constant term, prove that $\delta' \circ f = f'$, f' being the usual derivative of f.

20. Can I or D satisfy a linear equation $a \circ I = b$ or $a \circ D = b$, where a, b are nonzero real numbers? Generalize to higher degree equations.

21. Let a be a fixed real number, and take $\mathcal{R}_1$ to consist of all functions of the form $f_1(x) + f_2(x)e^{ax}$, where $f_1(x)$, $f_2(x)$ are polynomials with real coefficients. Endow $\mathcal{R}$ with addition and multiplication, addition being the usual pointwise addition of functions, and multiplication being convolution of functions.

(a) By imitating the proofs in the text, verify that $\mathcal{R}_1$ is a commutative ring. For closure under multiplication, first show by induction on k that both $\int_0^x t^k e^{at} dt$, $e^{ax} \int_0^x t^k e^{-at} dt$ have the form $f_1(x) + f_2(x) e^{ax}$. Conclude that $\mathcal{R}$ is a subring of $\mathcal{R}_1$.
(b) If $f(x) = f_1(x) + f_2(x) e^{ax}$, and if at least one of $f_1(x), f_2(x)$ is a nonzero polynomial, show that a positive number r exists such that $f(x)$ has constant nonzero sign for all x with $0 < x \leq r$. [Suggestion: If $f(x_k) = 0$ for a sequence of real numbers $x_1 > x_2 > x_3 > \cdots$ converging towards zero, use Rolle's theorem to show that a similar sequence exists for $f'(x)$. Observe that $f'(x) = f_1'(x) + f_3(x) e^{ax}$, with $\deg f_3(x) = \deg f_2(x)$. Repeatedly apply this and eventually obtain a similar sequence for a function of the form $f_4(x) e^{ax}$, with $\deg f_4(x) = \deg f_2(x)$. From this, obtain a contradiction.]
(c) Using (b), show that if $f(x), g(x) \in \mathcal{R}_1$, then $f(x) \circ g(x) = 0$ implies $f(x) = 0$ or $g(x) = 0$. [Suggestion: Imitate the proof in the text, to get (say) $f_1(x) + f_2(x) e^{ax} = 0$ for all x with $0 < x \leq r$. Then use (b) to show that $f_2(x)$, therefore $f_1(x)$ also, is the zero polynomial.]
(d) Conclude that $\mathcal{R}_1$ has a field of fractions, call it $\mathcal{K}_1$, containing $\mathcal{K}$ as a subfield.
(e) Verify that $(1 - ax) \circ e^{ax} = x$ by explicit calculation of the integral involved. Then multiply by $(1 - ax)^{-1}$ to get $e^{ax} = x/(1 - ax)$.
(f) Conclude that $e^{ax} \in \mathcal{K}$ for every constant a.
(g) Now take $\mathcal{R}_1$ to be all functions of the form $f_1(x) + f_2(x) \cos ax + f_3(x) \sin ax$, where f_1, f_2, f_3 are polynomials, and show in a similar manner that

$$\cos ax = x/(1 + \tfrac{1}{2}a^2x^2), \qquad \sin ax = \tfrac{1}{2}ax^2/(1 + \tfrac{1}{2}a^2x^2).$$

and hence that $\cos ax$ and $\sin ax$ are elements of $\mathcal{K}$.

By similar arguments, it can be shown that all functions of the form $f(x)e^{ax}$, $f(x)e^{ax} \cos bx$, $f(x)e^{ax} \sin bx$ are in $\mathcal{K}$, where $f(x)$ is a polynomial and a, b are constants.

22. Find expressions for the general element of $\mathcal{K}$ in terms of polynomials, δ functions, exponential, sine and cosine functions. [Suggestion: Express the typical element of $\mathbf{R}(s)$ in terms of partial fractions, then find elements of $\mathcal{K}$ mapping under L onto each type of partial fraction. Consulting a table of Laplace transforms, as in the Churchill book, may help.]

23. Using $L(x^m)L(x^n) = L(x^m \circ x^n)$, and the binomial theorem to expand $(x - t)^n$ into a finite series, prove this identity involving binomial coefficients:

$$\sum_{j=0}^{n} (-1)^j \binom{n}{j} \frac{1}{m + j + 1} = \frac{m!n!}{(m + n + 1)!}.$$

Each of the following discusses in greater depth one or more of the topics covered in the preceding chapters. Although advanced topics are discussed in several of the books, each has at least some elementary sections accessible to the readers of this text. The books listed below, as well as the journals, are available in most university and many college libraries.

Books

Bachman, P. *Introduction to p-adic Numbers and Valuation Theory* (Academic, 1964).
Beckenbach, E. F., editor. *Applied Combinatorial Mathematics* (Wiley, 1964).
Berg, L. *Introduction to the Operational Calculus* (North-Holland, 1967).
Berlekamp, E. *Algebraic Coding Theory* (McGraw-Hill, 1968).
Birkhoff, G. and T. C. Bartee. *Modern Applied Algebra* (McGraw-Hill, 1970).
Churchill, R. V. *Modern Operational Mathematics in Engineering* (McGraw-Hill, 1944).
Kinchin, A. Y. *Continued Fractions* (University of Chicago Press, 1964).
Korovokin, P. P. *Inequalities* (Blaisdell, 1961).
Lin, S. *An Introduction to Error Correcting Codes* (Prentice-Hall, 1970).
MacDuffee, C. C. *An Introduction to Abstract Algebra* (Dover, 1966).
McCoy, N. H. *Introduction to Modern Algebra* (Allyn and Bacon, 1968).
Mahler, K. *Introduction to p-adic Numbers and their Functions* (Cambridge University Press, 1973).
Marden, M. *Geometry of Polynomials* (American Mathematical Society, 1966).
Niven, I. *Irrational Numbers* (Mathematical Association of America, 1963); *Numbers: Rational and Irrational* (Random House, 1961).
Peterson, W. W. and E. J. Weldon. *Error Correcting Codes* (M.I.T. Press, 1972).
Phillips, E. G. *Functions of a Complex Variable* (Oliver and Boyd, 1951).
Pollard, H. *The Theory of Algebraic Numbers* (Mathematical Association of America, 1950).
Sawyer, W. W. *A Concrete Approach to Abstract Algebra* (Freeman, 1959).
Scheid, S. *Numerical Analysis* (McGraw-Hill, 1968).
Thurston, H. A. *The Number System* (Dover, 1968).
Uspensky, J. V. *Theory of Equations* (McGraw-Hill, 1948).
Uspensky, J. V. and M. A. Heaslot. *Elementary Number Theory* (McGraw-Hill, 1939).

Articles

Brown, J. L., Jr. and R. L. Duncan, "The Least Remainder Algorithm," *Fibonacci Quarterly*, 1971, pages 347–350, 401.
Collins, G. E., "Computer Algebra of Polynomials and Rational Functions," *American Mathematical Monthly*, vol. 80, 1973, pages 725–754.
Gregory, R. T., "The Use of Residue Arithmetic with Automatic Digital Computers," *Delta*, vol. 3, Fall 1972, pages 1–27.

ANSWERS TO SELECTED EXERCISES

When an exercise contains several questions, not individually numbered, the order of the answers corresponds to the order of the questions.

§1.2

1. $\varnothing$, {1}, {2}, {3}, {1, 2}, {1, 3}, {2, 3}, {1, 2, 3}. **2.** Yes.
3. {2, 4}; {1, 2, 3, 4, 7, 9, 10}; $\varnothing$; {1, 2, 4, 5, 6, 7}.
4. {1, 2}, {1, 3}, {1, 4}; complements are {3, 4}, {2, 4}, {2, 3}, respectively.
5. {2, 3}, {2, 4}, {3, 4}; complements are {1, 4}, {1, 3}, {1, 2}, respectively.
6. $V = \{5, 6, 8\}$. **7.** $V = \{2, 5, 6, 8\}$. **9.** No. **11.** No.

§1.3

1. 1 and 2. **2.** $-1, 0, 1$. **3.** $-1, 0, 1, 2$.
4. $-1, 0, 1$. **5.** No. No. No.
6. No. Many choices of a, b yield an example. One is $a = 2$, $b = -20{,}001$.
17. No. Yes.

§1.4

1. (a) $g = -1$; (b) $g = 2/3$; (c) $g(x) = -x + 5$, $g(x) = -x - 1$.
2. 17; 288. **3.** $-2, -3, -4, -5$; $1/2, 1/3, 1/4, 1/5$.
4. $g = 0$; $g = 1$ and $g = -1$; $g(x) = x$ and $g(x) = -x + b$ for any constant b.
5. $g = 0$; $g = 1$ and (for even n only) $g = -1$.
6. (a) $\mathcal{G}$ is a commutative semigroup, monoid, and group.
(b) $\mathcal{G}$ is a commutative semigroup, not a monoid, not a group.
(c) $\mathcal{G}$ is a noncommutative semigroup, not a monoid, not a group.
(d) $\mathcal{G}$ is a commutative semigroup and a commutative monoid, but not a group.
7. For all fractions:
(a) $\mathcal{G}$ is a commutative semigroup, monoid, and group.
(b) $\mathcal{G}$ is a commutative semigroup, not a monoid, not a group.
(c) $\mathcal{G}$ is a noncommutative semigroup, not a monoid, not a group.
(d) $\mathcal{G}$ is a commutative semigroup and monoid, but not a group.
For nonzero fractions:
(a) $\mathcal{G}$ is not a semigroup, not a monoid, not a group.
(b) $\mathcal{G}$ is a commutative semigroup, not a monoid, not a group.
(c) $\mathcal{G}$ is a noncommutative semigroup, not a monoid, not a group.
(d) $\mathcal{G}$ is a commutative semigroup, commutative monoid, and commutative group. **8.** No. **9.** Yes. No.
10. Under addition, the set of all even integers forms a commutative semigroup, monoid, and group, and the set of all odd integers does not form a semigroup, monoid, or group. The positive even integers alone form a commutative semigroup, not a monoid, not a group. The same is true of the negative even integers alone. Zero alone forms a commutative group.

§1.5

1. (a) Not a ring. (b) Ring with identity. (c) Ring, but not with identity. (d) Not a ring. **2.** Yes. No. Yes.
3. S is a commutative ring with identity. $-1 = 1$. S is still a commutative ring, but not with identity, and again $-1 = 1$.
4. All are commutative rings with identity. The rings (b), (e) are **Q**.
5. Subtraction is neither left nor right distributive with respect to addition. Division is right distributive, but not left distributive, with respect to addition.
6. (a) **Q**; (b) **Z**. **7.** The ring of Example 1.5.5. **Z**. **11.** Yes.

§1.6

1. None. None. The first ring in Exercise 3, §1.5 is both an integral domain and a field. The second is neither. **2.** Yes.
3. Yes. Yes. **4.** Only **Z** itself. **5.** Only **Q** itself.
6. Exactly the points (except the origin) on the X or Y axes.

§1.7

1. It is a field.
2. No. Still no, unless the ring is known to be free of zero divisors.
3. It is a field. **4.** **Q**, with its subring **Z**.
5. **Q**, and the field of Exercise 3 of §1.5.
6. **Z**, the even integers, the integral multiples of 3; ± 1, fractions with odd numerator and denominator, fractions with numerator and denominator not multiples of 3.
7. No. See Example 1.6.6. **8.** (a) **Q**; (b) **Z**.
9. The solution of $ax + b = 0$ is $x = -a^{-1}b$. No.

§2.2

1. $\frac{1}{2}(1 - 3^{-11})$. **2.** $\frac{16}{3}(4^7 - 1)$.
3. Many choices are possible. If $\alpha_1, \alpha_2, \alpha_3, \ldots$ and $\beta_1, \beta_2, \beta_3, \ldots$ are the sequences given in the text, possible choices are the sequences $\frac{1}{2}(\alpha_1 + \alpha_2)$, $\frac{1}{2}(\alpha_2 + \alpha_3)$, $\frac{1}{2}(\alpha_3 + \alpha_4), \ldots$ and $\frac{1}{2}(\beta_1 + \beta_2)$, $\frac{1}{2}(\beta_2 + \beta_3)$, $\frac{1}{2}(\beta_3 + \beta_4), \ldots$.
4. All real numbers x satisfying $0 \leq x \leq 1$. Yes.
5. All real numbers x for which $-4 \leq x \leq 1$. All real numbers. **6.** Yes.
7. Precisely those x for which $0 \leq x \leq 4$. Precisely those x satisfying $x \geq 4$.
8. Those x satisfying $1 \leq x \leq 2$. **9.** $\sqrt{8}$, $\sqrt{12}$ but not $\sqrt{4}$ are irrational.
15. If $x = 1$, the limit does not exist (or may be said to be $+\infty$).
16. Not necessarily. No. No. No.

§2.3

1. (a) Sum $= 1 + 4i$, difference $= 3 + 2i$, product $= -5 - i$.
   (b) Sum $= 1 + 2i$, difference $= -1$, product $= -1 + i$.
   (c) Sum $= 1 + 5i$, difference $= 5 - 3i$, product $= -10 + 10i$.
2. (a) $z = 1 - i$; (b) $z = -1 + 2i$; (c) $z = i$ and $z = -i$;
   (d) $z = 1$ and $z = -1$. **3.** $z = 2 + i$ and $z = i$.
4. Conjecture false: take $z_1 = z_2 = i$.
5. Conjecture false: take $z_1 = i$ and $z_2 = -1$.
6. Several squares are possible. One is obtained by adding $2 + 3i$ to the given vertices to get vertices $2 + 3i$, $3 + 3i$, $2 + 4i$, $3 + 4i$.
7. Several solutions are possible. One is $z_4 = z_1 + (z_2 - z_1) + (z_3 - z_1) = 5 - 2i$.
8. Yes. **9.** Not necessarily. **10.** Not necessarily.

§2.4

1. $2i = 2(\cos \pi/2 + i \sin \pi/2)$; $-7 = 7(\cos \pi + i \sin \pi)$;
   $1 + i = \sqrt{2}(\cos \pi/4 + i \sin \pi/4)$; $1 - \sqrt{3}i = 2(\cos(-\pi/3) + i \sin(-\pi/3))$.
2. $3i$; $\sqrt{2} - \sqrt{2}i$; -6.
3. $(2i)^4 = 16$; $(-7)^4 = 2401$; $(1 + i)^4 = -4$; $(1 - \sqrt{3}i)^4 = -8 + 8\sqrt{3}i$;
   $(3i)^4 = 81$; $(\sqrt{2} - \sqrt{2}i)^4 = -16$; $(-6)^4 = 1296$.
4. $z = i$. **5.** $\bar{z} = -4 + 5i$. **6.** $z = 0$.
7. z is purely imaginary (on the Y axis).
8. $z = 1, -1, i, -i, (1 + i)/\sqrt{2}, -(1 + i)/\sqrt{2}, (1 - i)/\sqrt{2}, -(1 - i)/\sqrt{2}$.
14. Yes.

§2.5

1. No. **2.** No.
3. If m carries 1 to -1 and i to i, then $m(z) = -iz - 1 + i$ or $m(z) = -\bar{z}$; i.e., $m = t_{-1+i}r_{-\pi/2}$ or $m = r_\pi c$. If m carries 1 to i and i to -1, then $m(z) = iz$ or $m(z) = \bar{z} - 1 + i$; i.e., $m = r_{\pi/2}$ or $m = t_{-1+i}c$.
4. $m(z) = iz$, that is, $m = r_{\pi/2}$.
5. $m(z) = iz + 1 - i$ or $m(z) = \bar{z}$; that is, $m = t_{1-i}r_{\pi/2}$ or $m = c$.
6. $m(1 + i) = -i$; $m^{-1}(1 + i) = 3 + 2i$; $m = t_{-1-2i}r_{\pi/2}c$.
   $m(1 + i) = 2 + 2i$; $m^{-1}(1 + i) = 0$; $m = t_{1+i}$.
10. Yes. Yes. Those w satisfying $|w| = 1$.

§2.6

1. **R**.
2. Two, i and $-i$. Four, $(1 + i)/\sqrt{2}$, $-(1 + i)/\sqrt{2}$, $(1 - i)/\sqrt{2}$, $-(1 - i)/\sqrt{2}$.
3. The roots are $(-a \pm (a^2 - 4b)^{1/2})/2$. See Exercise 16, §2.3, and Theorem 5.5.1 for methods for calculating square roots of complex numbers.

4. If $z_0 = x_0 + y_0 i$, take x_0, y_0 to be rational numbers near x, y respectively.
5. One formula is $\frac{1}{2}|$imaginary part of $(\bar{z}_2 - \bar{z}_1)(z_3 - z_1)|$.
6. If $z = x + yi$ is on the curve, then $y^2 = 2x - 1$.

§3.2 1. (a) $a + b = 3 + 2\lambda + \lambda^2 + 2\lambda^3$; $a - b = -1 + 2\lambda - \lambda^2$;
$ab = 2 + 4\lambda + \lambda^2 + 5\lambda^3 + 2\lambda^4 + \lambda^5 + \lambda^6$.
(b) $a + b = 3 + 2\lambda^3$; $a - b = 3 + 2\lambda - 2\lambda^3$; $ab = -3\lambda - \lambda^2 + 6\lambda^3 + 2\lambda^4$.
(c) $a + b = \lambda + 2\lambda^2 - \lambda^3$; $a - b = -\lambda - \lambda^3 - 2\lambda^4$; $ab = \lambda^3 - 2\lambda^5 - \lambda^7 - \lambda^8$.
2. 1.
3. Constant term $= -8$; leading term $= -\lambda^6$; leading coefficient $= -1$.
4. Polynomial a = nonzero constant; $a = 1$. 5. $a(\lambda) = \lambda^2 + 1$.
6. $10(\lambda + 3)^9$; $90(\lambda + 3)^8$; $720(\lambda + 3)^7$. 7. 5; 5; 8; 20; 19; no.
8. No. 9. No. Yes. No.
10. Yes. Consider λ^2, where the coefficient field is the field of Exercise 3, §1.5.
11. No; for an example take $\alpha_0 = \lambda$, $\alpha_1 = -1$, $m = 1$. Yes.
13. mn. 14. $c_0 = c_1 = \cdots = c_t = 0$; no; no. 18. Yes.
19. Only the nonzero constants, i.e., elements of $\mathfrak{F}$.

§3.3 1. With least nonnegative remainders:
(i) $q = 17, r = 6$; (ii) $q = -18, r = 1$; (iii) $q = -17, r = 6$;
(iv) $q = 18, r = 1$; (v) $q = 0, r = 7$; (vi) $q = -1, r = 118$.
With numerically least remainders:
(i) $q = 18, r = -1$; (ii) $q = -18, r = 1$; (iii) $q = -18, r = -1$;
(iv) $q = 18, r = 1$; (v) $q = 0, r = 7$; (vi) $q = 0, r = -7$.
2. With least nonnegative remainders:
(i) $q = 7, r = 4$; (ii) $q = -8, r = 5$; (iii) $q = -7, r = 4$;
(iv) $q = 8, r = 5$; (v) $q = 0, r = 9$; (vi) $q = -1, r = 58$.
With numerically least remainders:
(i) $q = 7, r = 4$; (ii) $q = -7, r = -4$; (iii) $q = -7, r = 4$;
(iv) $q = 7, r = -4$; (v) $q = 0, r = 9$; (vi) $q = 0, r = -9$.
3. (i) $q(\lambda) = \lambda^3 + \lambda$, $r(\lambda) = -2$; (ii) $q(\lambda) = 0$, $r(\lambda) = \lambda^2 + \lambda$;
(iii) $q(\lambda) = \frac{1}{2}\{(1 + i)\lambda - 1 + i\}$, $r(\lambda) = i$;
(iv) $q(\lambda) = \lambda^3 + \lambda^2 + \lambda + 1$, $r(\lambda) = 2$. 4. $a(2) = 68$. 5. $x = 1$.
6. Not without additional hypotheses. For example, with $a = 9$, $b = 5$, $a = 1 \cdot b + 4 = 2 \cdot b + (-1)$. 7. No. 8. No.
10. The remainder is necessarily even but the quotient may not be. Neither the quotient nor remainder are necessarily even or necessarily odd; however, precisely one of the quotient, remainder must be even.

§3.4 1. (a) $(a_1, a_2) = 14 = 3a_1 - 4a_2$; $\langle a_1, a_2 \rangle = 1232$.
(b) $(a_1, a_2, a_3) = 7 = -36a_1 + 48a_2 + a_3$; $\langle a_1, a_2, a_3 \rangle = 30800$.
(c) $(a_1, a_2, a_3, a_4) = 1 = -1872a_1 + 2496a_2 + 52a_3 - a_4$;
$\langle a_1, a_2, a_3, a_4 \rangle = 1016400$.
2. (a) $(a_1, a_2) = 10 = -11a_1 + 3a_2$; $\langle a_1, a_2 \rangle = 3700$.
(b) $(a_1, a_2, a_3) = 5 = 286a_1 - 78a_2 + a_3$; $\langle a_1, a_2, a_3 \rangle = 196100$.
(c) $(a_1, a_2, a_3, a_4) = 1 = 8866a_1 - 2418a_2 + 31a_2 - 2a_4 = 34a_3 - 117a_4$;
$\langle a_1, a_2, a_3, a_4 \rangle = 15099700$.
3. (a) $(a_1, a_2) = \lambda - 1 = 1 = -\frac{1}{2}(\lambda - 1)a_1 + \frac{1}{2}a_2$; $\langle a_1, a_2 \rangle = \lambda^4 - 1$.
(b) $(a_1, a_2, a_3) = 1 = \frac{1}{8}(\lambda^3 + \lambda^2 + \lambda - 3)a_1 - \frac{1}{8}(\lambda^2 + 2\lambda + 3)a_2 + \frac{1}{4}a_3$;
$\langle a_1, a_2, a_3 \rangle = \lambda^4 - 1$.
4. Take $a = 2$, $b = 2$, $c = 4$.
5. $x = -1 + 4t$, $y = 1 - 3t$, for arbitrary integral t; $x = 3 + 6t$, $y = -2 - 5t$, for arbitrary integral t; no x, y.
6. (a) $(a, b) = 7$, $\langle a, b \rangle = ab/7$. (b) $(a, b) = 8$, $\langle a, b \rangle = ab/8$.
(c) $(a, b) = 13$, $\langle a, b \rangle = ab/13$. (d) $(a, b) = 3$, $\langle a, b \rangle = ab/3$.
8. For $\mathbf{Z}$, $b = \pm a$, $c = \pm a$. 14. 1 or 2. 15. No.

§3.5

1. $a_1 = (\lambda - 1)(\lambda + 1)$; $a_2 = (\lambda^2 + 1)(\lambda - 1)$ in $\mathbf{Q}[\lambda]$ and $\mathbf{R}[\lambda]$, and $a_2 = (\lambda + i)(\lambda - i)(\lambda - 1)$ in $\mathbf{C}[\lambda]$; $a_3 = (\lambda^2 + 1)(\lambda + 1)$ in $\mathbf{Q}[\lambda]$ and $\mathbf{R}[\lambda]$, and $a_3 = (\lambda + i)(\lambda - i)(\lambda + 1)$ in $\mathbf{C}[\lambda]$.
2. 1001 and 10001 are not primes; 7 and 73 are factors respectively.
3. 2, 3, 5, 7, 11, 13, 17, 19, 23, 29, 31, 37, 41, 43, 47, 53, 59, 61, 67, 71, 73, 79, 83, 89, 97.
4. Yes. **5.** $1296 = 6 \cdot 6 \cdot 6 \cdot 6 = 81 \cdot 16$.

1. All except (viii) are valid. **2.** All except (ii) are valid.
3. Take $a = 1$, $b = 0$, $c = 2$, $m = 2$.
4. (i) $x = -1$; (ii) no solution; (iii) $x = 1$; (iv) $x = -1$; (v) no solutions.
5. (i) $x = 1$; (ii) $x = 1 + \frac{1}{2}\lambda$; $x = -1$.
6. (iv) $x = 1 + 5t$ for arbitrary integral t; (v) $x = 16 + 65t$ for arbitrary integral t.
7. No.

1. $(a, b) = 1$ in all cases; $\langle a, b\rangle = 1214 \cdot 1111$ for $\mathbf{Z}$, $\langle a, b\rangle = 1$ for $\mathbf{Q}[\lambda]$ and $\mathbf{C}[\lambda]$.
2. $(a, b) = 1$; when a, b, c are pairwise relatively prime.
3. Yes. **4.** $c = a(\alpha)$. **5.** $m = 1$.
7. Existence part valid, but not uniqueness part. **8.** 88; 500.
9. $42\lambda^5 + 60\lambda^4 + 80\lambda^3 + 2$. **10.** Yes. Yes. Not necessarily.

§4.2

1. $20747_{10} = 101000100001011_2 = 1130442_5 = 1000B_{12} = 510B_{16}$.
2. $BEFOD_{16} = 782093_{10}$; $51704_8 = 21444_{10}$; $51704_9 = 34105_{10}$.
3. $ABBA_{16} = 2401322_5$; $1234321_5 = 12100_{12}$.
4. $d_0 = 1$, $d_1 = 2$, $d_2 = -1$, $d_3 = 1$, all other digits zero.
5. $d_0 = \lambda + 1$, $d_1 = 2\lambda - 1$, $d_2 = 3\lambda - 2$.
6. $d_0 = 16$, $d_1 = 32$, $d_2 = 24$, $d_3 = 8$, $d_4 = 1$.
7. $d_0 = d_1 = d_2 = \cdots = 1$.
8. $a = 1(-1)0(-1)(-1)(-1)0_5 = 1331(-1)3_6$; $c = 49383_{10}$; $d = -8290_{10}$.
9. ...99997655; ...66664322; 9999994.

§4.3

1. Base five: $a + c = 22112$, $a - c = 11202$, $ac = 102112310$, $q = 3$, $r = 242$;
Base nine: $a + c = 16562$, $a - c = 11602$, $ac = 35562860$, $q = 5$, $r = 662$;
Base thirteen: $a + c = 16562$, $a - c = 11A02$, $ac = 309A7460$, $q = 7$, $r = A82$.
2. Base five, 31302; base nine, 21743; base thirteen, 1A73C.
3. $a + c = 2163_9$, $a - c = 1515_9$, $ac = 517515_9$.
4. Seven. **5.** Base two, 100000; base four, 200.
6. Using least nonnegative remainders:
(i) $q = 1110_2$, $r = 101_2$; (ii) $q = 320_4$, $r = 1021_4$; (iii) $q = 4200_8$, $r = 243_8$; (iv) $q = 115D_{16}$, $r = 22EA_{16}$. These are also the answers using numerically least remainders.
7. $a + c = 2422_8$; $a - c = 1(-1)(-3)2_8$; $ac = 134(-2)030_8$.
8. $a_1 + a_2 + a_3 = 36311$; $a_1 - a_2 = 43318$; $a_1a_2 = \ldots 999966528103$; $a_2a_3 = 4275771$.
9. 519514; 3820139; 998319. **13.** 5. **14.** $b = 4$; $b = 5$.

§4.4

2. $(-265, 370) = 5$; $(154, -112) = 14$; $(112, 175) = 7$.
3. Take 23 chips from the first pile.

§4.5

2. Any base b for which $5 \mid b$.
3. At most the degree of the lower degree polynomial.
4. $a = 21_{10}$; the exception is $b = 21_{10}$.

§5.2 1. For λ^2+1: no roots in **Q** or **R**, $\pm i$ in **C**.
For $\lambda^2+2\lambda+3$: no roots in **Q** or **R**, $-1\pm\sqrt{2}i$ in **C**.
For $\lambda-1$: roots in **Q**, **R**, or **C** are 1 (only).
For $2\lambda+5$: roots in **Q**, **R**, **C** are $-5/2$ (only).
For λ^2-2: no roots in **Q**, $\pm\sqrt{2}$ in both **R** and **C**.
For λ^3-1: only root in **Q** is 1, no other roots in **R**, but $(-1\pm\sqrt{3}i)/2$ in **C**.
For λ^3+1: only root in **Q** is -1, no other roots in **R**, but $(1\pm\sqrt{3}i)/2$ in **C**.
For $\lambda^4-4\lambda^2+4$: no roots in **Q**, only roots in **R** are $\pm\sqrt{2}$, each with multiplicity two.

2. 2 is a root of multiplicity two, other roots being $\pm i$. All roots of the derivative are 2, $(1\pm i)/2$ **4.** $\pm(1+\sqrt{3}i)$.

5. $(\lambda-2)^2$. **6.** Yes. Yes. **8.** Converse is false.

9. Derivative is the zero polynomial. **10.** No. **11.** Not necessarily. **12.** 1.

§5.3 1. (i) S is a cone, with the T axis as cone axis and the origin as cone apex.
(ii) S is a paraboloid of revolution, with the T axis as axis of revolution and the origin as apex.
(iii) S has the shape shown in Figure 5.3.3, with $P=0$ and $Q=1$, but "pointed" at both P and Q.
(iv) S has the shape shown in Figure 5.3.3, with $Q=0$ and $P=1$.

2. K_r is a circle of radius r^2 centered at 1, traversed twice as z covers C_r once.

3. $(\lambda^2+1)^{n/2}$.

§5.4 1. $u((\lambda-3)^2+4)$, where u is any real nonzero constant.

2. $f=[(\lambda-3-2i)(\lambda-3+2i)(\lambda-2+i)(\lambda-2-i)]^2$
$=[(\lambda^2-6\lambda+13)(\lambda^2-4\lambda+5)]^2$.

3. $4(\lambda+\frac{1}{2})(\lambda+\frac{3}{2})(\lambda^2+1)$ in **Q**$[\lambda]$ and **R**$[\lambda]$; $4(\lambda+\frac{1}{2})(\lambda+\frac{3}{2})(\lambda+i)(\lambda-i)$ in **C**$[\lambda]$.

4. $\lambda^3+2\lambda^2-3=(\lambda-1)(\lambda^2+3\lambda+3)$ in **Q**$[\lambda]$ and **R**$[\lambda]$,
$=(\lambda-1)(\lambda-\frac{1}{2}(-3+\sqrt{3}i))(\lambda-\frac{1}{2}(-3-\sqrt{3}i))$ in **C**$[\lambda]$.
$\lambda^3+2\lambda^2-1=(\lambda+1)(\lambda^2+\lambda-1)$ in **Q**$[\lambda]$,
$=(\lambda+1)(\lambda-\frac{1}{2}(-1+\sqrt{5}))(\lambda-\frac{1}{2}(-1-\sqrt{5}))$ in **R**$[\lambda]$ and **C**$[\lambda]$.

5. Yes. Not always. For p_1p_2, never. For **Z**, p_1+p_2 need not be a prime, and p_1p_2 never is prime.

6. $f(\lambda)=(\lambda^2+1)(\lambda^2+\lambda+1)$ in **Q**$[\lambda]$ and **R**$[\lambda]$,
$=(\lambda+i)(\lambda-i)(\lambda-\frac{1}{2}(-1+\sqrt{3}i))(\lambda-\frac{1}{2}(-1-\sqrt{3}i))$ in **C**$[\lambda]$.

7. $(\lambda^2+1)^2(\lambda^2+2)$. **8.** $2\lambda(\lambda^2+1)(3\lambda^2+5)$.

9. $(\lambda+1)(\lambda-1)(\lambda^2+2\lambda+2)=(\lambda+1)(\lambda-1)(\lambda+1+i)(\lambda+1-i)$;
$(2\lambda+1)(3\lambda^2-5\lambda+6)=3(2\lambda+1)(\lambda-\frac{1}{6}(5+\sqrt{47}i))(\lambda-\frac{1}{6}(5-\sqrt{47}i))$.

13. $f(\lambda)=(\lambda-1-i)(\lambda-1+i)(\lambda+2+\sqrt{2})(\lambda+2-\sqrt{2})$.

§5.5 1. (a) 3/2. (b) $-2, 3$. (c) $1+i, 1-i$. (d) $1, 1, -2$.
(e) $1+(\frac{5}{2}+\frac{1}{2}\sqrt{21})^{1/3}+(\frac{5}{2}-\frac{1}{2}\sqrt{21})^{1/3}, 1+(\frac{5}{2}+\frac{1}{2}\sqrt{21})^{1/3}\omega+(\frac{5}{2}-\frac{1}{2}\sqrt{21})^{1/3}\omega^2$, $1+(\frac{5}{2}+\frac{1}{2}\sqrt{21})^{1/3}\omega^2+(\frac{5}{2}-\frac{1}{2}\sqrt{21})^{1/3}\omega$. (f) Note that $(1+\frac{1}{3}\sqrt{3})^3=2+\frac{10}{9}\sqrt{3}$. Then the roots are $2, -1+i, -1-i$. (g) $0, i, -i$. (h) -2, $1+\sqrt{3}i$, $1-\sqrt{3}i$. (i) $-2^{1/3}-3^{1/3}$, $-2^{1/3}\omega-3^{1/3}\omega^2$, $-2^{1/3}\omega^2-3^{1/3}\omega$. (j) $2^{1/3}+5^{1/3}, 2^{1/3}\omega+5^{1/3}\omega^2, 2^{1/3}\omega^2+5^{1/3}\omega$. (k) $9^{1/3}+3^{1/3}, 9^{1/3}\omega+3^{1/3}\omega^2$, $9^{1/3}\omega^2+3^{1/3}\omega$. (l) $6, 3(-1+\sqrt{3}), 3(-1-\sqrt{3})$. (m) $\sqrt{2}, (\sqrt{3}-1)/\sqrt{2}$, $(-1-\sqrt{3})/\sqrt{2}$.
(n) $\pm1, \pm i, \pm(1+i)/\sqrt{2}$ for all choices of the $\pm$ signs.
(o) $\pm(1\pm i)2^{-1/4}$, for all choices of the $\pm$ signs.
(p) $(-1\pm\sqrt{3}i)/2$, $(1\pm\sqrt{7}i)/2$ (1 is a root of the auxiliary cubic).
(q) $-(1\pm\sqrt{5})/2, 2, -1$ (1 is a root of the auxiliary cubic).

2. $-2, -2, -1$.

3. i, $-(\sqrt{3}+i)/2$, $-(-\sqrt{3}+i)/2$; ±2, $\pm1\pm\sqrt{3}i$ for all choices of $\pm$ signs; $\pm\sqrt{2}\pm\sqrt{2}i$ for all choices of the $\pm$ signs.

4. $-1, i-i$. **5.** $1.7302050807<\sqrt{3}<1.7302050808$.

9. Product equals -1 if n is even, 1 if n is odd. **10.** -1.

§5.6

1. (i) One Jensen circle with line segment joining i, $-i$ as diameter. This line segment is also the convex hull of the roots.
   (ii) Two Jensen circles, the line segments joining i, $-i$ and joining $5+i$, $5-i$ being diameters. The convex hull of the roots is a rectangle with vertices i, $-i$, $5+i$, $5-i$.
   (iii) One Jensen circle, the line segment joining i, $-i$ being the diameter. The convex hull of the roots is a quadrilateral with 5, -6, i, $-i$ as vertices.
2. Begin with a polynomial $g(\lambda)$ having four real but not equally spaced roots, so that one minimum of g is strictly lower than the other. Then let $f(\lambda) = g(\lambda) + c$, where the constant c is chosen so that the higher minimum moves about the X axis, but the lower minimum stays below.
3. The sets in (i), (ii), (iv) only are convex. **4.** No.
5. The roots of f' are 0 with multiplicity $n-1$. The convex hull of the roots of f is an n-sided regular polygon inscribed in the unit circle, the vertices being the nth roots of 1.
6. Use Theorem 5.6.3 to give an overall estimate. Use Rolle's theorem to estimate the location of some real roots. Use the Jensen circles theorem to estimate the location of the complex roots. **10.** Yes. **11.** The union is usually not convex.

§5.7

1. (a) One real root only, between 0 and 1. (b) One real root only, between 0 and 1.
   (c) Three real roots, one between -3 and -2, one between 3 and 4, and one between 5 and 6. (d) No real roots.
   (e) No real roots. (f) The roots are 0, 2, $1+\sqrt{3}$, $1-\sqrt{3}$.
2. When $p^2 - 4q \geq 0$. **3.** $(\lambda^2+1)(\lambda-1)^2$.
4. For even n, $\lambda^n - 1$ has two real roots (± 1) and $\lambda^n + 1$ has no real roots. For odd n, $\lambda^n - 1$ has one real root $(+1)$ and $\lambda^n + 1$ has one real root (-1).

§5.8

1. $f(\lambda) = (\lambda - 2 - i)^2(\lambda - 2 + i)^2(\lambda - 3 - i)(\lambda - 3 + i)$. $f'(\lambda)$ has $2 \pm i$ as simple roots, and any nonreal roots are in the Jensen circle with $2 \pm i$ as diameter, or in the Jensen circle with $3 \pm i$ as diameter.
2. The first polynomial equals $(\lambda^2+\lambda+1)(\lambda^2+\lambda+2)$, the second $(\lambda^2+\lambda+2)(\lambda+1)$. The first has roots $(-1 \pm \sqrt{3}i)/2$ and $(-1 \pm \sqrt{7}i)/2$. The second has roots -1, $(-1 \pm \sqrt{7}i)/2$.
3. For even m, Newton's process converges to $-a^{1/m}$. For odd m, the process converges to $a^{1/m}$, apart from certain exceptional x_0 which lead to a horizontal tangent line. **4.** Once. Use $w = (z+i)(z-3i)$.
5. Precisely when it is a multiple root of f. **6.** Yes. No.

§6.2

1. The relations in (a), (b) only are equivalence relations.
   (a) There are three equivalence classes, represented by 0, 1, 2, and consisting respectively of the integers leaving remainder 0, remainder 1, remainder 2, when divided by 3.
   (b) There are infinitely many equivalence classes, the nonnegative integers being a system of distinct representatives. The equivalence class represented by 0 consists solely of 0, and the class represented by positive integer a consists of a and $-a$.
2. In the relation (e) of Exercise 1, the integer 0 fails to satisfy $0 \sim 0$.
3. The relation in (d) is reflexive and symmetric, but not transitive.
   The relation in (f) is reflexive and transitive, but not symmetric.
   The relation in (e) is symmetric and transitive, but not reflexive.
4. The relation on $\mathbf{Z}$ given by $a \sim b$ if and only if a is either b or $b+1$ is reflexive but neither symmetric nor transitive. The relation given by $a \sim b$ if and only if $a \neq b$ is symmetric but neither reflexive nor transitive. The relation given by $a \sim b$ if and only if $a > b$ is transitive but neither reflexive not symmetric. The relation given by $a \sim b$ if and only if $a = b + 1$ is not reflexive, not symmetric, and not

transitive. The reader may find it instructive to sketch the set R for each of the relations in the answers to Exercises 3, 4, as in Figures 6.2.5, 6.2.6, 6.2.7, then comparing with Figure 6.2.9.

5. Yes. An equivalence class consists of all positive integers with a fixed lowest digit. The integers 1, 2, 3, 4, 5, 6, 7, 8, 9, 10 constitute a system of distinct representatives. **6.** Given a, there may be no b satisfying $a \sim b$.

8. The origin itself constitutes one equivalence class. Otherwise, an equivalence class consists of all complex numbers on a ray ("half" a straight line) emanating from the origin, not including the origin. The complex numbers 0 and $e^{i\theta}$, for $0 \leq \theta < 2\pi$, constitute a system of distinct representatives.

10. The answer to Exercise 8, with the origin deleted from that answer.

§6.3

1. All pairs $(3t, 5t)$ for arbitrary nonzero integral t. All pairs $(3t, 4t)$ for arbitrary nonzero integral t.

2. All pairs $(t, t(\lambda - 1))$ for t an arbitrary nonzero polynomial with real coefficients. All pairs $(t(\lambda + 1), t(\lambda^2 + 1))$ for t an arbitrary nonzero polynomial with real coefficients.

3. (i) $\mathbf{Q}$; (ii) $\mathbf{Q}$; (iii) the fractions (quotients of polynomials) with coefficients from $\mathbf{Q}$ (see §6.4); (iv) $\mathbf{Q}$; (v) $\mathbf{R}$; (vi) $\mathbf{C}$.

4. Reflexivity and symmetry hold, but transitivity fails: $(1, 1) \sim (0, 0)$ and $(0, 0) \sim (2, 1)$, yet $(1, 1) \not\sim (2, 1)$. **5.** No.

§6.4

1. Not if α is a root of the polynomial in the denominator of $q(\lambda)$.

2. Let $q(\lambda) = \lambda$, $\alpha = -\frac{1}{2}$. If $|q(\alpha)|$ denotes the absolute value for rational functions applied to the constant rational function $q(\alpha)$, then we have $|q(\lambda)| = \rho^{-1}$, $|q(\alpha)| = 1$, so that $|q(\lambda)| \neq |q(\alpha)|$. If $|q(\alpha)|$ denotes the usual absolute value for rational numbers applied to the rational number $q(\alpha)$, then we have $|q(\lambda)| = \rho^{-1}$, $|q(\alpha)| = 1/2$, and again $|q(\lambda)| \neq |q(\alpha)|$.

3. No; take $q_1 = q_2 = 1$. Taking $q_1 = 2$, $q_2 = 1$ furnishes a counterexample for the principle of domination for the usual absolute value.

4. In the author's opinion, the last q_2 is reasonably near q_1. The four listed rational functions q_2 are progressively closer to q_1.

5. (i) Does not approach zero. (ii) Does approach zero.

6. $-5 \cdot 3 \cdot 2^{-2} \cdot 7$; $(\lambda^2 + 4)(\lambda + 2)^{-1}(\lambda + 3)^{-1}$; $(\lambda + 2i)(\lambda - 2i)(\lambda + 2)^{-1}(\lambda + 3)^{-1}$.

7. It is possible to have $|r| \leq \frac{1}{2}|b|$ if $\rho \leq \frac{1}{2}$. **8.** Take $q_2 = -q_1$.

9. $\mathcal{S}$ consists of rational functions for which the denominator has degree at least as high as that of the numerator. The corresponding fact for rational numbers is false. **10.** It does satisfy V4′.

§6.5

1. $1/3 = .\overline{01}_2 = .1\overline{0}_3 = .\overline{1}_4 = .\overline{13}_5 = .2\overline{0}_6$;
$1/2 = .1\overline{0}_2 = .\overline{1}_3 = .2\overline{0}_4 = .\overline{2}_5 = .3\overline{0}_6$;
$3/4 = .11\overline{0}_2 = .\overline{20}_3 = .3\overline{0}_4 = .\overline{3}_5 = .43\overline{0}_6$;
$1/6 = .0\overline{01}_2 = .0\overline{1}_3 = .0\overline{2}_4 = .\overline{04}_5 = .1\overline{0}_6$;
$13/3 = 100.\overline{01}_2 = 11.1\overline{0}_3 = 10.\overline{1}_4 = 4.\overline{13}_5 = 4.2\overline{0}_6$; $(b - 1)^{-1} = .\overline{1}_b$.

2. 7137/3100; 965/24. (These answers are in base ten.)

3. There are only b different digits possible to fill infinitely many digit positions. No. $\sqrt{2}$ is not a rational number. No.

4. $.75\overline{0} = .74\overline{9}$; $.9\overline{0}_{12} = .8\overline{B}_{12}$; no.

5. $1.01101_2 < \sqrt{2} < 1.01110_2$; $1.22524_6 < \sqrt{2} < 1.22525_6$; $1.32404_8 < \sqrt{2} < 1.32405_8$; $1.36485_9 < \sqrt{2} < 1.36486_9$; $1.6A09E_{16} < \sqrt{2} < 1.6A09F_{16}$; none of these representations are ultimately periodic.

6. $1/\lambda = \sum_{j=1}^{\infty} (\lambda + 1)^{-j}$; $\sum_{j=1}^{\infty} (z + 1)^{-j}$ converges to $1/z$ when $|z + 1| > 1$. The series for $\lambda/(\lambda - 1)^2$ is given in the text; when evaluated at z, it converges to $z/(z - 1)^2$ when $|z| > 1$.

7. $3/4 = \frac{3}{4}.\overline{0}$, $\sqrt{2} = \sqrt{2}.\overline{0}$.

8. Rational functions.

§6.6

1. $a + c = 43.022_5 = 41.522_6 = 37.522_9 = 37.522_{10} = 37.522_{12}$;
$a - c = 4.302_5 = 5.302_6 = 8.302_9 = 9.302_{10} = \mathrm{B}.302_{12}$;
$ac = 1003.00332_5 = 424.03132_6 = 342.55432_9 = 330.34332_{10} = 307.\mathrm{B}2132_{12}$;
$1.22_5 < a/c < 1.23_5$; $1.31_6 < a/c < 1.32_6$; $1.56_9 < a/c < 1.57_9$;
$1.65_{10} < a/c < 1.66_{10}$; $1.84_{12} < a/c < 1.85_{12}$.
2. $.3335_6 < a/c < .3340_6$. **3.** $1.01101_2 < \sqrt{2} < 1.01110_2$.
4. Exact value: 9.0888478. . .; approximate value: 9.04.

§6.7

1. Yes. No. **2.** Yes. **3.** $\mathbf{Q}$.
4. Yes. No. No. Yes. **5.** Use the fact that $\sqrt{pq}$ is not rational; no.
6. $1.02211_3 < (\alpha - 1)\beta/\gamma < 1.02212_3$; $0.34413_5 < (\alpha - 1)\beta/\gamma < 0.34414_5$.

§7.2

1. $\bar{2} + \bar{3} = \bar{1}$ in $\mathbf{Z}_4$, $= \bar{0}$ in $\mathbf{Z}_5$, $= \bar{5}$ in $\mathbf{Z}_6$; $\bar{2}\cdot\bar{3} = \bar{2}$ in $\mathbf{Z}_4$, $= \bar{1}$ in $\mathbf{Z}_5$, $= \bar{0}$ in $\mathbf{Z}_6$;
$\bar{3} = \bar{1}$ in $\mathbf{Z}_2$, $= \bar{0}$ in $\mathbf{Z}_3$.
2. $-\bar{2} = \bar{1}$ in $\mathbf{Z}_3$, $= \bar{2}$ in $\mathbf{Z}_4$, $= \bar{3}$ in $\mathbf{Z}_5$, $= \bar{4}$ in $\mathbf{Z}_6$, $= \bar{0}$ in $\mathbf{Z}_2$.
3. $\overline{\lambda + 2} + \overline{2\lambda + 3} = \overline{3\lambda + 5}$ in both cases; $\overline{\lambda + 2}\cdot\overline{2\lambda + 3} = \overline{7\lambda + 4}$ when $f(\lambda)$ is $\lambda^2 + 1$, and $= \overline{7\lambda + 6}$ when $f(\lambda) = \lambda^2$.
4. No solution possible in $\mathbf{Z}_6$ or $\mathbf{Z}_8$; $X = \bar{6}$ in $\mathbf{Z}_7$.
5. $X = \overline{-\lambda}$ when $f(\lambda) = \lambda^2 + 1$; no solution possible when $f(\lambda) = \lambda^2$.
6. In $\mathbf{Z}_2$: $\bar{0}, \bar{1}$; in $\mathbf{Z}_3$: $\bar{0}, \bar{2}$; in $\mathbf{Z}_4$: $\bar{2}, \bar{3}$; in $\mathbf{Z}_5$: $\bar{2}, \bar{3}$; in $\mathbf{Z}_6$: $\bar{0}, \bar{2}, \bar{3}, \bar{5}$.
8. $\bar{1}, \bar{3}, \bar{5}, \bar{7}$ are roots in $\mathbf{Z}_8$. **9.** $a = 3$ or 5; no; no. **10.** $|m|$.

§7.3

1. Over $\mathbf{Z}_2$, quotient $= \lambda^3 + \lambda^2 + \lambda$; over $\mathbf{Z}_3$, $\lambda^3 + 2\lambda^2 + \lambda$; and over $\mathbf{Z}_5$, $\lambda^3 + 4\lambda^2 + \lambda$. The remainder is 1 in all three cases.
2. In all three cases, the addition table is that displayed on page 312. The multiplication tables differ from the multiplication table displayed on page 312 only in the four elements occupying the last two rows and last two columns. These four elements are:

When $m(\lambda) = \lambda^2 + \lambda$

λ	0
0	$\lambda+1$

When $m(\lambda) = \lambda^2 + 1$

1	$\lambda+1$
$\lambda+1$	0

When $m(\lambda) = \lambda^2$

0	λ
λ	1

3. In both cases the addition table is:

	0	1	2	λ	$\lambda+1$	$\lambda+2$	2λ	$2\lambda+1$	$2\lambda+2$
0	0	1	2	λ	$\lambda+1$	$\lambda+2$	2λ	$2\lambda+1$	$2\lambda+2$
1	1	2	0	$\lambda+1$	$\lambda+2$	λ	$2\lambda+1$	$2\lambda+2$	2λ
2	2	0	1	$\lambda+2$	λ	$\lambda+1$	$2\lambda+2$	2λ	$2\lambda+1$
λ	λ	$\lambda+1$	$\lambda+2$	2λ	$2\lambda+1$	$2\lambda+2$	0	1	2
$\lambda+1$	$\lambda+1$	$\lambda+2$	λ	$2\lambda+1$	$2\lambda+2$	2λ	1	2	0
$\lambda+2$	$\lambda+2$	λ	$\lambda+1$	$2\lambda+2$	2λ	$2\lambda+1$	2	0	1
2λ	2λ	$2\lambda+1$	$2\lambda+2$	0	1	2	λ	$\lambda+1$	$\lambda+2$
$2\lambda+1$	$2\lambda+1$	$2\lambda+2$	2λ	1	2	0	$\lambda+1$	$\lambda+2$	λ
$2\lambda+2$	$2\lambda+2$	2λ	$2\lambda+1$	2	0	1	$\lambda+2$	λ	$\lambda+1$

§7.3 CONT.

The multiplication table in the first case is:

	0	1	2	λ	$\lambda+1$	$\lambda+2$	2λ	$2\lambda+1$	$2\lambda+2$
0	0	0	0	0	0	0	0	0	0
1	0	1	2	λ	$\lambda+1$	$\lambda+2$	2λ	$2\lambda+1$	$2\lambda+2$
2	0	2	1	2λ	$2\lambda+2$	$2\lambda+1$	λ	$\lambda+2$	$\lambda+1$
λ	0	λ	2λ	2	$\lambda+2$	$2\lambda+2$	1	$\lambda+1$	$2\lambda+1$
$\lambda+1$	0	$\lambda+1$	$2\lambda+2$	$\lambda+2$	2λ	1	$2\lambda+1$	2	λ
$\lambda+2$	0	$\lambda+2$	$2\lambda+1$	$2\lambda+2$	1	λ	$\lambda+1$	2λ	2
2λ	0	2λ	λ	1	$2\lambda+1$	$\lambda+1$	2	$2\lambda+2$	$\lambda+2$
$2\lambda+1$	0	$2\lambda+1$	$\lambda+2$	$\lambda+1$	2	2λ	$2\lambda+2$	λ	1
$2\lambda+2$	0	$2\lambda+2$	$\lambda+1$	$2\lambda+1$	λ	2	$\lambda+2$	1	2λ

The multiplication table in the second case agrees with that in the first case, except for the 36 elements in the last six rows and columns. These 36 entries are:

$2\lambda+1$	1	$\lambda+1$	$\lambda+2$	$2\lambda+2$	2
1	$\lambda+2$	2λ	2	λ	$2\lambda+1$
$\lambda+1$	2λ	2	$2\lambda+2$	1	λ
$\lambda+2$	2	$2\lambda+2$	$2\lambda+1$	$\lambda+1$	1
$2\lambda+2$	λ	1	$\lambda+1$	2	2λ
2	$2\lambda+1$	λ	1	2λ	$\lambda+2$

Both rings are fields. In the first case the element $\lambda + 1$ satisfies $(\lambda + 1)^2 + (\lambda + 1) + 2 = 0$, therefore acts as the element λ in the second case. Changing $\lambda + 1$ everywhere in the tables in the first case to λ (that is, changing λ to $\lambda + 2$) will therefore yield the tables in the second case.

4. 0; 0. **5.** $y^4 - y = (y - 0)(y - 1)(y - \lambda)(y - \lambda - 1)$.
6. λ, $\lambda + 1$, $\lambda^2 + \lambda + 1$, $\lambda^3 + \lambda + 1$, $\lambda^3 + \lambda^2 + 1$. **7.** $\mathbf{Z}_p$; $\mathbf{Z}_m$.
8. $\mathbf{Z}_2[\lambda]_{\lambda^2+\lambda+1}$; $\mathbf{Z}_2[\lambda]_{\lambda^3+\lambda+1}$; $\mathbf{Z}_5[\lambda]_{\lambda^2+2}$. **9.** $\mathbf{Z}_8$.
10. The multiplicative group of nonzero elements in the field $\mathbf{Z}_p$.

§7.4

1. Equal in (d), otherwise not equal.
2. f_1 is surjective in (b), (d), (g), injective in (d), (f), (g), and bijective in (d), (g); f_2 is surjective in (a), (b), (d), (e), injective in (a), (c), (d), (e), (f), bijective in (a), (d), (e).
3. $(f_1 + f_2)(r) = r^2 + r^3$, $(f_1 - f_2)(r) = r^2 - r^3$, $(f_1 f_2)(r) = (f_2 f_1)(r) = r^5$ (pointwise product), and $(f_1 f_2)(r) = (f_2 f_1)(r) = r^6$ (composition product).
4. Pointwise product: $f_1(x) = x$, $f_2(x) = x + 1$, for $x \in \mathbf{Z}_2$;
Composition product: $f_1(x) = x + 1$, $f_2(x) = 1$, for $x \in \mathbf{Z}_2$.
5. f is a surjection in (c), an injection in (b), (c), and a bijection in (c).

6. $g(2) = 4$, $g(3) = 1$, $g(4) = 2$; yes.
7. For any $a \neq 0$; $f^{-1}(x) = (x - b)/a$. **8.** Natural logarithm.
10. (ii) The distributive law $f_1(f_2 + f_3) = f_1 f_2 + f_1 f_3$ is the only ring axiom not satisfied; (iii) The zero polynomial function has no inverse—see Exercise 12. **14.** Deg $f = 1$.
15. Use the fundamental theorem of algebra; deg $f = 1$.

§7.5

1. When $m = 0$.
2. Automorphisms of $\mathbf{C}[\lambda]$ are the identity function and the function φ defined by $\varphi(a_0 + a_1\lambda + a_2\lambda^2 + \cdots) = \bar{a}_0 + \bar{a}_1\lambda + \bar{a}_2\lambda^2 + \cdots$, $\bar{a}$ denoting the complex conjugate of the complex number a. No.
3. Complex conjugation is its own inverse. **4.** φ is its own inverse.
5. No, for either choice of $m(\lambda)$ since $1 + 1 \neq 0$ in $\mathbf{Z}_4$ but $1 + 1 = 0$ in $\mathbf{Z}_2[\lambda]_{m(\lambda)}$.
6. $\varphi_0\varphi_1 \neq \varphi_1\varphi_0$. **7.** (i) The kernel is all polynomials for which 1 is a root, and the range in $\mathbf{Q}$; (ii) the kernel is 0 and the range is $\mathbf{Q}[\lambda]$; (iii) the kernel is zero and the range is $\mathbf{Q}[\lambda^2]$, i.e. polynomials in λ^2. $\mathbf{Q}$ is not an ideal in $\mathbf{Q}[\lambda]$. **9.** $\mathfrak{R}$.

§7.6

1. $(2\lambda - 1)(\lambda^2 - \lambda + 1)$; $2(\lambda - 1)(\lambda + 1)$. **2.** 4.
3. $\lambda^3 - p$ is a prime. **4.** All but $\lambda^n + \lambda$, $\lambda^n - 1$.
5. $(\lambda + 2)(\lambda + 3)$; prime; prime; prime; $(\lambda - 1)^2(\lambda + 1)^2$; $(\lambda^3 - 2)(\lambda^3 - 3)$.
7. No. **8.** No. Yes. No. **10.** No.

§7.7

1. For $\varphi_{2\lambda}\varphi_{\lambda+1}$: $2\lambda + 1$, $2\lambda + 2$, $(2\lambda + 3)/(4\lambda + 3)$;
for $\varphi_{\lambda+1}\varphi_{2\lambda}$: $2\lambda + 2$, $2\lambda + 3$, $(2\lambda + 4)/(4\lambda + 5)$. **2.** Yes; yes; yes; yes.
3. The six endomorphisms occur for $\alpha(\lambda) =: (\lambda + 1)/\lambda$, $\lambda + 1$, $\lambda/(\lambda + 1)$, $1/\lambda$, λ, $1/(\lambda + 1)$; yes; yes; no; yes.
4. Either $\alpha(\lambda) = \lambda$, or (for even n only) $\alpha(\lambda) = -\lambda + v$ for an arbitrary constant v. In the first case, φ_α is the identity function.
5. $f(\lambda) = ((\lambda - 3)/2)^3 + ((\lambda - 3)/2)^2 + (\lambda - 3)/2 + 1$;
$f(\lambda) = ((2\lambda - 1)/(1 - \lambda))^3 + ((2\lambda - 1)/(1 - \lambda))^2 + (2\lambda - 1)/(1 - \lambda) + 1$.
6. Ranges are isomorphic when φ_α, φ_β have equal kernels. For polynomials and rational functions, this holds when α, β are both nonconstant. In the polynomial case, the kernels are not the same when α, β are both constants, but the ranges are nevertheless isomorphic, both ranges being $\mathfrak{F}$.
8. Zero, and evaluation endomorphisms.
9. Zero, and evaluation endomorphisms; automorphisms are evaluation endomorphisms φ_α with deg $\alpha(\lambda) = 1$ in the case of $\mathbf{Z}_p[\lambda]$, and $\alpha(\lambda)$ a Mobius function in the case of $\mathbf{Z}_p(\lambda)$.

§7.8

1. $\mathbf{Z}_p$. **2.** No.
3. The isomorphism maps $x + y\lambda$ to $x + yi$, for rational x, y.
4. x is the congruence class represented by λ.
5. $\lambda^3 + 2\lambda + 2$ is a prime in $\mathbf{Z}_3[\lambda]$; $\lambda^3 + 2\lambda + 2 = (\lambda + 4)^2(\lambda + 2)$ in $\mathbf{Z}_5[\lambda]$;
$\lambda^3 + 2\lambda + 2 = (\lambda + 5)^2(\lambda + 4)$ in $\mathbf{Z}_7[\lambda]$.
6. The kernel must be 0 or all of $\mathbf{Q}(\lambda)$, since $\mathbf{Q}(\lambda)$ is a field. **7.** Yes; no.

§8.2

1. $|a|_2 = 2^{-3}$, $|a|_3 = 3^{-2}$, $|a|_5 = 5^2$; $|a|_2 = 2^{-3}$, $|a|_3 = 3^2$, $|a|_5 = 5^{-1}$.
2. $|a|_\lambda = \rho^2$, $|a|_{\lambda+1} = \rho^{-1}$, $|a|_{\lambda+1} = \rho^{-1}$. **3.** $|f(\lambda)|_p = \rho^e$.
4. $|q|_p = |1/q|_p = 1$ for $p \neq q$, $|q|_q = 1/q$, $|1/q|_q = q$.
5. $|a|_7 = 7^{-2}$; $|a|_2 = 2^{-3}$, $|a|_5 = 5^{-3}$.
6. All integers β for which $|\beta|_5 \leq 5^{-2}$ and $|\beta|_7 \leq 7^{-2}$; all polynomials $\beta(\lambda)$ for which $|\beta(\lambda)|_{\lambda-1} \leq \rho$, $|\beta(\lambda)|_{\lambda+1} \leq \rho$, and $|\beta(\lambda)|_{\lambda^2+1} \leq \rho$; $\beta = 3^2 \cdot 5^2 \cdot 7^2$.
11. 0.

§8.3 1. (a) $1/(\lambda - 1) + 2/(\lambda - 1)^2 + (2\lambda + 2)/(\lambda^2 + 1)$;
(b) $(\lambda + 1)/(\lambda^2 + 1) + 1/(\lambda^2 + 1)^2 + (\lambda - 1)/(\lambda^2 + 2)$;
(c) $1 + 1/\{2(\lambda - 1)\} + 1/(\lambda - 1)^2 + 1/\{2(\lambda + 1)\}$.

2. $-x - \ln|1 - x| + \ln|1 + x| +$ constant;
$-\frac{3}{4}\ln|1 - x| - \frac{1}{4}(1 - x)^{-1} + \frac{3}{4}\ln|1 + x| + \frac{1}{4}(1 + x)^{-1}$.

3. First obtain $1/3600 = 4/25 + 1/16 - 2/9$,
then $-599/3600 = -1 + 4/5^2 + 1/2 + 1/2^4 + 1/3^2$;
$82/63 = 1/3 + 1/3^3 + 6/7$. **4.** $|\alpha|_2 = 4$, $|\alpha|_3 = 9$, $|\alpha|_{11} = 121$.

5. $1/(\lambda^2 + \lambda + 1) + \lambda/(\lambda^2 + \lambda + 1)^2 + 1/\lambda^2$.

§8.4 1. $-1 = \bar{1}.0$ in 2-adic form; $-1 = \bar{2}.0$ in 3-adic form; $-1 = \bar{5}.0$ in 6-adic form.

2. $-111813/9900$; $-1049/140$. **3.** $-2 = \bar{9}8$; $-3 = \bar{9}7$.

4. $1599/13 = 123$; $674/9 = \bar{1}86$; $679/11 = \overline{27}89$. **5.** 3.

6. $1 - (\lambda - 1) + (\lambda - 1)^2 - \cdots = \sum_{j=0}^{\infty} (-1)^j(\lambda - 1)^j$; $|z - 1| < 1$.

7. Quotient $= 10^{10} + 3$; compare Exercise 5.

8. $d_0 = p$, all other digits zero. **9.** b is four.

§8.5 1. $37/10 = \langle 3, 1, 2, 3\rangle$; $-3/10 = \langle -1, 1, 2, 3\rangle$; $111/71 = \langle 1, 1, 1, 3, 2, 4\rangle$;
$22/7 = \langle 3, 7\rangle$; $73/28 = \langle 2, 1, 1, 1, 1, 5\rangle$.

2. $\sqrt{2} = \langle 1, 2, 2, 2, \ldots\rangle$ (2 repeats indefinitely);
$\sqrt{3} = \langle 1, 1, 2, 1, 2, 1, 2, \ldots\rangle$ (1, 2 repeats indefinitely);
$\sqrt{5} = \langle 2, 4, 4, 4, \ldots\rangle$ (4 repeats indefinitely);
$\sqrt{7} = \langle 2, 1, 1, 1, 4, 1, 1, 1, 4, \ldots\rangle$ (1, 1, 1, 4 repeats indefinitely).

3. $\alpha = (3 + \sqrt{15})/6$.

4. $682/305$ approximates $\sqrt{5}$ to about 2 units in the sixth decimal place.

5. Knowing q_5 implies knowledge of N_6, D_6, hence of N_6/D_6 which approximates π accurately to more than eight digits.

6. $e = \langle 2, 1, 2, 1, 1, 4, 1, 1, 6, 1, 1, 8, 1, \ldots\rangle$.

§8.6 1. (a) $y = \frac{2}{3}x^3 - \frac{1}{2}x^2 +$ constant; (b) $y = \frac{1}{4}x^4 - \frac{1}{3}x^3 +$ constant;
(c) $y = \frac{1}{2}x^2 + \frac{1}{6}x^3 + c_1x + c_2$, with c_1, c_2 constants;
(d) $y = 2x - 1$.

2. $\underbrace{x \cdot x \cdots x}_{k \text{ factors}} = x^{2k-1}/(2k - 1)!$. **3.** $y = 6x$.

4. No: $a^\circ f = a \times$ integral of f.

5. $D^\circ(x^2) = 2x$, $D^\circ x^\circ x = \frac{1}{2}x^2$; $I^\circ x^2 = \frac{1}{3}x^3$, $I^\circ x^\circ x = \frac{1}{24}x^4$;
$L(x^2) = 2/s^3$, $L(x^\circ x) = 1/s^4$.

TABLE OF NOTATION

$\mathbf{C}$	66	$\mathbf{Q}$	31	$\mathbf{R}$	56	$\mathbf{Z}$	30
$\mathcal{R}[\lambda]$	106	$\mathcal{R}[[\lambda]]$	111	$\mathcal{F}(\lambda)$	265	$\mathbf{Z}_m$	306
$\mathcal{R}_m$	306	$\mathcal{F}[\lambda]_m$	306	$\mathcal{R}^m$	337, 387	$\mid, \nmid$	121
$(\ )$	62, 98	$(\ )$	124	$\langle\ \rangle$	129, 429	$\geq$	7
$\equiv$	141	λ	103	$\lvert\ \rvert$	8, 72, 268	$\simeq$	343
$\lVert\ \rVert$	380	$\lvert\ \rvert_p$	381	!	34	$\binom{n}{m}$	158

Index

1 2 3 4 5 6 7 8 9 10 11 12 13 14 15 16 17 18 19 20 21 22 23 24 25 82 81 80 79 78 77 76 75 74 73